CAMBRIDGE TRACTS IN MATHEMATICS

General Editors

182 Nonlinear Markov Processes and Kinetic Equations

Nonlinear Markov Processes and Kinetic Equations

VASSILI N. KOLOKOLTSOV
University of Warwick

CAMBRIDGE
UNIVERSITY PRESS

CAMBRIDGE
UNIVERSITY PRESS

University Printing House, Cambridge CB2 8BS, United Kingdom

One Liberty Plaza, 20th Floor, New York, NY 10006, USA

477 Williamstown Road, Port Melbourne, VIC 3207, Australia

314-321, 3rd Floor, Plot 3, Splendor Forum, Jasola District Centre, New Delhi - 110025, India

103 Penang Road, #05-06/07, Visioncrest Commercial, Singapore 238467

Cambridge University Press is part of the University of Cambridge.

It furthers the University's mission by disseminating knowledge in the pursuit of education, learning and research at the highest international levels of excellence.

www.cambridge.org
Information on this title: www.cambridge.org/9780521111843

First published 2010

A catalogue record for this publication is available from the British Library

ISBN 978-0-521-11184-3 Hardback

Contents

Preface		*page* ix
Basic definitions, notation and abbreviations		xiv
1	**Introduction**	1
1.1	Nonlinear Markov chains	1
1.2	Examples: replicator dynamics, the Lotka–Volterra equations, epidemics, coagulation	6
1.3	Interacting-particle approximation for discrete mass-exchange processes	8
1.4	Nonlinear Lévy processes and semigroups	11
1.5	Multiple coagulation, fragmentation and collisions; extended Smoluchovski and Boltzmann models	13
1.6	Replicator dynamics of evolutionary game theory	24
1.7	Interacting Markov processes; mean field and kth-order interactions	28
1.8	Classical kinetic equations of statistical mechanics: Vlasov, Boltzmann, Landau	32
1.9	Moment measures, correlation functions and the propagation of chaos	34
1.10	Nonlinear Markov processes and semigroups; nonlinear martingale problems	39
Part I	**Tools from Markov process theory**	41
2	**Probability and analysis**	43
2.1	Semigroups, propagators and generators	43
2.2	Feller processes and conditionally positive operators	54

2.3	Jump-type Markov processes	64
2.4	Connection with evolution equations	67
3	**Probabilistic constructions**	73
3.1	Stochastic integrals and SDEs driven by nonlinear Lévy noise	73
3.2	Nonlinear version of Ito's approach to SDEs	82
3.3	Homogeneous driving noise	89
3.4	An alternative approximation scheme	90
3.5	Regularity of solutions	92
3.6	Coupling of Lévy processes	96
4	**Analytical constructions**	102
4.1	Comparing analytical and probabilistic tools	102
4.2	Integral generators: one-barrier case	104
4.3	Integral generators: two-barrier case	111
4.4	Generators of order at most one: well-posedness	114
4.5	Generators of order at most one: regularity	117
4.6	The spaces $(C^l_\infty(\mathbf{R}^d))^\star$	120
4.7	Further techniques: martingale problem, Sobolev spaces, heat kernels etc.	121
5	**Unbounded coefficients**	131
5.1	A growth estimate for Feller processes	131
5.2	Extending Feller processes	135
5.3	Invariant domains	138
Part II	**Nonlinear Markov processes and semigroups**	145
6	**Integral generators**	147
6.1	Overview	147
6.2	Bounded generators	149
6.3	Additive bounds for rates: existence	154
6.4	Additive bounds for rates: well-posedness	160
6.5	A tool for proving uniqueness	165
6.6	Multiplicative bounds for rates	169
6.7	Another existence result	170
6.8	Conditional positivity	173
7	**Generators of Lévy–Khintchine type**	175
7.1	Nonlinear Lévy processes and semigroups	175
7.2	Variable coefficients via fixed-point arguments	180

7.3 Nonlinear SDE construction 184
7.4 Unbounded coefficients 186

8 Smoothness with respect to initial data 188
8.1 Motivation and plan; a warm-up result 188
8.2 Lévy–Khintchine-type generators 192
8.3 Jump-type models 201
8.4 Estimates for Smoluchovski's equation 208
8.5 Propagation and production of moments for the Boltzmann equation 216
8.6 Estimates for the Boltzmann equation 219

Part III Applications to interacting particles 223

9 The dynamic law of large numbers 225
9.1 Manipulations with generators 225
9.2 Interacting diffusions, stable-like and Vlasov processes 232
9.3 Pure jump models: probabilistic approach 236
9.4 Rates of convergence for Smoluchovski coagulation 245
9.5 Rates of convergence for Boltzmann collisions 250

10 The dynamic central limit theorem 252
10.1 Generators for fluctuation processes 252
10.2 Weak CLT with error rates: the Smoluchovski and Boltzmann models, mean field limits and evolutionary games 263
10.3 Summarizing the strategy followed 267
10.4 Infinite-dimensional Ornstein–Uhlenbeck processes 268
10.5 Full CLT for coagulation processes (a sketch) 270

11 Developments and comments 275
11.1 Measure-valued processes as stochastic dynamic LLNs for interacting particles; duality of one-dimensional processes 275
11.2 Discrete nonlinear Markov games and controlled processes; the modeling of deception 279
11.3 Nonlinear quantum dynamic semigroups and the nonlinear Schrödinger equation 282
11.4 Curvilinear Ornstein–Uhlenbeck processes (linear and nonlinear) and stochastic geodesic flows on manifolds 293
11.5 The structure of generators 300
11.6 Bibliographical comments 310

Appendices 319

A Distances on measures 319

B Topology on càdlàg paths 324

C Convergence of processes in Skorohod spaces 329

D Vector-valued ODEs 334

E Pseudo-differential operator notation 337

F Variational derivatives 338

G Geometry of collisions 343

H A combinatorial lemma 347

I Approximation of infinite-dimensional functions 349

J Bogolyubov chains, generating functionals and Fock-space calculus 352

K Infinite-dimensional Riccati equations 355

References 360

Index 373

Preface

A nonlinear Markov evolution is a dynamical system generated by a measure-valued ordinary differential equation (ODE) with the specific feature that it preserves positivity. This feature distinguishes it from a general Banach-space-valued ODE and yields a natural link with probability theory, both in the interpretation of results and in the tools of analysis. However, nonlinear Markov evolution can be regarded as a particular case of measure-valued Markov processes. Even more important (and not so obvious) is the interpretation of nonlinear Markov dynamics as a dynamic law of large numbers (LLN) for general Markov models of interacting particles. Such an interpretation is both the main motivation for and the main theme of the present monograph.

The power of nonlinear Markov evolution as a modeling tool and its range of applications are immense, and include non-equilibrium statistical mechanics (e.g. the classical kinetic equations of Vlasov, Boltzmann, Smoluchovski and Landau), evolutionary biology (replicator dynamics), population and disease dynamics (Lotka–Volterra and epidemic models) and the dynamics of economic and social systems (replicator dynamics and games). With certain modifications nonlinear Markov evolution carries over to the models of quantum physics.

The general objectives of this book are: (i) to give the first systematic presentation of both analytical and probabilistic techniques used in the study of nonlinear Markov processes, semigroups and kinetic equations, thus providing a basis for future research; (ii) to show how the nonlinear theory is rooted in the study of the usual (linear) Markov semigroups and processes; and (iii) to illustrate general methods by surveying some applications to very basic nonlinear models from statistical (classical and quantum) physics and evolutionary biology.

The book addresses the most fundamental questions in the theory of nonlinear Markov processes: existence, uniqueness, constructions, approximation schemes, regularity, LLN limit and probabilistic interpretation. By a *probabilistic interpretation* of a nonlinear equation or of the corresponding evolution we mean specification of the underlying random process, whose marginal distributions evolve according to this equation, or in other words a path-integral representation for the solutions. This interpretation yields much more than just time dynamics, as it also specifies the correlations between various periods of evolution and suggests natural schemes for numerical solution such as nonlinear versions of the Markov chain Monte Carlo (MCMC) algorithm. Technically, a probabilistic interpretation is usually linked with an appropriate stochastic differential equation (SDE) underlying the given nonlinear dynamics.

Of course, many important issues are beyond the scope of this book. The most notable omissions are: (i) the long-term behavior of, and related questions about, stationary regimes and self-similar solutions; (ii) the effects of irregular behavior (e.g. gelation for the coagulation process); (iii) the DiPerna–Lions theory of generalized solutions; and (iv) numerical methods in general (though we do discuss approximation schemes). All these themes are fully addressed in the modern literature.

A particular feature of our exposition is the systematic combination of analytic and probabilistic tools. We use probability to obtain better insight into nonlinear dynamics and use analysis to tackle difficult problems in the description of random and chaotic behavior.

Whenever possible we adopt various points of view. In particular we present several methods for tackling the key results: analytic and probabilistic approaches to proving the LLN; direct and approximative schemes for constructing the solutions to SDEs; various approaches to the construction of solutions to kinetic equations, discussing uniqueness via duality, positivity and the Lyapunov function method; and the construction of Ornstein–Uhlenbeck semigroups via Riccati equations and SDEs.

An original aim of this book was to give a systematic presentation of all tools needed to grasp the proof of the central limit theorem (CLT) for coagulation processes from Kolokoltsov [136]. Putting this into a general framework required a considerable expansion of this plan. Apart from bringing together results and tools scattered through the journal literature, the main novelties are the following.

(i) The analysis of nonlinear Lévy processes, interacting degenerate stable-like processes and nonlinear Markov games is initiated (Sections 1.4, 7.1, 7.2, 11.2, 11.3).

(ii) A method of constructing linear and nonlinear Markov processes with general Lévy–Khintchine-type generators (including flows on manifolds such as curvilinear Ornstein–Uhlenbeck processes and stochastic geodesic flows) via SDEs driven by nonlinear distribution-dependent Lévy-type noise is put forward. In particular, a solution is suggested to the long-standing problem of identifying the continuity class of Lévy kernels for which the corresponding Lévy–Khintchine-type operators generate Feller processes: these kernels must be continuous in the Wasserstein–Kantorovich metric W_2 (Chapter 3). A modification of Feller semigroups suitable for the analysis of linear and nonlinear processes with unbounded coefficients is proposed (Chapter 5).

(iii) A class of pseudo-differential generators of "order at most one" is singled out, for which both linear and nonlinear theory can be developed by a direct analytic treatment (Sections 4.4, 7.2).

(iv) A class of infinite-dimensional Ornstein–Uhlenbeck semigroups and related infinite-dimensional Riccati equations, which arises as the limit of fluctuations for general mean field and kth-order interactions, is identified and its analysis is initiated (Chapter 10).

(v) A theory of smoothness with respect to initial data for a wide class of kinetic equations is developed (Chapter 8).

(vi) This smoothness theory is applied to obtain laws of large numbers (LLNs) and central limit theorems (CLTs) with rather precise convergence rates for Markov models of interactions with unbounded coefficients. These include nonlinear stable-like processes, evolutionary games, processes governed by Vlasov-type equations, Smoluchovski coagulation and Boltzmann collision models.

Readers and prerequisites

The book is aimed at researchers and graduate students in stochastic and functional analysis as applied to mathematical physics and systems biology (including non-equilibrium statistical and quantum mechanics, evolutionary games) as well as at natural scientists with strong mathematical backgrounds interested in nonlinear phenomena in dynamic modeling. The exposition is a step-by-step account that is intended to be accessible and comprehensible. A few exercises, mostly straightforward, are placed at the ends of some sections to illustrate or clarify points in the text.

The prerequisites for reading the book are (i) the basic notions of functional analysis (a superficial understanding of Banach and Hilbert spaces is sufficient; everything needed is covered in the early chapters of a standard treatise such as Reed and Simon [205] or Yosida [250]), (ii) abstract measure theory and the Lebesgue integral (including L_p-spaces and, preferably,

Fourier transforms) and (iii) probability theory and random processes (elementary distributions, characteristic functions, convergence of random variables, conditioning, Markov and Lévy processes, martingales; see e.g. Kallenberg [114], Shiryayev [220], Jacod and Protter [106], Applebaum [8], Kyprianou [153]).

The book is designed in such a way that, depending on their background and interests, readers may choose a selective path of reading. For instance, readers interested only in jump-type processes (including evolutionary games and spatially trivial Smoluchovski and Boltzmann models) do not need SDEs and ΨDO and so can read Sections 2.1, 2.3, 4.2, 4.3, Chapter 6 and Sections 8.3–8.6 and then look for sections that are relevant to them in Part III. However, readers interested in nonlinear Lévy processes, diffusions and stable-like processes should look at Chapters 2 and 3, Sections 4.4 and 4.7, Chapters 5 and 7, Sections 8.2, 9.1 and 9.2 and the relevant parts of Sections 10.1 and 10.2.

Plan of the book

In Chapter 1, the first four sections introduce nonlinear processes in the simplest situations, where either space or time is discrete (nonlinear Markov chains) or the dynamics has a trivial space dependence (the constant-coefficient case, describing nonlinear Lévy processes). The rest of this introductory chapter is devoted to an informal discussion of the limit of the LLN in Markov models of interaction. This limit is described by kinetic equations and its analysis can be considered as the main motivation for studying nonlinear Markov processes.

As the nonlinear theory is deeply rooted in the linear theory (since infinitesimal transformations are linear), Part I of the book is devoted to background material on the usual (linear) Markov processes. Here we systematically build the “linear basement” for the “nonlinear skyscrapers” to be erected later. Chapter 2 recalls some particularly relevant tools from the theory of Markov processes, stressing the connection between an analytical description (using semigroups and evolution equations) and a probabilistic description. Chapters 3 to 5 deal with methods of constructing Markov processes that serve as starting points for subsequent nonlinear extensions. The three cornerstones of our analysis – the concepts of positivity, duality and perturbation – are developed here in the linear setting.

Nonlinear processes *per se* are developed in Part II. Chapters 6 and 7 open with basic constructions and well-posedness results for nonlinear Markov semigroups and processes and the corresponding kinetic equations. Chapter 8, which is rather technical, is devoted to the regularity of nonlinear Markov semigroups with respect to the initial data. Though these results are of independent interest, the main motivation for their development here is to prepare a sound basis for the analytic study of the LLN undertaken later in the book.

In Part III we study the application of nonlinear processes to the dynamic LLN and the corresponding CLT for fluctuations (Chapters 9 and 10).

In Chapter 11 we sketch possible directions for the further development of the ideas presented here, namely the stochastic LLN and related measure-valued processes, nonlinear Markov games, nonlinear quantum dynamic semigroups and processes, linear and nonlinear processes on manifolds and, finally, the analysis of the generators of positivity-preserving evolutions. Section 11.6 concludes with historical comments and a short review of the (immense) literature on the subject and of related results.

The appendices collect together technical material used in the main text.

I am indebted to Diana Gillooly from CUP and to Ismael Bailleul, who devoted the time and energy to read extensively and criticize early drafts. I would also like to thank my colleagues and friends from all over the globe, from Russia to Mexico, for useful discussions that helped me to understand better the crucial properties of stochastic processes and interacting particles.

Basic definitions, notation and abbreviations

Kernels and propagators

Kernels and propagators are the main players in our story. We recall here the basic definitions. A *transition kernel* from a measurable space $(X, \mathcal{F})$ to a measurable space $(Y, \mathcal{G})$ is a function of two variables, $\mu(x, A)$, $x \in X$, $A \in \mathcal{G}$, which is $\mathcal{F}$-measurable as a function of x for any A and is a measure in $(Y, \mathcal{G})$ for any x. It is called a *transition probability kernel* or simply a *probability kernel* if all measures $\mu(x, .)$ are probability measures. In particular, a *random measure* on a measurable space $(X, \mathcal{F})$ is a transition kernel from a probability space to $(X, \mathcal{F})$. *Lévy kernels* from a measurable space $(X, \mathcal{F})$ to $\mathbf{R}^d$ are defined as above but now each $\mu(x, .)$ is a Lévy measure on $\mathbf{R}^d$, i.e. a (possibly unbounded) Borel measure such that $\mu(x, \{0\}) = 0$ and $\int \min(1, y^2)\mu(x, dy) < \infty$.

For a set S, a family of mappings $U^{t,r}$ from S to itself, parametrized by pairs of numbers $r \leq t$ (resp. $t \leq r$) from a given finite or infinite interval is called a *propagator* (resp. a *backward propagator*) in S if $U^{t,t}$ is the identity operator in S for all t and the following *chain rule*, or *propagator equation*, holds for $r \leq s \leq t$ (resp. for $t \leq s \leq r$): $U^{t,s}U^{s,r} = U^{t,r}$. A family of mappings T^t from S to itself parametrized by non-negative numbers t is said to form a *semigroup* (of the transformations of S) if T^0 is the identity mapping in S and $T^tT^s = T^{t+s}$ for all t, s. If the mappings $U^{t,r}$ forming a backward propagator depend only on the differences $r - t$ then the family $T^t = U^{0,t}$ forms a semigroup.

Basic notation

Sets and numbers

N, **Z**, **R**, **C** The sets of natural, integer, real and complex numbers;

$\mathbf{Z}_+$ The set $\mathbf{N} \cup \{0\}$

$\mathbf{R}_+$ (resp. $\bar{\mathbf{R}}_+$) The set of positive (resp. non-negative) numbers

$\mathbf{N}^\infty$, $\mathbf{Z}^\infty$, $\mathbf{R}^\infty$, $\mathbf{C}^\infty$ The sets of sequences from $\mathbf{N}$, $\mathbf{Z}$, $\mathbf{R}$, $\mathbf{C}$

$\mathbf{C}^d$, $\mathbf{R}^d$ The complex and real d-dimensional spaces

(x, y) or xy Scalar product of the vectors $x, y \in \mathbf{R}^d$

$|x|$ or $\|x\|$ Standard Euclidean norm $\sqrt{(x, x)}$ of $x \in \mathbf{R}^d$

$\mathrm{Re}\, a$, $\mathrm{Im}\, a$ Real and imaginary parts of a complex number a

$[x]$ Integer part of a real number x (the maximum integer not exceeding x)

S^d The d-dimensional unit sphere in $\mathbf{R}^{d+1}$

$B_r(x)$ (resp. B_r) The ball of radius r centred at x (resp. at the origin) in $\mathbf{R}^d$

$\bar{\Omega}$, $\partial\Omega$ Closure and boundary respectively of the subset Ω in a metric space

Functions

$C(S)$ (resp. $B(S)$) For a complete metric space (S, ρ) (resp. for a measurable space $(S, \mathcal{F})$), the Banach space of bounded continuous (resp. measurable) functions on S equipped with the sup norm $\|f\| = \|f\|_{C(S)} = \sup_{x\in S} |f(x)|$

$BUC(S)$ Closed subspace of $C(S)$ consisting of uniformly continuous functions

$C_f(S)$ (resp. $B_f(S)$) For a positive function f on X, the Banach space of continuous (resp. measurable) functions g on S with finite norm $\|g\|_{C_f(S)} = \|g/f\|_{C(S)}$ (resp. with B instead of C)

$C_{f,\infty}(S)$ (resp. $B_{f,\infty}(S)$) The subspace of $C_f(S)$ (resp. $B_f(S)$) consisting of functions g such that the ratio of g and f belongs to $C_\infty(S)$

$C_c(S) \subset C(S)$ Functions with a compact support

$C_{\mathrm{Lip}}(S) \subset C(S)$ Lipschitz continuous functions f, i.e. $|f(x) - f(y)| \leq \kappa\rho(x, y)$ with a constant κ

$C_{\mathrm{Lip}}(S)$ Banach space under the norm $\|f\|_{\mathrm{Lip}} = \sup_x |f(x)| + \sup_{x\neq y} |f(x) - f(y)|/|x - y|$

$C_\infty(S) \subset C(S)$ Functions f such that $\lim_{x\to\infty} f(x) = 0$, i.e. for all ϵ there exists a compact set $K : \sup_{x\notin K} |f(x)| < \epsilon$ (it is a closed subspace of $C(S)$ if S is locally compact)

$C^{\mathrm{sym}}(S^k)$ or $C_{\mathrm{sym}}(S^k)$ Symmetric continuous functions on X^k, i.e. functions invariant under any permutations of their arguments

$C^k(\mathbf{R}^d)$ (sometimes for brevity C^k) Banach space of k times continuously differentiable functions with bounded derivatives on $\mathbf{R}^d$ and for which the norm is the sum of the sup norms of the function itself and all its partial derivatives up to and including order k

$C^k_\infty(\mathbf{R}^d) \subset C^k(\mathbf{R}^d)$ Functions for which all derivatives up to and including order k are from $C_\infty(\mathbf{R}^d)$

$C^k_{\mathrm{Lip}}(\mathbf{R}^d)$ A subspace of $C^k(\mathbf{R}^d)$ whose derivative of order k is Lipschitz continuous; it is a Banach space equipped with the norm $\|f\|_{C^k_{\mathrm{Lip}}} = \|f\|_{C^{k-1}} + \|f^{(k)}\|_{\mathrm{Lip}}$

$C^k_{\mathrm{c}}(\mathbf{R}^d) = C_{\mathrm{c}}(\mathbf{R}^d) \cap C^k(\mathbf{R}^d)$

$\nabla f = (\nabla_1 f, \dots, \nabla_d f) = (\partial f/\partial x_1, \dots, \partial f/\partial x_d), \quad f \in C^1(\mathbf{R}^d)$

$L^p(\Omega, \mathcal{F}, \mu)$ or $L_p(\Omega, \mathcal{F}, \mu)$, $p \geq 1$ The usual Banach space of (equivalence classes of) measurable functions f on the measure space Ω such that $\|f\|_p = \left(\int |f|^p(x)\mu(dx)\right)^{1/p} < \infty$

$L_p(\mathbf{R}^d)$ The L_p-space that corresponds to Lebesgue measure

$L^\infty(\Omega, \mathcal{F}, P)$ Banach space of (equivalence classes of) measurable functions f on the measure space Ω with finite sup norm $\|f\| = \operatorname{ess\,sup}_{x\in\Omega} |f(x)|$

$W_1^l = W_1^l(\mathbf{R}^d)$ Sobolev Banach spaces of integrable functions on $\mathbf{R}^d$ whose derivatives up to and including order l (defined in the sense of distributions) are also integrable and equipped with the norms $\|f\|_{W_1^l} = \sum_{m=0}^l \|f^{(m)}\|_{L_1(\mathbf{R}^d)}$

$S(\mathbf{R}^d) = \{f \in C^\infty(\mathbf{R}^d) : \forall k, l \in \mathbf{N},\ |x|^k \nabla^l f \in C_\infty(\mathbf{R}^d)\}$ Schwartz space of fast-decreasing functions

Measures

$\mathcal{M}(S)$ (resp. $\mathcal{P}(S)$) The set of finite Borel measures (resp. probability measures) on a metric space S

$\mathcal{M}^{\mathrm{signed}}(S)$ Banach space of finite signed Borel measures on S; $\mu_n \to \mu$ weakly in $\mathcal{M}^{\mathrm{signed}}(S)$ means that $(f, \mu_n) \to (f, \mu)$ for any $f \in C(S)$

$\mathcal{M}_f(S)$ for a positive continuous function f on S The set of Radon measures on S with finite norm $\|\mu\|_{\mathcal{M}_f(S)} = \sup_{\|g\|_{C_f(S)} \leq 1}(g, \mu)$; $\mu_n \to \mu$ weakly in $\mathcal{M}_f(S)$ means that $(f, \mu_n) \to (f, \mu)$ for any $f \in C_f(S)$; if S is locally compact and $f \to \infty$ as $x \to \infty$ then $\mathcal{M}_f(S)$ is the Banach dual to $C_{f,\infty}(S)$, so that $\mu_n \to \mu$ $\star$-weakly in $\mathcal{M}_f(S)$ means that $(f, \mu_n) \to (f, \mu)$ for any $f \in C_{f,\infty}(S)$

$\mathcal{M}^+_{h\delta}(S)$ The set of finite linear combinations of Dirac's δ-measures on S with coefficients hk, $k \in \mathbf{N}$

μ^f The pushforward of μ by the mapping f: $\mu^f(A) = \mu(f^{-1}(A)) = \mu\{y : f(y) \in A\}$

$|\nu|$ For a signed measure ν, this is its (positive) total variation measure

$(f, g) = \int f(x)g(x)\,dx$ Scalar product for functions f, g on $\mathbf{R}^d$

$(f, \mu) = \int_S f(x)\mu(dx)$ Pairing of $f \in C(S)$, $\mu \in \mathcal{M}(S)$

Matrices and operators

$\mathbf{1}_M$ Indicator function of a set M (equals one or zero according to whether its argument is in M or otherwise)

$\mathbf{1}$ Constant function equal to one; also, the identity operator

$f = O(g)$ This means that $|f| \leq Cg$ for some constant C

$f = o(g)_{n\to\infty}$ This means that $\lim_{n\to\infty}(f/g) = 0$

A^T or A' Transpose of a matrix A

$A^\star$ or A' Dual or adjoint operator

Ker$A$, Sp A, tr A Kernel, spectrum and trace of the operator A

Probability

$\mathbf{E}$, $\mathbf{P}$ Expectation and probability of a function or event

$\mathbf{E}^x$, $\mathbf{P}^x$ for $x \in S$ (resp. $\mathbf{E}^\mu$, $\mathbf{P}^\mu$ for $\mu \in \mathcal{P}(S)$) Expectation and probability with respect to a process started at x (resp. with initial distribution μ)

Standard abbreviations

r.h.s.	right-hand side
l.h.s.	left-hand side
a.s.	almost surely
i.i.d.	independent identically distributed
BM	Brownian motion
CLT	central limit theorem
LLN	law of large numbers
ODE	ordinary differential equation
OU	Ornstein–Uhlenbeck
SDE	stochastic differential equation
ΨDO	pseudo-differential operator

1

Introduction

Sections 1.1–1.4 introduce nonlinear processes in the simplest situations, where either space or time is discrete (nonlinear Markov chains) or the dynamics has trivial space dependence (the constant-coefficient case describing nonlinear Lévy processes). The rest of the chapter is devoted to an informal discussion of the law of large numbers (LLN) for particles in Markov models of interaction. This limit is described by kinetic equations, and its analysis can be considered as the main motivation for the study of nonlinear Markov processes.

1.1 Nonlinear Markov chains

A discrete-time discrete-space *nonlinear Markov semigroup* Φ^k, $k \in \mathbf{N}$, is specified by an arbitrary continuous mapping $\Phi : \Sigma_n \to \Sigma_n$, where the simplex

$$\Sigma_n = \left\{ \mu = (\mu_1, \ldots, \mu_n) \in \mathbf{R}_+^n : \sum_{i=1}^n \mu_i = 1 \right\}$$

represents the set of probability laws on the finite state space $\{1, \ldots, n\}$. For a measure $\mu \in \Sigma_n$ the family $\mu^k = \Phi^k \mu$ can be considered as an evolution of measures on $\{1, \ldots, n\}$. But it does not yet define a random process, because finite-dimensional distributions are not yet specified. In order to obtain a process we have to choose a *stochastic representation* for Φ, i.e. to write it down in the form

$$\Phi(\mu) = \{\Phi_j(\mu)\}_{j=1}^n = \left\{ \sum_{i=1}^n P_{ij}(\mu)\mu_i \right\}_{j=1}^n, \tag{1.1}$$

where $P_{ij}(\mu)$ is a family of stochastic matrices[1] depending on μ (and so introducing nonlinearity!), whose elements specify the *nonlinear transition probabilities*. For any given $\Phi : \Sigma_n \mapsto \Sigma_n$ a representation (1.1) exists but is not unique. There exists a unique representation (1.1) given the additional condition that all matrices $P_{ij}(\mu)$ are one dimensional:

$$P_{ij}(\mu) = \Phi_j(\mu), \quad i, j = 1, \ldots, n. \tag{1.2}$$

Once a stochastic representation (1.1) for a mapping Φ is chosen we can naturally define, for any initial probability law $\mu = \mu^0$, a stochastic process i_l, $l \in \mathbf{Z}_+$, called a *nonlinear Markov chain*, on $\{1, \ldots, n\}$ in the following way. Starting with an initial position i_0, distributed according to μ, we then choose the next point i_1 according to the law $\{P_{i_0 j}(\mu)\}_{j=1}^n$. The distribution of i_1 now becomes $\mu^1 = \Phi(\mu)$:

$$\mu_j^1 = \mathbf{P}(i_1 = j) = \sum_{i=1}^n P_{ij}(\mu)\mu_i = \Phi_j(\mu).$$

Then we choose i_2 according to the law $\{P_{i_1 j}(\mu^1)\}_{j=1}^n$, and so on. The law governing this process at any given time k is $\mu^k = \Phi^k(\mu)$; that is, it is given by the semigroup. Now finite-dimensional distributions will be defined as well. Namely, for a function f of, say, two discrete variables, we have

$$\mathbf{E} f(i_k, i_{k+1}) = \sum_{i,j=1}^n f(i, j)\mu_i^k P_{ij}(\mu^k).$$

In other words, this process can be defined as a time-nonhomogeneous Markov chain with transition probabilities $P_{ij}(\mu^k)$ at time $t = k$.

Clearly the finite-dimensional distributions depend on the choice of representation (1.1). For instance, for the simplest representation (1.2) we have

$$\mathbf{E} f(i_0, i_1) = \sum_{i,j=1}^n f(i, j)\mu_i \Phi_j(\mu),$$

so that the discrete random variables i_0 and i_1 turn out to be independent.

Once a representation (1.1) is chosen, we can also define the transition probabilities P_{ij}^k at time $t = k$ recursively as

$$P_{ij}^k(\mu) = \sum_{m=1}^n P_{im}^{k-1}(\mu) P_{mj}(\mu^{k-1}).$$

[1] Recall that a $d \times d$ matrix Q is called stochastic if all its elements Q_{ij} are non-negative and such that $\sum_{j=1}^d Q_{ij} = 1$ for all i.

The semigroup identity $\Phi^{k+l} = \Phi^k \Phi^l$ implies that

$$\Phi_j^k(\mu) = \sum_{i=1}^n P_{ij}^k(\mu)\mu_i$$

and

$$P_{ij}^k(\mu) = \sum_{m=1}^n P_{im}^l(\mu) P_{mj}^{k-l}(\mu^l), \quad l < k.$$

Remark 1.1 In practical examples of the general model (1.1) the transition probabilities often depend on the law μ via such basic characteristics as its standard deviation or expectation. See e.g. Frank [79], where we can also find some elementary examples of *deterministic nonlinear Markov chains*, for which the transitions are certain once the distribution is known, i.e. where $P_{ij}(\mu) = \delta^j_{j(i,\mu)}$ for a given deterministic mapping $(i, \mu) \mapsto j(i, \mu)$.

We can establish nonlinear analogs of many results known for the usual Markov chains. For example, let us present the following simple fact about long-time behavior.

Proposition 1.2 *(i) For any continuous $\Phi : \Sigma_n \to \Sigma_n$ there exists a stationary distribution, i.e. a measure $\mu \in \Sigma_n$, such that $\Phi(\mu) = \mu$.*

(ii) If a representation (1.1) *for Φ is chosen in such a way that there exist $j_0 \in [1, n]$, time $k_0 \in \mathbf{N}$ and positive δ such that*

$$P_{ij_0}^{k_0}(\mu) \geq \delta, \tag{1.3}$$

for all i, μ, then $\Phi^m(\mu)$ converges to a stationary measure for any initial μ.

Proof Statement (i) is a consequence of the Brouwer fixed point principle. Statement (ii) follows from the representation (given above) of the corresponding nonlinear Markov chain as a time-nonhomogeneous Markov process. □

Remark 1.3 The convergence of $P_{ij}^m(\mu)$ as $m \to \infty$ can be shown by a standard argument. We introduce the bounds

$$m_j(t, \mu) = \inf_i P_{ij}^t(\mu), \qquad M_j(t, \mu) = \sup_i P_{ij}^t(\mu),$$

then we deduce from the semigroup property that $m_j(t, \mu)$ (resp. $M_j(t, \mu)$) is an increasing (resp. decreasing) function of t and finally we deduce from (1.3) that

$$M_j(t + k_0, \mu) - m_j(t + k_0, \mu) \leq (1 - \delta)\Big(M_j(t, \mu) - m_j(t, \mu)\Big),$$

implying the required convergence. (See e.g. Norris [193], Shiryayev [220], and Rozanov [210] for the time-homogeneous situation.)

We turn now to nonlinear chains in continuous time. A *nonlinear Markov semigroup* in continuous time and with finite state space $\{1, \ldots, n\}$ is defined as a semigroup Φ^t, $t \geq 0$, of continuous transformations of Σ_n. As in the case of discrete time the semigroup itself does not specify a process. A *continuous family of nonlinear transition probabilities* on $\{1, \ldots, n\}$ is a family $P(t, \mu) = \{P_{ij}(t, \mu)\}_{i,j=1}^n$ of stochastic matrices, depending continuously on $t \geq 0$ and $\mu \in \Sigma_n$, such that the *nonlinear Chapman–Kolmogorov equation* holds:

$$\sum_{i=1}^n \mu_i P_{ij}(t+s, \mu) = \sum_{k,i} \mu_k P_{ki}(t, \mu) P_{ij}\left(s, \sum_{l=1}^n P_{l.}(t, \mu)\mu_l\right). \tag{1.4}$$

This family is said to yield a *stochastic representation* for the Markov semigroup Φ^t whenever

$$\Phi_j^t(\mu) = \sum_i \mu_i P_{ij}(t, \mu), \qquad t \geq 0, \mu \in \Sigma_n. \tag{1.5}$$

If (1.5) holds, equation (1.4) simply represents the semigroup identity $\Phi^{t+s} = \Phi^t \Phi^s$.

Once a stochastic representation (1.5) for the semigroup Φ^k is chosen, we can define the corresponding stochastic process starting from $\mu \in \Sigma_n$ as a time-nonhomogeneous Markov chain with transition probabilities from time s to time t

$$p_{ij}(s, t, \mu) = P_{ij}(t - s, \Phi^s(\mu)).$$

To show the existence of a stochastic representation (1.5) we can use the same idea as in the discrete-time case and define $P_{ij}(t, \mu) = \Phi_j^t(\mu)$. However, this is not a natural choice from the point of view of stochastic analysis. A natural choice would arise from a generator that is reasonable from the point of view of the theory of Markov processes.

Namely, assuming the semigroup Φ^t is differentiable in t we can define the *(nonlinear) infinitesimal generator* of the semigroup Φ^t as the nonlinear operator on measures given by

$$A(\mu) = \frac{d}{dt}\Phi^t\Big|_{t=0}(\mu).$$

The semigroup identity for Φ^t implies that $\Phi^t(\mu)$ solves the Cauchy problem

$$\frac{d}{dt}\Phi^t(\mu) = A(\Phi^t(\mu)), \quad \Phi^0(\mu) = \mu. \tag{1.6}$$

As follows from the invariance of Σ_n under these dynamics, the mapping A is *conditionally positive*, in the sense that if $\mu_i = 0$ for a $\mu \in \Sigma_n$ then this implies $A_i(\mu) \geq 0$, and it is also *conservative* in the sense that A maps the measures from Σ_n to the space of signed measures

$$\Sigma_n^0 = \left\{\nu \in \mathbf{R}^n : \sum_{i=1}^n \nu_i = 0\right\}.$$

We shall say that such a generator A has a *stochastic representation* if it can be written in the form

$$A_j(\mu) = \sum_{i=1}^n \mu_i Q_{ij}(\mu) = (\mu Q(\mu))_j, \tag{1.7}$$

where $Q(\mu) = \{Q_{ij}(\mu)\}$ is a family of infinitesimally stochastic matrices depending on $\mu \in \Sigma_n$.[2] Thus in its stochastic representation the generator has the form of a usual Markov chain generator, though it depends additionally on the present distribution. The existence of a stochastic representation for the generator is not as obvious as for the semigroup but is not difficult to obtain, as shown by the following statement.

Proposition 1.4 *Given any nonlinear Markov semigroup Φ^t on Σ_n that is differentiable in t, its infinitesimal generator has a stochastic representation.*

An elementary proof can be obtained (see Stroock [227]) from the observation that as we are interested only in the action of Q on μ we can choose its action Σ_n^0 on the space transverse to μ in an arbitrary way. Instead of reproducing this proof we shall give in Section 6.8 a straightforward (and remarkably simple) proof of the generalization of this fact for nonlinear operators in general measurable spaces.

In practice, the converse problem is more important: the construction of a semigroup (a solution to (1.6)) from a given operator A, rather than the construction of the generator for a given semigroup. In applications, A is usually given directly in its stochastic representation. This problem will be one of our central concerns in this book, but in a quite general setting.

[2] A square matrix is called *infinitesimally stochastic* if it has non-positive (resp. non-negative) elements on the main diagonal (resp. off the main diagonal) and the sum of the elements of any row is zero. Such matrices are also called Q-matrices or Kolmogorov matrices.

1.2 Examples: replicator dynamics, the Lotka–Volterra equations, epidemics, coagulation

Nonlinear Markov semigroups abound among popular models in the natural and social sciences, so it is difficult to distinguish the most important examples. We shall discuss briefly here three biological examples (anticipating our future analysis of evolutionary games) and an example from statistical mechanics (anticipating our analysis of kinetic equations) illustrating the notions introduced above of stochastic representation, conditional positivity and so forth.

The *replicator dynamics* of the evolutionary game arising from the classical game of rock–paper–scissors (a hand game for two players) has the form

$$\begin{cases} \dfrac{dx}{dt} = (y-z)x, \\ \dfrac{dy}{dt} = (z-x)y, \\ \dfrac{dz}{dt} = (x-y)z, \end{cases} \tag{1.8}$$

(see e.g. Gintis [84], where a biological interpretation can be found also; the general equations of replicator dynamics are discussed in Section 1.6 of the present text). The generator of equations (1.8) clearly has a stochastic representation (1.7) with infinitesimal stochastic matrix

$$Q(\mu) = \begin{pmatrix} -z & 0 & z \\ x & -x & 0 \\ 0 & y & -y \end{pmatrix}, \tag{1.9}$$

where $\mu = (x, y, z) \in \Sigma_3$.

The famous *Lotka–Volterra equations* describing a biological system with two species, a predator and its prey, have the form

$$\begin{cases} \dfrac{dx}{dt} = x(\alpha - \beta y), \\ \dfrac{dy}{dt} = -y(\gamma - \delta x), \end{cases} \tag{1.10}$$

where $\alpha, \beta, \gamma, \delta$ are positive parameters. The generator of this model is conditionally positive but not conservative, as the total mass $x + y$ is not preserved. However, owing to the existence of the integral of motion $\alpha \log y - \beta y + \gamma \log x - \delta x$, the dynamics (1.10) is pathwise equivalent to the dynamics (1.8); i.e. there is a continuous mapping taking the phase portrait of system (1.10) to that of system (1.8).

One of the simplest deterministic *models of an epidemic* can be written as a system of four differential equations:

$$\begin{cases} \dot{X}(t) = -\lambda X(t)Y(t), \\ \dot{L}(t) = \lambda X(t)Y(t) - \alpha L(t), \\ \dot{Y}(t) = \alpha L(t) - \mu Y(t), \\ \dot{Z}(t) = \mu Y(t), \end{cases} \tag{1.11}$$

where $X(t)$, $L(t)$, $Y(t)$ and $Z(t)$ denote respectively the numbers of susceptible, latent, infectious and removed individuals at time t and the positive coefficients λ, α, μ (which may actually depend on X, L, Y, Z) reflect the rates at which susceptible individuals become infected, latent individuals become infectious and infectious individuals are removed. Written in terms of the proportions $x = X/\sigma$, $y = Y/\sigma$, $l = L/\sigma$, $z = Z/\sigma$, i.e. normalized to the total mass $\sigma = X + L + Y + Z$, this system becomes

$$\begin{cases} \dot{x}(t) = -\sigma\lambda x(t)y(t), \\ \dot{l}(t) = \sigma\lambda x(t)y(t) - \alpha l(t), \\ \dot{y}(t) = \alpha l(t) - \mu y(t), \\ \dot{z}(t) = \mu y(t), \end{cases} \tag{1.12}$$

with $x(t) + y(t) + l(t) + z(t) = 1$. Subject to the common assumption that $\sigma\lambda$, α and μ are constants, the r.h.s. is an infinitesimal generator of a nonlinear Markov chain in Σ_4. Again the generator depends quadratically on its variable and has an obvious stochastic representation (1.7) with infinitesimal stochastic matrix

$$Q(\mu) = \begin{pmatrix} -\lambda y & \lambda y & 0 & 0 \\ 0 & -\alpha & \alpha & 0 \\ 0 & 0 & -\mu & \mu \\ 0 & 0 & 0 & 0 \end{pmatrix}, \tag{1.13}$$

where $\mu = (x, l, y, z)$, yielding a natural probabilistic interpretation for the dynamics (1.12) as explained in the previous section. For a detailed deterministic analysis of this model and a variety of extensions we refer to the book by Rass and Radcliffe [201].

We turn now to an example from statistical mechanics, namely the dynamics of coagulation processes with a discrete mass distribution. Unlike the previous examples, the state space here is not finite but rather countable. As in the linear theory, the basic notions of finite nonlinear Markov chains presented above have a straightforward extension to the case of countable state spaces.

Let $x_j \in \mathbf{R}_+$ denote the amount of particles of mass $j \in \mathbf{N}$ present in the system. Assuming that the rate of coagulation of particles of masses i and j is proportional to the present amounts of particles x_i and x_j, with proportionality coefficients given by positive numbers K_{ij}, one can model the process by the system of equations

$$\dot{x}_j = \frac{1}{2} \sum_{i,k=1}^{\infty} K_{ik} x_i x_k (\delta_j^{i+k} - \delta_j^i - \delta_j^k), \qquad j = 1, 2, \ldots, \tag{1.14}$$

or equivalently

$$\dot{x}_j = \frac{1}{2} \sum_{i=1}^{j-1} K_{i,j-i} x_i x_{j-i} - \sum_{k=1}^{\infty} K_{kj} x_k x_j, \qquad j = 1, 2, \ldots \tag{1.15}$$

These are the much studied *Smoluchovski coagulation equations* for discrete masses. The r.h.s. is again an infinitesimal generator in the stochastic form (1.7) with quadratic dependence on the unknown variables, but now with a countable state space, the natural numbers, $\mathbf{N}$.

In the next section we introduce another feature (i.e. another probabilistic interpretation) of nonlinear Markov semigroups and processes. It turns out that they represent the dynamic law of large numbers (LLN) for Markov models of interacting particles. In particular, this representation will explain the frequent appearance of the quadratic r.h.s. in the corresponding evolution equations, as this quadratic dependence reflects the binary interactions that are most often met in practice. The simultaneous interactions of groups of k particles would lead to a polynomial of order k on the r.h.s.

1.3 Interacting-particle approximation for discrete mass-exchange processes

We now explain the natural appearance of nonlinear Markov chains as a *dynamic law of large numbers* in the case of *discrete mass-exchange processes*; these include coagulation, fragmentation, collision breakage and other mass-preserving interactions. Thus for the last time we will work with a discrete (countable) state space, trying to visualize the idea of the LLN limit for this easier-to-grasp situation. Afterwards we shall embark on our main journey, which is devoted to general (mostly locally compact) state spaces.

We denote by $\mathbf{Z}_+^\infty$ the subset of $\mathbf{Z}^\infty$ with non-negative elements that is equipped with the usual partial order: $N = \{n_1, n_2, \ldots\} \leq M = \{m_1, m_2, \ldots\}$ means that $n_j \leq m_j$ for all j. Let $\mathbf{R}_{+,\text{fin}}^\infty$ and $\mathbf{Z}_{+,\text{fin}}^\infty$ denote respectively the subsets of $\mathbf{R}_+^\infty$ and $\mathbf{Z}_+^\infty$ containing sequences with only a finite number of non-zero coordinates. We shall denote by $\{e_j\}$ the standard basis in $\mathbf{R}^\infty_{+,\mathrm{fin}}$ and will occasionally represent sequences $N = \{n_1, n_2, \ldots\} \in \mathbf{Z}^\infty_{+,\mathrm{fin}}$ as linear combinations $N = \sum_{j=1}^\infty n_j e_j$.

Suppose that a particle is characterized by its mass m, which can take only integer values. A collection of particles is then described by the vector $N = \{n_1, n_2, \ldots\} \in \mathbf{Z}^\infty_+$, where the non-negative integer n_j denotes the number of particles of mass j. The state space of our model is the set $\mathbf{Z}^\infty_{+,\mathrm{fin}}$ of finite collections of particles. We shall denote by $|N| = n_1 + n_2 + \cdots$ the number of particles in the state N, by $\mu(N) = n_1 + 2n_2 + \cdots$ the total mass of particles in this state and by $\mathrm{supp}\,(N) = \{j : n_j \neq 0\}$ the support of N considered as a measure on $\{1, 2, \ldots\}$.

Let Ψ be an arbitrary element of $\mathbf{Z}^\infty_{+,\mathrm{fin}}$. By its *mass-exchange* transformation we shall mean any transformation of Ψ into an element $\Phi \in \mathbf{Z}^\infty_{+,\mathrm{fin}}$ such that $\mu(\Psi) = \mu(\Phi)$. For instance, if Ψ consists of only one particle then this transformation is *pure fragmentation*, and if Φ consists of only one particle then this transformation is *pure coagulation* (not necessarily binary, of course). By a process of *mass exchange with a given profile* $\Psi = \{\psi_1, \psi_2, \ldots\} \in \mathbf{Z}^\infty_{+,\mathrm{fin}}$ we shall mean the Markov chain on $\mathbf{Z}^\infty_{+,\mathrm{fin}}$ specified by a Markov semigroup, on the space $B(\mathbf{Z}^\infty_{+,\mathrm{fin}})$ of bounded functions on $\mathbf{Z}^\infty_{+,\mathrm{fin}}$, whose generator is given by

$$G_\Psi f(N) = C_N^\Psi \sum_{\Phi:\mu(\Phi)=\mu(\Psi)} P_\Psi^\Phi \left[f(N - \Psi + \Phi) - f(N) \right]. \tag{1.16}$$

Here $C_N^\Psi = \prod_{i \in \mathrm{supp}\,(\Psi)} C_{n_i}^{\psi_i}$ (C_n^k denotes a binomial coefficient) and $\{P_\Psi^\Phi\}$ is any collection of non-negative numbers parametrized by $\Phi \in \mathbf{Z}^\infty_{+,\mathrm{fin}}$ such that $P_\Psi^\Phi = 0$ whenever $\mu(\Phi) \neq \mu(\Psi)$. It is understood that $G_\Psi f(N) = 0$ whenever $\Psi \leq N$ does not hold. Since mass is preserved this Markov chain is effectively a chain with a finite state space, specified by the initial condition, and hence it is well defined and does not explode in finite time. The behavior of the process defined by the generator (1.16) is as follows: (i) if $N \geq \Psi$ does not hold then N is a stable state; (ii) if $N \geq \Psi$ does hold then any randomly chosen subfamily Ψ of N can be transformed to a collection Φ with the rate P_Ψ^Φ. A subfamily Ψ of N consists of any ψ_1 particles of mass 1 from a given number n_1 of these particles, any ψ_2 particles of mass 2 from a given number n_2 etc. (notice that the coefficient C_N^Ψ in (1.16) is just the number of such choices).

More generally, if k is a natural number, a *mass-exchange process of order* k, or k-*ary mass-exchange process*, is a Markov chain on $\mathbf{Z}^\infty_{+,\mathrm{fin}}$ defined by the generator $G_k = \sum_{\Psi:|\Psi|\leq k} G_\Psi$. More explicitly,

$$G_k f(N) = \sum_{\Psi:|\Psi|\le k, \Psi\le N} C_N^\Psi \sum_{\Phi:\mu(\Phi)=\mu(\Psi)} P_\Psi^\Phi[f(N-\Psi+\Phi)-f(N)], \quad (1.17)$$

where P_Ψ^Φ is an arbitrary collection of non-negative numbers that vanish whenever $\mu(\Psi) \neq \mu(\Phi)$. As in the case of a single Ψ, for any initial state N this Markov chain lives on a finite state space of all M with $\mu(M) = \mu(N)$ and hence is always well defined.

We shall now perform a scaling that represents a discrete version of the general procedure leading to the law of large numbers for Markov models of interaction: this will be introduced in Section 1.5. The general idea behind such scalings is to make precise the usual continuous state space idealization of what is basically a finite model with an extremely large number of points (for example, water consists of a finite number of molecules but the general equation of thermodynamics treats it as a continuous medium).

Choosing a positive real h we shall consider, instead of a Markov chain on $\mathbf{Z}^\infty_{+,\mathrm{fin}}$, a Markov chain on $h\mathbf{Z}^\infty_{+,\mathrm{fin}} \subset \mathbf{R}^\infty$ with generator G_k^h given by

$$(G_k^h f)(hN) = \frac{1}{h} \sum_{\Psi:|\Psi|\le k, \Psi\le N} h^{|\Psi|} C_N^\Psi \sum_{\Phi:\mu(\Phi)=\mu(\Psi)} P_\Psi^\Phi \left[f(Nh - \Psi h + \Phi h) - f(Nh)\right]. \quad (1.18)$$

This generator can be considered to be the restriction to $B(h\mathbf{Z}^\infty_{+,\mathrm{fin}})$ of an operator in $B(\mathbf{R}^\infty_{+,\mathrm{fin}})$ which we shall again denote by G_k^h and which is defined by

$$(G_k^h f)(x) = \frac{1}{h} \sum_{\Psi:|\Psi|\le k} C_\Psi^h(x) \sum_{\Phi:\mu(\Phi)=\mu(\Psi)} P_\Psi^\Phi \left[f(x - \Psi h + \Phi h) - f(x)\right], \quad (1.19)$$

where the function C_Ψ^h is defined as

$$\prod_{j\in \mathrm{supp}\,(\Psi)} \frac{x_j(x_j - h)\cdots(x_j - (\psi_j - 1)h)}{\psi_j!}$$

when $x_j \ge (\psi_j - 1)h$ for all j; C_Ψ^h vanishes otherwise. Clearly, as $h \to 0$, the operator given by (1.19) converges on smooth enough functions f to the operator Λ_k on $B(\mathbf{R}^\infty_{+,\mathrm{fin}})$ given by

$$\Lambda_k f(x) = \sum_{\Psi:|\Psi|\le k} \frac{x^\Psi}{\Psi!} \sum_{\Phi:\mu(\Phi)=\mu(\Psi)} P_\Psi^\Phi \sum_{j=1}^\infty \frac{\partial f}{\partial x_j}(\phi_j - \psi_j), \quad (1.20)$$

where

$$x^\Psi = \prod_{j \in \mathrm{supp}\,(\Psi)} x_j^{\psi_j}, \qquad \Psi! = \prod_{j \in \mathrm{supp}\,(\Psi)} \psi_j!.$$

The operator Λ_k is an infinite-dimensional first-order partial differential operator. It is well known from the theory of stochastic processes that first-order partial differential operators generate deterministic Markov processes whose evolution is given by the characteristics of the partial differential operator Λ_k. These characteristics are described by the following infinite system of ordinary differential equations:

$$\dot{x}_j = \sum_{\Psi:|\Psi|\le k} \frac{x^\Psi}{\Psi!} \sum_{\Phi:\mu(\Phi)=\mu(\Psi)} P_\Psi^\Phi(\phi_j - \psi_j), \qquad j = 1, 2, \dots \tag{1.21}$$

This is the general system of kinetic equations describing the *dynamic law of large numbers* for k-ary mass-exchange processes with discrete mass distributions. In particular, in the case of binary coagulation, the P_Ψ^Φ are non-vanishing only for $|\Psi| = 2$ and $|\Phi| = 1$, and one can write $P_\Psi = K_{ij} = K_{ji}$ for a Ψ consisting of two particles of masses i and j (which coagulate to form a particle of mass $i + j$). Hence in this case (1.21) takes the form (1.14).

Let us stress again that the power of the polynomial x^Ψ in (1.21) corresponds to the number of particles taking part in each interaction. In particular, the most common, quadratic, dependence occurs when only binary interactions are taken into account.

Similarly, equations (1.10) and (1.12) can be deduced as the dynamic law of large numbers for the corresponding system of interacting particles (or species).

Of course, the convergence of the generator G_k^h to Λ_k for smooth functions does not necessarily imply the convergence of the corresponding semigroups or processes (especially for systems with an infinite state space). Additional arguments are required to justify this convergence. For the general discrete mass-exchange model (1.21), convergence was proved in Kolokoltsov [130], generalizing a long series of results for particular cases by various authors; see the detailed bibliography in Kolokoltsov [130]. We shall not develop this topic here, as we aim to work with more general models with continuous state spaces, including (1.21) as an easy special case.

1.4 Nonlinear Lévy processes and semigroups

To open our discussion of processes with uncountable state spaces, which generalize the nonlinear Markov chains introduced above, we shall define the simplest class of nonlinear processes having a Euclidean state space, namely nonlinear Lévy processes. Serious analysis of these processes will be postponed to Section 7.1.

A straightforward and constructive way to define a Lévy process is via its generator, given by the famous Lévy–Khintchine formula. Namely, a *Lévy process* X_t is a Markov process in $\mathbf{R}^d$ with a *generator* L *of Lévy–Khintchine form* given by

$$Lf(x) = \tfrac{1}{2}(G\nabla,\nabla)f(x) + (b,\nabla f(x)) + \int [f(x+y) - f(x) - (y,\nabla f(x))\mathbf{1}_{B_1}(y)]\nu(dy), \tag{1.22}$$

where $G = (G_{ij})$ is a symmetric non-negative matrix, $b = (b_i) \in \mathbf{R}^d$, $\nu(dy)$ is a Lévy measure[3] and

$$(G\nabla,\nabla)f(x) = \sum_{i,j=1}^{d} G_{ij}\frac{\partial^2 f}{\partial x_i \partial x_j}, \qquad (b,\nabla f(x)) = \sum_{i=1}^{n} b_i \frac{\partial f}{\partial x_i}.$$

In other words, X_t is a stochastic process such that

$$\mathbf{E}(f(X_t)|X_s = x) = (\Phi^{t-s}f)(x), \qquad f \in C(\mathbf{R}^d),$$

where Φ^t is the strongly continuous semigroup of linear contractions on $C_\infty(\mathbf{R}^d)$ generated by L. The generator L has an invariant domain $C^2_\infty(\mathbf{R}^d)$, so that for any $f \in C^2_\infty(\mathbf{R}^d)$ the function $\Phi^t f$ is the unique solution in $C^2_\infty(\mathbf{R}^d)$ of the Cauchy problem

$$\dot{f}_t = Lf_t, \qquad f_0 = f.$$

A Lévy process is not only a time-homogeneous Markov process (the transition mechanism from time s to time $t > s$ depends only on the difference $t - s$), but also a space-homogeneous Markov process, i.e.

$$\Phi^t T_\lambda = T_\lambda \Phi^t$$

for any $\lambda \in \mathbf{R}^d$, where T_λ is the translation operator $T_\lambda f(x) = f(x+\lambda)$. The existence (and essentially the uniqueness, up to a modification) of the Lévy process corresponding to a given L is a well-known fact.

Suppose now that a family A_μ of Lévy–Khintchine generators is given by

$$A_\mu f(x) = \tfrac{1}{2}(G(\mu)\nabla,\nabla)f(x) + (b(\mu),\nabla f)(x) + \int [f(x+y) - f(x) - (y,\nabla f(x))\mathbf{1}_{B_1}(y)]\nu(\mu,dy), \tag{1.23}$$

[3] I.e. a Borel measure in $\mathbf{R}^d$ such that $\nu\{0\} = 0$ and $\int \min(1,|y|^2)\nu(dy) < \infty$.

thus depending on $\mu \in \mathcal{P}(\mathbf{R}^d)$; for a Borel space X, $\mathcal{P}(X)$ denotes the set of probability measures on X. By the *nonlinear Lévy semigroup* generated by A_μ we mean the weakly continuous semigroup V^t of weakly continuous transformations of $\mathcal{P}(\mathbf{R}^d)$ such that, for any $\mu \in \mathcal{P}(\mathbf{R}^d)$ and any $f \in C^2_\infty(\mathbf{R}^d)$, the measure-valued curve $\mu_t = V^t(\mu)$ solves the problem

$$\frac{d}{dt}(f, \mu_t) = (A_{\mu_t} f, \mu_t), \qquad t \geq 0, \quad \mu_0 = \mu. \tag{1.24}$$

Once a Lévy semigroup has been constructed we can define the corresponding *nonlinear Lévy process* with initial law μ as the time-nonhomogeneous Lévy process generated by the family L_t given by

$$\begin{aligned} L_t f(x) = A_{V^t\mu} f(x) = \tfrac{1}{2}(G(V^t(\mu))\nabla, \nabla) f(x) + (b(V^t(\mu)), \nabla f)(x) \\ + \int [f(x+y) - f(x) \\ - (y, \nabla f(x)) \mathbf{1}_{B_1}(y)] \nu(V^t(\mu), dy), \end{aligned} \tag{1.25}$$

with law μ at $t = 0$.

We shall prove the existence of nonlinear Lévy semigroups in Section 7.1 under mild assumptions about the coefficients of L. Starting from this existence result we can obtain nonlinear analogs of many standard facts about Lévy processes, such as transience–recurrence criteria and local time properties, as presented in e.g. Bertoin [33]. We can also extend the theory to Hilbert- and Banach-space-valued Lévy processes (see e.g. Albeverio and Rüdiger [3] for the corresponding linear theory). However, we shall not go in this direction as our main objective here is to study nonlinear processes with variable coefficients.

1.5 Multiple coagulation, fragmentation and collisions; extended Smoluchovski and Boltzmann models

We shall now embark on an informal discussion of the methods and tools arising in the analysis of Markov models of interacting particles and their dynamic law of large numbers, in the case of an arbitrary (not necessarily countable) state space. Technical questions about the well-posedness of the evolutions and the justification of limiting procedures will be addressed in later chapters. Here the aim is (i) to get a brief idea of the kinds of equation that are worth studying and the limiting behaviors to expect, (ii) to develop some intuition about the general properties of these evolutions and about approaches to their analysis and (iii) to see how the analysis of various classical models in the natural sciences (most notably statistical and quantum physics and evolutionary biology) can be unified in a concise mathematical framework.

The ideas discussed go back to Boltzmann and Smoluchovski and were developed in the classical works of Bogolyubov, Vlasov, Leontovich, McKean, Katz, Markus, Lushnikov, Dobrushin and many others. The subject has attracted attention from both mathematicians and physicists and more recently from evolutionary biologists. The full model of kth-order interactions leading to measure-valued evolutions (see (1.73) below) was put forward in Belavkin and Kolokoltsov [25] and Kolokoltsov [132] and its quantum analog in Belavkin [21].

Let us stress that in this book our aim is to introduce the law of large numbers specified by nonlinear Markov processes in the mathematically most direct and unified way, without paying too much attention to the particulars of concrete physical models. A review of various relevant and physically meaningful scaling procedures (hydrodynamic, low-density, weak-coupling, kinetic etc.) can be found in the monograph by Spohn [225]; see also Balescu [16].

By a *symmetric function* of n variables we mean a function that is invariant under any permutation of these variables, and by a *symmetric operator* on the space of functions of n variables we mean an operator that preserves the set of symmetric functions.

We denote by X a locally compact separable metric space. Denoting by X^0 a one-point space and by X^j the j-fold product $X \times \cdots \times X$, considered with the product topology, we denote by $\mathcal{X}$ their disjoint union $\cup_{j=0}^{\infty} X^j$, which is again a locally compact space. In applications, X specifies the state space of one particle and $\mathcal{X} = \cup_{j=0}^{\infty} X^j$ stands for the state space of a random number of similar particles. We denote by $C_{\text{sym}}(\mathcal{X})$ the Banach spaces of symmetric bounded continuous functions on $\mathcal{X}$ and by $C_{\text{sym}}(X^k)$ the corresponding spaces of functions on the finite power X^k. The space of symmetric (positive finite Borel) measures is denoted by $\mathcal{M}_{\text{sym}}(\mathcal{X})$. The elements of $\mathcal{M}_{\text{sym}}(\mathcal{X})$ and $C_{\text{sym}}(\mathcal{X})$ are respectively the *(mixed) states and observables* for a Markov process on $\mathcal{X}$. We denote the elements of $\mathcal{X}$ by boldface letters, say $\mathbf{x}$, $\mathbf{y}$. For a finite subset $I = \{i_1, \ldots, i_k\}$ of a finite set $J = \{1, \ldots, n\}$, we denote by $|I|$ the number of elements in I, by $\bar{I}$ its complement $J \setminus I$ and by $\mathbf{x}_I$ the collection of variables $x_{i_1}, \ldots, x_{i_k}$.

Reducing the set of observables to $C_{\text{sym}}(\mathcal{X})$ means in effect that our state space is not $\mathcal{X}$ (resp. X^k) but rather the factor space $S\mathcal{X}$ (resp. SX^k) obtained by factorization with respect to all permutations, which allows the identification $C_{\text{sym}}(\mathcal{X}) = C(S\mathcal{X})$ (resp. $C_{\text{sym}}(X^k) = C(SX^k)$). Clearly

$S\mathcal{X}$ can be identified with the set of all finite subsets of X, the order being irrelevant.

A key role in the theory of measure-valued limits of interacting-particle systems is played by the inclusion $S\mathcal{X}$ to $\mathcal{M}(X)$ given by

$$\mathbf{x} = (x_1, \ldots, x_l) \mapsto \delta_{x_1} + \cdots + \delta_{x_l} = \delta_{\mathbf{x}}, \tag{1.26}$$

which defines a bijection between $S\mathcal{X}$ and the set $\mathcal{M}^+_\delta(X)$ of finite linear combinations of Dirac's δ-measures with natural coefficients. This bijection can be used to equip $S\mathcal{X}$ with the structure of a metric space (complete whenever X is complete) by "pulling back" any distance on $\mathcal{M}(X)$ that is compatible with its weak topology.

Clearly each $f \in C_{\text{sym}}(\mathcal{X})$ is defined by its components (restrictions) f^k on X^k, so that for $\mathbf{x} = (x_1, \ldots, x_k) \in X^k \subset \mathcal{X}$, say, we can write $f(\mathbf{x}) = f(x_1, \ldots, x_k) = f^k(x_1, \ldots, x_k)$. Similar notation will be used for the components of measures from $\mathcal{M}(\mathcal{X})$. In particular, the pairing between $C_{\text{sym}}(\mathcal{X})$ and $\mathcal{M}(\mathcal{X})$ can be written as

$$(f, \rho) = \int f(\mathbf{x})\rho(d\mathbf{x}) = f^0\rho_0 + \sum_{n=1}^{\infty} \int f(x_1, \ldots, x_n)\rho(dx_1 \cdots dx_n),$$
$$f \in C_{\text{sym}}(\mathcal{X}), \quad \rho \in \mathcal{M}(\mathcal{X}).$$

A useful class of measures (and mixed states) on $\mathcal{X}$ is given by *decomposable measures* of the form $Y^{\otimes}$; such a measure is defined for an arbitrary finite measure $Y(dx)$ on X by means of its components:

$$(Y^{\otimes})^n(dx_1 \cdots dx_n) = Y^{\otimes n}(dx_1 \cdots dx_n) = Y(dx_1) \cdots Y(dx_n).$$

Similarly, *decomposable observables* (multiplicative or additive) are defined for an arbitrary $Q \in C(X)$ as

$$(Q^{\otimes})^n(x_1, \ldots, x_n) = Q^{\otimes n}(x_1, \ldots, x_n) = Q(x_1) \cdots Q(x_n), \tag{1.27}$$

where

$$(Q^{\oplus})(x_1, \ldots, x_n) = Q(x_1) + \cdots + Q(x_n) \tag{1.28}$$

($Q^{\oplus}$ vanishes on X^0). In particular, if $Q = \mathbf{1}$, then $Q^{\oplus} = \mathbf{1}^{\oplus}$ is the number of particles: $\mathbf{1}^{\oplus}(x_1, \ldots, x_n) = n$.

In this section we are interested in pure jump processes on $\mathcal{X}$, whose semigroups and generators preserve the space C_{sym} of continuous symmetric functions and hence are given by symmetric transition kernels $q(\mathbf{x}; d\mathbf{y})$, which can thus be considered as kernels on the factor space $S\mathcal{X}$.

To specify a binary particle interaction of pure jump type we have to specify a continuous transition kernel

$$P^2(x_1, x_2; d\mathbf{y}) = \{P^2_m(x_1, x_2; dy_1 \cdots dy_m)\}$$

from SX^2 to $S\mathcal{X}$ such that $P^2(\mathbf{x}; \{\mathbf{x}\}) = 0$ for all $\mathbf{x} \in X^2$. By the *intensity* of the interaction we mean the total mass

$$P^2(x_1, x_2) = \int_{\mathcal{X}} P^2(x_1, x_2; d\mathbf{y}) = \sum_{m=0}^{\infty} \int_{X^m} P^2_m(x_1, x_2; dy_1 \cdots dy_m).$$

The intensity defines the rate of decay of any pair of particles x_1, x_2, and the measure $P^k(x_1, x_2; d\mathbf{y})$ defines the distribution of possible outcomes. If we suppose that any pair of particles, randomly chosen from a given set of n particles, can interact then this leads to a generator G_2 of binary interacting particles, defined in terms of the kernel P^2:

$$\begin{aligned}(G_2 f)(x_1, \dots, x_n) = &\sum_{I \subset \{1,\dots,n\}, |I|=2} \int [f(\mathbf{x}_{\bar{I}}, \mathbf{y}) - f(x_1, \dots, x_n)] \\ &\times P^2(\mathbf{x}_I, d\mathbf{y}) \\ = &\sum_{m=0}^{\infty} \sum_{I \subset \{1,\dots,n\}, |I|=2} \int [f(\mathbf{x}_{\bar{I}}, y_1, \dots, y_m) - f(x_1, \dots, x_n)] \\ &\times P^k_m(\mathbf{x}_I; dy_1 \dots dy_m).\end{aligned}$$

A probabilistic description of the evolution of a pure jump Markov process Z_t on $\mathcal{X}$ specified by this generator (if this process is well defined!) is the following.[4] Any two particles x_1, x_2 (chosen randomly and uniformly from n existing particles) wait for interaction to occur for a $P^2(x_1, x_2)$-exponential random time. The first pair in which interaction occurs produces in its place a collection of particles $y_1, \dots, y_m$ according to the distribution $P^2_m(x_1, x_2; dy_1 \cdots dy_m)/P^2(x_1, x_2)$. Then everything starts again from the new collection of particles thus obtained.

Similarly, a *k-ary interaction* or *kth-order interaction of pure jump type* is specified by a transition kernel

$$P^k(x_1, \dots, x_k; d\mathbf{y}) = \{P^k_m(x_1, \dots, x_k; dy_1 \cdots dy_m)\} \tag{1.29}$$

[4] See Theorem 2.34 for some background on jump-type processes, and take into account that the minimum of any finite collection of exponential random variables is again an exponential random variable.

from SX^k to $S\mathcal{X}$ such that $P^k(\mathbf{x}; \{\mathbf{x}\}) = 0$ for all $\mathbf{x} \in \mathcal{X}$, where the intensity is

$$P^k(x_1, \dots, x_k) = \int P^k(x_1, \dots, x_k; d\mathbf{y}) = \sum_{m=0}^{\infty} \int P_m^k(x_1, \dots, x_k; dy_1 \cdots dy_m). \tag{1.30}$$

This kernel defines a *generator G_k of k-ary interacting particles*:

$$(G_k f)(x_1, \dots, x_n) = \sum_{I \subset \{1,\dots,n\}, |I|=k} \int \left[f(\mathbf{x}_{\bar{I}}, \mathbf{y}) - f(x_1, \dots, x_n) \right] P^k(\mathbf{x}_I, d\mathbf{y}). \tag{1.31}$$

To model possible interactions of all orders up to a certain k, we can take the sum of generators of the type (1.31) for all $l = 1, 2, \dots, k$, leading to a model with generator given by

$$G_{\le k} f = \sum_{l=1}^{k} G_l f. \tag{1.32}$$

To ensure that operators of the type in (1.31) generate a unique Markov process we have to make certain assumptions. Physical intuition suggests that there should be conservation laws governing the processes of interaction. Precise criteria will be given in Part III.

Changing the state space according to the mapping (1.26) yields the corresponding Markov process on $\mathcal{M}_\delta^+(X)$. Choosing a positive parameter h, we now scale the empirical measures $\delta_{x_1} + \cdots + \delta_{x_n}$ by a factor h and the operator of k-ary interactions by a factor h^{k-1}.

Remark 1.5 Performing various scalings and analyzing scaling limits is a basic feature in the analysis of models in physics and biology. Scaling allows one to focus on particular aspects of the system under study. Scaling empirical measures by a small parameter h in such a way that the measure $h(\delta_{x_1} + \cdots + \delta_{x_n})$ remains finite when the number of particles n tends to infinity realizes the basic idea of a continuous limit, mentioned in Section 1.3 (when the number of molecules becomes large in comparison with their individual sizes, we observe and treat a liquid as a continuously distributed mass). Scaling kth-order interactions by h^{k-1} reflects the idea that, say, simultaneous ternary collisions are rarer events than binary collisions. Precisely this scaling makes these kth-order interactions neither negligible nor overwhelming in the approximation considered.

The above scaling leads to the generator Λ_k^h defined by

$$\Lambda_k^h F(h\delta_{\mathbf{x}}) = h^{k-1} \sum_{I \subset \{1,\ldots,n\}, |I|=k} \int_{\mathcal{X}} [F(h\delta_{\mathbf{x}} - h\delta_{\mathbf{x}_I} + h\delta_{\mathbf{y}}) - F(h\nu)] \times P(\mathbf{x}_I; d\mathbf{y}), \quad (1.33)$$

which acts on the space of continuous functions F on the set $\mathcal{M}_{h\delta}^+(X)$ of measures of the form $h\nu = h\delta_{\mathbf{x}} = h\delta_{x_1} + \cdots + h\delta_{x_n}$. Allowing interactions of order $\leq k$ leads to the generator $\Lambda_{\leq k}^h$ given by

$$\Lambda_{\leq k}^h F(h\delta_{\mathbf{x}}) = \sum_{l=1}^k \Lambda_l^h F(h\delta_{\mathbf{x}})$$
$$= \sum_{l=1}^k h^{l-1} \sum_{I \subset \{1,\ldots,n\}, |I|=l} \int_{\mathcal{X}} [F(h\delta_{\mathbf{x}} - h\delta_{\mathbf{x}_I} + h\delta_{\mathbf{y}}) - F(h\nu)] \times P(\mathbf{x}_I; d\mathbf{y}). \quad (1.34)$$

This generator defines our basic Markov model of exchangeable particles with (h-scaled) k-ary interactions of pure jump type. As we are aiming to take the limit $h \to 0$, with $h\delta_{\mathbf{x}}$ converging to a finite measure, the parameter h should be regarded as the inverse of the number of particles. There also exist important models with an input (i.e. with a term corresponding to $l = 0$ in (1.34)), but we shall not consider them here.

The scaling introduced before Remark 1.5, which is usual in statistical mechanics, is not the only reasonable one. In the theory of evolutionary games (see Section 1.6) or other biological models a more natural scaling is to normalize by the number of particles, i.e. a k-ary interaction term is divided by $n^{k-1} = (\|h\nu\|/h)^{k-1}$ (see e.g. the phytoplankton dynamics model in [9] or [211]). This leads to, instead of Λ_k^h, the operator $\tilde{\Lambda}_k^h$ given by

$$\tilde{\Lambda}_k^h F(h\delta_{\mathbf{x}}) = h^{k-1} \sum_{I \subset \{1,\ldots,n\}, |I|=k} \int_{\mathcal{X}} [F(h\nu - h\delta_{\mathbf{x}_I} + h\delta_{\mathbf{y}}) - F(h\nu)] \times \frac{P(\mathbf{x}_I; d\mathbf{y})}{\|h\delta_{\mathbf{x}}\|^{k-1}}, \quad (1.35)$$

or, more generally,

$$\tilde{\Lambda}_{\leq k}^h F(h\delta_{\mathbf{x}}) = \sum_{l=1}^k \tilde{\Lambda}_l^h F(h\delta_{\mathbf{x}}). \quad (1.36)$$

Applying the obvious relation

$$\sum_{I\subset\{1,\ldots,n\},|I|=2} f(\mathbf{x}_I) = \tfrac{1}{2}\iint f(z_1,z_2)\delta_{\mathbf{x}}(dz_1)\delta_{\mathbf{x}}(dz_2) - \tfrac{1}{2}\int f(z,z)\delta_{\mathbf{x}}(dz), \tag{1.37}$$

which holds for any $f \in C_{\text{sym}}(X^2)$ and $\mathbf{x} = (x_1,\ldots,x_n) \in X^n$, one observes that the expression for the operator Λ_2^h is

$$\begin{aligned}\Lambda_2^h F(h\delta_{\mathbf{x}}) = -\tfrac{1}{2}\int_{\mathcal{X}}\int_X &[F(h\delta_{\mathbf{x}} - 2h\delta_z + h\delta_{\mathbf{y}}) - F(h\delta_{\mathbf{x}})]\\ &\times P(z,z;d\mathbf{y})(h\delta_{\mathbf{x}})(dz)\\ + \tfrac{1}{2}h^{-1}\int_{\mathcal{X}}\int_{X^2} &[F(h\delta_{\mathbf{x}} - h\delta_{z_1} - h\delta_{z_2} + h\delta_{\mathbf{y}}) - F(h\delta_{\mathbf{x}})]\\ &\times P(z_1,z_2;d\mathbf{y})(h\delta_{\mathbf{x}})(dz_1)(h\delta_{\mathbf{x}})(dz_2).\end{aligned} \tag{1.38}$$

On the linear functions

$$F_g(\mu) = \int g(y)\mu(dy) = (g,\mu)$$

this operator acts according to

$$\begin{aligned}&\Lambda_2^h F_g(h\delta_{\mathbf{x}})\\ &\quad = \tfrac{1}{2}\int_{\mathcal{X}}\int_{X^2}[g^{\oplus}(\mathbf{y}) - g^{\oplus}(z_1,z_2)]P(z_1,z_2;d\mathbf{y})(h\delta_{\mathbf{x}})(dz_1)(h\delta_{\mathbf{x}})(dz_2)\\ &\qquad - \tfrac{1}{2}h\int_{\mathcal{X}}\int_X [g^{\oplus}(\mathbf{y}) - g^{\oplus}(z,z)]P(z,z;d\mathbf{y})(h\delta_{\mathbf{x}})(dz).\end{aligned}$$

It follows that if h tends to 0 and $h\delta_{\mathbf{x}}$ tends to some finite measure μ (in other words, if the number of particles tends to infinity but the "whole mass" remains finite owing to the scaling of each atom), the corresponding evolution equation $\dot{F} = \Lambda_2^h F$ on linear functionals $F = F_g$ tends to the equation

$$\begin{aligned}\frac{d}{dt}(g,\mu_t) &= \Lambda_2 F_g(\mu_t)\\ &= \tfrac{1}{2}\int_{\mathcal{X}}\int_{X^2}\left[g^{\oplus}(\mathbf{y}) - g^{\oplus}(\mathbf{z})\right] P^2(\mathbf{z};d\mathbf{y})\mu_t^{\otimes 2}(d\mathbf{z}), \qquad \mathbf{z} = (z_1,z_2),\end{aligned} \tag{1.39}$$

which is the *general kinetic equation for binary interactions of pure jump type* in *weak form*. "Weak" means that it must hold for all g belonging to $C_\infty(X)$ or at least to its dense subspace.

A similar procedure for the k-ary interaction operator in (1.33), based on the k-ary extension (H.4), (H.6) (see Appendix H) of (1.37), leads to the *general kinetic equation for k-ary interactions of pure jump type* in *weak form*:

$$\frac{d}{dt}(g,\mu_t) = \Lambda_k F_g(\mu_t)$$
$$= \frac{1}{k!}\int_{\mathcal{X}}\int_{X^k}\left[g^{\oplus}(\mathbf{y}) - g^{\oplus}(\mathbf{z})\right] P^k(\mathbf{z};d\mathbf{y})\mu_t^{\otimes k}(d\mathbf{z}), \qquad \mathbf{z}=(z_1,\ldots,z_k). \tag{1.40}$$

More generally, for interactions of order at most k, we start from the generator given in (1.32) and specified by the family of kernels $P = \{P(\mathbf{x}) = P^l(\mathbf{x}), \mathbf{x} \in X^l, l = 1,\ldots,k\}$ of type (1.30) and obtain the equation

$$\frac{d}{dt}(g,\mu_t) = \Lambda_{l\le k} F_g(\mu_t) = \sum_{l=1}^{k}\frac{1}{l!}\int_{\mathcal{X}}\int_{X^l}\left[g^{\oplus}(\mathbf{y}) - g^{\oplus}(\mathbf{z})\right] P^l(\mathbf{z};d\mathbf{y})\mu_t^{\otimes l}(d\mathbf{z}). \tag{1.41}$$

The same limiting procedure with the operator given by (1.35) yields the equation

$$\frac{d}{dt}\int_X g(z)\mu_t(dz)$$
$$= \frac{1}{k!}\int_{\mathcal{X}}\int_{X^k}\left[g^{\oplus}(\mathbf{y}) - g^{\oplus}(\mathbf{z})\right] P^k(\mathbf{z};d\mathbf{y})\left(\frac{\mu_t}{\|\mu_t\|}\right)^{\otimes k}(d\mathbf{z})\|\mu_t\|. \tag{1.42}$$

In the biological context the dynamics is traditionally written in terms of normalized (probability) measures. Because for positive μ the norm equals $\|\mu\| = \int_X \mu(dx)$ we see that, for positive solutions μ_t of (1.42), the norms of μ_t satisfy the equation

$$\frac{d}{dt}\|\mu_t\| = -\frac{1}{k!}\int_{X^k} Q(\mathbf{z})\left(\frac{\mu_t}{\|\mu_t\|}\right)^{\otimes k}(d\mathbf{z})\|\mu_t\|, \tag{1.43}$$

where

$$Q(\mathbf{z}) = -\int_{\mathcal{X}}\left[\mathbf{1}^{\oplus}(\mathbf{y}) - \mathbf{1}^{\oplus}(\mathbf{z})\right] P^k(\mathbf{z};d\mathbf{y}). \tag{1.44}$$

Consequently, rewriting equation (1.42) in terms of the normalized measure $\nu_t = \mu_t/\|\mu_t\|$ yields

$$\frac{d}{dt}\int_X g(z)\nu_t(dz) = \frac{1}{k!}\int_{\mathcal{X}}\int_{X^k}\left[g^{\oplus}(\mathbf{y}) - g^{\oplus}(\mathbf{z})\right] P^k(\mathbf{z}; d\mathbf{y})\nu_t^{\otimes k}(d\mathbf{z})$$
$$+\frac{1}{k!}\int_X g(z)\nu_t(dz)\int_{\mathcal{X}}\int_{X^k} Q(\mathbf{z})\nu_t^{\otimes k}(d\mathbf{z}). \qquad (1.45)$$

It is worth noting that the rescaling of interactions leading to (1.42) is equivalent to a time change in (1.40). A particular instance of this reduction in evolutionary biology is the well-known trajectory-wise equivalence of the Lotka–Volterra model and replicator dynamics; see e.g. [96].

We shall now consider some basic examples of interaction.

Example 1.6 (Generalized Smoluchovski coagulation model) The classical *Smoluchovski model* describes the mass-preserving binary coagulation of particles. In the more general context, often called cluster coagulation, see Norris [194], a particle is characterized by a parameter x from a locally compact state space X, where a mapping $E : X \to \mathbf{R}_+$, the *generalized mass*, and a transition kernel $P_1^2(z_1, z_2, dy) = K(z_1, z_2, dy)$, *the coagulation kernel*, are given such that the measures $K(z_1, z_2; .)$ are supported on the set $\{y : E(y) = E(z_1) + E(z_2)\}$. In this setting, equation (1.39) takes the form

$$\frac{d}{dt}\int_X g(z)\mu_t(dz)$$
$$= \frac{1}{2}\int_{X^3}[g(y) - g(z_1) - g(z_2)]K(z_1, z_2; dy)\mu_t(dz_1)\mu_t(dz_2). \qquad (1.46)$$

In the classical Smoluchovski model, we have $X = \mathbf{R}_+$, $E(x) = x$ and $K(x_1, x_2, dy) = K(x_1, x_2)\delta(x_1 + x_2 - y)$ for a certain symmetric function $K(x_1, x_2)$.

Using the scaling (1.35), which would be more appropriate in a biological context, we obtain, instead of (1.46), the equation

$$\frac{d}{dt}\int_X g(z)\mu_t(dz) = \int_{X^3}[g(y)-g(z_1)-g(z_2)]K(z_1, z_2, dy)\frac{\mu_t(dz_1)\mu_t(dz_2)}{\|\mu_t\|}. \qquad (1.47)$$

Example 1.7 (Spatially homogeneous Boltzmann collisions and beyond) Interpret $X = \mathbf{R}^d$ as the space of particle velocities, and assume that binary collisions $(v_1, v_2) \mapsto (w_1, w_2)$ preserve total momentum and energy:

$$v_1 + v_2 = w_1 + w_2, \qquad v_1^2 + v_2^2 = w_1^2 + w_2^2. \qquad (1.48)$$

These equations imply that

$$w_1 = v_1 - n(v_1 - v_2, n), \qquad w_2 = v_2 + n(v_1 - v_2, n)),$$
$$n \in S^{d-1}, \ (n, v_2 - v_1) \geq 0 \qquad (1.49)$$

(see Exercise 1.1 below). Assuming that the collision rates are shift invariant, i.e. they depend on v_1, v_2 only via the difference $v_2 - v_1$, the weak kinetic equation (1.39) describing the LLN dynamics for these collisions takes the form

$$\frac{d}{dt}(g, \mu_t) = \frac{1}{2} \int_{n \in S^{d-1}:(n, v_2 - v_1) \geq 0} \int_{\mathbf{R}^{2d}} \mu_t(dv_1)\mu_t(dv_2) \\ \times [g(w_1) + g(w_2) - g(v_1) - g(v_2)] B(v_2 - v_1, dn) \quad (1.50)$$

in which the *collision kernel* $B(v, dn)$ specifies a concrete physical collision model. In the most common models, the kernel B has a density with respect to Lebesgue measure on S^{d-1} and depends on v only via its magnitude $|v|$ and the angle $\theta \in [0, \pi/2]$ between v and n. In other words, one assumes $B(v, dn)$ has the form $B(|v|, \theta)dn$ for a certain function B. Extending B to angles $\theta \in [\pi/2, \pi]$ by

$$B(|v|, \theta) = B(|v|, \pi - \theta) \quad (1.51)$$

allows us finally to write the weak form of the Boltzmann equation as

$$\frac{d}{dt}(g, \mu_t) = \frac{1}{4} \int_{S^{d-1}} \int_{R^{2d}} [g(w_1) + g(w_2) - g(v_1) - g(v_2)] B(|v_1 - v_2|, \theta) \\ \times dn \mu_t(dv_1)\mu_t(dv_2), \quad (1.52)$$

where w_1 and w_2 are given by (1.49), θ is the angle between $v_2 - v_1$ and n and B satisfies (1.51).

Example 1.8 (Multiple coagulation, fragmentation and collision breakage) The processes combining the pure coagulation of no more than k particles, spontaneous fragmentation into no more than k pieces and collisions (or collision breakages) of no more than k particles are specified by the following transition kernels:

$$P_1^l(z_1, \ldots, z_l, dy) = K_l(z_1, \ldots, z_l; dy), \qquad l = 2, \ldots, k,$$

called *coagulation kernels*;

$$P_m^1(z; dy_1 \cdots dy_m) = F_m(z; dy_1 \cdots dy_m), \qquad m = 2, \ldots, k,$$

called *fragmentation kernels*; and

$$P_l^l(z_1, \ldots, z_l; dy_1 \cdots dy_l) = C_l(z_1, \ldots, z_l; dy_1 \cdots dy_2), \qquad l = 2, \ldots, k,$$

called *collision kernels*. The corresponding kinetic equation (cf. (1.39)) takes the form

$$\frac{d}{dt}\int g(z)\mu_t(dz)$$
$$= \sum_{l=2}^{k}\frac{1}{l!}\int_{z_1,\ldots,z_l,y}[g(y)-g(z_1)-\cdots-g(z_m)]K_l(z_1,\ldots,z_l;dy)\prod_{j=1}^{l}\mu_t(dz_j)$$
$$+\sum_{m=2}^{k}\int_{z,y_1,\ldots,y_m}[g(y_1)+\cdots+g(y_m)-g(z)]F_m(z;dy_1\ldots dy_m)\mu_t(dz)$$
$$+\sum_{l=2}^{k}\int[g(y_1)+\cdots+g(y_l)-g(z_1)-\cdots-g(z_l)]$$
$$\times C_l(z_1,\ldots,z_l;dy_1\cdots dy_l)\prod_{j=1}^{l}\mu_t(dz_j). \quad (1.53)$$

Exercise 1.1 Prove equations (1.49) and extend them to collisions in which k particles $\mathbf{v}=(v_1,\ldots,v_k)$ are scattered to k particles $\mathbf{w}=(w_1,\ldots,w_k)$ that preserve total energy and momentum, i.e.

$$v_1+\cdots+v_k=w_1+\cdots+w_k, \qquad v_1^2+\cdots+v_k^2=w_1^2+\cdots+w_k^2. \quad (1.54)$$

Deduce the following version of the Boltzmann equation for the simultaneous collision of k particles:

$$\frac{d}{dt}(g,\mu_t)=\frac{1}{k!}\int_{S_{\Gamma,\mathbf{v}}^{d(k-1)-1}}\int_{\mathbf{R}^{dk}}\left[g^{\oplus}(\mathbf{v}-2(\mathbf{v},\mathbf{n})\mathbf{n})-g^{\oplus}(\mathbf{v})\right]$$
$$\times B_k(\{v_i-v_j\}_{i,j=1}^{k};d\mathbf{n})\mu_t(dv_1)\cdots\mu_t(dv_l), \quad (1.55)$$

where

$$\Gamma=\{\mathbf{u}=(u_1,\ldots,u_k)\in\mathbf{R}^{dk}:u_1+\cdots+u_k=0\},$$
$$S_\Gamma^{d(k-1)-1}=\{\mathbf{n}\in\Gamma:\|\mathbf{n}\|=|n_1|^2+\cdots+|n_k|^2=1\},$$
$$S_{\Gamma,\mathbf{v}}^{d(k-1)-1}=\{\mathbf{n}\in S_\Gamma^{d(k-1)-1}:(\mathbf{n},\mathbf{v})\le 0\}.$$

Hint: In terms of $\mathbf{u}=\{u_1,\ldots,u_k\}$, defined by $\mathbf{w}=\mathbf{u}+\mathbf{v}$, conditions (1.54) mean that $\mathbf{u}\in\Gamma$ and

$$\|\mathbf{u}\|^2=\sum_{j=1}^{k}u_j^2=2\sum_{j=1}^{k}(w_j,u_j)=2(\mathbf{w},\mathbf{u})=-2\sum_{j=1}^{k}(v_j,u_j)=-2(\mathbf{v},\mathbf{u})$$

or, equivalently, that $\mathbf{u} = \|\mathbf{u}\|\mathbf{n}$, $\mathbf{n} \in S_\Gamma^{d(k-1)-1}$ and

$$\|\mathbf{u}\| = 2(\mathbf{w}, \mathbf{n}) = -2(\mathbf{v}, \mathbf{n}),$$

implying, in particular, that $(\mathbf{v}, \mathbf{n}) \leq 0$.

1.6 Replicator dynamics of evolutionary game theory

In this section we discuss examples of the kinetic equations that appear in *evolutionary game theory*. These models are often simpler, mathematically, than the basic models of statistical physics (at least from the point of view of justifying the dynamic law of large numbers), but we want to make precise their place in the general framework.

We start by recalling the notion of a game with a compact space of strategies, referring for general background to textbooks on game theory; see e.g. Kolokoltsov and Malafeev [139], [140] or Gintis [84]. A *k-person game (in its normal form)* is specified by a collection of k compact spaces $X_1, \ldots, X_k$ of possible pure strategies for the players and a collection of continuous payoff functions $H_1,\ldots,H_k$ on $X_1 \times \cdots \times X_k$. One step of such a game is played according to the following rule. Each player i, $i = 1, \ldots, k$, chooses independently a strategy $x_i \in X_i$ and then receives the payoff $H_i(x_1, \ldots, x_k)$, which depends on the choices of all k players. The collection of chosen strategies $x_1, \ldots, x_k$ is called a *profile (or situation) of the game*. In elementary models the X_i are finite sets. A game is called *symmetric* if the X_i do not depend on i, so that $X_i = X$, and the payoffs are symmetric in that they are specified by a single function $H(x; y_1, \ldots, y_{k-1})$ on X^k that is symmetric with respect to the last $k-1$ variables $y_1, \ldots, y_{k-1}$ via the formula

$$H_i(x_1, \ldots, x_k) = H(x_i, x_1, \ldots, x_{i-1}, x_{i+1}, \ldots, x_k).$$

Hence in symmetric games the label of the player is irrelevant; only the strategy is important.

By the *mixed-strategy extension* of a game with strategy spaces X_i and payoffs H_i, $i = 1, \ldots, k$, we mean a k-person game with strategy spaces $\mathcal{P}(X_i)$ (considered as compact in their weak topology), $i = 1, \ldots, k$, and payoffs

$$H_i^\star(P) = \int_{X^k} H_i(x_1, \ldots, x_k) P(dx_1 \cdots dx_k),$$
$$P = (p_1, \ldots, p_k) \in \mathcal{P}(X_1) \times \cdots \times \mathcal{P}(X_k).$$

Playing a *mixed strategy* p_i is interpreted as choosing the pure strategies randomly with probability law p_i. The key notion in the theory of games is that of Nash equilibrium. Let

$$H_i^\star(P\|x_i) = \int_{X_1\times\cdots\times X_{i-1}\times X_{i+1}\times\cdots\times X_n} H_i(x_1,\ldots,x_n)\,dp_1\cdots dp_{i-1}dp_{i+1}\cdots dp_n. \tag{1.56}$$

A situation $P = (p_1,\ldots,p_k)$ is called a *Nash equilibrium* if

$$H_i^\star(P) \geq H_i^\star(P\|x_i) \tag{1.57}$$

for all i and $x_i \in X_i$. For symmetric games and *symmetric profiles* $P = (p,\ldots,p)$, which are of particular interest for evolutionary games, the payoffs

$$H_i^\star(P) = H^\star(P) = \int_{X^k} H(x_1,\ldots,x_k)p^{\otimes k}(dx_1\cdots dx_k),$$

$$H_i^\star(P\|y) = H^\star(P\|y) = \int_{X^{k-1}} H(y,x_1,\ldots,x_{k-1})\,p^{\otimes(k-1)}(dx_1\cdots dx_{k-1})$$

do not depend on i, and the condition of equilibrium is

$$H^\star(P) \geq H^\star(P\|x), \qquad x \in X. \tag{1.58}$$

The *replicator dynamics (RD)* of evolutionary game theory is intended to model the process in which one approaches equilibrium from a given initial state by decreasing the losses produced by deviations from equilibrium (thus adjusting the strategy to the current situation). More precisely, assuming a mixed profile given by a density f_t with respect to a certain reference measure M on X (f_t can be interpreted as the fraction of a large population using strategy x), the replicator dynamics is defined by

$$\dot{f}_t(x) = f_t(x)\left[H^\star(f_tM\|x) - H^\star(f_tM)\right]. \tag{1.59}$$

The aim of this short section is to demonstrate how this evolution appears as a simple particular case of the LLN limit (1.43) of the scaled Markov model of type (1.31).

In the evolutionary biological context, particles become species of a certain large population, and the position of a particle $x \in X$ becomes the strategy of a species. A key feature distinguishing the evolutionary game setting from the general model developed in preceding sections is that the species produce new species of their own kind (with inherited behavioral patterns). In the usual model of evolutionary game theory it is assumed that any k randomly chosen species can occasionally meet and play a k-person symmetric game specified by a payoff function $H(x; y_1,\ldots,y_{k-1})$ on X^k, where the payoff measures fitness expressed in terms of the expected number of offspring.

Remark 1.9 Basic evolutionary models consider binary interactions ($k = 2$) with finite numbers of pure strategies, that is, elementary two-player games. However, arbitrary interaction laws and state spaces seem to be quite relevant in the biological context; see [201]. Such models would allow the analysis of animals living in groups or large families.

To specify a Markov model we need to specify the game a little further. We shall assume that X is a compact set and that the result of the game in which player x plays against $y_1, \dots, y_{k-1}$ is given by the probability rates $H^m(x; y_1, \dots, y_{k-1})$, $m = 0, 1, \dots$, of the number m of particles of the same type as x that would appear in place of x after this game (i.e. one interaction). To fit into the original model, the H^m can be chosen arbitrarily as long as the average change equals the original function H:

$$H(x; y_1, \dots, y_{k-1}) = \sum_{m=0}^{\infty} (m-1) H^m(x; y_1, \dots, y_{k-1}). \tag{1.60}$$

The simplest model is one in which a particle can either die or produce another particle of the same kind, with given rates H^0, H^2; the probabilities are therefore $H^0/(H^0 + H^2)$ and $H^2/(H^0 + H^2)$. Under these assumptions equation (1.60) reduces to

$$H(x; y_1, \dots, y_{k-1}) = H^2(x; y_1, \dots, y_{k-1}) - H^0(x; y_1, \dots, y_{k-1}). \tag{1.61}$$

In the general case, we have the model from Section 1.5 specified by transition kernels of the form

$$\begin{aligned} P_m^k(z_1, \dots, z_k; d\mathbf{y}) = {} & H^m(z_1; z_2, \dots, z_k) \delta_{z_1}(dy_1) \cdots \delta_{z_1}(dy_m) \\ & + H^m(z_2; z_1, \dots, z_k) \prod_{j=1}^{m} \delta_{z_2}(dy_j) + \cdots \\ & + H^m(z_k; z_1, \dots, z_{k-1}) \prod_{j=1}^{m} \delta_{z_k}(dy_j), \end{aligned} \tag{1.62}$$

so that

$$\begin{aligned} \int_{\mathcal{X}} [g^{\oplus}(\mathbf{y}) - g^{\oplus}(\mathbf{z})] \, P^k(\mathbf{z}; d\mathbf{y}) &= \sum_{m=0}^{\infty} (m-1) [g(z_1) H^m(z_1; z_2, \dots, z_k) + \cdots \\ & \qquad + g(z_k) H^m(z_k; z_1, \dots, z_{k-1})] \\ &= g(z_1) H(z_1; z_2, \dots, z_k) + \cdots \\ & \quad + g(z_k) H(z_k; z_1, \dots, z_{k-1}). \end{aligned}$$

Owing to the symmetry of H, equation (1.42) takes the form

$$\frac{d}{dt}\int_X g(x)\mu_t(dx)$$

$$= \frac{\|\mu_t\|}{(k-1)!}\int_{X^k} g(z_1)H(z_1; z_2, \dots, z_k)\left(\frac{\mu_t}{\|\mu_t\|}\right)^{\otimes k}(dz_1\cdots dz_k), \quad (1.63)$$

and hence for the normalized measure $\nu_t = \mu_t/\|\mu_t\|$ one obtains the evolution

$$\frac{d}{dt}\int_X g(x)\nu_t(dx) = \frac{1}{(k-1)!}\int_X \left[H^\star(\nu_t\|x) - H^\star(\nu_t)\right] g(x)\nu_t(dx), \quad (1.64)$$

which represents the *replicator dynamics* in weak form for a symmetric k-person game with an arbitrary compact space of strategies. It is obtained here as a simple particular case of (1.45).

If a reference probability measure M on X is chosen, equation (1.64) can be rewritten, in terms of the densities f_t of ν_t with respect to M, as (1.59).

Nash equilibria are connected with the replicator dynamics through the following result.

Proposition 1.10 *(i) If ν defines a symmetric Nash equilibrium for a symmetric k-person game specified by the payoff $H(x; y_1, \dots, y_{k-1})$ on X^k, where X is a compact space, then ν is a fixed point for the replicator dynamics* (1.64).

(ii) If ν is such that any open set in X has positive ν-measure (i.e. a "pure mixed" profile) then the converse to statement (i) holds.

Proof (i) By definition, ν defines a symmetric Nash equilibrium if and only if

$$H^\star(\nu\|x) \le H^\star(\nu)$$

for all $x \in X$. But the set $M = \{x : H^\star(\nu\|x) < H^\star(\nu)\}$ should have ν-measure zero (otherwise, integrating the above inequality would lead to a contradiction). This implies that

$$\int_X \left[H^\star(\nu\|x) - H^\star(\nu)\right] g(x)\nu_t(dx) = 0$$

for all g.

(ii) Conversely, assuming that the last equation holds for all g implies, because ν is a pure mixed profile, that

$$H^\star(\nu\|x) = H^\star(\nu)$$

on a open dense subset of X and hence everywhere, owing to the continuity of H. □

Exercise 1.2 Consider a mixed-strategy extension of a two-person symmetric game in which there is a compact space of pure strategies X for each player and a payoff specified by an antisymmetric function H on X^2, i.e. $H(x, y) = -H(y, x)$. Assume that there exists a positive finite measure M on X such that $\int H(x, y)M(dy) = 0$ for all x. Show that M specifies a symmetric Nash equilibrium and that, moreover, the function

$$L(f) = \int \ln f_t(x)M(dx)$$

is constant on the trajectories of the system (1.59). The function $L(f)$ is called the relative entropy of the measure M with respect to the measure $f_t M$.

Hint:

$$\frac{d}{dt}L(f_t) = \int H^\star(f_t M\|x)M(dx) - H^\star(f_t M),$$

and both terms on the r.h.s. vanish by the assumptions made.

1.7 Interacting Markov processes; mean field and kth-order interactions

Here we extend the models of Section 1.5 beyond pure jump interactions (analytically, beyond purely integral generators).

As shown in Belavkin and Kolokoltsov [25] under rather general assumptions, the parts of generators of Feller processes on $\mathcal{X}$ that are not of pure jump type can generate only processes that preserve the number of particles. Hence, a general Feller generator in $S\mathcal{X}$ has the form $B = (B^1, B^2, \ldots)$, where

$$\begin{aligned} &B^k f(x_1, \ldots, x_k) \\ &\quad = A^k f(x_1, \ldots, x_k) + \int_{\mathcal{X}} \left[f(\mathbf{y}) - f(x_1, \ldots, x_k)\right] P^k(x_1, \ldots, x_k, d\mathbf{y}); \end{aligned} \tag{1.65}$$

here P^k is a transition kernel from SX^k to $S\mathcal{X}$ and A^k generates a symmetric Feller process in X^k. However, with this generator the interaction of a subsystem of particles depends on the whole system: for the operator $(0, B^2, 0, \ldots)$, say, two particles will interact only in the absence of any other particle). To put all subsystems on an equal footing, one allows interactions between all subsets of k particles from a given collection of n particles. Consequently, instead of B^k one is led to a *generator of k-ary interactions* of the form

$$I_k[P^k, A^k]f(x_1, \dots, x_n)$$
$$= \sum_{I\subset\{1,\dots,n\},|I|=k} B_I f(x_1, \dots, x_n)$$
$$= \sum_{I\subset\{1,\dots,n\},|I|=k} \Big\{(A_I f)(x_1, \dots, x_n)$$
$$+ \int \big[f(\mathbf{x}_{\bar{I}}, \mathbf{y}) - f(x_1, \dots, x_n)\big] P^k(\mathbf{x}_I, d\mathbf{y})\Big\}, \tag{1.66}$$

where A_I (resp. B_I) is the operator $A^{|I|}$ (resp. $B^{|I|}$) acting on the variables $\mathbf{x}_I$. In quantum mechanics the transformation $B^1 \mapsto I_1$ is called the *second quantization* of the operator B^1. The transformation $B^k \mapsto I_k$ for $k > 1$ can be interpreted as the tensor power of this second quantization (see Appendix J for Fock-space notation).

The generators of interaction of order at most k have the form

$$\sum_{l=1}^{k} I_l[P^l, A^l]f(x_1, \dots, x_n)$$
$$= (G_{\le k} f)(x_1, \dots, x_n) + \sum_{I\subset\{1,\dots,n\}} (A_I f)(x_1, \dots, x_n), \tag{1.67}$$

where $G_{\le k} f$ is given by (1.32). The corresponding kth-order scaled evolution on $C(S\mathcal{X})$ is then governed by the equations

$$\dot{f}(t) = I^h[P, A]f(t), \qquad I^h[P, A] = \frac{1}{h}\sum_{l=1}^{k} h^l I_l[P^l, A^l]. \tag{1.68}$$

Scaling the state space by choosing f on $S\mathcal{X}$ to be of the form $f(\mathbf{x}) = F(h\delta_{\mathbf{x}})$, one defines the corresponding generators Λ_h^k on $C(\mathcal{M}_{h\delta}^+)$ by

$$\Lambda_h^l F(h\delta_{\mathbf{x}}) = h^{l-1} \sum_{I\subset\{1,\dots,n\},|I|=l} B_I^l F(h\delta_{\mathbf{x}}), \qquad \mathbf{x} = (x_1, \dots, x_n). \tag{1.69}$$

Using the combinatorial equation (H.4) for $F_g(h\nu)$ with $h\nu = h\delta_{x_1} + \cdots + h\delta_{x_n} = h\delta_{\mathbf{x}}$, one can write

$$\frac{1}{h}h^l \sum_{I\subset\{1,\dots,n\},|I|=l} (A_I F_g)(h\nu) = h^l \sum_{I\subset\{1,\dots,n\},|I|=l} (A^l g^{\oplus}(\mathbf{x}_I))$$
$$= \frac{1}{l!}\int (Ag^{\oplus})(z_1, \dots, z_l) \prod_{j=1}^{l} (h\nu)(dz_j) + O(h),$$

where the notation $(Ag^{\oplus})(z_1,\dots,z_l)=(A^l g^{\oplus}(z_1,\dots,z_l))$ is used. The same limiting procedure as in Section 1.5 now leads to, instead of (1.41), the more general equation

$$\frac{d}{dt}\int g(z)\mu_t(dz)$$
$$=\sum_{l=1}^{k}\frac{1}{l!}\int_{X^l}\left\{(Ag^{\oplus})(\mathbf{z})+\int_{\mathcal{X}}\left[g^{\oplus}(\mathbf{y})-g^{\oplus}(\mathbf{z})\right]P(\mathbf{z};d\mathbf{y})\right\}\mu_t^{\otimes l}(d\mathbf{z}). \quad (1.70)$$

More compactly, (1.70) can be written in terms of the operators B^k as

$$\frac{d}{dt}\int g(z)\mu_t(dz)=\sum_{l=1}^{k}\frac{1}{l!}\int_{X^l}(B^l g^{\oplus})(\mathbf{z})\mu_t^{\otimes l}(d\mathbf{z}) \quad (1.71)$$

or as

$$\frac{d}{dt}\int g(z)\mu_t(dz)=\int_{\mathcal{X}}(Bg^{\oplus})(\mathbf{z})\mu_t^{\tilde{\otimes}}(d\mathbf{z}), \quad (1.72)$$

where the convenient normalized tensor power of measures is defined, for arbitrary Y, by

$$(Y^{\tilde{\otimes}})_n(dx_1\cdots dx_n)=Y^{\tilde{\otimes}n}(dx_1\cdots dx_n)=\frac{1}{n!}Y(dx_1)\cdots Y(dx_n).$$

Finally, one can allow additionally for a *mean field interaction*, i.e. for the dependence of the family of operators A and transition kernels P in (1.65) on the current empirical measure $\mu=h\delta_{\mathbf{x}}$. In this case one obtains a generalized version of (1.70) in which A and P depend additionally on μ_t:

$$\frac{d}{dt}(g,\mu_t)=\int_{\mathcal{X}}\left\{(A[\mu_t]g^{\oplus})(\mathbf{z})+\int_{\mathcal{X}}\left[g^{\oplus}(\mathbf{y})-g^{\oplus}(\mathbf{z})\right]P(\mu_t,\mathbf{z};d\mathbf{y})\right\}\mu_t^{\tilde{\otimes}}(d\mathbf{z}), \quad (1.73)$$

or, more compactly,

$$\frac{d}{dt}(g,\mu_t)=\int_{\mathcal{X}}(B[\mu_t]g^{\oplus})(\mathbf{z})\mu_t^{\tilde{\otimes}l}(d\mathbf{z}). \quad (1.74)$$

This is the weak form of the general kinetic equation describing the dynamic LLN for Markov models of interacting particles, or interacting Markov processes, with mean field and kth-order interactions.

If the Cauchy problem for this equation is well posed, its solution μ_t for a given $\mu_0=\mu$ can be considered as a deterministic measure-valued Markov process. The corresponding semigroup is defined as $T_t F(\mu)=F(\mu_t)$. Using the variational derivatives (F.6) the evolution equation for this semigroup can be written as

$$\frac{d}{dt}F(\mu_t) = (\Lambda F)(\mu_t) = \int_{\mathcal{X}} B[\mu_t]\left(\frac{\delta F}{\delta \mu_t(.)}\right)^{\oplus}(\mathbf{z})\mu_t^{\tilde{\otimes}}(d\mathbf{z})$$
$$= \int_{\mathcal{X}} A[\mu_t]\left(\frac{\delta F}{\delta \mu_t(.)}\right)^{\oplus}(\mathbf{z})\mu_t^{\tilde{\otimes}}(d\mathbf{z})$$
$$+ \int_{\mathcal{X}^2}\left[\left(\frac{\delta F}{\delta \mu_t(.)}\right)^{\oplus}(\mathbf{y}) - \left(\frac{\delta F}{\delta \mu_t(.)}\right)^{\oplus}(\mathbf{z})\right] P(\mu_t, \mathbf{z}; d\mathbf{y})\mu_t^{\tilde{\otimes}}(d\mathbf{z}). \quad (1.75)$$

This is heuristic for the moment, as the assumptions of Lemma F.1 should be either checked for this case or modified appropriately. The kinetic equation (1.73) is simply a particular case of (1.75) for the linear functionals $F(\mu) = F_g(\mu) = (g, \mu)$.

For a linear Markov process it is instructive to study the analytic properties of its semigroup. For instance, knowledge about the domain or an invariant core is important. A core can often be identified with a certain class of smooth functions. A similar search for a core for the nonlinear semigroup $F(\mu_t)$ specified by (1.75) leads naturally to the question of the differentiability of μ_t with respect to the initial data μ_0. We shall explore this problem in Chapter 8, and later on its usefulness in relation to the LLN and CLT will be demonstrated.

If the transition kernels $P(\mu, \mathbf{z}; d\mathbf{y})$ preserve the number of particles, the components

$$B^l f(x_1, \ldots, x_l) = A^l f(x_1, \ldots, x_l) + \int_{X^l} [f(y_1, \ldots, y_l) - f(x_1, \ldots, x_l)] \times P^l(x_1, \ldots, x_l, dy_1 \cdots dy_l) \quad (1.76)$$

also preserve the number of particles, i.e. $B^l : C_{\mathrm{sym}}(X^l) \mapsto C_{\mathrm{sym}}(X^l)$. In particular, the integral transformations specified by P^k can be included in A^k. For instance, in the important case in which only binary interactions that preserve the number of particles are included, equation (1.74) takes the form

$$\frac{d}{dt}(g, \mu_t) = \int_X (B^1[\mu_t]g)(z)\mu_t(dz) + \tfrac{1}{2}\int_{X^2}(B^2[\mu_t]g^{\oplus})(z_1, z_2)\mu_t(dz_1)\mu_t(dz_2) \quad (1.77)$$

for conditionally positive operators B^1 and B^2 in $C(X)$ and $C^2_{\mathrm{sym}}(X^2)$ respectively.

It is worth stressing that the same kinetic equation can be obtained as the LLN dynamics for quite different Markov models of interaction. In particular, since A and P are symmetric, equation (1.73) (obtained as a limit for complicated interactions allowing for a change in the number of particles) can be written compactly as

$$\frac{d}{dt}(g, \mu_t) = (\tilde{A}[\mu_t]g, \mu_t) \tag{1.78}$$

for a certain conditionally positive operator $\tilde{A}$ in $C(X)$ depending on μ as on a parameter, which represents the mean field limit for a process preserving the number of particles. However, when analyzing the evolution (1.73) (or more specifically (1.70)) it is often convenient to keep track of the structure of the interaction and not to convert it to the concise form (1.78), in particular because some natural subcriticality conditions can be given in terms of this structure and may be lost in such a reduction. Having said this we stress that, when solving kinetic equations numerically using a scheme arising from particle approximations, we can try various approximations to find the most appropriate from the computational point of view.

In order to make all our heuristic calculations rigorous, we have to perform at least two (closely connected) tasks: first, to show the well-posedness of the Cauchy problem for equation (1.73) under certain assumptions and possibly in a certain class of measures; second, to prove the convergence of the Markov approximation processes in $\mathcal{M}^+_{h\delta}(X)$ to their solutions. These tasks will be dealt with in Parts II and III respectively. In the rest of this chapter we shall discuss briefly further classical examples. We shall also introduce certain tools and entities (moment measures, correlation functions, the nonlinear martingale problem) that are linked with evolutions of the type (1.73) and, as can be seen in the physical and mathematical literature, are used for practical calculations, for the analysis of the qualitative behavior of such evolutions and for the comparison of theoretical results with experiments and simulations.

1.8 Classical kinetic equations of statistical mechanics: Vlasov, Boltzmann, Landau

The kinetic equations (1.41),

$$\frac{d}{dt}(g, \mu_t) = \Lambda_{l\le k} F_g(\mu_t) = \sum_{l=1}^{k} \frac{1}{l!} \int_{\mathcal{X}} \int_{X^l} \left[g^{\oplus}(\mathbf{y}) - g^{\oplus}(\mathbf{z}) \right] P^l(\mathbf{z}; d\mathbf{y}) \mu_t^{\otimes l}(d\mathbf{z}),$$

usually appear in spatially homogeneous models of interaction. The richer model (1.66) allows us to include arbitrary underlying Markov motions, potential interactions, interdependence of volatility (diffusion coefficients) etc. We shall distinguish here some kinetic equations of the type (1.77) that are of particular interest for physics, showing how they fit into the general framework discussed above.

Example 1.11 (Vlasov's equation) As a classical particle is described by its position and momentum, let $X = \mathbf{R}^{2d}$ and both B^1 and B^2 be generators of deterministic dynamics (first-order differential operators). The *Vlasov equation* in a weak form,

$$\frac{d}{dt}(g, \mu_t) = \int_{R^2} \left(\frac{\partial H}{\partial p}\frac{\partial g}{\partial x} - \frac{\partial H}{\partial x}\frac{\partial g}{\partial p} \right) \mu_t(dx\, dp) + \int_{R^{4d}} \left(\nabla V(x_1 - x_2), \frac{\partial g}{\partial p_1}(x_1, p_1) \right) \times \mu_t(dx_1\, dp_1)\mu_t(dx_2\, dp_2) \tag{1.79}$$

is obtained when B^1 generates the Hamiltonian dynamics, i.e.

$$B^1 = \frac{\partial H}{\partial p}\frac{\partial}{\partial x} - \frac{\partial H}{\partial x}\frac{\partial}{\partial p}.$$

The function $H(x, p)$ is called the *Hamiltonian*; for example, $H = p^2/2 - U(x)$ for a given single-particle potential U. The generator B^2 specifies the potential interaction:

$$B^2 f(x_1, p_1, x_2, p_2) = \nabla V(x_1 - x_2)\frac{\partial f}{\partial p_1} + \nabla V(x_2 - x_1)\frac{\partial f}{\partial p_2}$$

for a given potential V.

Example 1.12 (Boltzmann's equation) Unlike the spatially homogeneous Boltzmann model leading to (1.50), the state space of the full model of collisions is $\mathbf{R}^{2d}$ as for the Vlasov equation. Assuming that away from collisions particles move according to the law of free motion and that collisions may occur when the distance between particles is small, we obtain the *mollified Boltzmann equation*

$$\frac{d}{dt}(g, \mu_t) = \left(v\frac{\partial}{\partial x}g, \mu_t \right) + \frac{1}{2}\int_{R^{4d}} \int_{n\in S^{d-1}:(n, v_1 - v_2)\geq 0} \eta(x_1 - x_2) \times B(v_1 - v_2, dn)\mu_t(dx_1\, dv_1)\mu_t(dx_2\, dv_2) \times \big[g(x_1, v_1 - n(v_1 - v_2, n)) + g(x_2, v_2 + n(v_1 - v_2, n)) - g(x_1, v_1) - g(x_2, v_2) \big], \tag{1.80}$$

where the *mollifier* η is a certain non-negative function with compact support.

More interesting, however, is the equation, obtained from (1.80) by a limiting procedure, that describes local collisions that occur only when the positions of the particles coincide. Namely, suppose that instead of η we have a family η^ϵ of functions converging weakly to the measure $\sigma(x_1)\delta_{x_1-x_2}dx_1$ with continuous σ:

$$\lim_{\epsilon\to 0}\int_{\mathbf{R}^{2d}} f(x_1,x_2)\eta^\epsilon(x_1-x_2)dx_1dx_2 = \int_{\mathbf{R}^d} f(x,x)\sigma(x)dx$$

for all $f \in C(\mathbf{R}^{2d})$. Assuming that $\mu_t(dx\,dv) = \mu_t(x,dv)dx$ (with an obvious abuse of notation) and taking the formal limit as $\epsilon \to 0$ in the kinetic equation (1.80) leads to the *Boltzmann equation* in weak form:

$$\begin{aligned}
&\frac{d}{dt}\int_{\mathbf{R}^d} g(x,v)\mu_t(x,dv)\\
&= \int_{\mathbf{R}^d} v\frac{\partial g}{\partial x}(x,v)\mu_t(x,dv)\\
&\quad + \frac{1}{2}\int_{R^{2d}}\int_{n\in S^{d-1}:(n,v_1-v_2)\geq 0} B(v_1-v_2,dn)\mu_t(x,dv_1)\mu_t(x,dv_2)\\
&\quad \times \big[g(x,v_1-n(v_1-v_2,n)) + g(x,v_2+n(v_1-v_2,n))\\
&\qquad - g(x,v_1) - g(x,v_2)\big]. \qquad (1.81)
\end{aligned}$$

A rigorous analysis of this limiting procedure and the question of the well-posedness of equation (1.81) are important open problems.

Example 1.13 (Landau–Fokker–Planck equation) The equation

$$\begin{aligned}
&\frac{d}{dt}(g,\mu_t)\\
&= \int_{R^{2d}}\big[\tfrac{1}{2}(G(v-v_\star)\nabla,\nabla)g(v) + (b(v-v_\star),\nabla g(v))\big]\mu_t(dv_\star)\mu_t(dv),
\end{aligned} \qquad (1.82)$$

for a certain non-negative matrix-valued field $G(v)$ and a vector field $b(v)$, is known as the Landau–Fokker–Planck equation. Physically this equation describes the limiting regime of Boltzmann collisions described by (1.50) when they become *grazing*, i.e. when v_1 is close to v_2. We will not describe this procedure in detail; for further information see Arseniev and Buryak [13], Villani [245] and references therein.

1.9 Moment measures, correlation functions and the propagation of chaos

For completeness, we shall introduce here the basic notion used in physics for describing interacting-particle systems, namely correlation functions, and the related effect of the propagation of chaos. Despite its wider importance, this material is not much used later in the book and is not crucial for understanding what follows; so this section could be skipped.

Let the measure $\rho = (\rho_0, \rho_1, \dots) \in \mathcal{M}_{\mathrm{sym}}(\mathcal{X})$. The important role in the analysis of interacting particles is played by the *moment measures* ν of ρ, defined as

$$\nu_n(dx_1 \dots dx_n) = \sum_{m=0}^{\infty} \int_{X^m} \frac{(n+m)!}{m!} \rho(dx_1 \cdots dx_n \dots dx_{n+m}), \tag{1.83}$$

where the integral in each term is taken over the variables $x_{n+1}, \dots, x_{n+m}$. If $X = \mathbf{R}^d$ and

$$\rho_n(dx_1 \cdots dx_n) = \rho_n(x_1, \dots, x_n) dx_1 \cdots dx_n,$$

then the densities of the ν_n with respect to Lebesgue measure, given by

$$\nu_n(x_1, \dots, x_n) = \sum_{m=0}^{\infty} \int_{X^m} \frac{(n+m)!}{m!} \rho(x_1, \dots, x_{n+m}) dx_{n+1} \cdots dx_{n+m},$$

are called the *correlation functions* of the family of densities $\{\rho_n\}$. For large-number asymptotics one normally uses the *scaled moment measures* ν^h of ρ, defined as

$$\nu_n^h(dx_1 \cdots dx_n) = h^n \sum_{m=0}^{\infty} \int_{X^m} \frac{(n+m)!}{m!} \rho(dx_1 \cdots dx_n \cdots dx_{n+m}) \tag{1.84}$$

for a positive parameter h.

Let us look at this transformation from the functional point of view. For any $g_k \in C(X^k)$ let us introduce an observable $S^h g_k \in C_{\mathrm{sym}}(\mathcal{X})$ defined by $S^h g_k(\mathbf{x}) = 0$ for $\mathbf{x} \in X^n$ with $n < k$ and by

$$S^h g_k(\mathbf{x}) = h^k \sum_{i_1, \dots, i_k \in \{1, \dots, n\}}' g(x_{i_1}, \dots, x_{i_k})$$

for $\mathbf{x} \in X^n$ with $n \geq k$; $\sum'$ means summation over all ordered k-tuples $i_1, \dots, i_k$ of different numbers from $\{1, \dots, n\}$ and $S^h g_0(\mathbf{x}) = g_0$. The function S^h has a clear combinatorial interpretation. For instance, if $g_1 = \mathbf{1}_A$ for a Borel set $A \subset X$ then $S^h g_1(\mathbf{x})$ is the number (scaled by h) of components of $\mathbf{x}$ lying in A. If $g_k = \mathbf{1}_{A_1} \cdots \mathbf{1}_{A_k}$ for pairwise disjoint Borel sets $A_i \subset X$ then

$$S^h g_k(\mathbf{x}) = S^h \mathbf{1}_{A_1}(\mathbf{x}) \cdots S^h \mathbf{1}_{A_k}(\mathbf{x})$$

is the number (scaled by h^k) of ways to choose one particle from A_1, one particle from A_2 and so on from a given collection $\mathbf{x}$. So S^h defines a functional extension of the well-known problem of counting the number of ways to choose a sequence of balls of prescribed colors, say green, blue, red, from a bag containing a given number of colored balls.

Clearly, if $g \in C_{\text{sym}}(X^k)$ then

$$S^h g_k(\mathbf{x}) = h^k k! \sum_{I \subset \{1,\dots,n\}:|I|=k} g(\mathbf{x}_I).$$

We have seen already that these observables appear naturally when deducing kinetic equations. The extension of the mapping $g_k \mapsto S^h g_k$ to the mapping $C(\mathcal{X}) \mapsto C_{\text{sym}}(\mathcal{X})$ by linearity can be expressed as

$$S^h g(\mathbf{x}) = \sum_{I \subset \{1,\dots,n\}} h^{|I|} |I|! g(\mathbf{x}_I).$$

Remark 1.14 It can be shown by induction that if $f = S^h g$ with $g \in C_{\text{sym}}(\mathcal{X})$ then g is unique and is given by

$$g(x_1, \dots, x_n) = \frac{1}{n!} h^{-n} \sum_{I \subset \{1,\dots,n\}} (-1)^{n-|I|} f(\mathbf{x}_I).$$

Lemma 1.15 *The mapping $\rho \mapsto \nu_k^h$ from $\mathcal{M}_{\text{sym}}(\mathcal{X})$ to $\mathcal{M}_{\text{sym}}(X^k)$ is dual to the mapping $g_k \mapsto S^h g_k$. That is, for ν^h defined by (1.84) and $g = (g_0, g_1, \dots) \in C(\mathcal{X})$,*

$$(S^h g_k, \rho) = (g_k, \nu_k^h), \qquad k = 0, 1, \dots \tag{1.85}$$

and

$$(S^h g, \rho) = (g, \nu^h) = \sum_{k=0}^{\infty} (g_k, \nu_k^h). \tag{1.86}$$

In particular, if $\mathbf{x}$ is a random variable in $\mathcal{X}$ that is distributed according to the probability law $\rho \in \mathcal{P}(\mathcal{X})$ then

$$\mathbf{E} S^h g_k(\mathbf{x}) = (g_k, \nu_k^h). \tag{1.87}$$

Proof We have

$$\begin{aligned}
(S^h g_k, \rho) &= \sum_{m=0}^{\infty} S^h g_k(x_1, \dots, x_{k+m}) \rho_{k+m}(dx_1 \cdots dx_{k+m}) \\
&= \sum_{m=0}^{\infty} h^k \sum_{i_1,\dots,i_k \in \{1,\dots,m+k\}}{}' g(x_{i_1}, \dots, x_{i_k}) \rho_{k+m}(dx_1 \cdots dx_{k+m}) \\
&= \sum_{m=0}^{\infty} h^k \frac{(m+k)!}{m!} g(x_1, \dots, x_k) \rho_{k+m}(dx_1 \cdots dx_{k+m}),
\end{aligned}$$

implying (1.85). □

The crucial property of the moment measures is obtained by observing that they represent essentially the moments of empirical measures.

Lemma 1.16 *Let* $\mathbf{x}$ *be a random variable in* $\mathcal{X}$ *that is distributed according to the probability law* $\rho \in \mathcal{P}(\mathcal{X})$*; then*

$$\mathbf{E}(g_k, (h\delta_{\mathbf{x}})^{\otimes k}) = (g_k, \nu_k^h) + O(h)\|f\| \max_{l<k} \|\nu_l^h\| \tag{1.88}$$

for $g_k \in C_{\text{sym}}(X^k)$*. In particular,*

$$\mathbf{E}(g_1, h\delta_{\mathbf{x}}) = (g_1, \nu_1^h), \tag{1.89}$$

$$\mathbf{E}(g_2, (h\delta_{\mathbf{x}})^{\otimes 2}) = (g_2, \nu_2^h) + h \int g_2(x, x)\nu_1^h(dx). \tag{1.90}$$

Proof We have

$$\begin{aligned}\mathbf{E}(g_1, h\delta_{\mathbf{x}}) = h &\int_X g_1(x_1)\rho_1(dx_1) \\ &+ h \int_{X^2} \left[g_1(x_1) + g_1(x_2)\right] \rho_2(dx_1 dx_2) + \cdots ,\end{aligned}$$

yielding (1.89). Similarly,

$$\begin{aligned}\mathbf{E}(g_2, (h\delta_{\mathbf{x}})^{\otimes 2}) = h^2 &\int_X g_2(x, x)\rho_1(dx) \\ &+ h^2 \int_{X^2} \left[g_2(x_1, x_1) + g_2(x_1, x_2)\right. \\ &\left. + g_2(x_2, x_1) + g_2(x_2, x_2)\right] \rho_2(dx_1 dx_2) + \cdots ,\end{aligned}$$

yielding (1.90). Then (1.88) is obtained by trivial induction. □

Suppose, now, we can prove that as $h \to 0$ the empirical measures $h\delta_{\mathbf{x}}$ of the approximating Markov process converge to the deterministic measure-valued process μ_t that solves the kinetic equation (1.73). By (1.89) the limiting measure μ_t coincides with the limit of the first moment measure ν_1^h. Moreover, once the convergence of $h\delta_{\mathbf{x}}$ to a deterministic limit is obtained, it is natural to expect (and often easy to prove) that the tensor powers $(h\delta_{\mathbf{x}})^{\otimes k}$ converge to the products $\mu_t^{\otimes k}$. Hence, by (1.88) we can then conclude that the moment measures ν_k^h converge to the products $\mu_t^{\otimes k}$. The possibility of expressing the moment measures as products is a manifestation of the so-called *propagation of chaos* property of the limiting evolution.

Exercise 1.3 Prove the following results. (i) If μ_t satisfies (1.71) then $\nu_t = (\mu_t)^{\tilde{\otimes}} \in \mathcal{M}_{\text{sym}}(\mathcal{X})$ satisfies the linear equation

$$\frac{d}{dt}\nu_t^l(dx_1\cdots dx_l) = \sum_{j=1}^{l}\sum_{k=1}^{K} C_{l+k-1}^l \int_{x_{l+1},\ldots,x_{l+k-1}} (B_k^{j,l+1,\ldots,l+k-1})^* \times \nu_t^{k+l-1}(dx_1\cdots dx_{l+k-1}), \quad (1.91)$$

where the C_m^l are the usual binomial coefficients, B_k^* is the dual to B_k and $(B_k^I)^*\nu_t(dx_1\cdots dx_m)$ specifies the action of B_k^* on variables with indexes from $I \subset \{1,\ldots,m\}$.

(ii) If the evolution of $\nu_t \in M_{\text{sym}}(\mathcal{X})$ is specified by (1.91) then the dual evolution on $C^{\text{sym}}(X)$ is given by the equation

$$\dot{g}(x_1,\ldots,x_l) = (\mathcal{L}_B g)(x_1,\ldots,x_l) = \sum_{I\subset\{1,\ldots,l\}}\sum_{j\notin I}(B_{|I|+1}^{j,I} g_{\bar{I}})(x_1,\ldots,x_l), \quad (1.92)$$

where $g_I(x_1,\ldots,x_l) = g(x_I)$ and $B_k^{j_1,\ldots,j_k}$ specifies the action of B_k on the variables $j_1,\ldots,j_k$. In particular,

$$(\mathcal{L}_B g^1)(x_1,\ldots,x_l) = (B_l(g^1)^+)(x_1,\ldots,x_l).$$

Hint: Observe that the strong form of (1.71) is

$$\dot{\mu}_t(dx) = \sum_{k=1}^{K}\frac{1}{(k-1)!}B_k^*(\mu_t\otimes\cdots\otimes\mu_t)(dx\,dy_1\cdots dy_{k-1}),$$

which implies (1.91) by straightforward manipulation. From (1.91) it follows that

$$\frac{d}{dt}(g,\nu_t) = \sum_{l=0}^{\infty}\sum_{k=1}^{K}\sum_{I\subset\{1,\ldots,l+k-1\},|I|=k-1}\sum_{j\notin I} B_k^{j,I} g_{\bar{I}}(x_1,\ldots,x_{l+k-1}) \times \nu_t(dx_1\cdots dx_{l+k-1})$$

$$= \sum_{m=0}^{\infty}\sum_{I\subset\{1,\ldots,m\}}\sum_{j\notin I} B_{|I|+1}^{j,I} g_{\bar{I}}(x_1,\ldots,x_m)\nu_t(dx_1\cdots dx_m),$$

which implies (1.92).

1.10 Nonlinear Markov processes and semigroups; nonlinear martingale problems

The aim of this chapter was to motivate the analysis of equations describing nonlinear positivity-preserving evolutions on measures. We can now identify, in a natural way, nonlinear analogs of the main notions from the theory of Markov processes and observe how the fundamental connection between Markov processes, semigroups and martingale problems is carried forward into the nonlinear setting.

Let $\tilde{\mathcal{M}}(X)$ be a dense subset of the space $\mathcal{M}(X)$ of finite (positive Borel) measures on a metric space X, considered in its weak topology. By a nonlinear *sub-Markov* (resp. *Markov*) *propagator* in $\tilde{\mathcal{M}}(X)$ we shall mean any propagator $V^{t,r}$ of possibly nonlinear transformations of $\tilde{\mathcal{M}}(X)$ that do not increase (resp. preserve) the norm. If $V^{t,r}$ depends only on the difference $t-r$ and hence specifies a semigroup, this semigroup is referred to as nonlinear or generalized *sub-Markov* or *Markov* respectively.

The usual, linear, Markov propagators or semigroups correspond to the case when all the transformations are linear contractions in the whole space $\mathcal{M}(X)$. In probability theory these propagators describe the evolution of averages of Markov processes, i.e. processes whose evolution after any given time t depends on the past $X_{\leq t}$ only via the present position X_t. Loosely speaking, to any nonlinear Markov propagator there corresponds a process whose behavior after any time t depends on the past $X_{\leq t}$ only through the position X_t of the process and its distribution at t. To be more precise, consider the nonlinear kinetic equation

$$\frac{d}{dt}(g, \mu_t) = (B[\mu_t]g, \mu_t), \tag{1.93}$$

where the family of operators $B[\mu]$ in $C(X)$ depend on μ as on a parameter and each $B[\mu]$ generates a Feller semigroup. (It was shown above that equations of this kind appear naturally as LLNs for interacting particles, and, as we shall see in Section 11.5, also that they arise from the mere assumption of positivity preservation.) Suppose that the Cauchy problem for equation (1.93) is well posed and specifies the weakly continuous Markov semigroup T_t in $\mathcal{M}(X)$. Suppose also that for any weakly continuous curve $\mu_t \in \mathcal{P}(X)$ the solutions to the Cauchy problem

$$\frac{d}{dt}(g, \nu_t) = (B[\mu_t]g, \nu_t) \tag{1.94}$$

define a weakly continuous propagator $V^{t,r}[\mu_.]$, $r \leq t$, of linear transformations in $\mathcal{M}(X)$ and hence a Markov process in X. Then, to any $\mu \in \mathcal{P}(X)$

there corresponds a Markov process X_t^μ in X with distributions $\mu_t = T_t(\mu)$ for all times t and with transition probabilities $p_{r,t}^\mu(x, dy)$ specified by equation (1.94) and satisfying the condition

$$\int_{X^2} f(y) p_{r,t}^\mu(x, dy) \mu_r(dx) = (f, V^{t,r} \mu_r) = (f, \mu_t). \tag{1.95}$$

We shall call the family of processes X_t^μ a *nonlinear Markov process.*

Thus a nonlinear Markov process is a semigroup of the transformations of distributions such that to each trajectory is attached a "tangent" Markov process with the same marginal distributions. The structure of these tangent processes is not intrinsic to the semigroup but can be specified by choosing a stochastic representation for the generator.

As in the linear case, the process X_t with càdlàg paths (or the corresponding probability distribution on the Skorohod space) solves the $(B[\mu], D)$-*nonlinear martingale problem* with initial distribution μ, meaning that X_0 is distributed according to μ and that the process

$$M_t^f = f(X_t) - f(X_0) - \int_0^t B[\mathcal{L}(X_s)] f(X_s)\, ds, \quad t \geq 0 \tag{1.96}$$

($\mathcal{L}(X_s)$ is the law of X_s) is a martingale for any $f \in D$, with respect to the natural filtration of X_t. This martingale problem is called well posed if, for any initial μ, there exists a unique X_t solving it.

PART I

Tools from Markov process theory

2

Probability and analysis

In this chapter we recall some particularly relevant tools from the theory of Markov processes and semigroups, stressing the connection between an analytical description in terms of evolution equations and a probabilistic description. To begin with, we introduce the duality between abstract semigroups and propagators of linear transformations, at the level of generality required for further applications.

2.1 Semigroups, propagators and generators

This section collects in a systematic way those tools from functional analysis that are most relevant to random processes; we consider these in the next section. Apart from recalling the notions of operator semigroups and their generators, we shall discuss their nonhomogeneous analogs, i.e. propagators, and use them to deduce a general well-posedness result for a class of nonlinear semigroups. For completeness we first recall the notion of unbounded operators (also fixing some notation), assuming, however, that readers are familiar with such basic definitions for Banach and Hilbert spaces as convergence, bounded linear operators and dual spaces and operators.

A *linear operator* A on a Banach space B is a linear mapping $A : D \mapsto B$, where D is a subspace of B called the *domain of* A. We say that the operator A is *densely defined* if D is dense in B. The operator A is called *bounded* if the norm $\|A\| = \sup_{x \in D} \|Ax\|/\|x\|$ is finite. If A is bounded and D is dense then A has a unique bounded extension (with the same norm) to an operator with the whole of B as its domain. It is also well known that a linear operator $A : B \to B$ is continuous if and only if it is bounded. For a continuous linear mapping $A : B_1 \to B_2$ between two Banach spaces its *norm* is defined as

$$\|A\|_{B_1 \mapsto B_2} = \sup_{x \neq 0} \frac{\|Ax\|_{B_2}}{\|x\|_{B_1}}.$$

The space of bounded linear operators $B_1 \to B_2$ equipped with this norm is a Banach space itself, often denoted by $\mathcal{L}(B_1, B_2)$.

A sequence of bounded operators A_n, $n = 1, 2, \ldots$, in a Banach space B is said to *converge strongly* to an operator A if $A_n f \to Af$ for any $f \in B$.

A linear operator on a Banach space is called a *contraction* if its norm does not exceed 1. A semigroup T_t of bounded linear operators on a Banach space B is called *strongly continuous* if $\|T_t f - f\| \to 0$ as $t \to 0$ for any $f \in B$.

Example 2.1 If A is a bounded linear operator on a Banach space then

$$T_t = e^{tA} = \sum_{n=0}^{\infty} \frac{t^n}{n!} A^n$$

defines a strongly continuous semigroup.

Example 2.2 The shifts $T_t f(x) = f(x + t)$ form a strongly continuous group of contractions on $C_\infty(\mathbf{R})$, $L^1(\mathbf{R})$ or $L^2(\mathbf{R})$. However, it is not strongly continuous on $C(\mathbf{R})$. Observe also that if f is an analytic function then

$$f(x + t) = \sum_{n=0}^{\infty} \frac{t^n}{n!} (D^n f)(x),$$

which can be written formally as $e^{tD} f(x)$.

Example 2.3 Let $\eta(y)$ be a complex-valued continuous function on $\mathbf{R}^d$ such that $Re\, \eta \leq 0$. Then

$$T_t f(y) = e^{t\eta(y)} f(y)$$

is a semigroup of contractions, on the Banach spaces $L^p(\mathbf{R}^d)$, $L^\infty(\mathbf{R}^d)$, $B(\mathbf{R}^d)$, $C(\mathbf{R}^d)$ or $C_\infty(\mathbf{R}^d)$, that is strongly continuous on $L^p(\mathbf{R}^d)$ and $C_\infty(\mathbf{R}^d)$ but not on the other three spaces.

An operator A with domain D is called *closed* if its graph is a closed subset of $B \times B$, that is, if $x_n \to x$ and $Ax_n \to y$ as $n \to \infty$ for a sequence $x_n \in D$ then $x \in D$ and $y = Ax$. The operator A is called *closable* if a closed extension of A exists, in which case the *closure of* A is defined as the minimal closed extension of A, that is, the operator whose graph is the closure of the graph of A. A subspace D of the domain D_A of a closed operator A is called a *core* for A if A is the closure of A restricted to D.

Let T_t be a strongly continuous semigroup of linear operators on a Banach space B. The *infinitesimal generator*, or simply the *generator*, of T_t is defined as the operator A given by

$$Af = \lim_{t \to 0} \frac{T_t f - f}{t}$$

and acting on the linear subspace $D_A \subset B$ (the *domain* of A) where this limit exists (in the topology of B). If the T_t are contractions then the *resolvent* of T_t (or of A) is defined for any $\lambda > 0$ as the operator R_λ given by

$$R_\lambda f = \int_0^\infty e^{-\lambda t} T_t f \, dt.$$

For example, the generator A of the semigroup $T_t f = e^{t\eta} f$ from Example 2.3 is given by the multiplication operator $Af = \eta f$ on functions f such that $\eta^2 f \in C_\infty(\mathbf{R}^d)$ or $\eta^2 f \in L^p(\mathbf{R}^d)$.

Theorem 2.4 (Basic properties of generators and resolvents) *Let T_t be a strongly continuous semigroup of linear contractions on a Banach space B and let A be its generator. Then the following hold:*

(i) $T_t D_A \subset D_A$ *for each* $t \geq 0$ *and* $T_t A f = A T_t f$ *for each* $t \geq 0$, $f \in D_A$.
(ii) $T_t f = \int_0^t A T_s f \, ds + f$ *for* $f \in D$.
(iii) R_λ *is a bounded operator on* B, *with* $\|R_\lambda\| \leq \lambda^{-1}$ *for any* $\lambda > 0$.
(iv) $\lambda R_\lambda f \to f$ *as* $\lambda \to \infty$.
(v) $R_\lambda f \in D_A$ *for any* f *and* $\lambda > 0$ *and* $(\lambda - A) R_\lambda f = f$, *i.e.* $R_\lambda = (\lambda - A)^{-1}$.
(vi) *If* $f \in D_A$ *then* $R_\lambda A f = A R_\lambda f$.
(vii) D_A *is dense in* B.
(viii) A *is closed on* D_A.

Proof (i) Observe that for, $\psi \in D_A$,

$$A T_t \psi = \left(\lim_{h \to 0} \frac{1}{h}(T_h - I) \right) T_t \psi = T_t \left(\lim_{h \to 0} \frac{1}{h}(T_h - I) \right) \psi = T_t A \psi.$$

(ii) This follows from (i).
(iii) We have $\|R_\lambda f\| \leq \int_0^\infty e^{-\lambda t} \|f\| \, dt = \lambda^{-1} \|f\|$.
(iv) This follows from the equation

$$\lambda \int_0^\infty e^{-\lambda t} T_t f \, dt = \lambda \int_0^\infty e^{-\lambda t} f \, dt + \lambda \int_0^\epsilon e^{-\lambda t} (T_t f - f) \, dt + \lambda \int_\epsilon^\infty e^{-\lambda t} (T_t f - f) \, dt,$$

observing that the first term on the r.h.s. is f and the second (resp. third) term is small for small ϵ (resp. for any ϵ and large λ).

(v) By definition

$$AR_\lambda f = \lim_{h\to 0}\frac{1}{h}(T_h - \mathbf{1})R_\lambda f = \lim_{h\to 0}\frac{1}{h}\int_0^\infty e^{-\lambda t}(T_{t+h}f - T_t f)\,dt$$

$$= \lim_{h\to 0}\left(\frac{e^{\lambda h}-1}{h}\int_0^\infty e^{-\lambda t}T_t f\,dt - \frac{e^{\lambda h}}{h}\int_0^h e^{-\lambda t}T_t f\,dt\right) = \lambda R_\lambda f - f.$$

(vi) This follows from the definitions and (ii).

(vii) This follows from (iv) and (v).

(viii) If $f_n \to f$ as $n\to\infty$ for a sequence $f_n \in D$ and A if $f_n \to g$ then

$$T_t f - f = \lim_{n\to\infty}\int_0^t T_s A f_n\,ds = \int_0^t T_s g\,ds.$$

Applying the fundamental theorem of calculus completes the proof. □

Remark 2.5 For all $\psi \in B$, the vector $\psi(t) = \int_0^t T_u\psi\,du$ belongs to D_A and $A\psi(t) = T_t\psi - \psi$. Moreover, $\psi(t) \to \psi$ as $t \to 0$ for all ψ and $A\psi(t) \to A\psi$ for $\psi \in D_A$. This observation yields another insightful proof of statement (vii) of Theorem 2.4 (bypassing the use of the resolvent).

Proposition 2.6 *Let an operator A with domain D_A generate a strongly continuous semigroup of linear contractions T_t. If D is a dense subspace of D_A that is invariant under all T_t then D is a core for A.*

Proof Let $\bar{D}$ be the domain of the closure of the operator A restricted to D. We have to show that for $\psi \in D_A$ there exists a sequence $\psi_n \in \bar{D}$, $n \in \mathbf{N}$, such that $\psi_n \to \psi$ and $A\psi_n \to A\psi$. By the remark above it is enough to show this for $\psi(t) = \int_0^t T_u\psi\,du$. As D is dense there exists a sequence $\psi_n \in D$ converging to ψ; hence $A\psi_n(t) \to A\psi(t)$. To complete the proof it remains to observe that $\psi_n(t) \in \bar{D}$ by the invariance of D. □

An important tool for the construction of semigroups is *perturbation theory*, which can be applied when a generator of interest can be represented as the sum of a well-understood operator and a term that is smaller (in some sense). Below we give the simplest result of this kind.

Theorem 2.7 *Let an operator A with domain D_A generate a strongly continuous semigroup T_t on a Banach space B, and let L be a bounded operator on B. Then $A + L$ with the same domain D_A also generates a strongly* *continuous semigroup Φ_t on B given by the series*

$$\Phi_t = T_t + \sum_{m=1}^{\infty} \int_{0\le s_1\le\cdots\le s_m\le t} T_{t-s_m} L T_{s_m-s_{m-1}} L \cdots T_{s_2-s_1} L \times T_{s_1} ds_1 \cdots ds_m \tag{2.1}$$

converging in the operator norm. Moreover, $\Phi_t f$ is the unique (bounded) solution of the integral equation

$$\Phi_t f = T_t f + \int_0^t T_{t-s} L \Phi_s f \, ds, \tag{2.2}$$

with a given $f_0 = f$. Finally, if D is an invariant core for A that is itself a Banach space under the norm $\|.\|_D$, if the T_t are uniformly (for t from a compact interval) bounded operators $D \to D$ and if L is a bounded operator $D \to D$ then D is an invariant core for $A + L$ and the Φ_t are uniformly bounded operators in D.

Proof On the one hand, clearly

$$\|\Phi_t\| \le \|T_t\| + \sum_{m=1}^{\infty} \frac{(\|L\|t)^m}{m!} \left(\sup_{s\in[0,t]} \|T_s\| \right)^{m+1},$$

implying the convergence of the series. Next, we have

$$\begin{aligned}
&\Phi_t \Phi_\tau f \\
&= \sum_{m=0}^{\infty} \int_{0\le s_1\le\cdots\le s_m\le t} T_{t-s_m} L T_{s_m-s_{m-1}} L \cdots T_{s_2-s_1} L T_{s_1} \, ds_1 \cdots ds_m \\
&\quad \times \sum_{n=0}^{\infty} \int_{0\le u_1\le\cdots\le u_n\le \tau} T_{\tau-u_n} L T_{u_n-u_{n-1}} L \cdots T_{u_2-u_1} L T_{u_1} \, du_1 \cdots du_n \\
&= \sum_{m,n=0}^{\infty} \int_{0\le u_1\le\cdots u_n\le\tau\le v_1\le\cdots\le v_m\le t+\tau} T_{t+\tau-v_m} L T_{v_m-v_{m-1}} L \cdots T_{v_1-u_n} L \cdots T_{u_2-u_1} L \\
&\qquad \times T_{u_1} dv_1 \cdots dv_m du_1 \cdots du_n \\
&= \sum_{k=0}^{\infty} \int_{0\le u_1\le\cdots\le u_k\le t+\tau} T_{t+\tau-u_k} L T_{u_k-u_{k-1}} L \cdots T_{u_2-u_1} L T_{u_1} \, du_1 \cdots du_k \\
&= \Phi_{t+\tau} f,
\end{aligned}$$

showing the main semigroup condition. Equation (2.2) is then a consequence of (2.1). On the other hand, if (2.2) holds then substituting the l.h.s. of this equation into its r.h.s. recursively yields

$$\begin{aligned}\Phi_t f &= T_t f + \int_0^t T_{t-s} L T_s f\, ds + \int_0^t ds_2 T_{t-s_2} L \int_0^{s_2} ds_1 T_{s_2-s_1} L \Phi_{s_1} f \\ &= T_t f + \sum_{m=1}^{N} \int_{0 \le s_1 \le \cdots \le s_m \le t} T_{t-s_m} L T_{s_m - s_{m-1}} L \cdots T_{s_2-s_1} L T_{s_1} f\, ds_1 \cdots ds_m \\ &+ \int_{0 \le s_1 \le \cdots \le s_{N+1} \le t} T_{t-s_{N+1}} L T_{s_{N+1}-s_N} L \cdots T_{s_2-s_1} L \Phi_{s_1} f\, ds_1 \cdots ds_m\end{aligned}$$

for arbitrary N. As the last term tends to zero, the series representation (2.1) follows.

Further, since the terms with $m > 1$ in (2.1) are $O(t^2)$ for small t,

$$\frac{d}{dt}\Big|_{t=0} \Phi_t f = \frac{d}{dt}\Big|_{t=0} \left(T_t f + \int_0^t T_{t-s} L T_s f\, ds \right) = \frac{d}{dt}\Big|_{t=0} T_t f + Lf,$$

so that $(d/dt)|_{t=0} \Phi_t f$ exists if and only if $(d/dt)|_{t=0} T_t f$ exists, and in this case

$$\frac{d}{dt}\Big|_{t=0} \Phi_t f = (A + L) f.$$

The last statement is obviously true, because the conditions on D ensure that the series (2.1) converges in the norm topology of D. □

For the analysis of time-nonhomogeneous and/or nonlinear evolutions we need to extend the notion of a generator to propagators. A backward propagator $U^{t,r}$ of uniformly (for t, r from a compact set) bounded linear operators on a Banach space B is called *strongly continuous* if the family $U^{t,r}$ depends strongly continuously on t and r. For a dense subspace D of B that is invariant under all $U^{t,r}$ we say that a family of linear operators A_t with common domain D is a *(nonhomogeneous) generator of the propagator $U^{t,r}$ on the common invariant domain D* if

$$\frac{d}{ds} U^{t,s} f = U^{t,s} A_s f, \quad \frac{d}{ds} U^{s,r} f = -A_s U^{s,r} f, \qquad t \le s \le r \tag{2.3}$$

for all $f \in D$, where the derivative exists in the topology of B and where for $s = t$ (resp. $s = r$) it is assumed to be only a right (resp. left) derivative.

Remark 2.8 The principle of uniform boundedness (well known in functional analysis) states that if a family T_α of bounded linear mappings from a Banach space X to another Banach space is such that the sets $\{\|T_\alpha x\|\}$ are bounded for each x then the family T_α is uniformly bounded. This implies that if $U^{t,r}$ is a strongly continuous propagator of bounded linear operators then the norms of $U^{t,r}$ are bounded uniformly for t, r from any compact interval. This fact is not of particular importance for our purposes, as we can include uniform boundedness on compact intervals in the definition. All our constructions of propagators yield this boundedness directly.

The next result extends Theorem 2.7 to propagators.

Theorem 2.9 *Let $U^{t,r}$ be a strongly continuous backward propagator of bounded linear operators in a Banach space B, a dense subspace $D \subset B$ is itself a Banach space under the norm $\|.\|_D$ and the $U^{t,r}$ are bounded operators $D \to D$. Suppose that a family of linear operators A_t generates this propagator on the common domain D (so that* (2.3) *holds). Let L_t be a family of bounded operators in both B and D that depend continuously on t in the strong topology as operators in B. Then $A_t + L_t$ generates a strongly continuous propagator $\Phi^{t,r}$ in B, on the same invariant domain D, where*

$$\Phi^{t,r} = U^{t,r} + \sum_{m=1}^{\infty} \int_{t \le s_1 \le \cdots \le s_m \le r} U^{t,s_1} L_{s_1} \cdots U^{s_{m-1},s_m} L_{s_m} \times U^{s_m,r} ds_1 \cdots ds_m. \tag{2.4}$$

This series converges in the operator norms of both B and D. Moreover, $\Phi^{t,r} f$ is the unique bounded solution of the integral equation

$$\Phi^{t,r} f = U^{t,r} f + \int_t^r U^{t,s} L_s \Phi^{s,r} f \, ds, \tag{2.5}$$

for a given $f_r = f$.

Proof This result is a straightforward extension of Theorem 2.7. The only difference to note is that, in order to obtain

$$\frac{d}{dt}\bigg|_{t=r} \int_t^r U^{t,s} L_s \Phi^{s,r} f \, ds = \frac{d}{dt}\bigg|_{t=r} \int_t^r U^{t,s} L_r \Phi^{s,r} f \, ds = -L_r f$$

one uses the continuous dependence of L_s on s (since the L_s are strongly continuous in s, the function $L_s \Phi^{s,r} f$ is continuous in s, because the family $\Phi^{s,r} f$ is compact as the image of a continuous mapping of the interval $[t, r]$). □

For a Banach space B or a linear operator A, one usually denotes by $B^\star$ or $A^\star$ respectively its *Banach dual*. The notation B', A' is also used.

Theorem 2.10 (Basic duality) *Let $U^{t,r}$ be a strongly continuous backward propagator of bounded linear operators in a Banach space B generated by a family of linear operators A_t on a common dense domain D invariant under* *all $U^{t,r}$. Let D be itself a Banach space with respect to a norm $\|.\|_D$ such that the A_t are continuous mappings $D \to B$. Then:*

(i) the family of dual operators $V^{s,t} = (U^{t,s})^\star$ forms a weakly continuous in s, t propagator of bounded linear operators in $B^\star$ (contractions if the $U^{t,r}$ are contractions) such that

$$\frac{d}{dt}V^{s,t}\xi = -V^{r,t}A_t^\star\xi, \quad \frac{d}{ds}V^{s,t}\xi = A_s^\star V^{s,t}\xi, \qquad t \le s \le r, \tag{2.6}$$

holds weakly in $D^\star$, i.e., the second equation, for example, means

$$\frac{d}{ds}(f, V^{s,t}\xi) = (A_s f, V^{s,t}\xi), \qquad t \le s \le r, \ f \in D; \tag{2.7}$$

(ii) $V^{s,t}\xi$ is the unique solution to the Cauchy problem (2.7), i.e. if $\xi_t = \xi$ for a given $\xi \in B^\star$ and ξ_s, $s \in [t, r]$, is a weakly continuous family in $B^\star$ satisfying

$$\frac{d}{ds}(f, \xi_s) = (A_s f, \xi_s), \qquad t \le s \le r, \ f \in D, \tag{2.8}$$

then $\xi_s = V^{s,t}\xi$ for $t \le s \le r$;

(iii) $U^{s,r}f$ is the unique solution to the inverse Cauchy problem of the second equation in (2.3), i.e. if $f_r = f$ and $f_s \in D$ for $s \in [t, r]$ satisfies the equation

$$\frac{d}{ds}f_s = -A_s f_s, \qquad t \le s \le r, \tag{2.9}$$

where the derivative exists in the norm topology of B, then $f_s = U^{s,r}f$.

Proof Statement (i) is a direct consequence of duality.

(ii) Let $g(s) = (U^{s,r}f, \xi_s)$ for a given $f \in D$. Writing

$$(U^{s+\delta,r}f, \ \xi_{s+\delta}) - (U^{s,r}f, \ \xi_s) \\ = (U^{s+\delta,r}f - U^{s,r}f, \ \xi_{s+\delta}) + (U^{s,r}f, \ \xi_{s+\delta} - \xi_s)$$

and using (2.3), (2.8) and the invariance of D, allows one to conclude that

$$\frac{d}{ds}g(s) = -(A_s U^{s,r}f, \xi_s) + (U^{s,r}f, A_s^\star\xi_s) = 0.$$

Hence $g(r) = (f, \xi_r) = g(t) = (U^{t,r}f, \xi_t)$, showing that ξ_r is uniquely defined. Similarly we can analyze any other point $r' \in (s, r)$.

(iii) In a similar way to the proof of (ii), this follows from the observation that

$$\frac{d}{ds}(f_s, V^{s,t}\xi) = 0.$$

□

The following simple stability result for these propagators is useful.

Theorem 2.11 *Suppose that we are given a sequence of propagators $U_n^{t,r}$, $n = 1, 2, \ldots$, with corresponding generators A_t^n, and a propagator $U^{t,r}$ with generator A_t. Suppose that all these propagators satisfy the same conditions as $U^{t,r}$ and A_t from Theorem 2.10 with the same D and B and with all bounds uniform in n. Moreover, let*

$$\|A_t^n - A_t\|_{D\to B} \leq \epsilon_n$$

uniformly for bounded times t, where $\epsilon_n \to 0$ as $n \to \infty$. Then:

$$\|U_n^{t,r}g - U^{t,r}g\|_B = O(1)\epsilon_n\|g\|_D$$

uniformly for bounded times t, r; $U_n^{t,r}$ converges to $U^{t,r}$ strongly in B; and the dual propagators $V_n^{r,t}$ converge to $V^{r,t}$ weakly in $B^\star$ and in the norm topology of $D^\star$.

Proof The required estimate follows from the obvious representation

$$(U_n^{t,r} - U^{t,r})g = U_n^{t,s}U^{s,r}g\Big|_{s=t}^{r} = \int_t^r U_n^{t,s}(A_s^n - A_s)U^{s,r}g\,ds, \qquad g \in D.$$

The other statements follow by the usual approximation argument and duality. □

The following result represents the basic tool (used in Chapter 6) allowing us to build nonlinear propagators from infinitesimal linear ones.[1] Recall that $V^{s,t}$ is the dual of $U^{t,s}$ given by Theorem 2.10.

Theorem 2.12 *Let D be a dense subspace of a Banach space B that is itself a Banach space such that $\|f\|_D \geq \|f\|_B$, and let $\xi \mapsto A[\xi]$ be a mapping from $B^\star$ to the bounded linear operator $A[\xi] : D \to B$, such that*

$$\|A[\xi] - A[\eta]\|_{D\to B} \leq c\|\xi - \eta\|_{D^\star}, \qquad \xi, \eta \in B^\star. \tag{2.10}$$

Let M be a bounded subset of $B^\star$ that is closed in the norm topologies of both $B^\star$ and $D^\star$ and, for a $\mu \in M$, let $C_\mu([0, r], M)$ be the metric space of the curves $\xi_s \in M$, $s \in [0, r]$, $\xi_0 = \mu$, that are continuous in the norm $D^\star$, with distance

$$\rho(\xi_., \eta_.) = \sup_{s\in[0,r]} \|\xi_s - \eta_s\|_{D^\star}.$$

Assume finally that, for any $\mu \in M$ and $\xi_. \in C_\mu([0, r], M)$, the operator curve $A[\xi_t] : D \to B$ generates a strongly continuous backward propagator

[1] The reader may choose to skip over this result until it is needed.

of uniformly bounded linear operators $U^{t,s}[\xi_.]$, $0 \le t \le s \le r$, *in* B *on the common invariant domain* D *(where, in particular,* (2.3) *holds), such that*

$$\|U^{t,s}[\xi_.]\|_{D\to D} \le c, \qquad t,s \le r, \tag{2.11}$$

for some constant $c > 0$*; the dual propagators* $V^{s,t}$ *preserve the set* M*. Then the weak nonlinear Cauchy problem*

$$\frac{d}{dt}(f,\mu_t) = (A[\mu_t]f,\mu_t), \qquad \mu_0 = \mu,\ f \in D, \tag{2.12}$$

is well posed in M*. More precisely, for any* $\mu \in M$ *it has a unique solution* $T_t(\mu) \in M$*, and the transformations* T_t *of* M *form a semigroup for* $t \in [0,r]$ *depending Lipschitz continuously on time* t *and on the initial data in the norm of* $D^\star$*, i.e.*

$$\|T_t(\mu) - T_t(\eta)\|_{D^\star} \le c(r,M)\|\mu-\eta\|_{D^\star}, \qquad \|T_t(\mu)-\mu\|_{D^\star} \le c(r,M)t. \tag{2.13}$$

Proof Since

$$(f,\ (V^{t,0}[\xi^1_.] - V^{t,0}[\xi^2_.])\mu) = (U^{0,t}[\xi^1_.]f - U^{0,t}[\xi^2_.]f,\ \mu)$$

and

$$\begin{aligned} U^{0,t}[\xi^1_.] - U^{0,t}[\xi^2_.] &= U^{0,s}[\xi^1_.]U^{s,t}[\xi^2_.]\Big|_{s=0}^{t} \\ &= \int_0^t U^{0,s}[\xi^1_.](A[\xi^1_s] - A[\xi^2_s])U^{s,t}[\xi^2_.]\,ds, \end{aligned}$$

and taking into account (2.10) and (2.11), one deduces that

$$\begin{aligned} \|(V^{t,0}[\xi^1_.] - V^{t,0}[\xi^2_.])\mu\|_{D^\star} &\le \|U^{0,t}[\xi^1_.] - U^{0,t}[\xi^2_.]\|_{D\to B}\|\mu\|_{B^\star} \\ &\le tc(r,M) \sup_{s\in[0,r]} \|\xi^1_s - \xi^2_s\|_{D^\star} \end{aligned}$$

(we have used the assumed boundedness of M). This implies that, for $t \le t_0$ with a small enough t_0, the mapping $\xi_t \mapsto V^{t,0}[\xi_.]$ is a contraction in $C_\mu([0,t],M)$. Hence by the contraction principle there exists a unique fixed point for this mapping. To obtain the unique global solution one just has to iterate the construction on the next interval $[t_0, 2t_0]$ and then on $[2t_0, 3t_0]$ etc. The semigroup property of T_t follows directly from uniqueness.

Finally, if $T_t(\mu) = \mu_t$ and $T_t(\eta) = \eta_t$ then

$$\begin{aligned} T_t(\mu) - T_t(\eta) &= V^{t,0}[\mu_.]\mu - V^{t,0}[\eta_.]\eta \\ &= (V^{t,0}[\mu_.] - V^{t,0}[\eta_.])\mu + V^{t,0}[\eta_.](\mu - \eta). \end{aligned}$$

Estimating the first term as above yields

$$\sup_{s\leq t} \|T_s(\mu) - T_s(\eta)\|_{D^\star} \leq c(r, M)\left[t \sup_{s\leq t} \|T_s(\mu) - T_s(\eta)\|_{D^\star} + \|\mu - \eta\|_{D^\star}\right],$$

which implies the first estimate in (2.13) for small times; this can then be extended to all finite times by iteration. The second estimate in (2.13) follows from (2.7). □

Remark 2.13 For our purposes, basic examples of the set M above are the following:

(i) One can take M to be the ball of fixed radius in $B^\star$;[2] it is natural to make this choice when all propagators are contractions;

(ii) if $B = C_\infty(\mathbf{R}^d)$ and $D = C^2_\infty(\mathbf{R}^d)$ or $D = C^1_\infty(\mathbf{R}^d)$ one can often take $M = \mathcal{P}(\mathbf{R}^d)$, which is closed in the norm topology of $D^\star$ because it is weakly (not $\star$-weakly) closed in $B^\star$ and hence also in $D^\star$.

Remark 2.14 If M is closed only in $B^\star$, but not in $D^\star$, the same argument as in the previous remark shows that there could exist at most one solution to (2.12); i.e. uniqueness holds.

We shall need also a stability result for the above nonlinear semigroups T_t with respect to small perturbations of the generator A.

Theorem 2.15 *Under the assumptions of Theorem 2.12 suppose that $\xi \mapsto \tilde{A}[\xi]$ is another mapping, from $B^\star$ to the bounded operator $\tilde{A}[\xi] : D \to B$, satisfying the same condition as A; the corresponding propagators $\tilde{U}^{t,s}$, $\tilde{V}^{s,t}$ satisfy the same conditions as $U^{t,s}$, $V^{s,t}$. Suppose that*

$$\|\tilde{A}[\xi] - A[\xi]\|_{D\to B} \leq \kappa, \qquad \xi \in M \tag{2.14}$$

for a constant κ. Then

$$\|\tilde{T}_t(\mu) - T_t(\eta)\|_{D^\star} \leq c(r, M)(\kappa + \|\mu - \eta\|_{D^\star}). \tag{2.15}$$

Proof As in the proof of Theorem 2.12, denoting $T_t(\mu)$ by μ_t and $\tilde{T}_t(\eta)$ by $\tilde{\eta}_t$ one can write

$$\mu_t - \tilde{\eta}_t = (V^{t,0}[\mu_.] - \tilde{V}^{t,0}[\tilde{\eta}_.])\mu + \tilde{V}^{t,0}[\tilde{\eta}](\mu - \eta)$$

and then

$$\sup_{s\leq t} \|\mu_s - \tilde{\eta}_s\|_{D^\star} \leq c(r, M)\left(t(\sup_{s\leq t} \|\mu_s - \tilde{\eta}_s\|_{D^\star} + \kappa) + \|\mu - \eta\|_{D^\star}\right),$$

[2] By the Banach–Alaoglu theorem the ball M is weakly closed in $B^\star$ and hence also in $D^\star$ since, for bounded subsets of $B^\star$, weak closures in $B^\star$ and $D^\star$ coincide and consequently M is closed in the norm of $D^\star$.

which implies (2.15) in the first place for small times and then for all finite times by iteration. □

2.2 Feller processes and conditionally positive operators

In this section we recall the basic properties of Feller processes and semigroups, fixing our notation and stressing the interplay between analytical and probabilistic interpretations.

A linear operator L on a functional space is called *positive* if $f \geq 0 \Longrightarrow Lf \geq 0$. A backward propagator (resp. a semigroup) of positive linear contractions in $B(S)$, $C(S)$ or $L_p(\mathbf{R}^d)$ is said to be a *sub-Markov backward propagator* (resp. a *sub-Markov semigroup*). It is called a *Markov (backward) propagator* (resp. a *Markov semigroup*) if additionally all these contractions are *conservative*, i.e. they take any constant function to itself. The connection with the theory of Markov processes is given by the following fundamental fact. For a Markov process X_t (defined on a probability space and taking values in a metric space), the transformations

$$\Phi^{s,t} f(x) = \mathbf{E}(f(X_t)|X_s = x) \tag{2.16}$$

form a Markov propagator in the space $B(S)$ of bounded Borel functions. In particular, if this Markov process is time homogeneous, the family

$$\Phi_t f(x) = \mathbf{E}(f(X_t)|X_0 = x) = \mathbf{E}_x f(X_t) \tag{2.17}$$

forms a Markov semigroup.

Usually Markov processes are specified by their *Markov transition probability families* $p_{r,t}(x, A)$ (which are the families of transition kernels from S to S parametrized by an ordered pair of real numbers $r \leq t$), so that

$$p_{s,t}(x, A) = (\Phi^{s,t} \mathbf{1}_A)(x) = P(X_t \in A|X_s = x)$$

or, equivalently,

$$(\Phi^{s,t} f)(x) = \int_S f(y) p_{s,t}(x, dy), \qquad f \in B(S).$$

The basic propagator equation $U^{t,s} U^{s,r} = U^{t,r}$ written in terms of the Markov transition families

$$p_{r,t}(x, A) = \int_S p_{s,t}(y, A) p_{r,s}(x, dy) \tag{2.18}$$

is called the *Chapman–Kolmogorov equation*.

A strongly continuous semigroup of positive linear contractions on $C_\infty(S)$ is called a *Feller semigroup*.

A (homogeneous) Markov process in a locally compact metric space S is called a *Feller process* if its Markov semigroup reduced to $C_\infty(S)$ is a Feller semigroup, i.e. if it preserves $C_\infty(S)$ and is strongly continuous there.

Theorem 2.16 *For an arbitrary Feller semigroup Φ_t in $C_\infty(S)$ there exists a (uniquely defined) family of positive Borel measures $p_t(x, dy)$ on S, with norm not exceeding* 1, *depending vaguely continuously on x, i.e.*

$$\lim_{x_n \to x} \int f(y) p_t(x_n, dy) = \int f(y) p_t(x, dy), \qquad f \in C_\infty(S)$$

and such that

$$\Phi_t f(x) = \int p_t(x, dy) f(y). \tag{2.19}$$

Proof Representation (2.19) follows from the Riesz–Markov theorem. The other properties of $p_t(x, dy)$ mentioned in the present theorem follow directly from the definition of a Feller semigroup. □

Formula (2.19) allows us to extend the operators Φ_t to contraction operators in $B(S)$. This extension clearly forms a sub-Markov semigroup in $B(S)$.

Let $K_1 \subset K_2 \subset \cdots$ be an increasing sequence of compact subsets of S exhausting S, i.e. $S = \cup_n K_n$. Let χ_n be any sequence of functions from $C_c(S)$ with values in [0, 1] and such that $\chi(x) = 1$ for $|x| \in K_n$. Then, for any $f \in B(S)$, one has (by monotone or dominated convergence)

$$\begin{aligned} \Phi_t f(x) = \int p_t(x, dy) f(y) &= \lim_{n \to \infty} \int p_y(x, dy) \chi_n(y) f(y) \\ &= \lim_{n \to \infty} (\Phi_t(\chi_n f))(x) \end{aligned} \tag{2.20}$$

(for positive f the limit is actually the supremum over n). This simple equation is important, as it allows us to define the *minimal extension of Φ_t to $B(S)$* directly via Φ_t, avoiding explicit reference to $p_t(x, dy)$.

Theorem 2.17 *If Φ_t is a Feller semigroup then uniformly, for x from a compact set,*

$$\lim_{t \to 0} \Phi_t f(x) = f(x), \qquad f \in C(\mathbf{R}^d),$$

where Φ_t denotes the extension (2.20).

Proof By linearity and positivity it is enough to show this for $0 \le f \le 1$. In this case, for any compact set K and a non-negative function $\phi \in C_\infty(\mathbf{R}^d)$ that equals 1 in K,

$$(f - \Phi_t f)\mathbf{1}_K \leq [f\phi - \Phi_t(f\phi)]\mathbf{1}_K,$$

and similarly

$$[1 - f - \Phi_t(1 - f)]\mathbf{1}_K \leq [(1 - f)\phi - \Phi_t((1 - f)\phi)]\mathbf{1}_K.$$

The latter inequality implies that

$$\begin{aligned}(\Phi_t f - f)\mathbf{1}_K &\leq [\Phi_t \mathbf{1} - \mathbf{1} + \mathbf{1} - f - \Phi_t((1 - f)\phi)]\mathbf{1}_K \\ &\leq [(1 - f)\phi - \Phi_t((1 - f)\phi)]\mathbf{1}_K.\end{aligned}$$

Consequently

$$|f - \Phi_t f|\mathbf{1}_K \leq |f\phi - \Phi_t(f\phi)|\mathbf{1}_K + |(1 - f)\phi - \Phi_t((1 - f)\phi)|\mathbf{1}_K,$$

which implies the required convergence on the compact set K by the strong continuity of Φ_t. □

Corollary 2.18 *If Φ is a Feller semigroup then the dual semigroup $\Phi_t^\star$ on $\mathcal{M}(X)$ is a positivity-preserving semigroup of contractions depending continuously on t in both the vague and weak topologies.*

Proof Everything is straightforward from the above definitions except weak continuity, which follows from the previous theorem since

$$\begin{aligned}(f, \Phi_t^\star \mu - \mu) &= (\Phi_t f - f, \mu) \\ &= \int_{|x|<K} (\Phi_t f - f)(x)\mu(dx) + \int_{|x|\geq K} (\Phi_t f - f)(x)\mu(dx);\end{aligned}$$

for $f \in C(\mathbf{R}^d)$ the second integral can be made arbitrarily small by choosing large enough K, and the first integral is then small for small t by Theorem 2.17. □

A Feller semigroup Φ_t is called *conservative* if all measures $p_t(x, .)$ in the representation (2.19) are probability measures or, equivalently, if the natural extension of Φ_t to $B(S)$ given by (2.20) preserves constants and hence forms a Markov semigroup in $B(S)$.

Another useful link between the Markov property and continuity is stressed in the following modification of the Feller property. A *C-Feller semigroup* in $C(S)$ is a sub-Markov semigroup in $C(S)$, i.e. it is a semigroup of contractions Φ_t in $C(S)$ such that $0 \leq u \leq 1$ implies $0 \leq \Phi_t u \leq 1$. Note that on the one hand this definition does not include strong continuity and on the other hand it applies to any topological space S, not necessarily one that is locally compact or even metric. Of course, a Feller semigroup Φ_t is *C-Feller* if the space $C(S)$ is invariant under the natural extension (2.20), and a C-Feller semigroup Φ_t is

Feller if $C_\infty(S)$ is invariant under all Φ_t and the corresponding restriction is strongly continuous. It is worth stressing that a Feller semigroup may not be C-Feller, and vice versa; see the exercises at the end of Section 2.4.

Feller semigroups arising from Markov processes are obviously conservative. Conversely, any conservative Feller semigroup is the semigroup of a certain Markov process; this follows from representation (2.19) for the kernels p_t and a basic construction of Markov processes based on Kolmogorov's existence theorem.

Proposition 2.19 *A Feller semigroup is C-Feller if and only if Φ_t applied to a constant is a continuous function. In particular, any conservative Feller semigroup is C-Feller.*

Proof By Proposition A.5 the vague continuity and weak continuity of $p_t(x, dy)$ with respect to x coincide under the condition of continuous dependence of the total mass $p_t(x, S)$ on x. □

Theorem 2.20 *If X_t^x is a Feller process in $\mathbf{R}^d$ whose starting point is denoted by x then (i) $X_t^x \to X_t^y$ weakly as $x \to y$ for any t, and (ii) $X_t^x \to x$ in probability as $t \to 0$.*

Proof Proposition A.5 implies statement (i) and the weak convergence $X_t^x \to x$ as $t \to 0$. In particular, the family of distributions of X_t, $t \in [0, 1]$, is tight (see the definition before Theorem A.12). Taking this into account, in order to show the convergence in probability one has to show that, for any $K > \epsilon > 0$,

$$\lim_{t\to 0} \mathbf{P}(\epsilon < |X_t^x - x| < K) = 0.$$

Now choosing an arbitrary non-negative function $f(y) \in C_\infty(\mathbf{R}^d)$ that vanishes at x and equals 1 for $\epsilon < \|x - y\| < K$ yields

$$\mathbf{P}(\epsilon < |X_t^x - x| < K) \leq \mathbf{E} f(X_t^x) \to f(x) = 0,$$

as required. □

Theorem 2.21 *Let X_t be a Lévy process with characteristic exponent*

$$\eta(u) = i(b, u) - \tfrac{1}{2}(u, Gu) + \int_{\mathbf{R}^d} [e^{i(u,y)} - 1 - i(u, y)\mathbf{1}_{B_1}(y)]\nu(dy). \quad (2.21)$$

Then X_t is a Feller process with semigroup Φ_t such that

$$\Phi_t f(x) = \int f(x + y) p_t(dy), \qquad f \in C(\mathbf{R}^d), \quad (2.22)$$

where p_t is the law of X_t. This semigroup is translation invariant, i.e.

$$(\Phi_t f)(x+z) = (\Phi_t f(.+z))(x).$$

Proof Formula (2.22) follows from the definition of Lévy processes as time-homogeneous and translation-invariant Markov process. Notice that any $f \in C_\infty(\mathbf{R}^d)$ is uniformly continuous. For any such f,

$$\begin{aligned}\Phi_t f(x) - f(x) &= \int [f(x+y) - f(x)] p_t(dy) \\ &= \int_{|y|>K} [f(x+y) - f(x)] p_t(dy) \\ &\quad + \int_{|y|\le K} [f(x+y) - f(x)] p_t(dy);\end{aligned}$$

the first (resp. the second) term is small for small t and any K by the stochastic continuity of X (resp. for small K and arbitrary t by the uniform continuity of f). Hence $\|\Phi_t f - f\| \to 0$ as $t \to 0$. To see that $\Phi_t f \in C_\infty(\mathbf{R}^d)$ for $f \in C_\infty(\mathbf{R}^d)$ one writes similarly

$$\Phi_t f(x) = \int_{|y|>K} f(x+y) p_t(dy) + \int_{|y|\le K} f(x+y) p_t(dy)$$

and observes that the second term clearly belongs to $C_\infty(\mathbf{R}^d)$ for any K and that the first can be made arbitrarily small by choosing large enough K. □

Remark 2.22 A Fourier transform takes the semigroup Φ_t to a multiplication semigroup,

$$\Phi_t f(x) = F^{-1}(e^{t\eta} F f), \qquad f \in S(\mathbf{R}^d),$$

because

$$\begin{aligned}(F\Phi_t f)(p) &= \frac{1}{(2\pi)^{d/2}} \int e^{-ipx} \int f(x+y) p_t(dy) \\ &= \frac{1}{(2\pi)^{d/2}} \int\int e^{-ipz+ipy} \int f(z) p_t(dy) \\ &= (Ff)(p) e^{t\eta(p)}.\end{aligned}$$

This yields another proof of the Feller property of the semigroup Φ_t.

Theorem 2.23 *If X_t is a Lévy process with characteristic exponent* (2.21) *then its generator L is given by*

$$Lf(x) = \sum_{j=1}^{d} b_j \frac{\partial f}{\partial x_j} + \frac{1}{2} \sum_{j,k=1}^{d} G_{jk} \frac{\partial^2 f}{\partial x_j \partial x_k}$$

$$+ \int_{\mathbf{R}^d} \left(f(x+y) - f(x) - \sum_{j=1}^{d} y_j \frac{\partial f}{\partial x_j} \mathbf{1}_{B_1}(y) \right) \nu(dy) \qquad (2.23)$$

on the Schwartz space S of fast-decreasing smooth functions. Moreover, the Lévy exponent is expressed in terms of the generator by the formula

$$\eta(u) = e^{-iux} L e^{iux}. \qquad (2.24)$$

Each space $C_\infty^k(\mathbf{R}^d)$ with $k \geq 2$ is an invariant core for L.

Proof Let us first check (2.23) for exponential functions. Namely, for $f(x) = e^{i(u,x)}$,

$$\Phi_t f(x) = \int f(x+y) p_t(dy) = e^{i(u,x)} \int e^{i(u,y)} p_t(dy) = e^{i(u,x)} e^{t\eta(u)}.$$

Hence

$$Lf(x) = \frac{d}{dt}\bigg|_{t=0} \Phi_t f(x) = \eta(u) e^{i(u,x)}$$

is given by (2.23), owing to the elementary properties of exponents. By linearity this extends to functions of the form $f(x) = \int e^{i(u,x)} g(u) du$ with $g \in S$. But this class coincides with S by Fourier's theorem. To see that $C_\infty^k(\mathbf{R}^d)$ is invariant under Φ_t for any $k \in \mathbf{N}$ it is enough to observe that the derivative $\nabla_l \Phi_t f$ for a function $f \in C_\infty^1(\mathbf{R}^d)$ satisfies the same equation as $\Phi_t f$ itself. Finally, $Lf \in C_\infty(\mathbf{R}^d)$ for any $f \in C_\infty^2(\mathbf{R}^d)$. □

By a straightforward change of variable, one finds that the operator $L^\star$ is given by

$$L^\star f(x) = -\sum_{j=1}^{d} b_j \frac{\partial f}{\partial x_j} + \frac{1}{2} \sum_{j,k=1}^{d} G_{jk} \frac{\partial^2 f}{\partial x_j \partial x_k}$$

$$+ \int_{\mathbf{R}^d} \left(f(x-y) - f(x) + \sum_{j=1}^{d} y_j \frac{\partial f}{\partial x_j} \mathbf{1}_{B_1}(y) \right) \nu(dy) \qquad (2.25)$$

is adjoint to (2.23) in the sense that

$$\int Lf(x) g(x)\, dx = \int f(x) L^\star g(x)\, dx,$$

for f, g from the Schwartz space S.

Remark 2.24 The operator (2.23) is a ΨDO (see Appendix E) with the symbol $\eta(p)$, where η is the characteristic exponent (2.21). In fact, by (E.3) one simply has to check that $(FLf)(p) = \eta(p)(Ff)(p)$. Since

$$(FLf)(p) = \frac{1}{(2\pi)^{d/2}}(e^{-ip\cdot}, Lf) = \frac{1}{(2\pi)^{d/2}}(L^\star e^{-ip\cdot}, f),$$

this follows from the equation

$$L^\star e^{-ipx} = \eta(p)e^{-ipx},$$

which in its turn is a direct consequence of the properties of the exponent function.

The following are the basic definitions relating to the generators of Markov processes. One says that an operator A in $C(\mathbf{R}^d)$ defined on a domain D_A (i) is *conditionally positive*, if $Af(x) \geq 0$ for any $f \in D_A$ such that $f(x) = 0 = \min_y f(y)$; (ii) satisfies the *positive maximum principle (PMP)* if $Af(x) \leq 0$ for any $f \in D_A$ such that $f(x) = \max_y f(y) \geq 0$; (iii) is *dissipative* if $\|(\lambda - A)f\| \geq \lambda\|f\|$ for $\lambda > 0$, $f \in D_A$; (iv) is *local* if $Af(x) = 0$ whenever $f \in D_A \cap C_c(\mathbf{R}^d)$ vanishes in a neighborhood of x; (v) is *locally conditionally positive* if $Af(x) \geq 0$ whenever $f(x) = 0$ and has a local minimum there; (vi) satisfies a *local PMP* if $Af(x) \leq 0$ for any $f \in D_A$ having a local non-negative maximum at x.

For example, the multiplication operator taking $u(x)$ to $c(x)u(x)$, for some function $c \in C(\mathbf{R}^d)$, is always conditionally positive but it satisfies the PMP only in the case of non-negative c.

The importance of these notions lies in the following fact.

Theorem 2.25 *Let A be a generator of a Feller semigroup Φ_t. Then*

(i) A is conditionally positive,
(ii) A satisfies the PMP on D_A,
(iii) A is dissipative.

If moreover A is local and D_A contains C_c^∞ then A is locally conditionally positive and satisfies the local PMP on C_c^∞.

Proof This is very simple. For (i), note that

$$Af(x) = \lim_{t\to 0}\frac{\Phi_t f(x) - f(x)}{t} = \lim_{t\to 0}\frac{\Phi_t f(x)}{t} \geq 0$$

by positivity preservation. For (ii) note that if $f(x) = \max_y f(y)$ then, by Exercise 2.2 (see below), $\Phi_t f(y) \leq f(x)$ for all y, t, implying that

$Af(x) \le 0$. For (iii) choose x to be the maximum point of $|f|$. By passing to $-f$ if necessary we can consider $f(x)$ to be positive. Then

$$\|(\lambda - A)f\| \ge \lambda\|f\| \ge \lambda f(x) - Af(x) \ge \lambda f(x)$$

by the PMP. □

Let us observe that if S is compact and a Feller semigroup in $C(S)$ is conservative then the constant unit function $\mathbf{1}$ belongs to the domain of its generator A, and $A\mathbf{1} = 0$. Hence it is natural to call such generators *conservative*. In the case of noncompact $S = \mathbf{R}^d$, we shall say that a generator of a Feller semigroup A is *conservative* if $A\phi_n(x) \to 0$ for any x as $n \to \infty$, where $\phi_n(x) = \phi(x/n)$ and ϕ is an arbitrary function from $C_c^2(\mathbf{R}^d)$ that equals 1 in a neighborhood of the origin and has values in $[0, 1]$. We shall see at the end of the next section that the conservativity of a semigroup implies the conservativity of the generator with partial inverse to be given in Theorem 2.40.

We recall now the basic structural result about generators of Feller processes by formulating the following fundamental fact.

Theorem 2.26 (Courrège) *If the domain of a conditionally positive operator L (in particular, the generator of a Feller semigroup) in $C_\infty(\mathbf{R}^d)$ contains the space $C_c^2(\mathbf{R}^d)$ then it has the following Lévy–Khintchine form with variable coefficients:*

$$Lf(x) = \tfrac{1}{2}(G(x)\nabla, \nabla)f(x) + (b(x), \nabla f(x)) + c(x)f(x) + \int [f(x+y) - f(x) - (\nabla f(x), y)\mathbf{1}_{B_1}(y)]\nu(x, dy), \quad f \in C_c^2(\mathbf{R}^d). \tag{2.26}$$

Here $G(x)$ is a symmetric non-negative matrix and $\nu(x, .)$ is a Lévy measure on $\mathbf{R}^d$, i.e.

$$\int_{\mathbf{R}^n} \min(1, |y|^2)\nu(x; dy) < \infty, \qquad \nu(\{0\}) = 0, \tag{2.27}$$

that depends measurably on x. If additionally L satisfies the PMP then $c(x) \le 0$ everywhere.

The proof of this theorem is based only on standard calculus, though it requires some ingenuity (the last statement in the theorem being of course obvious). It can be found in [52], [43] and [105] and will not be reproduced here. We will simply indicate the main strategy, showing how the Lévy kernel comes into play. Namely, as follows from conditional positivity, $Lf(x)$ for any x is a positive linear functional on the space of continuous functions with support in $\mathbf{R}^d \setminus \{0\}$. Hence, by the Riesz–Markov theorem for these functions,

$$Lf(x) = \tilde{L}f(x) = \int f(y)\tilde{\nu}(x, dy) = \int f(x+y)\nu(x, dy)$$

for some kernel ν such that $\nu(x, \{x\}) = 0$. Next, we can deduce from conditional positivity that L should be continuous as a mapping from $C_c^2(\mathbf{R}^d)$ to the bounded Borel functions. This in turn allows us to deduce the basic moment condition (2.27) on ν. We then observe that the difference between L and $\tilde{L}$ must be a second-order differential operator. Finally, one shows that this differential operator must also be conditionally positive.

Remark 2.27 Actually, when proving Theorem 2.26 (see [43]) one arrives at a characterization not only for conditionally positive operators but also for conditionally positive linear functionals obtained by fixing the arguments. Namely, it is shown that if a linear functional $(Ag)(x) : C_c^2 \mapsto \mathbf{R}^d$ is conditionally positive at x, i.e. if $Ag(x) \geq 0$ whenever a non-negative g vanishes at x, then $Ag(x)$ is continuous and has the form (2.26) (irrespective of the properties of $Ag(y)$ at other points y).

Corollary 2.28 *If the domain of the generator L of a conservative Feller semigroup Φ_t in $C_\infty(\mathbf{R}^d)$ contains C_c^2 then it has the form* (2.26) *with vanishing $c(x)$. In particular, L is conservative.*

Proof By Theorems 2.25 and 2.26, L has the form (2.26) on $C_c^2(\mathbf{R}^d)$ with non-positive $c(x)$. The conservativity of L means that $L\phi_n(x) \to 0$ for any x as $n \to \infty$, where $\phi_n(x) = \phi(x/n)$ and ϕ is an arbitrary function from $C_c^2(\mathbf{R}^d)$ that equals 1 in a neighborhood of the origin and has values in $[0, 1]$. Clearly $\lim_{n\to\infty} L\phi_n(x) = c(x)$. So conservativity is equivalent to the property that $c(x) = 0$ identically. Since Φ_t is a conservative Feller semigroup it corresponds to a certain Markov (actually Feller) process X_t. □

The inverse question whether a given operator of the form (2.26) (or its closure) actually generates a Feller semigroup, which roughly speaking means having regular solutions to the equation $\dot{f} = Lf$ (see the next section), is non-trivial and has attracted much attention. We shall deal with it in the next few chapters.

We conclude this section by recalling Dynkin's formula connecting Markov processes and martingales.

Theorem 2.29 (Dynkin's formula) *Let $f \in D$, the domain of the generator L of a Feller process X_t. Then the process*

$$M_t^f = f(X_t) - f(X_0) - \int_0^t Lf(X_s)\, ds, \qquad t \geq 0, \tag{2.28}$$

is a martingale (with respect to the filtration for which X_t is a Markov process) under any initial distribution ν. It is often called Dynkin's martingale.

Proof

$$\mathbf{E}(M^f_{t+h}|\mathcal{F}_t) - M^f_t$$
$$= \mathbf{E}\Big(f(X_{t+h}) - \int_0^{t+h} Lf(X_s)\,ds\,\Big|\,\mathcal{F}_t\Big) - \Big(f(X_t) - \int_0^t Lf(X_s)\,ds\Big)$$
$$= \Phi_h f(X_t) - \mathbf{E}\Big(\int_t^{t+h} Lf(X_s)\,ds\,\Big|\,\mathcal{F}_t\Big) - f(X_t)$$
$$= \Phi_h f(X_t) - f(X_t) - \int_0^h L\Phi_s f(X_t)\,ds = 0. \qquad \square$$

This result motivates the following definition. Let L be a linear operator given by $L : D \to B(\mathbf{R}^d)$, $D \in C(\mathbf{R}^d)$. One says that a process X_t with càdlàg paths (or the corresponding probability distribution on the Skorohod space) solves the (L, D)-*martingale problem* for initial distribution μ if X_0 is distributed according to μ and the process (2.28) is a martingale for any $f \in D$. This martingale problem is called *well posed* if, for any initial μ, there exists a unique X_t solving it. The following result is a direct consequence of Theorem 2.29. It will be used later on in the construction of Markov semigroups.

Proposition 2.30 *(i) A Feller process X_t solves the (L, D)-martingale problem, where L is the generator of X_t and D is any subspace of its domain.*

(ii) If the (L, D)-martingale problem is well posed, there can exist no more than one Feller process whose generator is an extension of L.

Exercise 2.1 Let X_t be a Markov chain on $\{1, \ldots, n\}$ with transition probabilities $q_{ij} > 0$, $i \neq j$, which are defined via the semigroup of stochastic matrices Φ_t with generator by given

$$(Af)_i = \sum_{j \neq i} (f_j - f_i) q_{ij}.$$

Let $N_t = N_t(i)$ denote the number of transitions during time t of a process starting at some point i. Show that $N_t - \int_0^t q(X_s)\,ds$ is a martingale, where $q(l) = \sum_{j \neq l} q_{lj}$ denotes the intensity of jump l. Hint: To check that $\mathbf{E}N_t = \mathbf{E}\int_0^t q(X_s)\,ds$ show that the function $\mathbf{E}N_t$ is differentiable and that

$$\frac{d}{dt}\mathbf{E}(N_t) = \sum_{j=1}^n P(X_t = j) q_j.$$

Exercise 2.2 Show that if Φ is a positive contraction in $B(S)$, where S is a metric space, then $a \leq f \leq b$ for $f \in B(S)$ and $a, b \in \mathbf{R}$ implies that $a \leq \Phi f \leq b$. Hint: First settle the case when either a or b vanishes.

2.3 Jump-type Markov processes

In this section we consider in more detail the bounded conditionally positive operators that correspond probabilistically to pure jump processes.

Proposition 2.31 *Let S be a locally compact metric space and on the one hand let L be a bounded conditionally positive operator from $C_\infty(S)$ to $B(S)$. Then there exists a bounded transition kernel $\nu(x, dy)$ in S with $\nu(x, \{x\}) = 0$ for all x, and a function $a(x) \in B(S)$, such that*

$$Lf(x) = \int_S f(z)\nu(x, dz) - a(x)f(x). \tag{2.29}$$

On the other hand, if L is of this form then it is a bounded conditionally positive operator $C(S) \mapsto B(S)$.

Proof If L is conditionally positive in $C_\infty(S)$ then $Lf(x)$ is a positive functional on $C_\infty(S \setminus \{x\})$ and hence, by the Riesz–Markov theorem, there exists a measure $\nu(x, dy)$ on $S \setminus \{x\}$ such that $Lf(x) = \int_S f(z)\nu(x, dz)$ for $f \in C_\infty(S \setminus \{x\})$. As L is bounded, these measures are uniformly bounded. As any $f \in C_\infty(S)$ can be written as $f = f(x)\chi + [f - f(x)\chi]$, where χ is an arbitrary function with a compact support and $\chi(x) = 1$, it follows that

$$Lf(x) = f(x)L\chi(x) + \int [f - f(x)\chi](z)\nu(x, dz),$$

which clearly has the form (2.29). The inverse statement is obvious. □

Remark 2.32 The condition $\nu(x, \{x\}) = 0$ is natural for a probabilistic interpretation (see below). From the analytical point of view it makes representation (2.29) unique.

We shall now describe analytical and probabilistic constructions of pure jump processes, focusing our attention on the most important case, that of continuous kernels.

Theorem 2.33 *Let $\nu(x, dy)$ be a weakly continuous uniformly bounded transition kernel in a complete metric space S such that $\nu(x, \{x\}) = 0$ and also $a \in C(S)$. Then the operator defined in (2.29) has $C(S)$ as its domain and* *generates a strongly continuous semigroup T_t in $C(S)$ that preserves positivity and is given in terms of certain transition kernels $p_t(x, dy)$:*

$$T_t f(x) = \int p_t(x, dy) f(y).$$

In particular, if $a(x) = \|\nu(x, .)\|$ then $T_t \mathbf{1} = \mathbf{1}$ and T_t is the Markov semigroup of a Markov process that we shall call a pure jump or jump-type Markov process.

Proof Since L is bounded, it generates a strongly continuous semigroup. As it can be given by the integral form

$$Lf(x) = \int_S f(z)\tilde{\nu}(x, dz),$$

where the signed measure $\tilde{\nu}(x, .)$ coincides with ν outside $\{x\}$ and $\tilde{\nu}(x, \{x\}) = -a(x)$, it follows from the convergence in norm of the exponential series for $T_t = e^{tL}$ that the T_t are integral operators. To see that these operators are positive we can observe that the T_t are bounded from below by the resolving operators of the equation $\dot{f}(x) = -a(x) f(x)$, which are positive. Application of the standard construction for Markov processes (via Kolmogorov's existence theorem) yields the existence of the corresponding Markov process. □

Remark 2.34 An alternative analytical proof can be given by perturbation theory (Theorem 2.7), considering the integral term in (2.29) as a perturbation. This approach leads directly to the representation (2.32) obtained below probabilistically. From this approach, to obtain positivity is straightforward.

A characteristic feature of pure jump processes is the property that their paths are a.s. piecewise constant, as is shown by the following result on the probabilistic interpretation of these processes.

Theorem 2.35 *Let $\nu(x, dy)$ be a weakly continuous uniformly bounded transition kernel in a metric space S such that $\nu(x, \{x\}) = 0$. Let $a(x) = \nu(x, S)$. Define the following process X_t^x. Starting at a point x, the process remains there for a random $a(x)$-exponential time τ, i.e. this time is distributed according to $\mathbf{P}(\tau > t) = \exp[-ta(x)]$, and then jumps to a point $y \in S$ distributed according to the probability law $\nu(x, .)/a(x)$. Then the procedure is repeated, now starting from y, etc. Let N_t^x denote the number of jumps of this process during a time t when starting from a point x. Then*

$$\mathbf{P}(N_t^x = k) = \int_{0<s_1<\cdots<s_k<t} \int_{S^k} e^{-a(y_k)(t-s_k)} \nu(y_{k-1}, dy_k)$$
$$\times e^{-a(y_{k-1})(s_k-s_{k-1})} \nu(y_{k-2}, dy_{k-1}) \cdots e^{-a(y_1)(s_2-s_1)} \nu(x, dy_1)$$
$$\times e^{-s_1 a(x)} ds_1 \cdots ds_k, \qquad (2.30)$$

$$\mathbf{P}(N_t^x > k) = \int_{0<s_1<\cdots<s_k<t} \int_{S^k} [1 - e^{-a(y_k)(t-s_k)}] \nu(y_{k-1}, dy_k)$$
$$\times e^{-a(y_{k-1})(s_k-s_{k-1})} \nu(y_{k-2}, dy_{k-1}) \cdots e^{-a(y_1)(s_2-s_1)} \nu(x, dy_1)$$
$$\times e^{-s_1 a(x)} ds_1 \cdots ds_k, \qquad (2.31)$$

and N_t^x is a.s. finite. Moreover, for a bounded measurable f,

$$\mathbf{E} f(X_t^x) = \sum_{k=0}^{\infty} \mathbf{E} f(X_t^x) \mathbf{1}_{N_t^x = k}$$
$$= \sum_{k=0}^{\infty} \int_{0<s_1<\cdots<s_k<t} \int_{S^k} e^{-a(y_k)(t-s_k)} \nu(y_{k-1}, dy_k)$$
$$\times e^{-a(y_{k-1})(s_k-s_{k-1})} \nu(y_{k-2}, dy_{k-1}) \cdots e^{-a(y_1)(s_2-s_1)} \nu(x, dy_1)$$
$$\times e^{-s_1 a(x)} f(y_k) ds_1 \cdots ds_k, \qquad (2.32)$$

and there exists (in the sense of the sup norm) the derivative

$$\left.\frac{d}{dt}\right|_{t=0} \mathbf{E} f(X_t^x) = \int_S f(z) \nu(x, dz) - a(x) f(x).$$

Proof Let $\tau_1, \tau_2, \ldots$ denote the (random) sequence of the jump times. By the definition of the exponential waiting time,

$$\mathbf{P}(N_t^x = 0) = P(\tau_1 > t) = e^{-a(x)t}.$$

Next, by conditioning,

$$\mathbf{P}(N_t^x = 1) = \mathbf{P}(\tau_2 > t - \tau_1, \ \tau_1 \le t)$$
$$= \int_0^t \mathbf{P}(\tau_2 > t - \tau_1 | \tau_1 = s) a(x) e^{-sa(x)} \, ds$$
$$= \int_0^t \int_S \mathbf{P}(\tau_2 > t - s | \tau_1 = s, X(s) = y) \nu(x, dy) e^{-sa(x)} \, ds$$
$$= \int_0^t \int_S e^{-a(y)(t-s)} \nu(x, dy) e^{-sa(x)} \, ds$$

and

$$\mathbf{P}(N_t^x > 1) = \mathbf{P}(\tau_2 \le t - \tau_1, \tau_1 \le t)$$
$$= \int_0^t \int_S [1 - e^{-a(y)(t-s)}]\nu(x, dy) e^{-sa(x)}\, ds;$$

similarly, one obtains (2.30), (2.31) for arbitrary k. Denoting $M = \sup_x a(x)$ and taking into account the elementary inequality $1 - e^{-a} \le a$, $a > 0$, one obtains from (2.31)

$$\mathbf{P}(N_t^x > k) \le M^{k+1} t \int\!\!\int_{0<s_1<\cdots<s_k<t} ds_1 \cdots ds_k \le \frac{(Mt)^{k+1}}{k!},$$

implying the convergence of the series $\sum_{k=0}^{\infty} \mathbf{P}(N_t^x > k)$. Hence, by the Borel–Cantelli lemma, N_t^x is a.s. finite. In particular, the first equation in (2.32) holds. Next, we have

$$\mathbf{E} f(X_t^x) \mathbf{1}_{N_t^x = 1} = \int_0^t \int_S f(y)\nu(x, dy) e^{-sa(x)} P(\tau_2 > t - s | X_s = y)\, ds$$
$$= \int_0^t \int_S e^{-a(y)(t-s)} f(y)\nu(x, dy) e^{-sa(x)} ds$$

The other terms of the series (2.32) are computed similarly. The equation for the derivative then follows straightforwardly, as only the first two terms of the series (2.32) contribute to the derivative, the rest of the terms being of order at least t^2. □

Remark 2.36 The deduction of the expansion (2.32) given above shows clearly its probabilistic meaning. As mentioned earlier, it can be obtained by analytical methods (perturbation theory). We shall consider this approach further in Section 4.2 when analyzing pure jump processes with unbounded rates $a(x)$.

Exercise 2.3 If S in Theorem 2.33 is locally compact and a bounded ν (depending weakly continuously on x) is such that $\lim_{x\to\infty} \int_K \nu(x, dy) = 0$ for any compact set K, then an operator L of the form (2.29) preserves the space $C_\infty(S)$ and hence generates a Feller semigroup.

2.4 Connection with evolution equations

From the definition of the generator and the invariance of its domain it follows that if Φ_t is the Feller semigroup of a process X_t with a generator L and domain D_L then $\Phi_t f(x)$ solves the Cauchy problem

$$\frac{d}{dt} f_t(x) = L f_t(x), \qquad f_0 = f, \tag{2.33}$$

whenever $f \in D_L$, the derivative being taken in the sense of the sup norm of $C(\mathbf{R}^d)$. Formula (2.17) yields the probabilistic interpretation of this solution and an explicit formula.

In the theory of linear differential equations, the solution $G(t, x, x_0)$ of (2.33) with $f_0 = \delta_{x_0} = \delta(. - x_0)$, i.e. satisfying (2.33) for $t > 0$ and the limiting condition in the weak form

$$\lim_{t \to 0} G(t, ., x_0, g) = \lim_{t \to 0} \int G(t, x, x_0) g(x)\, dx = g(x_0)$$

for any $g \in C_c^\infty$, is called the *Green function* or *heat kernel* of the problem (2.33) (whenever it exists, of course, which may not be the case in general). In probability language the Green function $G(t, x, x_0)$ is the density at x_0 of the distribution of the process X_t that started at x.

Consequently, if the distribution of a Lévy process X_t has a density $\omega(t, y)$ then $\Phi_t \delta_{x_0}(x) = \omega(t, x_0 - x)$, as follows from (2.22), so that $\omega(t, x_0 - x)$ is the Green function $G(t, x, x_0)$ in this case. The density of the probability law of X_t can be found as the Fourier transform of its characteristic function.

In particular, the Green function for the pseudo-differential (fractional parabolic) equation

$$\frac{\partial u}{\partial t} = (A, \nabla u(x)) - a|\nabla u|^\alpha$$

(see Appendix E for fractional derivatives) is given by the so-called *stable density*

$$S(x_0 - At - x; \alpha, at) = (2\pi)^{-d} \int_{\mathbf{R}^d} \exp[-at|p|^\alpha + ip(x + At - x_0)]\, dp.$$

Together with the existence of a solution one is usually interested in its uniqueness. The next statement shows how naturally this issue is settled via conditional positivity.

Theorem 2.37 *Let a subspace $D \subset C(\mathbf{R}^d)$ contain constant functions, and let an operator $L : D \mapsto C(\mathbf{R}^d)$ satisfying the PMP be given. Let $T > 0$ and $u(t, x) \in C([0, T] \times R^d)$. Assume that $u(0, x)$ is everywhere non-negative and that $u(t, .) \in C_\infty(\mathbf{R}^d) \cap D$ for all $t \in [0, T]$ is differentiable in t for $t > 0$ and satisfies the evolution equation*

$$\frac{\partial u}{\partial t} = Lu, \qquad t \in (0, T].$$

Then $u(t, x) \geq 0$ everywhere.

Proof Suppose that $\inf u = -\alpha < 0$. For a $\delta < \alpha/T$ consider the function

$$v_\delta = u(t, x) + \delta t.$$

Clearly this function also has a negative infimum. Since v tends to a positive constant δt as $x \to \infty$, v has a global negative minimum at some point (t_0, x_0) which lies in $(0, T] \times \mathbf{R}^d$. Hence $(\partial v/\partial t)(t_0, x_0) \leq 0$ and, by the PMP, $Lv(t_0, x_0) \geq 0$. Consequently we have, on the one hand,

$$\left(\frac{\partial v}{\partial t} - Lv\right)(t_0, x_0) \leq 0.$$

On the other hand, from the evolution equation and PMP we deduce that

$$\left(\frac{\partial v}{\partial t} - Lv\right)(t_0, x_0) \geq \left(\frac{\partial u}{\partial t} - Lu\right)(t_0, x_0) + \delta = \delta.$$

This contradiction completes the proof. □

Corollary 2.38 *Under the same conditions on D and L as in the above theorem, assume that $f \in C([0, T] \times \mathbf{R}^d)$ and $g \in C_\infty(\mathbf{R}^d)$. Then the Cauchy problem*

$$\frac{\partial u}{\partial t} = Lu + f, \qquad u(0, x) = g(x), \tag{2.34}$$

can have at most one solution $u \in C([0, T] \times \mathbf{R}^d)$ such that $u(t, .) \in C_\infty(\mathbf{R}^d)$ for all $t \in [0, T]$.

Now we shall touch upon the problem of reconstructing a Feller semigroup from a rich enough class of solutions to the Cauchy problem (2.33).

Theorem 2.39 *Let L be a conditionally positive operator in $C_\infty(\mathbf{R}^d)$ satisfying the PMP, and let D be a dense subspace of $C_\infty(\mathbf{R}^d)$ containing $C_c^2(\mathbf{R}^d)$ and belonging to the domain of L. Suppose that U_t, $t \geq 0$, is a family of bounded (uniformly for $t \in [0, T]$ for any $T > 0$) linear operators in $C_\infty(\mathbf{R}^d)$ such that U_t preserves D and that $U_t f$ for any $f \in D$ is a classical solution of* (2.33) *(i.e. it holds for all $t \geq 0$, the derivative being taken in the sense of the sup norm of $C(\mathbf{R}^d)$). Then U_t is a strongly continuous semigroup of positive operators in C_∞ defining a unique classical solution $U_t \in C_\infty(\mathbf{R}^d)$ of* (2.33) *for any $f \in D$.*

Proof Uniqueness and positivity follow from the previous theorem, if one takes into account that by Courrége's theorem, Theorem 2.26, the operator L naturally extends to constant functions preserving the PMP. However, uniqueness implies the semigroup property, because U_{t+s} and $U_t U_s$ solve the same

Cauchy problem. Finally, to prove strong continuity observe that if $\phi \in D$ then (since L and U_s commute by Theorem 2.4)

$$U_t\phi - \phi = \int_0^t LU_s\phi\, ds = \int_0^t U_s L\phi\, ds,$$

and

$$\|U_t\phi - \phi\| \leq t \sup_{s\leq t} \|U_s\| \|L\phi\|.$$

Since D is dense, the case of arbitrary ϕ is dealt with by the standard approximation procedure. □

The next result gives a simple analytical criterion for conservativity. It also introduces the very important formula (2.35) for the solution of nonhomogeneous equations, which is sometimes called the *du Hamel principle.*

Theorem 2.40 *(i) Under the assumptions of the previous theorem assume in addition that D is a Banach space itself, under a certain norm* $\|\phi\|_D \geq \|\phi\|$ *such that L is a bounded operator taking D to* $C_\infty(\mathbf{R}^d)$ *and the operators* U_t *are bounded (uniformly for t from compact sets) as operators in D. Then the function*

$$u = U_t g + \int_0^t U_{t-s} f_s\, ds \tag{2.35}$$

is the unique solution to equation (2.34) *in* $C_\infty(\mathbf{R}^d)$.

(ii) Let L be uniformly conservative in the sense that $\|L\phi_n\| \to 0$ *as* $n \to \infty$ *for* $\phi_n(x) = \phi(x/n)$, $n \in \mathbf{N}$, *and for any* $\phi \in C_c^2(\mathbf{R}^d)$ *that equals* 1 *in a neighborhood of the origin and has values in* [0, 1]. *Then* U_t *is a conservative Feller semigroup.*

Proof (i) Uniqueness follows from Theorem 2.37. Since the U_t are uniformly bounded in D it follows that the function u from (2.35) is well defined and belongs to D for all t. Next, straightforward formal differentiation shows that u satisfies (2.34). To prove the existence of the derivative one writes

$$\frac{\partial g}{\partial t} = LU_f + \lim_{\delta\to 0}\frac{1}{\delta}\int_0^t (U_{t+\delta-s} - U_{t-s})\phi_s\, ds + \lim_{\delta\to 0}\frac{1}{\delta}\int_t^{t+\delta} U_{t+\delta-s}\phi_s\, ds.$$

The first limit here exists and equals $L\int_0^t U_{t-s}\phi_s\, ds$. However,

$$\lim_{\delta\to 0}\frac{1}{\delta}\int_t^{t+\delta} U_{t+\delta-s}\phi_s\, ds = \phi_t + \lim_{\delta\to 0}\frac{1}{\delta}\int_t^{t+\delta}(U_{t+\delta-s}\phi_s - \phi_t)\, ds,$$

and so the second limit is seen to vanish.

(ii) Clearly the function ϕ_n solves the problem

$$\frac{\partial u}{\partial t} = Lu - L\phi_n, \qquad u(0,x) = \phi(x),$$

and hence by (i)

$$\phi_n(x) = U_t\phi_n + \int_0^t U_{t-s}L\phi_n\, ds.$$

As $n \to \infty$ the integral on the r.h.s. of this equation tends to zero in $C_\infty(\mathbf{R}^d)$ and $\phi_n(x)$ tends to 1 for each x. Hence

$$\lim_{n\to\infty} U_t\phi_n(x) = 1, \qquad x \in \mathbf{R}^d,$$

implying that in a representation of the type (2.19) for U_t (which exists due to the positivity of U_t) all measures $p_t(x, dy)$ are probability measures. This completes the proof. □

We conclude this section with some simple exercises illustrating various versions of the Feller property.

Exercise 2.4 Let X_t be a deterministic process in $\mathbf{R}$ solving the ODE $\dot{x} = x^3$. Show that (i) the solution to this equation with initial condition $X(0) = x$ is

$$X_x(t) = \operatorname{sgn}(x)\left(\frac{1}{-2t + x^{-2}}\right)^{1/2}, \qquad |x| < \frac{1}{\sqrt{2t}},$$

(ii) the corresponding semigroup has the form

$$\Phi_t f(x) = \begin{cases} (X_x(t)), & |x| < \dfrac{1}{\sqrt{2t}}, \\ 0, & |x| \geq \dfrac{1}{\sqrt{2t}}, \end{cases} \tag{2.36}$$

in $C_\infty(\mathbf{R})$ and is Feller and (iii) the corresponding measures from representation (2.19) are

$$p_t(x, dy) = \begin{cases} \delta(X_x(t) - y)), & |x| < \dfrac{1}{\sqrt{2t}}, \\ 0, & |x| \geq \dfrac{1}{\sqrt{2t}}, \end{cases} \tag{2.37}$$

implying that this Feller semigroup is not conservative, as its minimal extension takes the constant 1 to the indicator function of the interval $(-1/\sqrt{2t}, 1/\sqrt{2t})$. (It is instructive to see where the criterion of conservativity from Theorem 2.40 breaks down in this example.)

Exercise 2.5 Let X_t be a deterministic process in $\mathbf{R}$ solving the ODE $\dot{x} = -x^3$. Show that (i) the solution to this equation with initial condition $X(0) = x$ is

$$X_x(t) = \operatorname{sgn}(x) \left(\frac{1}{2t + 1/x^2} \right)^{1/2},$$

(ii) the corresponding semigroup is conservative and C-Feller but not Feller, as it does not preserve the space $C_\infty(\mathbf{R}^d)$.

Exercise 2.6 Let X_t be a deterministic process in $\mathbf{R}_+$ solving the ODE $\dot{x} = -1$ and "killed" at the boundary $\{x = 0\}$, i.e. it vanishes at the boundary at the moment it reaches it. Show that the corresponding semigroup on $C_\infty(\mathbf{R}_+)$ (which is the space of continuous functions on $\mathbf{R}_+$ tending to zero both for $x \to \infty$ and $x \to 0$) is given by (2.19) with

$$p_t(x, dy) = \begin{cases} \delta(x - t - y), & x > t, \\ 0, & x \leq t, \end{cases} \tag{2.38}$$

and is Feller but not conservative, as its minimal extension to $C(\mathbf{R}_+)$ (which stands for killing at the boundary) takes the constant 1 to the indicator $\mathbf{1}_{[t,\infty)}$. However, if, instead of a killed process, one defines a corresponding stopped process that remains at the boundary $\{x = 0\}$ once it reaches it, the corresponding semigroup is given on $C_\infty(\bar{\mathbf{R}}_+)$ by (2.19) with

$$p_t(x, dy) = \begin{cases} \delta(x - t - y), & x > t, \\ \delta(y), & x \leq t. \end{cases} \tag{2.39}$$

This is a conservative Feller semigroup on $C_\infty(\bar{\mathbf{R}}_+)$ that constitutes an extension (but not a minimal one) of the previously constructed semigroup for the killed process.

Exercise 2.7 This exercise is designed to show that the stopped process from the previous exercise does not give a unique extension of a Feller semigroup on $C_\infty(\mathbf{R}_+)$ to $C_\infty(\bar{\mathbf{R}}_+)$. Namely, consider a mixed "stopped and killed" process in which a particle moves according to the equation $\dot{x} = -1$ until it reaches the boundary, where it remains for a θ-exponential random time and then vanishes. Show that such a process specifies a non-conservative Feller semigroup on $C_\infty(\bar{\mathbf{R}}_+)$ given by

$$\Phi_t f(x) = \begin{cases} f(x - t), & x > t, \\ f(0)e^{-\theta(t-x)}, & x \leq t. \end{cases} \tag{2.40}$$

3

Probabilistic constructions

We develop here the theory of stochastic differential equations (SDEs) driven by nonlinear Lévy noise, aiming at applications to Markov processes. To make the basic ideas clearer, we start with symmetric square-integrable Lévy processes and then extend the theory to more general cases. One tool we use is the coupling of Lévy processes. To avoid interruptions to the exposition, the relevant results on coupling are collected in Section 3.6.

3.1 Stochastic integrals and SDEs driven by nonlinear Lévy noise

Suppose that $Y_s(\eta)$ is a family of symmetric square-integrable Lévy processes in $\mathbf{R}^d$ with càdlàg paths, depending on a parameter $\eta \in \mathbf{R}^n$ and specified by their generators L_η, where

$$L_\eta f(x) = \frac{1}{2}(G(\eta)\nabla, \nabla)f(x) + \int [f(x+y) - f(x) - (y, \nabla)f(x)]\nu(\eta, dy) \tag{3.1}$$

and

$$\nu(\eta)(\{0\}) = 0, \qquad \sup_\eta \left(\operatorname{tr} G(\eta) + \int |y|^2 \nu(\eta, dy) \right) = \kappa_1 < \infty. \tag{3.2}$$

Our first objective is to define the stochastic integral $\int_0^t \alpha_s dY_s(\xi_s)$ for random processes α and ξ. We start with piecewise-constant α and ξ. To simplify the notation we assume that they are constant on intervals with binary rational bounds. More precisely, suppose that $(\Omega, \mathcal{F}, P)$ is a filtered probability space with a filtration $\mathcal{F}_t$ that satisfies the usual conditions of completeness and right continuity. Let $\tau_k = 2^{-k}$. Processes of the form

$$\alpha_t = \sum_{j=0}^{[t/\tau_k]} \alpha^j \mathbf{1}_{(j\tau_k,(j+1)\tau_k]}, \qquad \xi_t = \sum_{j=0}^{[t/\tau_k]} \xi^j \mathbf{1}_{(j\tau_k,(j+1)\tau_k]}, \tag{3.3}$$

where α^j and ξ^j are $\mathcal{F}_{j\tau_k}$-measurable $\mathbf{R}^d$- and $\mathbf{R}^n$-valued random variables, will be called *simple*. Our *stochastic integral* for such α, ξ is defined as

$$\int_0^t \alpha_s dY_s(\xi_s) = \sum_{j=0}^{[t/\tau_k]} \alpha^j (Y_{\min(t,(j+1)\tau_k)} - Y_{j\tau_k})(\xi^j). \tag{3.4}$$

However, for this formula to make sense for random ξ_t some measure-theoretic reasoning is required, since a natural question arises: on which probability space is this process defined? Everything would be fine if all Lévy processes $Y_s(x)$ were defined on a single probability space and depended measurably on x. Can this be done? Of course each $Y_s(\eta)$ exists on, say, $D(\mathbf{R}_+, \mathbf{R}^d)$. But if one uses the normal Kolmogorov construction and defines $Y_s(x)$ on the infinite product space $\prod_{x\in\mathbf{R}^n} D(\mathbf{R}_+, \mathbf{R}^d)$ then what is the mechanism that ensures measurability with respect to x?

To move ahead, we apply the following *randomization* and *conditional independence and randomization* lemmas (see Lemma 3.22 and Proposition 6.13 respectively in Kallenberg [114]):

Lemma 3.1 *Let $\mu(x, dz)$ be a probability kernel from a measurable space X to a Borel space Z. Then there exists a measurable function $f : X\times[0,1] \to Z$ such that if θ is uniformly distributed on* $[0, 1]$ *then $f(X, \theta)$ has distribution $\mu(x, .)$ for every $x \in X$.*

Lemma 3.2 *Let ξ, η, ζ be random variables with values in measurable spaces Z, X, U respectively, where Z is Borel. Then ξ is η-conditionally independent on ζ if and only if $\xi = f(\eta, \theta)$ a.s. for some measurable function $f : X \times [0, 1] \to Z$ and some random variable θ, uniformly distributed on* $[0, 1]$, *that is independent of η and ζ.*

In order to apply these results we need to compare the Lévy measures. To this end, we introduce an extension of the Wasserstein–Kantorovich distance to unbounded measures. Namely, let $\mathcal{M}_p(\mathbf{R}^d)$ denote the class of Borel measures μ on $\mathbf{R}^d \setminus \{0\}$, not necessarily finite but with finite pth moment (i.e. such that $\int |y|^p \mu(dy) < \infty$). For a pair of measures ν_1, ν_2 in $\mathcal{M}_p(\mathbf{R}^d)$ we define the distance $W_p(\nu_1, \nu_2)$ using (A.2):

$$W_p(\nu_1, \nu_2) = \left(\inf_\nu \int |y_1 - y_2|^p \nu(dy_1 dy_2)\right)^{1/p},$$

where inf is now taken over all $\nu \in \mathcal{M}_p(\mathbf{R}^{2d})$ such that condition (A.1),

$$\int_{S\times S} [\phi_1(x) + \phi_2(y)]\, \nu(dxdy) = (\phi_1, \nu_1) + (\phi_2, \nu_2),$$

holds for all ϕ_1, ϕ_2 satisfying $\phi_i(.)/|.|^p \in C(\mathbf{R}^d)$. It is easy to see that for finite measures this definition coincides with the usual definition.

Remark 3.3 Although the measures ν_1 and ν_2 are infinite, the distance $W_p(\nu_1, \nu_2)$ is finite. In fact, let a decreasing sequence of positive numbers ϵ_n^1 be defined by writing $\nu_1 = \sum_{n=1}^{\infty} \nu_1^n$, where the probability measures ν_1^n have their support in the closed shells $\{x \in \mathbf{R}^d : \epsilon_n^1 \le |x| \le \epsilon_{n-1}^1\}$ (with $\epsilon_0^1 = \infty$). The quantities ϵ_2^n and ν_2^n are defined similarly. Then the sum $\sum_{n=1}^{\infty} \nu_1^n \otimes \nu_2^n$ is a coupling of ν_1 and ν_2 having a finite value of $\int |y_1 - y_2|^p \nu_1(dy_1)\nu_2(dy_2)$.

Moreover, by the same argument as for finite measures (see [201], [246] or Proposition A.13), we can show that whenever the distance $W_p(\nu_1, \nu_2)$ is finite the infimum in (A.2) is achieved, i.e. there exists a measure $\nu \in \mathcal{M}_p(\mathbf{R}^{2d})$ such that

$$W_p(\mu_1, \mu_2) = \left(\int |y_1 - y_2|^p \nu(dy_1 dy_2) \right)^{1/p}. \tag{3.5}$$

We now make the following crucial assumption about the family $Y_s(x)$:

$$\left\{ \mathrm{tr} \left[\sqrt{G(x_1)} - \sqrt{G(x_2)} \right]^2 \right\}^{1/2} + W_2(\nu(x_1, .), \nu(x_2, .)) \le \kappa_2 \|x_1 - x_2\| \tag{3.6}$$

for some constant κ_2 and any $x_1, x_2 \in \mathbf{R}^d$. By Proposition 3.19 below the mapping from $x \in \mathbf{R}^n$ to the law of the Lévy process $Y_s(x)$ is then continuous and hence measurable. Consequently, by Lemma 3.1 (with Z the complete metric space $D(\mathbf{R}_+, \mathbf{R}^d)$, a Borel space) one can define the processes $Y_s(x)$ as measurable functions of x living on the standard probability space $[0, 1]$ with Lebesgue measure. This makes expression (3.4) well defined. However, this is still not quite satisfactory for our purposes as this construction does not take into account the dependence of the natural filtration of $Y_s(x)$ on x. To settle this issue, let $f_s^k(x, \omega) \in D([0, \tau_k], \mathbf{R}^d)$, $x \in \mathbf{R}^d$, $\omega \in [0, 1]$, $s \in [0, \tau_k]$, be the function from Lemma 3.1 constructed for the parts of the Lévy processes $Y_s(x)$ lying on an interval of length τ_k. In particular, for each $x \in \mathbf{R}^d$ the random process $f_s^k(x, .)$, defined on the standard probability space $([0, 1], \mathcal{B}([0, 1]))$ with Lebesgue measure, is a Lévy process on the time interval $s \in [0, \tau_k]$. As the basic probability space for the construction of the integral in (3.5) we choose the space $\Omega \times [0, 1]^\infty$ with product σ-algebra and product measure (each interval $[0, 1]$ being equipped with Lebesgue measure).

Let us now define the process (3.4) as the random process on the probability space $\Omega \times [0,1]^\infty$ (with points denoted by $(\omega, \delta_1, \delta_2, \dots)$) given by the formula

$$\int_0^t \alpha_s dY_s(\xi_s)(\omega, \{\delta_j\}_1^\infty) = \sum_{j=0}^{[t/\tau_k]} \alpha^j(\omega) f^k_{\min(\tau_k, t-j\tau_k)}(\xi^j(\omega), \delta_j), \tag{3.7}$$

and let $\mathcal{F}_t^{\alpha,\xi}$ be the filtration on $\Omega \times [0,1]^\infty$ that is generated by ξ_τ, α_τ, $\int_0^\tau \alpha_s dY_s(\xi_s)$, $\tau \le t$. The following statement summarizes the basic properties of the simple integral in (3.7).

Theorem 3.4 *Let* (3.2), (3.6) *hold for the family of Lévy generators given by* (3.1), *let α, ξ be simple processes of the form* (3.3) *and let α be bounded by a constant A. Then* (3.7) *defines a càdlàg process on the probability space $\Omega \times [0,1]^\infty$ having the following properties.*

(i) It is adapted to the filtration $\mathcal{F}_t^{\alpha,\xi}$.

(ii) The random process $f^k_{\cdot}(\xi^j(\omega), \delta_j)$ conditioned either on ξ^j or on the σ-algebra $\mathcal{F}_{j\tau_k}^{\alpha,\xi}$ is distributed in the same way as the Lévy process $Y_{\cdot}(\xi^j)$.

(iii) The process given in (3.7) *is a square-integrable $\mathcal{F}_t^{\alpha,\xi}$-martingale (note that it is not assumed to be a martingale with respect to its own natural filtration), and*

$$\mathbf{E}\left(\int_0^t \alpha_s dY_s(\xi_s)\right)^2 \le A^2 t \kappa_1. \tag{3.8}$$

(iv) Definition (3.7) *is unambiguous with respect to the choice of partition length τ_k. Namely, if one writes the processes* (3.3) *as*

$$\begin{aligned}\alpha_t &= \sum_{j=0}^{[t/\tau_k]} \alpha^j (\mathbf{1}_{(2j\tau_{k+1},(2j+1)\tau_{k+1}]} + \mathbf{1}_{((2j+1)\tau_{k+1},2(j+1)\tau_{k+1}]}),\\ \xi_t &= \sum_{j=0}^{[t/\tau_k]} \xi^j (\mathbf{1}_{(2j\tau_{k+1},(2j+1)\tau_{k+1}]} + \mathbf{1}_{((2j+1)\tau_{k+1},2(j+1)\tau_{k+1}]},\end{aligned} \tag{3.9}$$

then the integral (3.7) *has the same distribution as a similar integral for the processes* (3.9) *defined with respect to partitions of length τ_{k+1}.*

(v) Let $\tilde\alpha, \tilde\xi$ be another pair of simple processes with the same bound A as for $\tilde\alpha_t$. Then

$$\begin{aligned}&W^2_{2,t,\mathrm{un}}\left(\int_0^t \alpha_s dY_s(\xi_s), \int_0^t \tilde\alpha_s dY_s(\tilde\xi_s)\right)\\ &\qquad \le 4A^2\kappa_2^2 W^2_{2,t,\mathrm{un}}(\xi,\tilde\xi) + 4\kappa_1 W^2_{2,t,\mathrm{un}}(\alpha,\tilde\alpha),\end{aligned} \tag{3.10}$$

where the distance $W_{2,t,\mathrm{un}}$ *is defined in* (A.4), *and where distances in* $\mathbf{R}^d$ *are Euclidean.*

Proof (i) This is obvious.
(ii) On the one hand, $f^k_.(\xi^j(\omega), \delta_j)$ conditioned on ξ^j is distributed in the same way as the Lévy process $Y_.(\xi^j)$ by construction but, on the other hand, $f^k_.(\xi^j(\omega), \delta_j)$ is ξ^j-conditionally independent of α_j and of $\int_0^{j\tau_k} \alpha_s dY_s(\xi_s)$ by Lemma 3.2, implying that the distributions of $f^k_.(\xi^j(\omega), \delta_j)$ conditioned on either ξ^j or $\mathcal{F}^{\alpha,\xi}_{j\tau_k}$ are the same.
(iii) To prove the martingale property it is enough to show that

$$\mathbf{E}\left(\int_0^t \alpha_s dY_s(\xi_s)\middle|\, \mathcal{F}^{\alpha,\xi}_\tau\right) = \int_0^\tau \alpha_s dY_s(\xi_s), \qquad \tau < t,$$

and it is sufficient to show this for $j\tau_k \leq \tau < t \leq (j+1)\tau_k$, for any j. But the latter follows from (ii), because all these Lévy processes have zero expectation. Next, by conditioning and statement (ii) one has

$$\mathbf{E}\left(\int_0^t \alpha_s dY_s(\xi_s)\right)^2 = \sum_{j=0}^{[t/\tau_k]} \mathbf{E}\left(\alpha^j,\ f^k_{\min(\tau_k, t-j\tau_k)}(\xi_j, .)\right)^2,$$

implying (3.8) by assumption (3.2) and also by statement (ii).
(iv) By conditioning with respect to the collection of the random variables α^j, ξ^j the statement reduces to the i.i.d. property of increments of Lévy processes.
(v) By (iv) we can define both processes using partitions of the same length, say τ_k. Suppose first that the curves α_t, ξ_t are not random. Then the terms in the sum (3.7) are independent. Doob's maximum inequality allows us to estimate the l.h.s. of (3.10) as follows:

$$\begin{aligned}
2\sum_{j=0}^{[t/\tau_k]} \inf \mathbf{E}&\left(\alpha^j f^k_{\min(\tau_k,t-j\tau_k)}(\xi^j, .) - \tilde{\alpha}^j f^k_{\min(\tau_k,t-j\tau_k)}(\tilde{\xi}^j, .)\right)^2 \\
&\leq 4\sum_{j=0}^{[t/\tau_k]} \left(\alpha^j - \tilde{\alpha}^j,\ \mathbf{E}[f^k_{\min(\tau_k,t-j\tau_k)}(\xi^j, .)]\right)^2 \\
&\quad + 4A^2 \sum_{j=0}^{[t/\tau_k]} \inf \mathbf{E}[f^k_{\min(\tau_k,t-j\tau_k)}(\xi^j, .) - f^k_{\min(\tau_k,t-j\tau_k)}(\tilde{\xi}^j, .)]^2,
\end{aligned}$$

where the infimum is taken over all couplings of $f^k_s(\xi^j, .)$ and $f^k_s(\tilde{\xi}^j, .)$ that yield a Lévy process in $\mathbf{R}^{2d}$ (and hence a martingale), so that Doob's maximum inequality is applicable. This implies that

$$W_{2,t,\text{un}}^2\left(\int_0^t \alpha_s dY_s(\xi_s), \int_0^t \tilde{\alpha}_s dY_s(\tilde{\xi}_s)\right)$$
$$\leq 4A^2\kappa_2^2 \int_0^t (\xi_s - \tilde{\xi}_s)^2\, ds + 4\kappa_1 \int_0^t (\alpha_s - \tilde{\alpha}_s)^2\, ds$$
$$\leq 4A^2\kappa_2^2 t \sup_{s\leq t}(\xi_s - \tilde{\xi}_s)^2 + 4\kappa_1 t \sup_{s\leq t}(\alpha_s - \tilde{\alpha}_s)^2$$

by assumption (3.6) and estimate (3.8). This in turn implies the general estimate (3.10) by conditioning with respect to the collection of the random variables α^j, ξ^j. □

Recall now that any left-continuous square-integrable adapted process can be approximated in L^2 (on each bounded interval $[0, t]$) by simple left-continuous processes. We can now define the *stochastic integral driven by nonlinear Lévy noise*, $\int_0^t \alpha_s dY_s(\xi_s)$, for any left-continuous adapted (more generally, predictable) square-integrable processes α, ξ with bounded α: it is the limit, in the sense of distribution on the Skorohod space of càdlàg paths, of the corresponding integral over the simple approximations of α and ξ. We summarize below the basic properties of this integral.

Theorem 3.5 (Stochastic integral driven by nonlinear Lévy noise) *Let* (3.2), (3.6) *hold for the family Y_s defined by* (3.1) *and* (3.2). *Then the above limit exists and does not depend on the approximation sequence, there exists a filtered probability space on which the processes $\xi_t, \alpha_t, \int_0^t \alpha_s dY_s(\xi_s)$ are defined as adapted processes, the integral $\int_0^t \alpha_s dY_s(\xi_s)$ is a square-integrable martingale with càdlàg paths such that estimate* (3.8) *holds and, for any s, the increments $\int_s^t \alpha_\tau dY_\tau(\xi_\tau)$ and $\int_0^s \alpha_\tau dY_\tau(\xi_\tau)$ are $\sigma\{\xi_\tau, \alpha_\tau, \tau \leq s\}$-conditionally independent. Finally, for any other pair of processes $\tilde{\alpha}, \tilde{\xi}$ with the same bound A for $\tilde{\alpha}_t$, estimate* (3.10) *holds.*

Proof (i) The existence of the limit and its independence of the approximation follows from Theorem 3.4 and estimates (3.8), (3.10). Notice that the weak convergence deduced from convergence with respect to the metric W_2 is stronger than Skorohod convergence. As the required σ-algebras $\mathcal{F}_t$, one then can choose the σ-algebras generated by the limiting processes $\xi_\tau, \alpha_\tau, \int_0^\tau \alpha_s dY_s(\xi_s)$ for $\tau \leq t$. □

Remark 3.6 For the purpose of constructing Markov processes, assuming the existence of a finite second moment of the relevant Lévy measure is satisfactory because, using perturbation theory, we can always reduce any given generator to a generator whose Lévy measure has compact support. It is natural, however, to ask whether the above theory extends to general Lévy measures. In fact it does. The main idea is to substitute the metric W_2 by an appropriate equivalent metric on $\mathbf{R}^d$. The natural choice for Lévy measures is either of the metrics ρ_β or $\tilde{\rho}_\beta$, to be introduced in Section 3.6. Alternatively, one can treat the part of a Lévy process having a finite Lévy measure separately, since it is defined as a Lebesgue integral, in the usual way.

Let us now consider an SDE driven by *nonlinear Lévy noise* and having the form

$$X_t = x + \int_0^t a(X_{s-})dY_s(g(X_{s-})) + \int_0^t b(X_{s-})\,ds. \tag{3.11}$$

Theorem 3.7 (SDE driven by nonlinear Lévy noise) *Let* (3.2), (3.6) *hold for the family* (3.1). *Let b, g, a be bounded Lipschitz-continuous functions, the first two of which are from $\mathbf{R}^n$ to $\mathbf{R}^n$ and the third of which is from $\mathbf{R}^n$ to $n \times d$ matrices, with a common Lipschitz constant κ. Let x be a random variable that is independent of the $Y_s(z)$. Then the solution to* (3.11) *exists in the sense of distribution*[1] *and is unique.*

Proof This is based on the contraction principle in the complete metric space $M_2(t)$ of the distributions on the Skorohod space of càdlàg paths $\xi \in D([0,t], \mathbf{R}^d)$ with finite second moment $W_{2,t,\text{un}}(\xi, 0) < \infty$ and with metric $W_{2,t,\text{un}}$. For any $\xi \in M_2(t)$, let

$$\Phi(\xi)_t = x + \int_0^t a(\xi_{s-})\,dY_s(g(\xi_{s-})) + \int_0^t b(\xi_{s-})\,ds.$$

By Theorem 3.5, for an arbitrary coupling of the pair of processes ξ^1, ξ^2, we have

$$\begin{aligned}
&W^2_{2,t,\text{un}}(\Phi(\xi^1), \Phi(\xi^2)) \\
&\quad \le \mathbf{E}\int_0^t \Big\{8A^2\kappa_2^2[g(\xi_s^1) - g(\xi_s^1)]^2 \\
&\qquad\qquad + 8\kappa_1[a(\xi_s^1) - a(\xi_s^2)]^2 + 2[b(\xi_s^1) - b(\xi_s^2)]^2\Big\}\,ds \\
&\quad \le \kappa t(8A^2\kappa_2^2 + 8\kappa_1 + 2)\,\mathbf{E}\sup_{s\le t}(\xi_s^1 - \xi_s^2)^2,
\end{aligned}$$

implying that

$$W^2_{2,t,\text{un}}(\Phi(\xi^1), \Phi(\xi_2)) \le \kappa t(8A^2\kappa_2^2 + 8\kappa_1 + 2)W_2^2(\xi^1, \xi_2).$$

Thus the mapping $\xi \mapsto \Phi(\xi)$ is a contraction in $M_2(t)$ for $t\kappa(8A^2\kappa_2^2 + 8\kappa_1 + 2) < 1$. This implies the existence and uniqueness of a fixed point and hence of a solution to (3.11) for this t. For large t this construction can be extended by iteration. □

[1] Thus the equation implies the coincidence of the distributions.

Our main motivation for analyzing equation (3.11) lies in the fact that the solution to the particular case

$$X_t = x + \int_0^t dY_s(X_{s-}) + \int_0^t b(X_{s-})\,ds \tag{3.12}$$

specifies a Markov process with generator given by

$$\begin{aligned} Lf(x) &= \tfrac{1}{2}(G(x)\nabla, \nabla)f(x) + (b(x), \nabla f(x)) \\ &\quad + \int [f(x+y) - f(x) - (y, \nabla)f(x)]\nu(x, dy), \end{aligned} \tag{3.13}$$

yielding not only the existence but also a construction of such a process. This is not difficult to see. We shall prove it in the next section as a by-product of a constructive approach (using Euler-type approximations) to the analysis of the SDE (3.11).

To conclude this section, let us consider some basic examples for which the assumptions of the above theorems hold.

To begin with, we observe that the assumption on the Lévy kernel ν in (3.6) is satisfied if we can decompose the Lévy measures $\nu(x; .)$ into countable sums $\nu(x; .) = \sum_{n=1}^{\infty} \nu_n(x; .)$ of probability measures such that $W_2(\nu_i(x; .), \nu_i(z; .)) \leq a_i|x - z|$ and the series $\sum a_i^2$ converges. It is well known that the optimal coupling of probability measures (the Kantorovich problem) cannot always be realized using mass transportation (i.e. a solution to the Monge problem), since this could lead to examples for which the construction of the process via standard stochastic calculus would not work. However, non-degeneracy is not built into such examples and this leads to serious difficulties when one tries to apply analytical techniques in these circumstances.

Another particularly important situation occurs when all members of a family of measures $\nu(x; .)$ have the same star shape, i.e. they can be represented as

$$\nu(x, dy) = \nu(x, s, dr)\,\omega(ds), \qquad y \in \mathbf{R}^d,\ r = |y| \in \mathbf{R}_+,\ s = y/r \in S^{d-1}, \tag{3.14}$$

with a certain measure ω on S^{d-1} and a family of measures $\nu(x, s, dr)$ on $\mathbf{R}_+$. This allows us to reduce the general coupling problem to a much more easily handled one-dimensional problem, because evidently if $\nu_{x,y,s}(dr_1 dr_2)$ is a coupling of $\nu(x, s, dr)$ and $\nu(y, s, dr)$ then $\nu_{x,y,s}(dr_1 dr_2)\omega(ds)$ is a coupling of $\nu(x, .)$ and $\nu(y, .)$. If one-dimensional measures have no atoms then their coupling can be naturally organized by "pushing" along a certain mapping. Namely, the measure ν^F is the pushing forward of a measure ν on $\mathbf{R}_+$ by a mapping $F : \mathbf{R}_+ \mapsto \mathbf{R}_+$ whenever

$$\int f(F(r))\nu(dr) = \int f(u)\nu^F(du),$$

for a sufficiently rich class of test functions f, say, for the indicators of intervals. Suppose that we are looking for a family of monotone continuous bijections $F_{x,s} : \mathbf{R}_+ \mapsto \mathbf{R}_+$ such that $\nu^{F_{x,s}} = \nu(x, s, .)$. Choosing $f = \mathbf{1}_{[F(z),\infty)}$ as a test function in the above definition of pushing forward yields

$$G(x, s, F_{x,s}(z)) = \nu([z, \infty)) \tag{3.15}$$

for $G(x, s, z) = \nu(x, s, [z, \infty)) = \int_z^\infty \nu(x, s, dy)$. Clearly if the $\nu(x, s, .)$ and ν, although technically unbounded, are in fact bounded on any interval separated from the origin and if they have no atoms and do not vanish on any open interval then this equation defines a unique continuous monotone bijection $F_{x,s} : \mathbf{R}_+ \mapsto \mathbf{R}_+$ with an inverse that is also continuous. Hence we arrive at the following criterion.

Proposition 3.8 *Suppose that the Lévy kernel $\nu(x, .)$ can be represented in the form* (3.14) *and that ν is a Lévy measure on $\mathbf{R}_+$ such that all $\nu(x, s, .)$ and ν are unbounded, have no atoms and do not vanish on any open interval. Then the family $\nu(x, .)$ depends Lipshitz continuously on x in W_2 whenever the unique continuous solution $F_{x,s}(z)$ to* (3.15) *is Lipschitz continuous in x with a constant $\kappa_F(z, s)$ for which the condition*

$$\int_{\mathbf{R}_+} \int_{S^{d-1}} \kappa_F^2(r, s)\omega(ds)\nu(dr) < \infty \tag{3.16}$$

holds.

Proof By the above discussion, the solution F specifies the coupling $\nu_{x,y}(dr_1\, dr_2\, ds_1\, ds_2)$ of $\nu(x, .)$ and $\nu(y, .)$ via

$$\int f(r_1, r_2, s_1, s_2)\nu_{x,y}(dr_1\, dr_2\, ds_1\, ds_2)$$
$$= \int f(F_{x,s}(r), F_{y,s}(r), s, s)\omega(ds)\nu(dr),$$

so that for Lipschitz continuity of the family $\nu(x, .)$ it is sufficient to have

$$\int_{\mathbf{R}_+} \int_{S^{d-1}} \left[F_{x,s}(r) - F_{y,s}(r)\right]^2 \omega(ds)\nu(dr) \le c(x - y)^2,$$

which is clearly satisfied whenever (3.16) holds. □

It is worth mentioning that a coupling for the sum of Lévy measures can be organized separately for each term, allowing the use of the above statement for the star shape components and, say, some discrete methods for the discrete components.

If ν has a density, a more explicit criterion can be given; see Stroock [227]. The following simple example is worth mentioning.

Corollary 3.9 *Let*

$$\nu(x; dy) = a(x, s) r^{-(1+\alpha(x,s))} \, dr \, \omega(ds), \qquad y \in \mathbf{R}^d, \; r = |y| \in \mathbf{R}_+, \; s = y/r \in S^{d-1}, \tag{3.17}$$

where a, α are $C^1(\mathbf{R}^d)$ functions of the variable x, depend continuously on s and take values in $[a_1, a_2]$ and $[\alpha_1, \alpha_2]$ respectively for $0 < a_1 \leq a_2$, $0 < \alpha_1 \leq \alpha_2 < 2$. Then the family of measures $\mathbf{1}_{B_K}(y)\nu(x, dy)$ depends Lipschitz continuously on x in W_2.

Proof Choose $\nu((z, K]) = 1/z - 1/K$. Since then

$$G(x, s, z) = \int_z^K a(x, s) r^{-(1+\alpha(x,s))} \, dr = \frac{a(x, s)}{\alpha(x, s)} (z^{-\alpha(x,s)} - K^{-\alpha(x,s)}),$$

it follows that the solution to (3.15) is given by

$$F_{x,s}(z) = \left[K^{-\alpha} + \frac{\alpha}{a} \left(\frac{1}{z} - \frac{1}{K} \right) \right]^{-1/\alpha} (x, s)$$

implying that $F(1) = 1$, $F_{x,s}(z)$ is of order $(az/\alpha)^{1/\alpha}$ for small z and $|\nabla_x F|$ is bounded by $O(1) z^{1/\alpha} \log z$. Hence condition (3.16) can be rewritten as a condition for the integrability around the origin of the function

$$z^{2(\alpha_2^{-1} - 1)} \log^2 z,$$

and clearly it holds true. □

Processes whose generators have Lévy measures of the form (3.17) are often called *stable-like*.

3.2 Nonlinear version of Ito's approach to SDEs

We shall develop now a constructive approach to the proof of Theorem 3.7. It bypasses the results on the stochastic integral in the previous section, yields a process satisfying the equation strongly (i.e. not only in the sense of the coincidence of distributions) and makes explicit the underlying Markovian structure, which is of major importance for our purposes.

Our approach is a nonlinear version of Ito's method (as detailed in Stroock [227]) and can be regarded as a stochastic version of the Euler approximation scheme for solving differential equations.

We shall deal again with equation (3.11), restricting our attention for simplicity to the case $g(x) = x$ and $a(x) = 1$. On the one hand this case is sufficient for the application we have in mind and on the other hand it captures the main difficulties, so that the extension to a general version of (3.11) is more or less straightforward (though much heavier in notation). For notational convenience, we shall include now the drift in the noise term. Thus we will deal with the equation

$$X_t = x + \int_0^t dY_s(X_{s-}), \tag{3.18}$$

in which $Y_s(\eta)$ is a family of Lévy processes with generators given by

$$\begin{aligned} L_\eta f(x) &= \tfrac{1}{2}(G(\eta)\nabla, \nabla) f(x) + (b(\eta), \nabla f(x)) \\ &\quad + \int [f(x+y) - f(x) - (y, \nabla) f(x)] \nu(\eta, dy), \end{aligned} \tag{3.19}$$

where

$$\nu(\eta, \{0\}) = 0, \qquad \sup_\eta \left(\operatorname{tr} G(\eta) + |b(\eta)| + \int |y|^2 \nu(\eta, dy) \right) = \kappa_1 < \infty \tag{3.20}$$

and

$$\begin{aligned} \left\{ \operatorname{tr} \left[\sqrt{G(x_1)} - \sqrt{G(x_2)} \right]^2 \right\}^{1/2} + |b(x_1) - b(x_2)| + W_2(\nu(x_1, .), \nu(x_2, .)) \\ \leq \kappa_2 \|x_1 - x_2\|. \end{aligned} \tag{3.21}$$

As mentioned above the solutions will be constructed from an Euler-type approximation scheme. Namely, let $Y_\tau^l(x)$ be the collection (depending on $l = 0, 1, 2, \ldots$) of independent families of the Lévy processes $Y_\tau(x)$, depending measurably on x, that were constructed in Lemma 3.1. We define the approximations $X^{\mu,\tau}$ by

$$X_t^{\mu,\tau} = X_{l\tau}^{\mu,\tau} + Y_{t-l\tau}^l(X_{l\tau}^{\mu,\tau}), \qquad \mathcal{L}(X_\mu^\tau(0)) = \mu, \tag{3.22}$$

for $l\tau < t \leq (l+1)\tau$, where $\mathcal{L}(X)$ means the law of X. Clearly these approximation processes are càdlàg. If $x \in \mathbf{R}^d$ then for brevity we shall write $X_t^{x,\tau}$ for $X_t^{\delta_x,\tau}$.

Remark 3.10 Clearly, if the limit of X_t^{μ,τ_k} as $k \to \infty$ exists in the metric $W_{2,t,\mathrm{un}}$ then it solves equation (3.18) in the sense of the previous section.

To avoid any appeal to the previous theory, we can simply define the weak solution to (3.11) as the weak limit of X_t^{μ,τ_k}, $\tau_k = 2^{-k}$, $k \to \infty$, in the sense of the distributions on the Skorohod space of càdlàg paths (this definition is of course implied by the convergence in distribution of the distance $W_{2,t,\mathrm{un}}$). This definition, as a limit of approximations, is constructive and sufficient for the applications we have in mind. It is not, however, very attractive aesthetically. Alternatively, we can define a solution to (3.11) in terms of martingales, i.e. as a process X_t^μ for which the processes (3.26) are martingales for smooth functions f.

Theorem 3.11 (SDEs driven by nonlinear Lévy noise revisited) *Suppose that the assumptions of Theorem 3.7 hold, i.e. that* (3.2), (3.6) *hold for* ν *and* G *and that* b *is a bounded Lipschitz-continuous function* $\mathbf{R}^n \to \mathbf{R}^n$. *Then:*

(i) for any $\mu \in \mathcal{P}(\mathbf{R}^d) \cap \mathcal{M}_2(\mathbf{R}^d)$ *there exists a limit process* X_t^μ *for the approximations* $X_t^{\mu,\tau}$ *such that*

$$\sup_\mu \sup_{s\in[0,t]} W_2^2\left(X^{\mu,\tau_k}_{[s/\tau_k]\tau_k}, X_t^\mu\right) \le c(t)\tau_k, \tag{3.23}$$

and, more strongly,

$$\sup_\mu W_{2,t,\mathrm{un}}^2\left(X^{\mu,\tau_k}, X^\mu\right) \le c(t)\tau_k; \tag{3.24}$$

(ii) the distributions $\mu_t = \mathcal{L}(X_t^\mu)$ *depend* $1/2$*-Hölder continuously on* t *in the metric* W_2 *and Lipschitz continuously on the initial condition:*

$$W_2^2(X_t^\mu, X_t^\eta) \le c(T) W_2^2(\mu, \eta); \tag{3.25}$$

(iii) the processes

$$M(t) = f(X_t^\mu) - f(x) - \int_0^t Lf(X_s^\mu)\, ds \tag{3.26}$$

are martingales for any $f \in C^2(\mathbf{R}^d)$, *where*

$$\begin{aligned} Lf(x) &= \tfrac{1}{2}(G(x)\nabla, \nabla) f(x) + (b(x), \nabla f(x)) \\ &\quad + \int [f(x+y) - f(x) - (y, \nabla) f(x)]\nu(x, dy), \end{aligned}$$

in other words, the process X_t^μ *solves the corresponding martingale problem;*

(iv) the operators given by $T_t f(x) = \mathbf{E} f(X_t^x)$ *form a conservative Feller semigroup preserving the space of Lipschitz-continuous functions and having the domain of the generator containing* $C_\infty^2(\mathbf{R}^d)$.

Proof *Step 1* (*Continuity of approximations with respect to the initial data*) By the definition of the distance one has

$$W_2^2(x_1 + Y_\tau(x_1),\ x_2 + Y_\tau(x_2)) \le \mathbf{E}(\xi_1 - \xi_2)^2$$

for any random variables (ξ_1, ξ_2) with the projections $\xi_i = x_i + Y_\tau(x_i)$, $i = 1, 2$. Choosing the coupling given in Proposition 3.19 below yields

$$\mathbf{E}(\xi_1 - \xi_2)^2 \le (1 + c\tau)(x_1 - x_2)^2.$$

Hence, taking the infimum over all couplings yields

$$W_2^2(x_1 + Y_\tau(x_1),\ x_2 + Y_\tau(x_2)) \le (1 + c\tau) W_2^2(\mathcal{L}(x_1), \mathcal{L}(x_2)). \qquad (3.27)$$

Applying this inequality inductively, we obtain

$$W_2^2(X_{k\tau}^{\mu,\tau}, X_{k\tau}^{\eta,\tau}) \le e^{1+2ck\tau} W_2^2(\mu, \eta). \qquad (3.28)$$

Step 2 (*Subdivision*) We want to estimate the W_2-distance between the random variables

$$\xi_1 = x + Y_\tau(x) = x' + (Y_\tau - Y_{\tau/2})(x), \qquad \xi_2 = z' + (Y_\tau - Y_{\tau/2})(z'),$$

where

$$x' = x + Y_{\tau/2}(x), \qquad z' = z + Y_{\tau/2}(z)$$

and in (3.28) $\mu = \mathcal{L}(x)$, $\eta = \mathcal{L}(z)$, $\eta' = \mathcal{L}(z')$. We shall couple ξ_1 and ξ_2 using Proposition 3.19 twice. By (3.60), see Section 3.6, we have

$$W_2^2(\xi_1, \xi_2) \le \mathbf{E}(x' - z')^2 + c\tau[\mathbf{E}(x' - z')^2 + \mathbf{E}(x - z')^2].$$

Hence, by (3.27) and (A.3), $W_2^2(\xi_1, \xi_2)$ does not exceed

$$W_2^2(x, z)(1 + 2c\tau)(1 + c\tau) + c\tau \mathbf{E}(x - z')^2$$

or, consequently,

$$W_2^2(x, z)(1 + c\tau) + c\tau \mathbf{E}(Y_{\tau/2}(z))^2$$

(with another constant c). Hence

$$W_2^2(\xi_1, \xi_2) \le W_2^2(x, z)(1 + c\tau) + c\tau^2$$

(with yet another c). We arrive at

$$W_2^2(X_{k\tau}^{\mu,\tau}, X_{k\tau}^{\mu,\tau/2}) \le c\tau^2 + (1 + c\tau) W_2^2\left(X_{(k-1)\tau}^{\mu,\tau}, X_{(k-1)\tau}^{\mu,\tau/2}\right). \qquad (3.29)$$

Step 3 (*Existence of the limits of marginal distributions*) By induction one estimates the l.h.s. of the previous inequality using the relation

$$\tau^2[1+(1+c\tau)+(1+c\tau)^2+\cdots+(1+c\tau)^{(k-1)}] \le c^{-1}\tau(1+c\tau)^k \le c(t)\tau.$$

Repeating this subdivision and using the triangle inequality for distances yields

$$W_2^2(X_{k\tau}^{\mu,\tau}, X_{k\tau}^{\mu,\tau/2^m}) \le c(t)\tau.$$

This implies the existence of the limit $X_x^{\tau_k}([t/\tau_k]\tau_k)$, as $k \to \infty$, in the sense of (3.23).

Step 4 (*Improving convergence*) For $f \in C^2(\mathbf{R}^d)$ the processes

$$M_\tau(t) = f(X_t^{\mu,\tau}) - f(x) - \int_0^t L\left[X_{[s/\tau]\tau}^{\mu,\tau}\right] f(X_s^{\mu,\tau})\,ds, \qquad \mu = \mathcal{L}(x), \tag{3.30}$$

are martingales, by Dynkin's formula applied to the Lévy processes $Y_\tau(z)$. Our aim is to pass to the limit $\tau_k \to 0$ to obtain a martingale characterization of the limiting process. But first we have to strengthen our convergence result.

Observe that the step-by-step inductive coupling of the trajectories $X^{\mu,\tau}$ and $X^{\eta,\tau}$ used above to prove (3.28) actually defines the coupling between the distributions of these random trajectories in the Skorohod space $D([0,T],\mathbf{R}^d)$, i.e. it defines a random trajectory $(X^{\mu,\tau}, X^{\eta,\tau})$ in $D([0,T],\mathbf{R}^{2d})$. We can construct the Dynkin martingales for this coupled process in the same way as for $X^{\mu,\tau}$. Namely, for a function f of two variables with bounded second derivatives, the process

$$M_\tau(t) = f(X_t^{\mu,\tau}, X_t^{\eta,\tau}) - \int_0^t \tilde{L}_s f(X_s^{\mu,\tau}, X_s^{\eta,\tau})\,ds, \qquad \mu = \mathcal{L}(x_\mu),\ \eta = \mathcal{L}(x_\eta),$$

is a martingale; here $\tilde{L}_t$ is the coupling operator (see (3.58) below) constructed from the Lévy processes Y with parameters $X_{[t/\tau]\tau}^{\mu,\tau}$ and $X_{[t/\tau]\tau}^{\eta,\tau}$. These martingales are very useful for comparing different approximations. For instance, choosing $f(x,y) = (x-y)^2$ leads to a martingale of the form

$$(X_t^{\mu,\tau} - X_t^{\eta,\tau})^2 + \int_0^t O(1)(X_s^{\mu,\tau} - X_s^{\eta,\tau})^2\,ds.$$

Applying the martingale property in conjunction with Gronwall's lemma yields

$$\sup_{s\le t} \mathbf{E}(X_s^{\tau,\mu} - X_s^{\tau,\eta})^2 \le c(t)\mathbf{E}(x_\mu - x_\eta)^2,$$

giving another proof of (3.28). Moreover, applying Doob's maximal inequality (with $p = 2$) to the martingale $\tilde{M}_\tau$ constructed from $f(x, y) = x - y$ implies that

$$\mathbf{E} \sup_{s \le t} |\tilde{M}_\tau(s)|^2 \le 4\,\mathbf{E}(x_\mu - x_\eta)^2$$

and, consequently,

$$\mathbf{E} \sup_{s \le t} \left(X_s^{\mu,\tau} - X_s^{\eta,\tau} + \int_0^s O(1)|X_v^{\mu,\tau} - X_v^{\eta,\tau}|\,dv \right) \le 6\,\mathbf{E}(x_\mu - x_\eta)^2.$$

Applying Gronwall's lemma yields

$$\mathbf{E} \sup_{s \le t} (X_s^{\mu,\tau} - X_s^{\eta,\tau})^2 \le c(t)\,\mathbf{E}(x_\mu - x_\eta)^2,$$

which allows us to improve (3.28) to the estimate of the distance between paths:

$$W_{2,t,\mathrm{un}}^2(X^{\mu,\tau}, X^{\eta,\tau}) \le c(t) W_2^2(\mu, \eta). \tag{3.31}$$

Similarly, we can strengthen the estimates for subdivisions leading to (3.24).

Using the Skorohod representation theorem for the weak converging sequence of random trajectories X^{μ,τ_k} (let us stress again that convergence with respect to the distance $W_{2,t,\mathrm{un}}$ implies the weak convergence of the distributions in the sense of the Skorohod topology), we can put the processes X^{μ,τ_k} on a single probability space, forcing them to converge to X^μ almost surely in the sense of the Skorohod topology.

Step 5 (*Solving the martingale problem and obtaining the Markov property*) Passing to the limit $\tau = \tau_k \to 0$ in (3.30) and using the continuity and boundedness of f and Lf and the dominated convergence theorem allows us to conclude that these martingales converge, a.s. and in L^1, to the martingale

$$M(t) = f(X_t^\mu) - f(x) - \int_0^t (Lf)(X_s^\mu)\,ds, \tag{3.32}$$

in other words that the process X_t^μ solves the corresponding martingale problem.

Passing to the limit $\tau_k \to 0$ in the Markov property for the approximations, i.e. in

$$\mathbf{E}\left(f(X_t^{\mu,\tau_k}) \big| \sigma(X_u^{\mu,\tau_k})\big|_{u \le j\tau_k} \right) = \mathbf{E}\left(f(X_t^{\mu,\tau_k}) \big| X_{j\tau_k}^{\mu,\tau_k} \right),$$

yields the Markov property for the limit X_t^μ.

Step 6 (*Completion*) Observe now that (3.28) implies (3.25). Moreover, the mapping $T_t f(x) = \mathbf{E} f(X_t^x)$ preserves the set of Lipschitz-continuous functions. In fact, if f is Lipschitz with constant h then

$$\left|\mathbf{E}f\left(X^{x,\tau}_{[t/\tau]\tau}\right) - \mathbf{E}f\left(X^{z,\tau}_{[t/\tau]\tau}\right)\right| \le h\,\mathbf{E}\left\|X^{x,\tau}_{[t/\tau]\tau} - X^{z,\tau}_{[t/\tau]\tau}\right\|$$
$$\le h\left(\mathbf{E}\left\|X^{x,\tau}_{[t/\tau]\tau} - X^{z,\tau}_{[t/\tau]\tau}\right\|^2\right)^{1/2}.$$

Taking the infimum yields

$$\left|\mathbf{E}f\left(X^{x,\tau}_{[t/\tau]\tau}\right) - \mathbf{E}f\left(X^{z,\tau}_{[t/\tau]\tau}\right)\right| \le hc(t_0)W_2(x,z).$$

Similarly, one may show (first for Lipschitz-continuous f and then for all $f \in C(\mathbf{R}^d)$ using the standard approximation) that

$$\sup_{t\in[0,t_0]}\sup_x\left|\mathbf{E}f\left(X^{x,\tau_k}_{[t/\tau_k]\tau_k}\right) - \mathbf{E}f(X^x_t)\right| \to 0, \qquad k\to\infty, \tag{3.33}$$

for all $f \in C(\mathbf{R}^d)$. From this convergence one deduces that T_t, given by $T_tf(x) = \mathbf{E}f(X^x_t)$, is a positivity-preserving family of contractions in $C(\mathbf{R}^d)$ that also preserves constants. Moreover, as the dynamics of averages of the approximation process clearly preserves the space $C_\infty(\mathbf{R}^d)$, the same holds for the limiting mappings T_t. Consequently, the operators T_t form a conservative Feller semigroup.

From the inequality

$$W_2^2\left(X^{\mu,\tau}_{l\tau}, X^{\mu,\tau}_{(l-1)\tau}\right) \le \mathbf{E}\left[Y_\tau\left(X^{\mu,\tau}_{(l-1)\tau}\right)\right]^2 \le c\tau$$

it follows that the curve μ_t depends 1/2-Hölder continuously on t in W_2.

Finally, it follows from the martingale property of (3.32) and the continuity of $Lf(X^\mu_s)$ that, for $f \in C^2(\mathbf{R}^d)$,

$$\frac{1}{t}(T_tf - f) = Lf + o_{t\to 0}(1),$$

implying that f belongs to the generator of L. □

It is worth noting that, in the simpler case of generators of order at most one, the continuity of Lévy measures with respect to a more easily handled metric W_1 is sufficient, as the following result shows. We omit the proof because it is just a simplified version of the proof of Theorem 3.11.

Theorem 3.12 *For the operator L given by*

$$Lf(x) = (b(x), \nabla f(x)) + \int [f(x+z) - f(x)]\nu(x,\mu;dz), \qquad \nu(x,.) \in \mathcal{M}_1(\mathbf{R}^d), \tag{3.34}$$

where

$$\|b(x) - b(z)\| + W_1(\nu(x,.), \nu(z,.)) \le \kappa\|x - z\| \tag{3.35}$$

holds true for a constant κ, there exists a unique Feller process X_t^μ solving (3.18) *such that*

$$\sup_\mu W_{1,t,\mathrm{int}}\left(X_\mu^{\tau_k}, X_\mu\right) \le c(t)\tau_k; \tag{3.36}$$

the distributions $\mu_t = \mathcal{L}(X(t))$ depend 1/2-Hölder continuously on t and Lipschitz continuously on the initial condition in the metric W_1.

3.3 Homogeneous driving noise

The usual stochastic calculus based on Wiener noise and Poisson random measures yields a construction of Markov processes such that the Lévy measures of the generator $\nu(x, dy)$ are connected by a family of sufficiently regular transformations, i.e.

$$\nu(x, dy) = \nu^{F_x}(dy) \qquad \Longleftrightarrow \qquad \int f(y)\nu(x, dy) = \int f(F_x(y))\nu(dy)$$

for some given Lévy measure ν and a family of measurable mappings F_x. As we noted above, such transformations yield a natural coupling of Lévy measures via

$$\int f(y_1, y_2)\nu_{x_1,x_2}(dy_1\, dy_2) = \int f(F_{x_1}(y), F_{x_2}(y))\nu(dy).$$

Writing down the conditions of Theorems 3.11 or 3.7 in terms of this coupling yields the standard conditions on F, allowing one to solve the corresponding SDE. An Ito-type construction for this case is presented in detail in [227]. Let us discuss this matter briefly.

Let Y be a Lévy process with generator given by

$$Lf(x) = \tfrac{1}{2}(G\nabla, \nabla)f(x) + \int [f(x+y) - f(x) - (y, \nabla)f(x)]\nu(dy),$$

with $\int |y|^2\nu(dy) < \infty$, and let $\tilde{N}(ds\, dx)$ be the corresponding compensated Poisson measure of jumps. We are interested now in a stochastic equation, of standard form

$$X_t = x + \int_0^t \sigma(X_{s-})\, dB_s^G + \int_0^t b(X_{s-})\, ds + \int_0^t\int F(X_{s-}, z)\tilde{N}(ds\, dz), \tag{3.37}$$

where F is a measurable mapping from $\mathbf{R}^n \times \mathbf{R}^d$ to $\mathbf{R}^n$ and σ maps $\mathbf{R}^n$ to $n \times d$ matrices. The analysis of the previous section suggests the Ito–Euler approximation scheme (3.22) for the solutions of (3.37), with

$$Y_t(z) = \sigma(z)B_t^G + b(z)t + \int_0^t \int F(z,y)\tilde{N}(ds\,dy). \tag{3.38}$$

Clearly $Y_t(z)$ is a Lévy process with generator given by

$$L_z f(x) = \frac{1}{2}(\sigma(z)\sigma^T(z)\nabla,\ \nabla)f(x) + (b(z), \nabla f(x))$$
$$+ \int \left[f(x + F(z,y)) - f(x) - (F(z,y), \nabla f(x)) \right] \nu(dy).$$

Proposition 3.13 *Let Y_t be a Lévy process as introduced above. Let*

$$|b(y_1) - b(y_2)|^2 + \|\sigma(y_1) - \sigma(y_2)\|^2 + \int |F(y_1, w) - F(y_2, w)|^2 \nu(dw)$$
$$\leq \kappa |y_1 - y_2|^2 \tag{3.39}$$

and

$$\sup_y \left(|b(y)| + \|\sigma(y)\| + \int |F(y,z)|^2 \nu(dz) \right) < \infty.$$

Then the approximations X_t^{μ,τ_k} converge to the solution of (3.37) *in the norm* $(\mathbf{E} \sup_{s \leq t} |Y_s|^2)^{1/2}$. *The limiting process is Feller with generator L acting on $C^2(\mathbf{R}^d)$ as follows:*

$$Lf(x) = \frac{1}{2}(\sigma(x)G\sigma^T(x)\nabla, \nabla)f(x) + (b(x), \nabla f(x))$$
$$+ \int \left[f(x + F(x,y)) - f(x) - (F(x,y), \nabla f(x)) \right] \nu(dy). \tag{3.40}$$

Proof This is a consequence of Theorem 3.7. It is worth noting that the present case enjoys an important simplification, both technical and ideological, compared with Theorem 3.7. Namely, now all Lévy processes are directly defined on the same probability space. This allows us to avoid technical complications and to construct a solution strongly on the same probability space, for which convergence holds in the usual L^2-sense (without reference to Wasserstein–Kantorovich metrics). □

Remark 3.14 It is not difficult to allow for linear growth of the coefficients. One can also include a compound Poisson component; see Stroock [227].

3.4 An alternative approximation scheme

Let us discuss briefly an alternative approximation scheme for constructing Markov processes based on nonlinear random integrals.

For a process X_t and a measurable function f, let a *nonlinear random integral* based on the noise dX_t be defined as a limit in probability:

$$\int_0^t f(dX_s) = \lim_{\max_i (s_{i+1}-s_i)\to 0} \sum_{i=1}^n f(X_{s_{i+1}} - X_{s_i}) \tag{3.41}$$

(the limit is taken over finite partitions $0 = s_0 < s_1 < \cdots < s_n = t$ of the interval $[0, t]$) when it exists. In particular, $\int_0^t (dY_s)^2$ is the *quadratic variation* of Y.

Proposition 3.15 *Let Y be a compound Poisson process and $N(dsdz)$ be the corresponding Poisson random measure with intensity $\lambda\, dsdz$, so that*

$$Y_t = Z(1) + \cdots + Z(N_t) = \int_0^t \int zN(ds\, dz)$$

with i.i.d. $Z(i)$ and Poisson process N_t. Let $\tilde{Y}$, $\tilde{N}$ denote the corresponding compensated processes.

(i) If the integral (3.41) *is defined (as a finite or infinite limit) for $X = Y$ (in particular, this is the case for either bounded or positive f) then*

$$\int_0^t f(dY_s) = \int_0^t \int f(z)N(ds\, dz) = f(Z(1)) + \cdots + f(Z(N_t)); \tag{3.42}$$

as a special case, note that the quadratic variation of a compound Poisson process equals the sum of the squares of its jumps;

(ii) If $f \in C^1(\mathbf{R}^d)$ then

$$\begin{aligned}\int_0^t f(d\tilde{Y}_s) &= \int_0^t \int f(z)N(dt\, dz) - t\lambda(\nabla f(0), \mathbf{E}Z(1)) \\ &= \int_0^t \int f(z)\tilde{N}(dt\, dz) + t\lambda[\mathbf{E}f(Z(1)) - (\nabla f(0), \mathbf{E}Z(1))].\end{aligned} \tag{3.43}$$

Proof Statement (i) is obvious. For (ii) observe that, since the number of jumps of $Y(t)$ is a.s. finite on each finite time interval, for partitions with small enough $\max_i(s_{i+1} - s_i)$ any interval $s_{i+1} - s_i$ contains not more than one jump, implying that $\int_0^t f(d\tilde{Y})$ equals $\int_0^t f(dY)$ plus the limit of the sums $\sum_{i=1}^n f(-\lambda \mathbf{E}Z(1)(s_{i+1}) - (s_i))$. □

An alternative approximation scheme for constructing a Markov process with generator equation (3.40) can be obtained by considering stochastic equations with nonlinear increments:

$$X = x + \int_0^t \sigma(X_{s-})\, dB_s^G + \int_0^t b(X_{s-})\, ds + \int g(X_s, dY_s^{\text{red}}),$$

where g is a measurable function and the reduced Lévy process Y^{red} is generated by the integral part of the full generator of Y. A natural approximation is, of course,

$$X_t^{\mu,\tau} = X_{l\tau}^{\mu,\tau} + \sigma(X_{l\tau}^{\mu,\tau})(B_t^G - B_{l\tau}^G) \\ + b(X_{l\tau}^{\mu,\tau})(t - l\tau) + g(X_{l\tau}^{\mu,\tau}, Y_t - Y_{l\tau}) \qquad (3.44)$$

for $l\tau < t \leq (l+1)\tau$.

Proposition 3.16 *Let g satisfy the same conditions as F in Proposition 3.13 and additionally let it have a bounded second derivative with respect to the second variable. Then the scheme (3.44) converges to a Feller process with generator given by*

$$Lf(x) = \tfrac{1}{2}(\sigma(x)G\sigma^T(x)\nabla, \nabla)f(x) \\ + \left(b(x) + \int_{B_1}\left[g(x,z) - \left(z, \frac{\partial g}{\partial z}(x,0)\right)\right]\nu(dz), \nabla f(x)\right) \\ + \int \left[f(x+g(x,y)) - f(x) - (g(x,y), \nabla f(x))\right]\nu(dy). \qquad (3.45)$$

Proof This follows from Propositions 3.13 and 3.15. □

3.5 Regularity of solutions

We shall discuss the regularity of the solutions to SDEs, focusing our attention on equation (3.37). Regularity for the more general equation (3.18) is discussed in [137].

Recall that we denote by C^k_{Lip} (resp. C^k_∞) the subspace of functions from $C^k(\mathbf{R}^d)$ with a Lipschitz-continuous derivative of order k (resp. with all derivatives up to order k vanishing at infinity).

Theorem 3.17 *Assume that the conditions of Proposition 3.13 hold and put $G = 1$ for simplicity.*

(i) Let $b, \sigma \in C^1_{\text{Lip}}(\mathbf{R}^d)$ and let

$$\sup_z \int \left\| \frac{\partial}{\partial z} F(z,w) \right\|^\beta \nu(dw) < \infty \qquad (3.46)$$

hold for $\beta = 2$.

Then the approximations $X_t^{x,\tau}$ to (3.37) are a.s. differentiable with respect to x and, for $\beta = 2$,

$$\mathbf{E}\left\| \frac{\partial X_{\tau l}^{x,\tau}}{\partial x} \right\|^\beta \leq (1 + c\tau)^l. \qquad (3.47)$$

(ii) Assume further that $b, \sigma \in C^2_{\mathrm{Lip}}(\mathbf{R}^d)$,

$$\sup_z \int \left\| \frac{\partial^2}{\partial z^2} F(z, w) \right\|^2 \nu(dw) < \infty, \tag{3.48}$$

and (3.46) *holds for* $\beta = 4$. *Then* (3.47) *holds for* $\beta = 4$, *the approximations* $X_t^{x,\tau}$ *to* (3.37) *are a.s. twice differentiable with respect to* x *and*

$$\mathbf{E} \left\| \frac{\partial^2 X^{x,\tau}_{\tau l}}{\partial x^2} \right\|^2 \le c(t), \qquad l\tau \le t. \tag{3.49}$$

Moreover, the solutions X_t^x *of* (3.37) *are a.s. differentiable with respect to* x *and the spaces* C^1_{Lip} *and* $C^1_{\mathrm{Lip}} \cap C^1_\infty$ *are invariant under the semigroup* T_t.

(iii) Assume further that $b, \sigma \in C^3_{\mathrm{Lip}}(\mathbf{R}^d)$,

$$\sup_z \int \left\| \frac{\partial^3}{\partial z^3} F(z, w) \right\|^2 \nu(dw) < \infty \tag{3.50}$$

and (3.46) *holds for* $\beta = 6$. *Then* (3.47) *holds for* $\beta = 6$, *the approximations* $X_t^{x,\tau}$ *to* (3.37) *are a.s. three times differentiable with respect to* x *and*

$$\mathbf{E} \left\| \frac{\partial^3 X^{x,\tau}_{\tau l}}{\partial x^3} \right\|^2 \le c(t), \qquad l\tau \le t. \tag{3.51}$$

Moreover, the solutions X_t^x *of* (3.37) *are a.s. twice differentiable with respect to* x,

$$\sup_{s \le t} \mathbf{E} \left\| \frac{\partial^2 X^x_s}{\partial x^2} \right\|^2 \le c(t), \tag{3.52}$$

the spaces C^2_{Lip} *and* $C^2_{\mathrm{Lip}} \cap C^2_\infty$ *are invariant under the Markov semigroup* T_t *and the latter space represents an invariant core for* T_t. *Moreover, in this case* T_t *and the corresponding process are uniquely defined by the generator* L.

Proof (i) Differentiating (3.38) yields

$$\frac{\partial}{\partial z} Y_t(z) = \frac{\partial}{\partial z} \sigma(z) B_t + \frac{\partial}{\partial z} b(z) t + \int_0^t \int \frac{\partial}{\partial z} F(z, y) \tilde{N}(ds\, dy). \tag{3.53}$$

Under our assumptions this expression is well defined and specifies a Lévy process for any z. Consequently the approximations $X_t^{x,\tau}$ are differentiable with respect to x a.s. and, by the chain rule,

$$\frac{\partial X^{x,\tau}_{\tau l}}{\partial x} = \left\{ 1 + \frac{\partial}{\partial z} [Y_{\tau l}(z) - Y_{(l-1)\tau}(z)] \right\} \Bigg|_{z = X^{x,\tau}_{\tau(l-1)}} \frac{\partial X^{x\tau}_{\tau(l-1)}}{\partial x}.$$

Consequently, by Exercise 3.1 below,

$$\mathbf{E}\left\|\frac{\partial X_{\tau l}^{x,\tau}}{\partial x}\right\|^2 \le (1+c\tau)\,\mathbf{E}\left\|\frac{\partial X_{\tau(l-1)}^{x,\tau}}{\partial x}\right\|^2,$$

implying (3.47) for $\beta = 2$ by induction.
(ii) Similarly, assumption (3.46) for any even β (we need $\beta = 4$ or $\beta = 6$) implies the corresponding estimate (3.47). Next,

$$\frac{\partial^2 X_{l\tau}^{x,\tau}}{\partial x^2} = \left\{1 + \frac{\partial}{\partial z}[Y_{\tau l}(z) - Y_{(l-1)\tau}(z)]\right\}\Bigg|_{z=X_{\tau(l-1)}^{x,\tau}} \frac{\partial^2 X_{\tau(l-1)}^{x\tau}}{\partial x^2}$$
$$+ \left(\left\{\frac{\partial^2}{\partial z^2}[Y_{\tau l}(z) - Y_{(l-1)\tau}(z)]\right\}\Bigg|_{z=X_{\tau(l-1)}^{x,\tau}} \frac{\partial X_{\tau(l-1)}^{x\tau}}{\partial x}, \frac{\partial X_{\tau(l-1)}^{x\tau}}{\partial x}\right),$$

so that this derivative exists and is continuous in x a.s. for all l, and

$$\mathbf{E}\left\|\frac{\partial^2 X_x^\tau(\tau l)}{\partial x^2}\right\|^2 \le (1+c\tau)\,\mathbf{E}\left\|\frac{\partial X_x^\tau(\tau(l-1))}{\partial x}\right\|^2 + c\tau(1+c\tau)^{l-1},$$

where we have used (3.47) with $\beta = 4$ and the estimate

$$\mathbf{E}\left(1 + \frac{\partial Y_t(z)}{\partial z}\right)\frac{\partial^2 Y_t(z)}{\partial z^2} = O(t)$$

that follows from the well-known formula

$$\mathbf{E}\left(\int_0^t\int_{\{|x|\le 1\}} f(x)\tilde{N}(ds\,dx)\int_0^t\int_{\{|x|\le 1\}} g(x)\tilde{N}(ds\,dx)\right)$$
$$= t\int_{\{|x|\le 1\}} f(x)g(x)\nu(dx) \qquad (3.54)$$

for stochastic integrals over random Poisson measures.

By induction, one then obtains the estimate

$$\mathbf{E}\left\|\frac{\partial^2 X_x^\tau(\tau l)}{\partial x^2}\right\|^2 \le lc\tau(1+c\tau)^{l-1} \le c(t)$$

for $l\tau \le t$.

Consequently the family of first derivatives of the approximations is Lipschitz continuous uniformly in finite time, so that we can choose a converging subsequence as $\tau_k \to 0$, the limit being of course $\partial X_t^x/\partial x$, that satisfies the same estimate (3.47) as the approximations. Furthermore, if $f \in C^1_{\mathrm{Lip}}$ then

$$\left|\frac{\partial}{\partial x}\mathbf{E}f(X_s^{x,\tau}) - \frac{\partial}{\partial x}\mathbf{E}f(X_s^{z,\tau})\right| \le \mathbf{E}\left|\left(\frac{\partial f}{\partial x}(X_s^{x,\tau}) - \frac{\partial f}{\partial x}(X_s^{z,\tau})\right)\frac{\partial X_s^{x,\tau}}{\partial x}\right| + \mathbf{E}\left|\frac{\partial f}{\partial x}(X_s^{z,\tau})\left(\frac{\partial X_s^{x,\tau}}{\partial x} - \frac{\partial X_s^{z,\tau}}{\partial z}\right)\right| \le c(t)\|x - z\|.$$

Hence, from the sequence of the uniformly Lipschitz-continuous functions $(\partial/\partial x)\mathbf{E}f(X_s^{x,\tau_k})$, $k = 1, 2, \dots$, we can choose a convergence subsequence, whose limit is clearly $(\partial/\partial x)\mathbf{E}f(X_t^x)$, showing that $\mathbf{E}f(X_t^x) \in C^1_{\mathrm{Lip}}$. From the uniform convergence it also follows that $\mathbf{E}f(X_t^x) \in C^1_{\mathrm{Lip}} \cap C^1_\infty$ whenever the same holds for f.

(iii) Similarly,

$$\begin{aligned}
\frac{\partial^3 X_{l\tau}^{x,\tau}}{\partial x^3}
&= \left\{1 + \frac{\partial}{\partial z}[Y_{\tau l}(z) - Y_{(l-1)\tau}(z)]\right\}\bigg|_{z=X^{x,\tau}_{\tau(l-1)}} \frac{\partial^3 X^{x\tau}_{\tau(l-1)}}{\partial x^3} \\
&\quad + 3\left(\left\{\frac{\partial^2}{\partial z^2}[Y_{\tau l}(z) - Y_{(l-1)\tau}(z)]\right\}\bigg|_{z=X^{x,\tau}_{\tau(l-1)}} \frac{\partial^2 X^{x\tau}_{\tau(l-1)}}{\partial x^2}, \frac{\partial X^{x\tau}_{\tau(l-1)}}{\partial x}\right) \\
&\quad + \left\{\frac{\partial^3}{\partial z^3}[Y_{\tau l}(z) - Y_{(l-1)\tau}(z)]\right\}\bigg|_{z=X^{x,\tau}_{\tau(l-1)}} \frac{\partial X^{x\tau}_{\tau(l-1)}}{\partial x}\frac{\partial X^{x\tau}_{\tau(l-1)}}{\partial x}\frac{\partial X^{x\tau}_{\tau(l-1)}}{\partial x},
\end{aligned}$$

leading to (3.51) and to the invariance of the space $C^2_{\mathrm{Lip}} \cap C^2_\infty$.

Finally, regularity implies uniqueness by Theorem 2.10. □

Consider now an example describing a possibly degenerate diffusion combined with a possibly degenerate *stable-like process*. Namely, let

$$\begin{aligned}
Lf(x) &= \tfrac{1}{2}\operatorname{tr}\left[\sigma(x)\sigma^T(x)\nabla^2 f(x)\right] + (b(x), \nabla f(x)) + \int [f(x+y) - f(x)]\,\nu(x, dy) \\
&\quad + \int_P (dp)\int_0^K d|y| \int_{S^{d-1}} a_p(x,s)\frac{f(x+y) - f(x) - (y, \nabla f(x))}{|y|^{\alpha_p(x,s)+1}} \\
&\quad \times d|y|\omega_p(ds),
\end{aligned} \tag{3.55}$$

where $s = y/|y|$, $K > 0$, (P, dp) is a Borel space with a finite measure dp and the ω_p are certain finite Borel measures on S^{d-1}.

Proposition 3.18 *(i) Let σ, b be Lipschitz continuous, and let a_p, α_p be $C^1(\mathbf{R}^d)$ functions of the variable x (uniformly in s, p) that depend* *continuously on s, p and take values in compact subintervals of* $(0, \infty)$ *and* $(0, 2)$ *respectively. Finally, let* ν *be a uniformly bounded measure depending weakly continuously on x and such that* $\int_{|y-x|\le A} \nu(x, dy) \to 0$ *as* $x \to \infty$ *for any A. Then a generator L of the form (3.55) generates a Feller process for which the domain of the generator of the corresponding Feller semigroup* T_t *contains* $C_c^2(\mathbf{R}^d)$.

(ii) Suppose additionally that for some $k > 2$ *one has* $\sigma, b \in C^k_{\text{Lip}}(\mathbf{R}^d)$, *that* a, α *are of the class* $C^k(\mathbf{R}^d)$ *as functions of x uniformly in s and that the kernel* ν *is k times differentiable in x, with*

$$\int_{B_a} \sum_{l=1}^{k} \left| \frac{\partial^l \nu}{\partial x^l} \right| (x, dy) \to 0$$

as $x \to \infty$ *for any a. Then, for* $l = 2, \ldots, k-1$, *the space* $C^l_{\text{Lip}} \cap C^l_\infty$ *is an invariant domain for the Feller semigroup and this semigroup is uniquely defined.*

Proof Perturbation theory reduces the problem to the case of vanishing ν, when the result follows from Theorem 3.17 taking into account Corollary 3.9. □

3.6 Coupling of Lévy processes

We describe here the natural *coupling of Lévy processes*, and this leads in particular to the analysis of their weak derivatives with respect to a parameter. Recall that by C^k_{Lip} we denote the subspace of functions from $C^k(\mathbf{R}^d)$ with a Lipschitz-continuous derivative of order k.

Proposition 3.19 *Let* Y^i_s, $i = 1, 2$, *be two Lévy processes in* $\mathbf{R}^d$ *specified by their generators* L_i, *which are given by*

$$L_i f(x) = \frac{1}{2}(G_i \nabla, \nabla) f(x) + (b_i, \nabla f(x)) + \int \left[f(x+y) - f(x) - (\nabla f(x), y) \right] \nu_i(dy), \tag{3.56}$$

with $\nu_i \in \mathcal{M}_2(\mathbf{R}^d)$. *Let* $\nu \in \mathcal{M}_2(\mathbf{R}^{2d})$ *be a coupling of* ν_1, ν_2, *i.e. let*

$$\int\int [\phi_1(y_1) + \phi_2(y_2)] \nu(dy_1 dy_2) = (\phi_1, \nu_1) + (\phi_2, \nu_2) \tag{3.57}$$

hold for all ϕ_1, ϕ_2 *satisfying* $\phi_i(.)/|.|^2 \in C(\mathbf{R}^d)$. *Then the operator*

$$
\begin{aligned}
&Lf(x_1, x_2) \\
&\quad = \left[\tfrac{1}{2}(G_1\nabla_1, \nabla_1) + \tfrac{1}{2}(G_2\nabla_2, \nabla_2) + \left(\sqrt{G_2}\sqrt{G_1}\nabla_1, \nabla_2\right)\right] f(x_1, x_2) \\
&\qquad + (b_1, \nabla_1 f(x_1, x_2)) + (b_2, \nabla_2 f(x_1, x_2)) \\
&\qquad + \int \{f(x_1 + y_1, x_2 + y_2) - f(x_1, x_2) \\
&\qquad\qquad - [(y_1, \nabla_1) + (y_2, \nabla_2)] f(x_1, x_2)\}\, \nu(dy_1 dy_2) \qquad (3.58)
\end{aligned}
$$

(where ∇_i means the gradient with respect to x_i) specifies a Lévy process Y_s in $\mathbf{R}^{2d}$, with characteristic exponent

$$
\begin{aligned}
\eta_{x_1,x_2}(p_1, p_2) = -\tfrac{1}{2}\left[\sqrt{G(x_1)}p_1 + \sqrt{G(x_2)}p_2\right]^2 + ib(x_1)p_1 + ib(x_2)p_2 \\
+ \int \left[e^{iy_1p_1 + iy_2p_2} - 1 - i(y_1p_1 + y_2p_2)\right] \nu(dy_1 dy_2),
\end{aligned}
$$

that is a coupling of Y_s^1, Y_s^2 in the sense that the components of Y_s have the distributions of Y_s^1 and Y_s^2 respectively. Moreover, if $f(x_1, x_2) = h(x_1 - x_2)$ for a function $h \in C^2(\mathbf{R}^d)$ then

$$
\begin{aligned}
Lf(x_1, x_2) = \tfrac{1}{2}((\sqrt{G_1} - \sqrt{G_2})^2\nabla, \nabla)h(x_1 - x_2) + (b_1 - b_2, \nabla h)(x_1 - x_2) \\
+ \int \left[h(x_1 - x_2 + y_1 - y_2) - h(x_1 - x_2)\right. \\
\left. - (y_1 - y_2, \nabla h(x_1 - x_2))\right] \nu(dy_1 dy_2). \qquad (3.59)
\end{aligned}
$$

Finally,

$$
\begin{aligned}
&\mathbf{E}(\xi + Y_t^1 - Y_t^2)^2 \\
&\quad = [\xi + t(b_1 - b_2)]^2 + t\left[\operatorname{tr}\left(\sqrt{G_1} - \sqrt{G_2}\right)^2 + \iint (y_1 - y_2)^2 \nu(dy_1 dy_2)\right]. \qquad (3.60)
\end{aligned}
$$

Proof This is straightforward. The second moment (3.60) is found by twice differentiating the characteristic function. □

One can extend this result to Lévy measures without a finite second moment, using the equivalent metric on $\mathbf{R}^d$ with varying order for large and small distances. We shall demonstrate this possibility for the case of Lévy measures with a finite "outer moment" of at least first order. Let ρ be any continuous increasing concave function $\mathbf{R}_+ \mapsto \mathbf{R}_+$ such that $\rho(0) = 0$. As one may easily see that the function $\rho(|x - y|)$ specifies a metric in any $\mathbf{R}^d$, the triangle inequality in particular holds:

$$\rho(|x+y|) \le \rho(|x|) + \rho(|y|).$$

This metric is equivalent to (i.e. specifies the same topology as) the Euclidean metric (*Exercise: check this!*). The natural choice for dealing with Lévy processes is the function

$$\rho_\beta(r) = \min(r, r^{\beta/2}), \qquad \beta \in [0,2].$$

However, the fact that ρ_β is not smooth is technically inconvenient. Thus in intermediate calculations we shall often use a smooth approximation to it, $\tilde\rho$, defined as a smooth (at least twice continuously differentiable) increasing concave function $\mathbf{R}_+ \mapsto \mathbf{R}_+$ such that $\tilde\rho_\beta(r) \ge \rho_\beta(r)$ everywhere and $\tilde\rho_\beta(r) = \rho_\beta(r)$ for $r \le 1$ and $r \ge 2$.

Proposition 3.20 *Let $\beta \in [1,2]$ and let Y_s^i, $i = 1, 2$, be two Lévy processes in $\mathbf{R}^d$ specified by their generators* (3.56) *with*

$$\int \rho_\beta^2(|y|)\nu_i(dy) = \int \min(|y|^\beta, |y|^2)\nu_i(dy) < \infty.$$

Let a Lévy measure ν on $\mathbf{R}^{2d}$ have a finite "mixed moment"

$$\int \rho_\beta^2(|(y_1,y_2)|)\nu(dy_1dy_2) = \int \min(|(y_1,y_2)|^\beta, |(y_1,y_2)|^2)\nu(dy_1dy_2) < \infty$$

and let it be a coupling of ν_1, ν_2, i.e. (A.1) *holds for positive ϕ_1, ϕ_2. Then the operator equation* (3.58) *specifies a Lévy process Y_s in $\mathbf{R}^{2d}$ that is a coupling of Y_s^1, Y_s^2 such that*

$$\begin{aligned}\mathbf{E}\rho_\beta^2(|Y_t^1 - Y_t^2|) &= \mathbf{E}\min(|Y_t^1 - Y_t^2|^\beta, |Y_t^1 - Y_t^2|^2)\\ &\le tc(t)\Big[\|\sqrt{G_1} - \sqrt{G_2}\|^2 + |b_1 - b_2|^2\\ &\qquad + \iint \min(|y_1 - y_2|^\beta, |y_1 - y_2|^2)\nu(dy_1dy_2)\Big]\end{aligned} \tag{3.61}$$

and moreover

$$\begin{aligned}\mathbf{E}\tilde\rho_\beta^2(|x + Y_t^1 - Y_t^2|) \le \tilde\rho_\beta^2(|x|) + tc(t)\Big[&\|\sqrt{G_1} - \sqrt{G_2}\|^2 + |b_1 - b_2|^2\\ &+ \iint \rho_\beta^2(|y_1 - y_2|)\nu(dy_1dy_2)\Big],\end{aligned} \tag{3.62}$$

with a constant $c(t)$ that is bounded for finite t.

Proof Clearly (3.61) follows from (3.62). To prove the latter formula observe that by Dynkin's formula

$$\mathbf{E}\tilde{\rho}_\beta^2(|x+Y_t^1-Y_t^2|)=\tilde{\rho}_\beta^2(|x|)+\mathbf{E}\int_0^t Lf(Y_s^1,Y_s^2)\,ds,$$

where $f(x,y)=\tilde{\rho}_\beta^2(|x-y|)$. If $\beta=1$ then the first and second derivatives of ρ_β^2 are uniformly bounded. Consequently, by (3.59) one has

$$\begin{aligned}
&\mathbf{E}\tilde{\rho}_1^2(|x+Y_t^1-Y_t^2|)\\
&\le \tilde{\rho}_1^2(|x|)+ct\|\sqrt{G_1}-\sqrt{G_2}\|^2+ct|b_1-b_2|\,\left|\nabla\tilde{\rho}_1^2(|x+Y_t^1-Y_t^2|)\right|\\
&\quad+ct\int_{\{|y_1-y_2|\le 1\}}(y_1-y_2)^2\nu(dy_1dy_2)+ct\int_{\{|y_1-y_2|>1\}}|y_1-y_2|\nu(dy_1dy_2),
\end{aligned}$$

implying (3.62) by Gronwall's lemma, and also the estimate

$$\begin{aligned}
|b_1-b_2|\,\left|\nabla\tilde{\rho}_1^2(|x+Y_t^1-Y_t^2|)\right| &\le 2|b_1-b_2|^2+2\left|\nabla\tilde{\rho}_1^2(|x+Y_t^1-Y_t^2|)\right|^2\\
&\le 2|b_1-b_2|^2+c\tilde{\rho}_1^2(|x+Y_t^1-Y_t^2|).
\end{aligned}$$

If $\beta>1$, only the second derivative is bounded and we have

$$\begin{aligned}
&\mathbf{E}\tilde{\rho}_\beta^2(|x+Y_t^1-Y_t^2|)\\
&\le \tilde{\rho}_\beta^2(|x|)+ct\left\|\sqrt{G_1}-\sqrt{G_2}\right\|^2+ct|b_1-b_2|\,\left|\nabla\tilde{\rho}_1^2(|x+Y_t^1-Y_t^2|)\right|\\
&\quad+ct\int_{\{|y_1-y_2|\le 1\}}(y_1-y_2)^2\nu(dy_1dy_2)+c\mathbf{E}\int_0^t ds\int_{\{|y_1-y_2|>1\}}\nu(dy_1dy_2)\\
&\quad\times\Big[\tilde{\rho}_\beta^2(|x+Y_s^1-Y_s^2+y_1-y_2|)-\tilde{\rho}_\beta^2(|x+Y_s^1-Y_s^2|)\\
&\quad-\Big(y_1-y_2,\nabla\rho_\beta^2(|x+Y_s^1-Y_s^2|)\Big)\Big].
\end{aligned}$$

Taking into account that

$$\int_{\{|y_1-y_2|>1\}}\nu(dy_1dy_2)\le\int_{\{\min(|y_1|,|y_2|)>1/2\}}\nu(dy_1dy_2)<\infty,$$

$$\tilde{\rho}_\beta^2(|x+Y_s^1-Y_s^2+y_1-y_2|)\le 2\tilde{\rho}_\beta^2(x+Y_s^1-Y_s^2)+2\rho_\beta^2(y_1-y_2)$$

and (owing to the Hölder inequality)

$$\begin{aligned}
\left|\Big(y_1-y_2,\nabla\tilde{\rho}_\beta^2(|x+Y_s^1-Y_s^2|)\Big)\right| &\le \frac{1}{\beta}|y_1-y_2|^\beta+\frac{\beta-1}{\beta}\left|\nabla\tilde{\rho}_\beta^2(|x+Y_1-Y_2|)^{\beta/(\beta-1)}\right|\\
&\le \frac{1}{\beta}|y_1-y_2|^\beta+c\frac{\beta-1}{\beta}\tilde{\rho}_\beta^2(|x+Y_1-Y_2|)
\end{aligned}$$

we can conclude that

$$\mathbf{E}\tilde{\rho}_\beta^2(|x+Y_t^1-Y_t^2|) \le \tilde{\rho}_\beta^2(|x|) + ct(\|\sqrt{G_1}-\sqrt{G_2}\|^2 + |b_1-b_2|^2)$$
$$+ ct\int \min(|y_1-y_2|^\beta, |y_1-y_2|^2)\nu(dy_1dy_2)$$
$$+ c\mathbf{E}\int_0^t \tilde{\rho}_\beta^2(|x+Y_s^1-Y_s^2|)\,ds,$$

implying (3.62) by Gronwall's lemma. □

Similarly one obtains

Proposition 3.21 *Let Y_s^i, $i=1,2$, be two Lévy processes in $\mathbf{R}^d$ specified by their generators L_i, which are given by*

$$L_if(x) = (b_i, \nabla f(x)) + \int [f(x+y) - f(x)]\nu_i(dy) \tag{3.63}$$

with $\nu_i \in \mathcal{M}_1(\mathbf{R}^d)$. Let $\nu \in \mathcal{M}_1(\mathbf{R}^{2d})$ be a coupling of ν_1, ν_2, i.e. let (A.1) *hold for all ϕ_1, ϕ_2 satisfying $\phi_i(.)/|.| \in C(\mathbf{R}^d)$. Then the operator L given by*

$$Lf(x_1,x_2) = (b_1, \nabla_1 f(x_1,x_2)) + (b_2, \nabla_2 f(x_1,x_2))$$
$$+ \int [f(x_1+y_1, x_2+y_2) - f(x_1,x_2)]\nu(dy_1dy_2) \tag{3.64}$$

specifies a Lévy process Y_s in $\mathbf{R}^{2d}$ that is a coupling of Y_s^1, Y_s^2 such that, for all t,

$$\mathbf{E}\|\xi + Y_t^1 - Y_t^2\| \le \|\xi\| + t\left(\|b_1-b_2\| + \int\int \|y_1-y_2\|\nu(dy_1dy_2)\right). \tag{3.65}$$

Proof One approximates $|y|$ by a smooth function, applies Dynkin's formula and then passes to the limit. □

We add here a couple of simple exercises on estimates of Lévy processes.

Exercise 3.1 Let Y_t be a Lévy process with generator L given by

$$Lf(x) = \tfrac{1}{2}(G\nabla,\nabla)f(x) + (b, \nabla f(x))$$
$$+ \int [f(x+y) - f(x) - (\nabla f(x), y)]\nu(dy). \tag{3.66}$$

(i) Show that

$$\mathbf{E}(\xi+Y_t)^2 = (\xi+tb)^2 + t\left(\operatorname{tr} G + \int y^2\nu(dy)\right) \tag{3.67}$$

for any $\xi \in \mathbf{R}^d$. Hint: use characteristic functions.

(ii) Show that if $\int_{|y|>1} |y|^k \nu(dy) < \infty$ then $\mathbf{E}|Y_t|^k = O(t)$ for any integer $k > 1$ and small t.

(iii) Show that

$$\mathbf{E}\rho_\beta^2(|Y_t|) \leq tc(t)\left(\|G\| + \int \rho_\beta^2(|y|)\nu(dy)\right).$$

Hint: Proceed as in the proof of Proposition 3.20.

Exercise 3.2 Let Y_s be a Lévy process with generator L given by

$$Lf(x) = \int [f(x+y) - f(x) - (y, \nabla f(x))]\nu(dy).$$

Then

$$\mathbf{E}|Y_t| \leq 2t \int |y|\nu(dy).$$

Hint: Use Dynkin's formula.

4

Analytical constructions

Chapter 3 was devoted to the construction of Markov processes by means of SDEs. Here we shall discuss analytical constructions. In Section 4.1 we sketch the content of the chapter, making, in passing, a comparison between these two approaches.

4.1 Comparing analytical and probabilistic tools

Sections 4.2 and 4.3 deal with the integral generators corresponding probabilistically to pure jump Markov processes. The basic series expansion (4.3), (4.4) is easily obtained analytically via the du Hamel principle, and probabilistically it can be obtained as the expansion of averages of terms corresponding to a fixed number of jumps; see Theorem 2.35. Thus for bounded generators both methods lead to the same, easily handled, explicit formula for such processes. In the less trivial situation of unbounded rates the analytical treatment given below rapidly yields the general existence result and eventually, subject to the existence of a second bound, uniqueness and non-explosion. However, if the process does explode in finite time, leading to non-uniqueness, specifying the various processes that arise (i.e. solutions to the evolution equation) requires us to fix "boundary conditions at infinity", and this is most naturally done probabilistically by specifying the behavior of a process after it reaches infinity (i.e. after explosion). We shall not develop the theory in this direction; see, however, Exercise 2.7.

In Section 4.4 we analyze generators of "order at most one", which can be described in this way because they are defined on first-order differentiable functions. We have chosen to give a straightforward analytical treatment in detail, though a probabilistic analysis based on SDEs driven by nonlinear Lévy noise (Theorem 3.12) would lead to the same results. Having a domain containing $C^1_\infty(\mathbf{R}^d)$ (and not just $C^2_\infty(\mathbf{R}^d)$, as in the general case, or $C_\infty(\mathbf{R}^d)$, as in the case of integral generators), the class of operators of "order at most one" is naturally singled out as an intermediate link between the integral and general Lévy–Khintchine generators. Its nonlinear counterpart, which we consider later, contains distinctive models including Vlasov evolution, the mollified Boltzmann equation and interacting stable-like processes with index $\alpha < 1$. Because an operator of "order at most one" is defined on $C^1_\infty(\mathbf{R}^d)$, in sufficiently regular situations the corresponding evolution can be proved to have $C^1_\infty(\mathbf{R}^d)$ as an invariant core. In this case the dual evolution on measures can be lifted to the dual Banach space $(C^1_\infty(\mathbf{R}^d))^\star$. Consequently we might expect that the continuity properties of this evolution, which are crucial for nonlinear extensions, can be expressed in terms of the topology of the space $(C^1_\infty(\mathbf{R}^d))^\star$. This is indeed the case. To visualize this space, we supply in Section 4.6 a natural representation of its elements in terms of the usual measures.

Methods and results for the construction of Markov processes are numerous and their full exposition is beyond the scope of this book. In Section 4.7 we sketch briefly various analytical techniques used for dealing with general Lévy–Khintchine-type generators with variable coefficients. First we discuss basic results for the martingale problem, which lies at the crossroads of analytical and probabilistic techniques (these are linked by Dynkin's formula); it also plays a crucial role in the extensions to unbounded coefficients discussed later. Then we give some results on decomposable generators as well as on heat kernel estimates for stable-like processes.

Comparing in general terms the analytical and probabilistic approaches to Lévy–Khintchine-type generators, we observe that in the analytical approaches some non-degeneracy should be assumed (either for the diffusion coefficients or for the Lévy measure), whereas probabilistic constructions often require more regularity.

To illustrate this point in part, let us assess the analytical meaning of the SDE-based construction of Feller semigroups given in Section 3.2. Lifting the one-step approximation $x \mapsto x + Y_\tau(x)$ to the averages yields the transformation

$$\Phi_\tau f(x) = \mathbf{E} f(x + Y_\tau(x)),$$

so that the semigroup is constructed as the limit of the approximations

$$T^n_t = \Phi_{t/n} \cdots \Phi_{t/n} \qquad (n \text{ times})$$

as $n \to \infty$. Such a limit is often the *T-product* or *chronological product* of the infinitesimal transformations Φ_τ (see [173] and also Section 6.2, (6.7)).

Analytically the approximations Φ_τ can be written as

$$(\Phi_\tau f)(x) = \left(e^{tL_x} f\right)(x),$$

where L_x denotes the generator L of the limiting semigroup T_t with coefficients fixed at point x, i.e. L is the generator of the Lévy process $Y_\tau(x)$. Thus from the analytic point of view the choice of the approximations Φ_τ corresponds to a version of the method of frozen coefficients, which is well established in the theory of partial differential equations. The lack of a non-degeneracy assumption makes it difficult to prove the convergence of this T-product analytically, but a probabilistic analysis does the job by means of the coupling method. However, the conditions of Theorem 4.23, formulated below, contain some non-degeneracy assumptions on the corresponding Lévy measure. This allows us to prove convergence by utilizing T-products in a purely analytical manner.

4.2 Integral generators: one-barrier case

This section starts with an analytical study of positive evolutions with unbounded *integral generators*. Under the general assumption that $(X, \mathcal{F})$ is a measurable space, consider the problem

$$\begin{aligned} \dot{u}_t(x) = A_t u_t(x) = \int u_t(z)\nu_t(x; dz) - a_t(x)u_t(x), \\ u_r(x) = \phi(x), \ t \geq r \geq 0, \end{aligned} \tag{4.1}$$

defining the operator A_t. Here $a_t(x)$ is a measurable non-negative function of two variables that is locally bounded in t for any x such that the integral $\xi_t(x) = \int_0^t a_s(x)ds$ is well defined and is continuous in t; $\nu_t(x, \cdot)$ is a transition kernel from $\mathbf{R}_+ \times X$ to X (i.e. a family of finite measures on X depending measurably on $t \geq 0$, $x \in X$); and ϕ is a given measurable function.

By the du Hamel principle (see e.g. Theorem 2.9), equation (4.1) is formally equivalent to the integral equation

$$u_t(x) = I_\phi^r(u)_t = e^{-[\xi_t(x)-\xi_r(x)]}\phi(x) + \int_r^t e^{-[\xi_t(x)-\xi_s(x)]} L_s u_s(x)\, ds, \tag{4.2}$$

where $L_t v(x) = \int v(z)\nu_t(x, dz)$.

An easy observation about (4.2) is the following. The iterations of the mapping I^r_ϕ from (4.2) are connected with the partial sums

$$S_m^{t,r}\phi = e^{-(\xi_t-\xi_r)}\phi + \sum_{l=1}^m \int_{r\le s_l\le\cdots\le s_1\le t} e^{-(\xi_t-\xi_{s_1})}L_{s_1}\cdots e^{-(\xi_{s_{l-1}}-\xi_{s_l})}L_{s_l} \times e^{-(\xi_{s_l}-\xi_r)}\phi\, ds_1\cdots ds_l \tag{4.3}$$

(where $e^{-\xi_t}$ designates the operator for multiplication by $e^{-\xi_t(x)}$) of the perturbation series solution

$$S^{t,r} = \lim_{m\to\infty} S_m^{t,r} \tag{4.4}$$

to (4.2) by the equations

$$(I^r_\phi)^m(u)_t = S_{m-1}^{t,r}\phi + \int_{r\le s_m\le\cdots\le s_1\le t} e^{-(\xi_t-\xi_{s_1})}L_{s_1}\cdots e^{-(\xi_{s_{m-1}}-\xi_{s_m})}L_{s_m} \times u\, ds_1\cdots ds_m \tag{4.5}$$

and

$$I^r_\phi(S_m^{\cdot,r}\phi) = S_{m+1}^{\cdot,r}\phi$$

(see again Theorem 2.9).

For a measurable positive function f on X we denote by $B_f(X)$ (resp. $B_{f,\infty}(X)$) the space of measurable functions g on X such that $g/f \in B(X)$ (resp. g/f is bounded and tends to zero as f goes to infinity) with norm $\|g\|_{B_f} = \|g/f\|$. In the case when S is a Borel space, the corresponding subspaces of continuous functions are denoted by $C_f(X)$ and $C_{f,\infty}(X)$. By $\mathcal{M}_f(X)$ we denote the set of (positive, but not necessarily finite) measures μ on $(X,\mathcal{F})$ such that $(f,\mu)<\infty$. The corresponding space of signed measures ξ is a Banach space if it is equipped with the norm $(f,|\xi|)$.

Theorem 4.1 *Suppose that $\nu_t(x,.)\in\mathcal{M}_\psi(X)$ and*

$$A_t\psi(x)\le c\psi(x), \qquad t\in[0,T], \tag{4.6}$$

for a strictly positive measurable function ψ on X and a constant $c=c(T)$. Then the following statements hold.

(i) For all $0\le r\le t\le T$,

$$(I^r_\psi)^m(\psi)_t \le \left(1+c(t-r)+\cdots+\frac{1}{m!}c^m(t-r)^m\right)\psi, \tag{4.7}$$

and consequently $S^{t,r}\psi(x)$ is well defined as a convergent series for each t, x; furthermore

$$S^{t,r}\psi(x) \le e^{c(t-r)}\psi(x). \tag{4.8}$$

(ii) For an arbitrary $\phi \in B_\psi(X)$ the perturbation series $S^{t,r}\phi = \lim_{m\to\infty} S_m^{t,r}\phi$ is absolutely convergent for all t, x and the function $S^{t,r}\phi$ solves (4.2) *and represents its minimal solution, i.e. $S^{t,r}\phi \le u$ pointwise for any other solution u to* (4.2).

(iii) The family $S^{t,r}$ forms a propagator in $B_\psi(X)$, with norm

$$\|S^{t,r}\|_{B_\psi(X)} \le e^{c(t-r)} \tag{4.9}$$

depending continuously on t in the following sense:

$$\sup_{\|\phi\|_{B_\psi(X)}\le 1} \|(S^{t,r} - S^{\tau,r})\phi \mathbf{1}_{M_k}\|_{B_\psi(X)} \to 0, \qquad t \to \tau, \tag{4.10}$$

for any k, where

$$M_k = \left\{x : \sup_{s\in[0,T]} a_s(x) \le k\right\}.$$

(iv) For any $\phi \in B_\psi(X)$ the function $S^{t,r}\phi(x)$ is differentiable in t for each x and satisfies equation (4.1) *pointwise, i.e. for any x.*

Proof (i) This is given by induction on m. Suppose that (4.7) holds for a particular m. Since (4.6) implies that

$$L_t\psi(x) \le [c + a_t(x)]\,\psi(x) = \left[c + \dot{\xi}_t(x)\right]\psi(x),$$

it follows that

$$\begin{aligned}(I_\psi^r)^{m+1}(\psi)_t \le{}& e^{-[\xi_t(x)-\xi_r(x)]}\psi(x)\\ &+ \int_r^t e^{-(\xi_t(x)-\xi_s(x))}(c + \dot{\xi}_s(x))\\ &\quad\times \left(1 + c(s-r) + \cdots + \frac{1}{m!}c^m(s-r)^m\right)\psi(x)\,ds.\end{aligned}$$

Consequently, as

$$\int_r^t e^{-(\xi_t-\xi_s)}\dot{\xi}_s \frac{1}{l!}(s-r)^l\,ds = \frac{1}{l!}(t-r)^l - \frac{1}{(l-1)!}\int_r^t e^{-(\xi_t-\xi_s)}(s-r)^{l-1}\,ds$$

for $l > 0$, it remains to show that

$$\sum_{l=1}^{m} c^l \left(\frac{1}{l!}(t-r)^l - \frac{1}{(l-1)!} \int_r^t e^{-(\xi_t-\xi_s)}(s-r)^{l-1} ds \right)$$
$$+ \sum_{l=0}^{m} c^{l+1} \frac{1}{l!} \int_r^t e^{-(\xi_t-\xi_s)}(s-r)^l ds$$
$$\leq c(t-r) + \cdots + \frac{1}{(m+1)!} c^{m+1}(t-r)^{m+1}.$$

But this inequality holds because the l.h.s. equals

$$\sum_{l=1}^{m} \frac{c^l}{l!}(t-r)^l + \frac{c^{m+1}}{m!} \int_r^t e^{-(\xi_t-\xi_s)}(s-r)^m \, ds.$$

(ii) Applying statement (i) separately to the positive and negative parts of ϕ one obtains the convergence of the series $S^{t,r}\phi$ and the estimate (4.9). From the definitions it follows that

$$I_\phi^r(S_m^{\cdot,r}\phi)_t = S_{m+1}^{t,r}\phi,$$

implying that $S^{t,r}\phi$ satisfies (4.2) and is minimal, since any solution u of this equation satisfies the equation $u_t = (I_\phi^r)^m(u)_t$ and hence (owing to (4.5)) also the inequality $u_t \geq S_{m-1}^{t,r}\phi$.

(iii) Once the convergence of the series $S^{t,r}$ is proved, the propagator (or Chapman–Kolmogorov) equation follows from the standard manipulations of integrals given in the proof of Theorem 2.7.

The continuity of $S^{t,r}$ in t follows from the formula

$$S^{t,r}\phi - S^{\tau,r}\phi = (e^{-(\xi_t-\xi_\tau)} - 1)e^{-(\xi_\tau-\xi_r)}\phi$$
$$+ \int_r^\tau (e^{-(\xi_t-\xi_\tau)} - 1)e^{-(\xi_\tau-\xi_s)} L_s S^{s,r}\phi \, ds$$
$$+ \int_\tau^t e^{-(\xi_t-\xi_s)} L_s S^{s,r}\phi \, ds \tag{4.11}$$

for $r \leq \tau \leq t$. By the propagator property, it is enough to show (4.10) for $r = \tau$. But, as follows from (4.11),

$$|(S^{t,r}\phi - \phi)| \leq (1 - e^{-(\xi_t-\xi_r)})\psi + e^{c(t-r)} \int_r^t e^{-(\xi_t-\xi_s)}(c\psi + a_s\psi) \, ds. \tag{4.12}$$

Consequently,

$$|(S^{t,r}\phi - \phi)\mathbf{1}_{M_k}| \leq e^{c(t-r)}(2k + c)(t-r)\psi, \tag{4.13}$$

implying (4.10) for $r = \tau$.

(iv) The differentiability of $S^{t,r}\phi(x)$ for each x follows from (4.11). Differentiating equation (4.2), one sees directly that $S^{t,r}\phi$ satisfies (4.1) and that all the required formulas hold pointwise. □

The propagator $S^{t,r}$ is called the *minimal propagator* associated with the family A_t. It can be used to define the so-called *minimal jump sub-Markov process* specified by the generator A_t. However, for this and other applications to time-nonhomogeneous stochastic processes one usually needs equation (4.1) in inverse time, i.e. the problem

$$\dot{u}_t(x) = -A_t u_t(x) = -\int u_t(z)\nu_t(x; dz) + a_t(x)u_t(x),$$
$$u_r(x) = \phi(x),\ 0 \le t \le r, \quad (4.14)$$

for which the corresponding integral equation takes the form

$$u_t(x) = I^r_\phi(u)_t = e^{\xi_t(x)-\xi_r(x)}\phi(x) + \int_t^r e^{\xi_t(x)-\xi_s(x)} L_s u_s(x)\, ds. \quad (4.15)$$

The statements of Theorem 4.1 (and their proofs) obviously hold for the perturbation series $S^{t,r}$ constructed from (4.15), with the same estimate (4.9), but which now of course form a backward propagator, called the *minimal backward propagator* associated with A_t. Let us denote by $V^{t,r} = (S^{r,t})^\star$ the dual propagator on the space $\mathcal{M}_\psi(X)$. By the above explicit construction and Fubbini's theorem, the inverse propagator $S^{r,t}$ consists of integral operators specified by a family of transition kernels, implying that the operators $V^{t,r}$ actually act on measures (and not just on the dual space to $B_\psi(X)$). In particular, if $\psi = 1$ and $c = 0$ in Theorem 4.1, the minimal backward propagator $S^{r,t}$ is sub-Markov and specifies a sub-Markov process on X. Remarkably, the following holds in full generality.

Proposition 4.2 *Under the assumptions of Theorem 4.1 let $S^{r,t}$ denote the corresponding minimal backward propagator. Then the dual propagator $V^{t,r}$ is strongly continuous in the norm topology of $\mathcal{M}_\psi(X)$.*

Proof Since $V^{t,r}$ is a propagator, it is sufficient to show that $V^{t,r}\mu - \mu$ tends to zero as $t \to r$, $t \ge r$, in the norm topology of $\mathcal{M}_\psi(X)$. Let $0 \le f \le \psi$. One has

$$(f,\ V^{t,r}\mu - \mu) = (S^{r,t}f - f,\ \mu).$$

For any $\epsilon > 0$ there exists a k such that $\int_{\bar{M}_k} \psi(x)\mu(dx) < \epsilon$, where $\bar{M}_k = X \setminus M_k$. Consequently, by (4.12) and (4.13),

$$\sup_{0\le f\le\psi} |(f,\ V^{t,r}\mu - \mu)| \le e^{c(t-r)}[(2\epsilon + (2k+c)(t-r)(\psi,\mu)],$$

which can be made arbitrary small for small enough $t - r$. □

Remark 4.3 Strong continuity of the dual propagator is a strong property. The operator $e^{t\Delta}$ on $\mathcal{P}(\mathbf{R}^d)$ does not possess this property.

Corollary 4.4 *Under the assumptions of Proposition 4.2, $\mu_t = V^{t,r}\mu$ represents the minimal solution of the equation*

$$\mu_t = e^{-(\xi_t-\xi_r)}\mu + \int_r^t e^{-(\xi_t-\xi_s)} L'_s \mu_s \, ds, \tag{4.16}$$

where $L'_s\mu(.) = \int \mu(dx)\nu_s(x,.)$.

Proof The series expansion for μ_t solving (4.16) is dual to the series defining the backward propagator $S^{r,t}$. □

Equation (4.16) is sometimes called the *mild form* of the equation

$$\frac{d}{dt}\mu_t = -a_t\mu_t + L'_t\mu_t, \tag{4.17}$$

which is dual to (4.14). It is of course natural to ask when equation (4.17) itself is satisfied. The condition to be introduced below is reminiscent of the property of infinity of not being an entrance boundary point to the process corresponding to (4.17) (see [130]), as it ensures that the unbounded part of the kernel $\nu_s(y,dz)$ is directed towards infinity and not within the domains M_k.

Proposition 4.5 *Under the assumptions of Proposition 4.2, assume additionally that for any k*

$$\sup_{t\in[0,T]} \|\nu_t(., M_k)\|_{B_\psi(X)} < \infty. \tag{4.18}$$

Then the minimal solutions μ_t to (4.16) *satisfy equation* (4.17) *in the sense of "convergence in M_k", i.e. for any k we have*

$$\frac{d}{dt}\mathbf{1}_{M_k}\mu_t = \mathbf{1}_{M_k}(-a_t\mu_t + L'_t\mu_t), \tag{4.19}$$

where the derivative on the l.h.s. is defined in the sense of the Banach topology of $\mathcal{M}_\psi(X)$.

Proof Let us show (4.19) for $r = t = 0$, which is sufficient by the propagator property of μ_t. To this end one observes that, for any $f \in B_\psi(X)$, equation (4.16) implies that

$$\begin{aligned}(f\mathbf{1}_{M_k}, \mu_t - \mu) &= ((e^{-\xi_t} - 1)f\mathbf{1}_{M_k}, \mu)\\ &\quad + \int_{y\in M_k}\int_X \mu(dz)\int_0^t e^{-(\xi_t-\xi_s)(y)}\nu(z,dy)f(y)\,ds\\ &\quad + \int_{y\in M_k}\int_X\int_0^t e^{-(\xi_t-\xi_s)(y)}\nu(z,dy)(\mu_s-\mu)(dz)f(y)\,ds,\end{aligned}$$

yielding (4.19) for $t = 0$, since the last term is of order $o(t)$ as $t \to 0$ by the norm continuity of μ_t and assumption (4.18). □

In the time-nonhomogeneous case we can now fully describe the analytic properties of the corresponding semigroup on measures.

Proposition 4.6 *Under the assumptions of Proposition 4.5 assume additionally that neither ν_t nor a_t depends explicitly on t. Then the domain $D_{A'}$ of the generator A' of the semigroup $V^{t,0}$ (which is strongly continuous in $\mathcal{M}_\psi(X)$) consists precisely of those $\mu \in \mathcal{M}_\psi(X)$ for which the measure $A'\mu = -a\mu + L'\mu$ (which is σ-finite for any $\mu \in \mathcal{M}_\psi(X)$, as it is finite on each M_k by (4.18)) belongs to $\mathcal{M}_\psi(X)$. In particular, for any $\mu \in D_{A'}$ the whole curve $\mu_t = V^{t,0}\mu$ belongs to $D_{A'}$ and constitutes the unique solution in $D_{A'}$ of equation* (4.17) *(the derivative being defined with respect to the Banach topology of $\mathcal{M}_\psi(X)$).*

Proof The first statement is easily deduced from (4.19) and the definition of the generator. It implies the second statement by duality (see Theorem 2.10, applied to V rather than U). □

Finally let us give a criterion for the minimal propagator to preserve the set of continuous functions.

Proposition 4.7 *Under the assumptions of Theorem 4.1 assume that X is a metric space equipped with its Borel sigma algebra, $a_t(x)$ is a continuous function of two variables and that $\nu_t(x,.)$ depends continuously on x in the norm topology of $\mathcal{M}_\psi(X)$. Then the propagator $S^{t,r}$ from Theorem 4.1 preserves the set of continuous functions. In particular, if $\psi = 1$, $c = 0$ and all objects involved are time independent then the propagator $S^{t,r}$ specifies a C-Feller semigroup.*

Proof If $u_t(x) = S^{t,r}\phi(x)$ then

$$\begin{aligned} u_t(x) - u_t(y) &= e^{-[\xi_t(x)-\xi_r(x)]}\phi(x) - e^{-[\xi_t(y)-\xi_r(y)]}\phi(y) \\ &\quad + \int_r^t \int_X \Big(e^{-[\xi_t(x)-\xi_s(x)]}\nu_s(x,dz) \\ &\qquad - e^{-[\xi_t(y)-\xi_s(y)]}\nu_s(y,dz)\Big)\, u_s(z)\, ds. \end{aligned} \tag{4.20}$$

□

Exercise 4.1 Suppose that $A_t\psi \le c\psi + \phi$ for positive functions ϕ and ψ and all $t \in [0,T]$, where A_t is from (4.1). Then

$$S^{t,r}\psi \le e^{c(t-r)}\left(\psi + \int_r^t S^{t,\tau}\, d\tau\phi\right).$$

Hint: using (4.7) yields

$$I^r_\psi(\psi)_t \le [1 + c(t-r)]\,\psi + \int_r^t e^{-(\xi_t-\xi_s)}\phi\,ds,$$

$$(I^r_\psi)^2(\psi)_t \le \left(1 + c(t-r) + \frac{c^2}{2}(t-r)^2\right)\psi + \int_r^t e^{-(\xi_t-\xi_s)}\,[1 + c(s-r)]\,\phi\,ds$$
$$+ \int_r^t e^{-(\xi_t-\xi_s)} L_s \int_r^s e^{-(\xi_s-\xi_\tau)}\phi\,d\tau\,ds,$$

etc. and hence

$$(I^r_\psi)^m(\psi)_t \le e^{c(t-r)}\left(\psi + \int_r^t e^{-(\xi_t-\xi_s)}\phi\,ds\right.$$
$$\left. + \int_r^t e^{-(\xi_t-\xi_s)} L_s \int_r^s e^{-(\xi_s-\xi_\tau)}\phi\,d\tau\,ds + \cdots\right)$$
$$= e^{c(t-r)}\left[\psi + \int_r^t d\tau \left(e^{-(\xi_t-\xi_\tau)} + \int_\tau^t e^{-(\xi_t-\xi_s)} L_s e^{-(\xi_s-\xi_\tau)}\,ds + \cdots\right)\phi\right],$$

and the proof is completed by noting that

$$S^{t,r}\psi = \lim_{m\to\infty} S^{t,r}_{m-1}\psi \le \lim_{m\to\infty}(I^r_\psi)^m(\psi)_t.$$

4.3 Integral generators: two-barrier case

We shall develop the ideas of this section further in the general discussion of unbounded coefficients given in Chapter 5.

To obtain the strong continuity of the backward propagator $S^{t,r}$ itself (not just of its dual, as in Proposition 4.2 above), the existence of a second bound for A_t (see (4.1)) can be helpful.

Theorem 4.8 *Suppose that two functions ψ_1, ψ_2 on X are given, both satisfying* (4.6) *and such that $0 < \psi_1 \le \psi_2$, $\psi_1 \in B_{\psi_2,\infty}(X)$ and a_t is bounded on any set where ψ_2 is bounded. Then $S^{t,r}$, $t \le r$ (constructed above from* (4.14), (4.15)*) is a strongly continuous backward propagator in $B_{\psi_2,\infty}(X)$.*

Proof By Theorem 4.1, the $S^{t,r}$ are bounded in $B_{\psi_2}(X)$. Moreover, as $S^{t,r}\phi$ tends to ϕ as $t \to r$ uniformly on the sets where ψ_2 is bounded, it follows that

$$\|S^{t,r}\phi - \phi\|_{B_{\psi_2}(X)} \to 0, \qquad t \to r,$$

for any $\phi \in B_{\psi_1}(X)$, and hence also for any $\phi \in B_{\psi_2,\infty}(X)$ since $B_{\psi_1}(X)$ is dense in $B_{\psi_2,\infty}(X)$. □

As the convergence of the above perturbation series for $S^{t,r}$ is point-wise, it does not provide a mechanism for preserving continuity. An exercise at the end of this section settles this matter under strong continuity assumptions on ν.

From now on we will assume that X is a locally compact metric space equipped with its Borel sigma algebra. We shall discuss the conservativity, uniqueness and smoothness of the propagators constructed.

Theorem 4.9 *Under the assumptions of Theorem 4.8, assume that X is a locally compact metric space equipped with its Borel sigma algebra, the functions ψ_1, ψ_2 are continuous and such that ψ_2 and ψ_2/ψ_1 tend to infinity as x goes to infinity, a_t is a continuous mapping from t to $C_{\psi_2/\psi_1,\infty}$ and L_t is a continuous mapping from t to bounded operators that take C_{ψ_1} to $C_{\psi_2,\infty}$.*

Then B_{ψ_1} is an invariant core for the propagator $S^{t,r}$ in the sense of the definition given before Theorem 2.9, i.e.

$$
\begin{aligned}
A_r\phi &= \lim_{t\to r, t\le r} \frac{S^{t,r}\phi - \phi}{r-t} = \lim_{s\to r, s\ge r} \frac{S^{r,s}\phi - \phi}{s-r}, \\
\frac{d}{ds} S^{t,s}\phi &= S^{t,s}A_s\phi, \quad \frac{d}{ds} S^{s,r}\phi = -A_s S^{s,r}\phi, \qquad t < s < r,
\end{aligned}
\tag{4.21}
$$

for all $\phi \in B_{\psi_1}(X)$; all these limits exist in the Banach topology of $B_{\psi_2,\infty}(X)$. Moreover B_{ψ_1} and $B_{\psi_2,\infty}$ are invariant under $S^{t,r}$, so that C_{ψ_1} is an invariant core of the strongly continuous propagator $S^{t,r}$ in $C_{\psi_2,\infty}$. In particular, if a_t, L_t do not depend on t then A generates a strongly continuous semigroup on $C_{\psi_2,\infty}$ for which C_{ψ_1} is an invariant core. Finally, if $\psi_1 = 1$, $A\psi_1 = 0$ then the propagator $S^{t,r}$ preserves constants and specifies a Markov process on X.

Proof By Theorem 4.1, $S^{t,r}\phi$ satisfies equations (4.14) pointwise. To show that these equations hold in the topology of $B_{\psi_2,\infty}$ one needs to show that the operators $A_t(\phi)$ are continuous as functions from t to $B_{\psi_2,\infty}$ for each $\phi \in B_{\psi_1}$. But this follows directly from our continuity assumptions on a_t and L_t. To show that the space C_{ψ_1} is invariant (and this implies the remaining statements), we shall approximate $S^{t,r}$ by evolutions with bounded intensities. Let χ_n be a measurable function $X \to [0,1]$ such that $\chi_n(x) = 1$ for $\psi_2(x) \le n$ and $\chi_n(x) = 0$ for $\psi_2(x) > n+1$. Write $\nu_t^n(x,dz) = \chi_n(x)\nu_t(x,dz)$ and $a_t^n = \chi_n a_t$, and let $S_n^{t,r}$ and A_t^n denote the propagators constructed as above, but with ν_t^n and a_t^n instead of ν_t and a_t. Then the propagators $S_n^{t,r}$ converge strongly in the Banach space $B_{\psi_2,\infty}$ to the propagator $S^{t,r}$. In fact, as $S^{t,r}$ and $S_n^{t,r}$ are uniformly bounded, it is enough to show the convergence for the elements ϕ of the invariant core B_{ψ_1}. For such a ϕ one has

$$(S^{t,r} - S_n^{t,r})(\phi) = \int_t^r \frac{d}{ds} S^{t,s} S_n^{s,r} \phi \, ds = \int_t^r S^{t,s}(A_s - A_s^n) S_n^{s,r} \phi \, ds. \quad (4.22)$$

By invariance $S_n^{s,r}\phi \in B_{\psi_1}$, implying that $(A_s - A_s^n)S_n^{s,r}\phi \in B_{\psi_2}$ and tends to zero in the norm of B_{ψ_2} as $n \to \infty$; hence the r.h.s. of (4.22) tends to zero in B_{ψ_2} as $n \to \infty$.

To complete the proof it remains to observe that, as the generators of $S_n^{t,r}$ are bounded, the corresponding semigroups preserve continuity (since they can be constructed as a convergent exponential series). Hence $S^{t,r}$ preserves continuity as well, since $S^{t,r}\phi$ is a (uniform) limit of continuous functions.

If $A\mathbf{1} = 0$, the constant functions solve equation (4.14). By uniqueness (which can be deduced from either Theorem 2.10 or Theorem 2.37), it follows that $S^{t,r}\mathbf{1} = \mathbf{1}$. □

We can also encounter situations when the a_t have a higher growth rate than ψ_2. To get uniqueness in this case, one has to introduce the martingale problem. For simplicity, we shall discuss this extension for locally compact X and time-homogeneous kernels only.

Theorem 4.10 *Under the assumptions of Theorem 4.8, let X be locally compact, $\psi_1 = 1$, $A\psi_1 = 0$, a_t and L_t be independent of t and $\psi_2(x) \to \infty$ as $x \to \infty$. Then the martingale problem for the operator A is well posed in $C_c(X)$ and its solution defines a (pure jump) Markov process in X such that*

$$\mathbf{P}\left(\sup_{0 \le s \le t} \psi_2(X_s^x) > r\right) \le \frac{c(t, \psi_2(x))}{r}. \quad (4.23)$$

Proof The proof is the same as that for the more general case, given in Theorem 5.2 below (it uses ψ_2 instead of f_L and the approximation $A_n(x) = \chi_q(a(x)/n)A(x))$ and hence is omitted. □

Finally, we shall discuss additional regularity for the dual to the evolution on measures (4.14). To this end we shall introduce the notion of the dual transition kernel. Recall that by a *transition kernel* in X one usually means the family $\nu(x, .)$ of measures from $\mathcal{M}(X)$, depending weakly measurably on $x \in X$, such that $\int f(y)\nu(x, dy)$ is measurable for any bounded measurable f. Let X be $\mathbf{R}^d$ or an open subset of $\mathbf{R}^d$. A transition kernel $\nu'(x, dy)$ will be called *dual* to ν if the measures $dx\, \nu(x, dy)$ and $\nu'(y, dx)\, dy$ coincide as Borel measures in $X \times X$. Clearly, if a dual exists then it is uniquely defined up to a natural equivalence. Though in the usual examples (such as the Boltzmann equation) with which we shall deal the dual kernels are given explicitly, let us state for completeness the basic existence result.

Proposition 4.11 *For a transition kernel $\nu(x,.)$ in X (where X is $\mathbf{R}^d$ or its open subset) a dual exists if and only if the projection of the measure $dx\,\nu(x,dy) \in \mathcal{M}(X \times X)$ onto the second variable is absolutely continuous with respect to Lebesgue measure.*

Proof This is a direct consequence of the well-known measure-theoretic result on the disintegration of measures (see e.g. [86]), stating that if ν is a probability measure on $X \times X$ with a projection onto (i.e. a marginal of) the first variable μ, then there exists a measurable mapping $x \mapsto \nu(x,.)$ from X to $\mathcal{P}(X)$ uniquely determined μ-a.s. such that

$$\int_{X\times X} f(x,y)\nu(dxdy) = \int_X \left(\int_X f(x,y)\nu(x,dy)\right)\mu(dx), \quad f \in C(X\times X).$$

□

In the next proposition the continuity of the transitional kernel implies weak continuity in $\mathcal{M}_{\psi_2}(X)$, i.e. that $\int f(y)\nu(x,dy)$ is continuous (though it may be unbounded) for any $f \in C_{\psi_2}(X)$.

Proposition 4.12 *Under the assumptions of Theorem 4.9 assume that X is either $\mathbf{R}^d$ or an open subset of $\mathbf{R}^d$ and that there exists a dual continuous transition kernel $\nu'_t(x,.)$ to $\nu_t(x,.)$ satisfying the same conditions as ν. Then the (forward) propagator $V^{t,s}$ on measures which is dual to the backward evolution specified by* (4.14) *preserves the space of absolutely continuous measures. When reduced to densities it is bounded both in $C_{\psi_1}(X)$ and $C_{\psi_2}(X)$ and, for any $g \in C_{\psi_1}$, yields the unique solution $g_s = V^{s,t}g$ in C_{ψ_1} to the Cauchy problem*

$$\frac{d}{ds}g_s = A'_s g_s, \qquad g_t = g, \tag{4.24}$$

for $s \geq t$. Here

$$A'_t g(x) = \int g(y)\nu'_t(x,dy) - a_t(x)g(x),$$

and the derivative in s in (4.24) *is defined pointwise and uniformly for x from any subset where ψ_2 is bounded.*

Proof This follows directly from Theorems 4.9 and 2.10. □

4.4 Generators of order at most one: well-posedness

This section is devoted to a purely analytical construction of processes generated by integro-differential (or pseudo-differential) operators *of order at most one*, i.e. the operators given by

$$Lf(x) = (b(x), \nabla f(x)) + \int_{\mathbf{R}^d\setminus\{0\}} \left[f(x+y) - f(x)\right] \nu(x, dy) \tag{4.25}$$

with Lévy measures $\nu(x, .)$ having finite first moment $\int_{B_1} |y|\nu(x, dy)$.

Theorem 4.13 *Assume that $b \in C^1(\mathbf{R}^d)$ and that $\nabla\nu(x, dy)$, the gradient with respect to x, exists in the weak sense as a signed measure and depends weakly continuously on x. Moreover, assume that*

$$\sup_x \int \min(1, |y|)\nu(x, dy) < \infty, \qquad \sup_x \int \min(1, |y|)|\nabla\nu(x, dy)| < \infty \tag{4.26}$$

and that for any $\epsilon > 0$ there exists a $K > 0$ such that

$$\sup_x \int_{\mathbf{R}^d\setminus B_K} \nu(x, dy) < \epsilon, \qquad \sup_x \int_{\mathbf{R}^d\setminus B_K} |\nabla\nu(x, dy)| < \epsilon, \tag{4.27}$$

$$\sup_x \int_{B_{1/K}} |y|\nu(x, dy) < \epsilon. \tag{4.28}$$

Then L generates a conservative Feller semigroup T_t in $C_\infty(\mathbf{R}^d)$ with invariant core $C^1_\infty(\mathbf{R}^d)$. Moreover, T_t reduced to $C^1_\infty(\mathbf{R}^d)$ is also a strongly continuous semigroup in the Banach space $C^1_\infty(\mathbf{R}^d)$.

Proof Notice first that (4.26) implies that, for any $\epsilon > 0$,

$$\sup_x \int_{\mathbf{R}^d\setminus B_\epsilon} \nu(x, dy) < \infty, \qquad \sup_x \int_{\mathbf{R}^d\setminus B_\epsilon} |\nabla\nu(x, dy)| < \infty. \tag{4.29}$$

Next, since the operator given by

$$\int_{\mathbf{R}^d\setminus B_1} \left[f(x+y) - f(x)\right] \nu(x, dy) \tag{4.30}$$

is bounded in the Banach spaces $C(\mathbf{R}^d)$ and $C^1(\mathbf{R}^d)$ (owing to (4.26)) and also in the Banach spaces $C_\infty(\mathbf{R}^d)$ and $C^1_\infty(\mathbf{R}^d)$ (owing to (4.27)), by the standard perturbation argument (see e.g. Theorem 2.7) we can reduce the situation to the case when all the $\nu(x, dy)$ have support in B_1, which we shall assume from now on.

Let us introduce the approximation

$$L_\epsilon f(x) = (b(x), \nabla f(x)) + \int_{\mathbf{R}^d\setminus B_\epsilon} \left[f(x+y) - f(x)\right] \nu(x, dy). \tag{4.31}$$

For any $\epsilon > 0$ the operator L_ϵ generates a conservative Feller semigroup T_t^ϵ in $C_\infty(\mathbf{R}^d)$, with invariant cores $C^1_\infty(\mathbf{R}^d)$, because the first term in (4.31) does this and the second term corresponds to a bounded operator in the Banach spaces $C_\infty(\mathbf{R}^d)$ and $C^1_\infty(\mathbf{R}^d)$ (owing to (4.29)), so that perturbation theory

(Theorem 2.7) applies; conservativity also follows from the perturbation series representation. Differentiating the equation $\dot{f}(x) = L_\epsilon f(x)$ with respect to x yields

$$\frac{d}{dt}\nabla_k f(x) = L_\epsilon \nabla_k f(x) + (\nabla_k b(x), \nabla f(x)) + \int_{B_1 \setminus B_\epsilon} \big[f(x+y) - f(x)\big] \nabla_k \nu(x, dy). \tag{4.32}$$

Considering (4.32) as an evolution equation for $g = \nabla f$ in the Banach space $C_\infty(\mathbf{R}^d \times \{1, \ldots, d\}) = C_\infty(\mathbf{R}^d) \times \cdots \times C_\infty(\mathbf{R}^d)$, observe that the r.h.s. is represented as the sum of a diagonal operator that generates a Feller semigroup and of two bounded (uniformly in ϵ by (4.26)) operators. Hence this evolution is well posed. To show that the derivative of $f(x)$ is given by the semigroup generated by (4.32), we would first approximate b, ν by a sequence of the twice continuously differentiable objects b_n, ν_n, for $n \to \infty$. The corresponding approximating generator of type (4.32) has an invariant core $C^1_\infty(\mathbf{R}^d)$ and hence the uniqueness of the solutions to the corresponding evolution equation would hold (by Theorem 2.10), implying that this solution coincides with the derivative of the corresponding $(T^\epsilon_t)_n f$. Passing to the limit $n \to \infty$ would then complete the argument.

Hence $\nabla_k T^\epsilon_t f$ is uniformly bounded for all $\epsilon \in (0, 1]$ and for t from any compact interval, whenever $f \in C^1_\infty(\mathbf{R}^d)$. Therefore, writing

$$(T^{\epsilon_1}_t - T^{\epsilon_2}_t) f = \int_0^t T^{\epsilon_2}_{t-s}(L_{\epsilon_1} - L_{\epsilon_2}) T^{\epsilon_1}_s \, ds$$

for arbitrary $\epsilon_1 > \epsilon_2$ and making the estimate

$$|(L_{\epsilon_1} - L_{\epsilon_2}) T^{\epsilon_1}_s f(x)| \le \int_{B_{\epsilon_1} \setminus B_{\epsilon_2}} |(T^{\epsilon_1}_s f)(x+y) - (T^{\epsilon_1}_s f)(x)| \nu(x, dy)$$
$$\le \int_{B_{\epsilon_1}} \|\nabla T^{\epsilon_1}_s f\| |y| \nu(x, dy) = o(1) \|f\|_{C^1_\infty}, \qquad \epsilon_1 \to 0,$$

from (4.28) yields

$$\|(T^{\epsilon_1}_t - T^{\epsilon_2}_t) f\| = o(1) t \|f\|_{C^1_\infty}, \qquad \epsilon_1 \to 0. \tag{4.33}$$

Therefore the family $T^\epsilon_t f$ converges to a family $T_t f$ as $\epsilon \to 0$. Clearly the limiting family T_t specifies a strongly continuous semigroup in $C_\infty(\mathbf{R}^d)$. Writing

$$\frac{T_t f - f}{t} = \frac{T_t f - T^\epsilon_t f}{t} + \frac{T^\epsilon_t f - f}{t}$$

and noting that by (4.33) the first term is $o(1)\|f\|_{C^1_\infty}$ as $\epsilon \to 0$ allows us to conclude that $C^1_\infty(\mathbf{R}^d)$ belongs to the domain of the generator of the semigroup T_t in $C_\infty(\mathbf{R}^d)$ and that it is given on that domain by (4.25).

Applying to T_t the same procedure as was applied above to T_t^ϵ (i.e. differentiating the evolution equation with respect to x) shows that T_t also defines a strongly continuous semigroup in $C^1_\infty(\mathbf{R}^d)$, as its generator differs from the diagonal operator with all entries on the diagonal equal to L only by a bounded additive term. □

The conditions of the above result were designed to obtain a Feller semigroup and a Feller process. However, a Markov process with a C-Feller semigroup can be constructed under weaker assumptions.

Theorem 4.14 *Assume that the conditions of Theorem 4.13 hold apart from* (4.27) *and* (4.28). *Then there exists a unique Markov process in* $\mathbf{R}^d$ *whose Markov semigroup reduces to a conservative C-Feller semigroup T_t in $C(\mathbf{R}^d)$ of the form*

$$T_t f(x) = \int f(y) p_t(x, dy),$$

with probability transition kernels p_t, and such that

(i) $T_t f(x) \to f(x)$ *as* $t \to 0$ *uniformly on* x *from compact subsets,*
(ii) the space $C^1(\mathbf{R}^d)$ *is invariant under* T_t *and*
(iii) $\dot{T}_t f(x) = L f(x)$ *for any* $f \in C^1(\mathbf{R}^d)$, $t \geq 0$ *and* $x \in R^d$.

Proof The statement of the theorem clearly holds for approximating operators L_ϵ given by (4.31); this follows from the explicit form of the perturbation series representation for the corresponding semigroup (the integral part of L_ϵ is considered as a perturbation, all terms of the perturbation series are integral operators and hence so is the limit T_t^ϵ). The existence of the Markov process then follows from the standard construction of Markov processes. Passing to the limit as in the previous theorem (with all uniform limits substituted by a limit that is uniform on compact sets) yields the required properties for T_t. Uniqueness follows by the same duality arguments as in Theorem 2.10. □

4.5 Generators of order at most one: regularity

In Theorem 4.14 strong continuity, which is a very convenient property for analysis, was lost. In such situations a natural space in which strong continuity holds is the space $BUC(X)$ of bounded uniformly continuous functions on X.

Proposition 4.15 *The space $BUC(\mathbf{R}^d)$ is a closed subspace in $C(\mathbf{R}^d)$, and $C^\infty(\mathbf{R}^d)$ is dense in $BUC(\mathbf{R}^d)$.*

Proof The first statement is obvious. For the second, approximate $f \in BUC(\mathbf{R}^d)$ by the usual convolution

$$f_n(x) = \int f(y)\phi_n(x-y)dy = \int f(x-y)\phi_n(y)dy, \qquad n \to \infty,$$

where $\phi_n(x) = n\phi(nx)$ and ϕ is a non-negative infinitely differentiable function with a compact support and with $\int \phi(y)dy = 1$; the uniform continuity of f ensures that $f_n \to f$ as $n \to \infty$ in $C(\mathbf{R}^d)$. □

It is clear that under the conditions of Theorem 4.14 the semigroup T_t is strongly continuous in $BUC(\mathbf{R}^d)$, because the space $C^1(\mathbf{R}^d)$ is invariant under T_t and dense in $BUC(\mathbf{R}^d)$ and T_t is strongly continuous in $C^1(\mathbf{R}^d)$ (in the topology of $BUC(\mathbf{R}^d)$). However, as Lf need not belong to $BUC(\mathbf{R}^d)$ for $f \in C^1(\mathbf{R}^d)$, we cannot state that $C^1(\mathbf{R}^d)$ is an invariant domain. Further regularity is needed to obtain an easy-to-handle invariant domain. For instance, the following holds.

Theorem 4.16 *Assume that $b \in C^2(\mathbf{R}^d)$, $\nu(x, dy)$ is twice weakly continuously differentiable in x and*

$$\sup_{x,i,j} \int \min(1, |y|) \left(\nu(x, dy) + \left| \frac{\partial}{\partial x^i} \nu(x, dy) \right| + \left| \frac{\partial^2}{\partial x^i \partial x^j} \nu(x, dy) \right| \right) < \infty. \tag{4.34}$$

Then L of the form (4.25) generates a strongly continuous semigroup in $BUC(\mathbf{R}^d)$ with invariant core $C^2(\mathbf{R}^d)$.

Proof The invariance of $C^2(\mathbf{R}^d)$ is proved in the same way as the invariance of $C^1(\mathbf{R}^d)$ in Theorem 4.13. Moreover, $Lf \in C^1(\mathbf{R}^d) \subset BUC(\mathbf{R}^d)$ for any $f \in C^2(\mathbf{R}^d)$. Finally, $C^2(\mathbf{R}^d)$ is dense in $BUC(\mathbf{R}^d)$ by Proposition 4.15. □

For applications to nonlinear semigroups a nonhomogeneous extension of Theorems 4.13 or 4.14 is needed. As a proof can be obtained by a straightforward extension of the above, it is omitted here.

Theorem 4.17 *Assume that the family of operators L_t has the form given by (4.25), b and ν depending continuously on time t (ν weakly) and satisfying the conditions of the previous theorem as functions of x with all estimates being uniform on compact time intervals. Then the corresponding family of operators L_t generates a strongly continuous backward propagator $U^{t,s}$ of* *linear contractions in $C_\infty(\mathbf{R}^d)$ with invariant domain $C^1_\infty(\mathbf{R}^d)$ (in the sense of the definition given before Theorem 2.9, so that equations* (2.3) *are satisfied by L_t instead of A_t). Moreover $U^{t,s}$ is also a bounded strongly continuous propagator in $C^1_\infty(\mathbf{R}^d)$.*

Suppose that the Levy measures ν in (4.25) have densities, i.e. $\nu(x, dy) = \nu(x, y)dy$; then the operator dual (in the sense of the natural duality between measures and functions) to the integral part of L is clearly given, on functions g, by

$$\int \big[g(x-y)\nu(x-y, y) - g(x)\nu(x, y)\big]\, dy.$$

In particular, if

$$\sup_x \int \big[\nu(x-y, y) - \nu(x, y)\big]\, dy < \infty, \tag{4.35}$$

the dual to L can be written in the form

$$L^\star g(x) = -(b(x), \nabla g(x)) + \int_{\mathbf{R}^d} \big[g(x-y) - g(x)\big]\, \nu(x, y)\, dy \\ + \int g(x-y)\big[\nu(x-y, y) - \nu(x, y)\big]\, dy - \nabla \cdot b(x) g(x). \tag{4.36}$$

We shall now obtain a regularity result for both the initial and the dual problems.

Theorem 4.18 *Let $k \in \mathbf{N}$, $k \geq 2$. Suppose that $\nu_t(x, dy) = \nu_t(x, y)\, dy$, that $b_t(.) \in C^k(\mathbf{R}^d)$ uniformly in t, that ν_t, b_t depend continuously on t and that $\nu_t(x, y)$ is k times continuously differentiable in x and satisfies*

$$\int \min(1, |y|) \sup_{t \leq T, x \in \mathbf{R}^d} \left(\nu_t(x, y) + \left| \frac{\partial \nu_t}{\partial x}(x, y) \right| + \cdots \right. \\ \left. + \left| \frac{\partial^k \nu_t}{\partial x^k}(x, y) \right| \right) dy < \infty. \tag{4.37}$$

Then the corresponding family of operators L_t generates a strongly continuous backward propagator $U^{t,s}$ of linear contractions in $C_\infty(\mathbf{R}^d)$ with the following properties.

(i) Each space $C^l_\infty(\mathbf{R}^d)$, $l = 1, \ldots, k$, is invariant and $U^{t,s}$ is strongly continuous in each of these Banach spaces.

(ii) Its dual propagator $V^{s,t}$ on $\mathcal{M}(X)$ preserves absolutely continuous measures and, when reduced to densities, represents a strongly continuous propagator both in $L^1(\mathbf{R}^d)$ and $C_\infty(\mathbf{R}^d)$.

(iii) The spaces $C^l_\infty(\mathbf{R}^d)$ and the Sobolev spaces $W^l_1(\mathbf{R}^d)$ (up to lth order, derivatives defined in the sense of distributions are integrable functions), $l = 0, \ldots, k-1$, are invariant under $V^{s,t}$ and the family $V^{s,t}$ forms a bounded strongly continuous propagator in each Banach space $W^l_1(\mathbf{R}^d)$. The dual propagator $V^{s,t}$ extends to the $\star$-weakly continuous propagator in $(C^l_\infty(\mathbf{R}^d))^\star$, $l = 1, \ldots, k$, which for any $l < k$ specifies the unique solution to the Cauchy problem for the equation $\dot{m}u_t = L^\star \mu_t$ understood weakly in $(C^{l+1}_\infty(\mathbf{R}^d))^\star$, i.e. in the sense that

$$\frac{d}{dt}(g, \mu_t) = (L_t, \mu_t) \qquad \text{for all } g \in C^{l+1}_\infty(\mathbf{R}^d).$$

Proof (i) First observe that (4.37) trivially implies (4.26)–(4.28). Differentiating $\dot{f} = L_t f$ shows that the evolution of the derivatives is governed by the same generator as that given by (4.25), up to a bounded additive term, as long as (4.37) holds.

(ii) and (iii) Notice that the assumption (4.37) ensures that $L^\star$, given by (4.36), has the same form as L up to an additive term which is bounded in $W^1_l(\mathbf{R}^d)$, $l = 1, \ldots, k-1$. Hence we can apply Theorem 4.17 to $L^\star$. Results for Sobolev spaces are obtained in the same way as for the spaces $C^l_\infty(\mathbf{R}^d)$ (this is possible because the derivatives satisfy the same equation as the function itself up to an additive bounded term). Note only that to get strong continuity in some integral norm, say in $L_1(\mathbf{R}^d)$, one observes first that such a norm is strongly continuous when reduced to $L_1(\mathbf{R}^d) \cap C^1_\infty(\mathbf{R}^d)$ and then (by the density argument) in the whole space $L_1(\mathbf{R}^d)$.

(iv) One has

$$\frac{d}{dt}(g, \mu_t) = \frac{d}{dt}(U^{0,t} g, \mu) = (U^{0,t} L_t g, \mu) = (L_t g, \mu_t).$$

Uniqueness follows from Theorem 2.10 with $B = C^l_\infty(\mathbf{R}^d)$, $D = C^{l+1}_\infty(\mathbf{R}^d)$. □

Exercise 4.2 State and prove a version of Theorem 4.17 for propagators in $C(\mathbf{R}^d)$ (i.e. a nonhomogeneous version of Theorem 4.14).

4.6 The spaces $(C^l_\infty(\mathbf{R}^d))^\star$

Though this is not strictly necessary for our purposes, we would like to visualize briefly the dual spaces $(C^l_\infty(\mathbf{R}^d))^\star$ that are playing an important role in our analysis.

Proposition 4.19 *If* $\nu \in (C^1_\infty(\mathbf{R}^d))^\star$ *then there exist* $d+1$ *finite signed Borel measures* $\nu_0, \nu_1, \dots, \nu_d$ *on* $\mathbf{R}^d$ *such that*

$$\nu(f) = (f, \nu) = (f, \nu_0) + \sum_{i=1}^d (\nabla_i f, \nu_i).$$

Conversely, any expression of this form specifies an element ν *from* $(C^1_\infty(\mathbf{R}^d))^\star$.

Proof Let us define an embedding $C^1_\infty(\mathbf{R}^d)$ to $C_\infty(\mathbf{R}^d) \times \cdots \times C_\infty(\mathbf{R}^d)$ ($d+1$ terms) by $f \mapsto (f, \nabla_1 f, \dots, \nabla_d f)$. It is evident that this mapping is injective and continuous and that its image is closed. Hence by the Hahn–Banach theorem any continuous linear functional on $C^1_\infty(\mathbf{R}^d)$ can be extended to a linear functional on the product space $(C_\infty(\mathbf{R}^d))^{d+1}$, yielding the required representation. □

Similar representations hold for other spaces $(C^l_\infty(\mathbf{R}^d))^\star$. However, this result does not give a full description of the space $(C^1_\infty(\mathbf{R}^d))^\star$, as the measures ν_i are not unique (integration by parts allows to transform ν_0 to other ν_i and vice versa). In particular, this makes unclear (unlike for the space $C_\infty(\mathbf{R}^d)$) the natural extension of a measure $\nu \in (C^1_\infty(\mathbf{R}^d))^\star$ to a functional on $C^1(\mathbf{R}^d)$, because such an extension depends on the choice of the ν_i.

We shall use two natural topologies in $(C^1_\infty(\mathbf{R}^d))^\star$: the *Banach topology*, given by the norm

$$\|\nu\|_{(C^1_\infty(\mathbf{R}^d))^\star} = \sup_{\|g\|_{C^1_\infty(\mathbf{R}^d)} \le 1} (g, \nu),$$

and the $\star$-*weak topology* with the convergence $\xi_n \to \xi$ as $n \to \infty$, meaning the convergence $(g, \xi_n) \to (g, \xi)$ for any $g \in C^1_\infty(\mathbf{R}^d)$.

4.7 Further techniques: martingale problem, Sobolev spaces, heat kernels etc.

In this and previous chapters we have touched upon several methods of constructing Markov (in particular Feller) semigroups from a given formal (pre)generator. Our exposition has been far from exhaustive. Here we briefly review other methods, only sketching the proofs and referring to the literature for the full story.

Our main topic here is the martingale problem approach. Having in mind the basic connection between this problem and Feller processes, see Proposition 2.30, it is natural to suggest that solving a martingale problem could serve as a useful intermediate step for the construction of a Markov process. As the following fact shows, we can obtain the existence of a solution to a martingale problem under rather general assumptions.

Theorem 4.20 *Assume that L is defined on $C_c^2(\mathbf{R}^d)$ by the usual Lévy–Khintchine form, i.e.*

$$Lu(x) = \operatorname{tr}\left[G(x)\frac{\partial^2}{\partial x^2}\right]u(x) + (b(x), \nabla)u(x) + \int \left[u(x+y) - u(x) - \mathbf{1}_{B_1}(y)(y, \nabla)u(x)\right]\nu(x, dy) \tag{4.38}$$

(as usual G is a non-negative matrix and ν is a Lévy measure) and that the symbol $p(x, \xi)$ of this pseudo-differential operator, where

$$p(x,\xi) = -(G(x)\xi, \xi) + i(b(x), \xi) + \int \left[e^{i\xi y} - 1 - i\mathbf{1}_{|y|\le 1}(y)(\xi, y)\right]\nu(x, dy),$$

is continuous. Moreover, let

$$\sup_x \left(\frac{\|G(x)\|}{1+|x|^2} + \frac{|b(x)|}{1+|x|} + \frac{1}{(1+|x|)^2}\int_{B_1}|y|^2\nu(x, dy) + \int_{\{|y|>1\}}\nu(x, dy)\right) < \infty. \tag{4.39}$$

Then the martingale problem has a solution for any initial probability μ.

This theorem was proved in Kolokoltsov [130] and extended the previous result of Hoh [97], where this existence was obtained under a stronger assumption of bounded coefficients, i.e. for

$$\sup_x \left(\|G(x)\| + |b(x)| + \int \min(1, |y|^2)\,\nu(x, dy)\right) < \infty. \tag{4.40}$$

Referring to these papers for a full proof, let us indicate its main idea. There is a very general result (see Ethier and Kurtz [74]) implying the existence of a solution to our martingale problem with sample paths in $D(\mathbf{R}_+, \dot{\mathbf{R}}^d)$, where $\dot{\mathbf{R}}^d$ is a one-point compactification of $\mathbf{R}^d$, so one needs only to find an appropriate Lyapunov function to ensure that, under the assumption of the theorem, the solution process cannot actually reach infinity in any finite time.

An important input from the martingale problem approach is the following localization procedure, allowing one to show the well-posedness of a martingale problem if it is well posed in a neighborhood of any point.

Let X_t be a process with sample paths in $D([0,\infty),\mathbf{R}^d)$ and initial distribution μ. For an open subset $U \subset \mathbf{R}^d$, define the exit time from U as

$$\tau_U = \inf\{t \geq 0 : X_t \notin U \text{ or } X(t-) \notin U\}. \tag{4.41}$$

We write τ_U^x when stressing that the initial point is x. Let L be an operator in $C(\mathbf{R}^d)$ with domain D. One says that X_t solves the *stopped martingale problem* for L in U starting with μ if $X_t = X_{\min(t,\tau)}$ a.s. and

$$f(X_t) - \int_0^{\min(t,\tau)} Lf(X_s)\,ds$$

is a martingale for any $f \in D$.

Theorem 4.21 *Suppose that the martingale problem is well posed for (L, D). Then for any initial μ there exists a unique solution to the stopped martingale problem that does not depend on the definition of $Lf(x)$ for $x \notin \bar{U}$ (provided that for this extension the martingale problem is well posed).*

We refer to Chapter 4, Section 6, of Ethier and Kurtz [74] for a complete proof. Let us sketch, however, a simple argument that works in the most important case, i.e. when $D \subset C^2(\mathbf{R}^d)$ is a core of L and L generates a Feller semigroup T_t. In this case we can approximate L on D by bounded L_n generating pure jump processes with semigroups T_t^n converging to T_t as $n \to \infty$. In view of the probabilistic construction of T^n (see Theorem 2.35) the corresponding stopped processes do not depend on the definition of $L_n f(x)$ for x outside U. Hence the same is true for their limit.

Theorem 4.22 *Suppose that L and L_k, $k = 1, 2, \ldots$, are operators, of the type* (4.38), *defined on $D = C_c^2(\mathbf{R}^d)$ and that (U_k), $k = 1, 2, \ldots$, is an open covering of $\mathbf{R}^d$ such that $L_k f(x) = Lf(x)$ for $x \in U_k$, $f \in D$. Assume that for any initial distribution μ the martingale problem for L has a solution and that the martingale problems for all the L_k are well posed. Then the martingale problem for L is also well posed.*

We refer to Ethier and Kurtz [74] or Stroock and Varadhan [229] for a rigorous proof, whose idea is quite transparent: in any domain U_k a solution to a martingale problem for L coincides with that for L_k (and is unique, by Theorem 4.21). When leaving U_k the process finds itself in some other set $U_{k'}$ and its behavior is uniquely specified by the solution to the martingale problem for $L_{k'}$, etc. This result is very useful since, by Proposition 2.30, uniqueness for a Feller problem can be settled by obtaining the well-posedness of the corresponding martingale problem.

The construction and analysis of the martingale problem and the corresponding Markov semigroups can be greatly facilitated by using advanced functional analysis techniques, in particular Fourier analysis. As these methods work more effectively in Hilbert spaces and the original Feller semigroups act in the Banach space $C_\infty(\mathbf{R}^d)$, one looks for auxiliary Hilbert spaces where the existence of a semigroup can be shown as a preliminary step. For these auxiliary spaces it is natural to use the *Sobolev spaces* $H^s(\mathbf{R}^d)$ defined as the completions of the Schwartz space $S(\mathbf{R}^d)$ with respect to the Hilbert norm

$$\|f\|_{H^s}^2 = \int f(x)(1+\Delta)^s f(x)\,dx.$$

In particular, H^0 coincides with the usual L^2. The celebrated *Sobolev imbedding lemma* states that H^s is continuously imbedded in $(C_\infty \cap C^l)(\mathbf{R}^d)$ whenever $s > l + d/2$. Consequently, if we can show the existence of a semigroup in H^s, this lemma supplies automatically an invariant dense domain (and hence a core) for its extension to C_∞. For a detailed discussion of Fourier analysis and Sobolev spaces in the context of Markov processes we refer to Jacob [103].

As an example of the application of the techniques mentioned above, we shall discuss Markov semigroups with so-called decomposable generators. Let ψ_n, $n = 1, \ldots, N$, be a finite family of generators of Lévy processes in $\mathbf{R}^d$, i.e. for each n

$$\psi_n f(x) = \operatorname{tr}\left[G^n \frac{\partial^2}{\partial x^2}\right] f(x) + \left(b^n, \frac{\partial}{\partial x}\right) f(x) + \int \left[f(x+y) - f(x) - (\nabla f(x), y)\mathbf{1}_{B_1}(y)\right] \nu^n(dy),$$

where $G^n = (G^n_{ij})$ is a non-negative symmetric $d \times d$ matrix and ν^n is a Lévy measure. Recall that the function

$$p_n(\xi) = -(G^n\xi, \xi) + i(b^n, \xi) + \int \left[e^{i\xi y} - 1 - i(\xi, y)\mathbf{1}_{B_1}(y)\right] \nu^n(dy)$$

is called the *symbol* of the operator ψ_n. We denote by p_n^ν the corresponding integral terms, so that

$$p_n^\nu(\xi) = \int \left[e^{i\xi y} - 1 - i(\xi, y)\mathbf{1}_{B_1}(y)\right] \mu^n(dy)$$

and by $\tilde{p}_n^\nu$ the part corresponding to ν_n reduced to the unit ball:

$$\tilde{p}_n^\nu(\xi) = \int_{B_1} \left[e^{i\xi y} - 1 - i(\xi, y)\right] \nu^n(dy).$$

We also write $p_0 = \sum_{n=1}^N p_n$.

Let a_n be a family of positive continuous functions on $\mathbf{R}^d$. By a *decomposable generator* we mean an operator of the form $\sum_{n=1}^N a_n(x)\psi_n$. These operators are simpler to deal with analytically, and at the same time their properties capture the major qualitative features of the general case. Decomposable generators appear in many applications, for example in connection with interacting-particle systems (see e.g. Kolokoltsov [129]), where the corresponding functions a_n are usually unbounded but smooth.

Theorem 4.23 *Suppose that there exist constants $c > 0$ and $\alpha_n > 0$, $\beta_n < \alpha_n$, $n = 1, \ldots, N$, such that for all n*

$$|\mathrm{Im}\, p_n^\nu(\xi)| \le c|p_0(\xi)|, \qquad |\mathrm{Re}\, \tilde{p}_n^\nu(\xi)| \ge c^{-1}|\mathrm{pr}_n(\xi)|^{\alpha_n},$$
$$|\nabla \tilde{p}_n^\nu(\xi)| \le c|\mathrm{pr}_n(\xi)|^{\beta_n}, \tag{4.42}$$

where pr_n is the orthogonal projection operator on the minimal subspace containing the support of the measure $\mathbf{1}_{B_1}\nu^n$. The a_n are positive s-times continuously differentiable functions, with $s > 2 + d/2$, such that $a_n(x) = O(1 + |x|^2)$ for those n where $G^n \neq 0$ or $\mathbf{1}_{B_1}\nu^n \neq 0$, $a_n(x) = O(|x|)$ for those n where $\beta^n \neq 0$ and $a_n(x)$ is bounded whenever $\mathbf{1}_{\mathbf{R}^d \setminus B_1}\nu^n \neq 0$. Then there exists a unique extension of the operator $L = \sum_{n=1}^N a_n(x)\psi_n$ (with initial domain $C_c^2(\mathbf{R}^d)$) that generates a Feller semigroup in $C_\infty(\mathbf{R}^d)$.

In practice, condition (4.42) is not very restrictive. It allows, in particular, any α-stable measures ν (whatever their degeneracy). Moreover, if $\int |\xi|^{1+\beta_n}\nu_n(d\xi) < \infty$ then the last condition in (4.42) holds, because

$$\nabla \tilde{p}_n^\nu(\xi) = \int iy(e^{i\xi y} - 1)\nu^n(dy)$$

and $|e^{ixy} - 1| \le c|xy|^\beta$ for any $\beta \le 1$ and some $c > 0$. In particular, the last condition in (4.42) always holds for $\beta_n = 1$. As no restrictions on the differential part of p_n are imposed, all (possibly degenerate) diffusion processes with smooth symbols are covered by our assumptions.

We refer to Kolokoltsov [130] for a detailed, rather lengthy, proof, the idea being as follows. By the localization theorem, Theorem 4.22, and the basic existence theorem, Theorem 4.20, we can reduce the situation to the case when $a_n \in C^s(\mathbf{R}^d)$ and $|a_n(x) - a_n(x_0)|$ are small compared with $a_n(x_0)$ for some x_0. And this case is dealt with using non-stationary perturbation theory in Sobolev spaces for vanishing drifts, supplemented further by analysis in the "interaction representation". In Hoh [97] the same result is proved using the analysis of resolvents in Sobolev spaces, under the additional assumption that the a_n are bounded, the symbols p_n are real and $|p_0(\xi)| \ge c|\xi|^\alpha$ (thus excluding the usual degenerate diffusions). In Jacob *et al.* [105] it was noted that the proof of Hoh

[97] can be generalized to a slightly weaker condition, where the assumption of the reality of the symbols p_n is replaced by the assumption that their imaginary parts are bounded by their real parts.

One should also mention an important aspect, that of proving the existence and obtaining estimates of transition probability densities (also called heat kernels) for Feller processes that specify Green functions for the Cauchy problems of the corresponding evolution equations. Though heat-kernel estimates for diffusion equations are well established, for Feller processes with jumps much less is known. We shall formulate here a result on *heat kernels for stable-like processes*, i.e. for the equation

$$\frac{\partial u}{\partial t} = -a(x)|-i\nabla|^{\alpha(x)}u, \qquad x \in \mathbf{R}^d,\ t \geq 0, \tag{4.43}$$

referring for the details to the original paper, Kolokoltsov [125]).

If $a(x)$ and $\alpha(x)$ are constants, the Green function for (4.43) is given by the stable density (see Section 2.4)

$$S(x_0 - x; \alpha, at) = (2\pi)^{-d} \int_{\mathbf{R}^d} \exp\left[-at|p|^\alpha + ip(x - x_0)\right] dp. \tag{4.44}$$

In the theory of pseudo-differential operators, equation (4.43) is written in pseudo-differential form as

$$\frac{\partial u}{\partial t} = \Phi(x, -i\nabla)u(x) \tag{4.45}$$

with symbol

$$\Phi(x, p) = -a(x)|p|^{\alpha(x)}. \tag{4.46}$$

As follows from direct computations (see equation (E.1)), an equivalent form of (4.43) is the following integro-differential form, of Lévy–Khintchine type:

$$\frac{\partial u}{\partial t} = -a(x)c(\alpha) \int_0^\infty \left[u(x + y) - u(x) - (y, \nabla u)\right] \frac{d|y|}{|y|^{1+\alpha}} \tag{4.47}$$

with some constant $c(\alpha)$. We shall not need this form much, but it is important to have in mind that the operator on the r.h.s. of (4.43) satisfies the PMP, which is clear from the representation given in (4.47) but is not so obvious from (4.43).

Naturally, one might expect that, for small times, the Green function of equation (4.43) with varying coefficients can be approximated by the Green function of the corresponding problem with constant coefficients, i.e. by the function

$$G_0(t, x, x_0) = S(x - x_0, \alpha(x_0), a(x_0)t). \tag{4.48}$$

This is in fact true, as the following result shows (see Kolokoltsov [125] for the proof and a discussion of the related literature).

Theorem 4.24 *Let $\beta \in (0, 1]$ be arbitrary and let $\alpha \in [\alpha_d, \alpha_u]$, $a \in [a_d, a_u]$ be β-Hölder continuous functions on $\mathbf{R}^d$ with values in compact subsets of $(0, 2)$ and $(0, \infty)$ respectively. Then the Green function u_G for equation (4.43) exists in the sense that it is a function, continuous for $t > 0$, defining a solution to equation (4.43) as a distribution. That is, for any $f \in C(\mathbf{R}^d)$ the function*

$$f_t(x) = \int u_G(t, x, y) f(y)\, dy \tag{4.49}$$

satisfies the equation

$$(f_t, \phi) = (f, \phi) + \int_0^t (f_s, L'\phi)\, ds \tag{4.50}$$

for any $\phi \in C^2(\mathbf{R}^d)$ with compact support, where L' is the operator dual to the operator L on the r.h.s. of equation (4.43). Moreover, for $t \leq T$, with any given T,

$$\begin{aligned} u_G(t, x, x_0) &= S(x - x_0, \alpha(x_0), a(x_0)t) \left[1 + O(t^{\beta/\alpha_u})(1 + |\log t|)\right] \\ &\quad + O(t) f^d_{\alpha_d}(x - x_0) \end{aligned} \tag{4.51}$$

and the resolving operator $f \mapsto \int u_G(t, x, x_0) f(x_0)\, dx_0$ of the Cauchy problem for (4.43) specifies a conservative Feller semigroup. If the functions α, a are of the class $C^2(\mathbf{R}^2)$ then $u_G(t, x, x_0)$ solves (4.43) classically for $t > 0$ and the corresponding Feller semigroup preserves the space $C^2(\mathbf{R}^2)$ and is bounded with respect to the Banach norm of this space.

Of course, real-life processes are often nonhomogeneous in time. However, as we shall see, nonhomogeneous processes arise as natural tools in the analysis of nonlinear Markov evolutions. Having this in mind we present below a time-nonhomogeneous version of the above results, also including drifts and sources (see Kolokoltsov [134]).

Theorem 4.25 *Suppose that $\alpha(x) \in [\alpha_d, \alpha_u] \subset (1, 2)$ and that $a_t(x)$, $a_t^{-1}(x)$, A_t, $B_t(x)$, $f_t(x, z)$ are bounded and twice continuously differentiable functions of $t \in [0, T]$, $x, z \in \mathbf{R}^d$, such that a_t, f_t are non-negative and $f_t(x, z) \leq B(1 + |z|^{\beta+d})^{-1}$ where β, $B > 0$ are constants. Then the following hold.*

(i) The equation

$$\frac{d}{dt}u_t(x) = -a_t(x)|\nabla|^{\alpha(x)}u_t(x) + (A_t(x), \nabla u_t(x)) + B_t(x)u_t(x) \\ + \int [u_t(x+z) - u_t(x)]\, f_t(x,z)\, dz \qquad (4.52)$$

has a corresponding Green function $u_G(t,s,x,y)$, $T \geq t \geq s \geq 0$, *i.e. a solution for the initial condition* $u_G(s,s,x,y) = \delta(x-y)$, *such that*

$$u_G(t,s,x,y) = G_0(t,s,x,y)\left[1 + O(1)\min(1,|x-y|) + O(t^{1/\alpha})\right] \\ + O(t)(1+|x-y|^{d+\min(\alpha,\beta)})^{-1}. \qquad (4.53)$$

Here $O(1)$ *and* $O(t^{1/\alpha})$ *each depend only on* T *and* $C(T)$*;*

$$G_0(t,s,x,y) \\ = (2\pi)^{-d}\int \exp\left[-\int_s^t a_r(y)\,dr\,|p|^\alpha + i\left(p,\ x - y - \int_s^t A_r(y)dr\right)\right]dp \qquad (4.54)$$

is a shifted stable density; and the last term in (4.53) *can be omitted whenever* $\beta \geq \alpha$.

(ii) The Green function $u_G(t,s,x,y)$ *is everywhere non-negative and satisfies the Chapman–Kolmogorov equation; moreover, in the case* $B_t = 0$ *one has* $\int u_G(t,s,x,y)dy = 1$ *for all* x *and* $t > s$.

(iii) If the coefficient functions are from the class $C^2(\mathbf{R}^d)$ *then* $u_G(t,s,x,y)$ *is continuously differentiable in* t, s *and satisfies* (4.52) *classically; for any* $u_s \in C_\infty(X)$ *there exists a unique (classical) solution* u_t *in* $C_\infty(X)$ *to the Cauchy problem* (4.52) *(i.e. a continuous mapping* $t \to u_t \in C_\infty(X)$ *that solves* (4.52) *for* $t > s$ *and coincides with* u_s *at* $t = s$*); we have* $u_t \in C^1(X)$ *for all* $t > s$ *with*

$$\|u_t\|_{C^1(X)} = O(t-s)^{-1/\alpha}\|u_s\|_{C(X)};$$

and if $a_t, A_t, B_t, f_t \in C^k(X)$, $k > 0$, *the mapping* $u_s \mapsto u_t$ *is a bounded operator in* $C^k(X)$ *uniformly for* $t \in [0,T]$.

Remark 4.26 If the coefficients are not assumed to be differentiable but only Hölder continuous, then statements (i) and (ii) of the theorem remain valid only if one understands the solutions to (4.52) in the sense of operators defined on distributions (as in Theorem 4.24).

Corollary 4.27 *Under the assumptions of Theorem 4.25 the mapping $u_s \mapsto u_t$ extends to the bounded linear mapping $\mathcal{M}(X) \mapsto \mathcal{M}(X)$ that is also continuous in the vague topology and is such that its image always has a density (with respect to Lebesgue measure) that solves equation* (4.52) *for $t > s$.*

To conclude this lengthy yet rather sketchy section we mention some other relevant methods whose exposition is beyond the scope of this book.

1. Dirichlet forms To a linear operator in a Banach space B there corresponds a bilinear form (Ax, y), $x \in B$, $y \in B^\star$ (the Banach dual to B), which if B is a Hilbert space H (so that H and $H^\star$ are naturally identified) can be reduced to a bilinear form on $H \times H$. The forms arising from the generators of Markov processes are usually referred to as Dirichlet forms. We can characterize this class of forms rather explicitly (in the famous Beuring–Deny formula), in a way that is similar to the Courrége characterization of the generators. In some situations it turns out to be easier to analyze Markov semigroups using their Dirichlet forms rather than their generators. We refer to the monographs by Fukushima *et al.* [81] and Ma and Röckner [167] for the basic theory of Dirichlet forms and their applications.

2. Resolvent and Hille–Yosida theorem There is a general result characterizing the generators of bounded (in particular contraction) semigroups, namely the *Hille–Yosida theorem.* Specifically for Feller semigroups, it states that a linear operator A in $C_\infty(\mathbf{R}^d)$ defined on a dense domain D is closable and its closure generates a Feller semigroup if and only if it satisfies the PMP and the range $\lambda - A$ is dense in $C_\infty(\mathbf{R}^d)$ for some $\lambda > 0$. This last condition is of course the most difficult to check, but we can get nontrivial results on the existence of Feller semigroups using this approach by Fourier analysis of the resolvent $(\lambda - A)^{-1}$ in Sobolev spaces; see Jacob [103] and references therein.

3. Subordination Changing the time scales of processes in a random way yields a remarkable method of constructing new processes, not only Markovian processes but also those with a memory that can be described by differential equations that are fractional in time (see Evans and Jacob [75], Kolokoltsov, Korolev and Uchaikin [138], Kolokoltsov [134] and Meerschaert and Scheffler [184] and references therein).

4. Semiclassical asymptotics Scaling the derivatives of a differential or pseudo-differential equation often yields an approximation in terms of the solutions of certain ODE, called the characteristic equation for the initial problem. The interpretation as a quasi-classical or semiclassical approximation arises naturally from the models of quantum mechanics. In the theory of diffusion, a similar approximation is often called the *small-diffusion* approximation. This technique allows one to obtain effective two-sided estimates and small-time asymptotics for the transition probabilities of stochastic processes. We refer to Kolokoltsov [124] for an extensive account and to Kolokoltsov [126] for the semiclassical approach to nonlinear diffusions arising in the theory of superprocesses.

5. Malliavin calculus This was originally developed by Paul Malliavin in an attempt to obtain a probabilistic proof of the famous Hörmander result on the characterization of degenerate diffusions having a smooth Green function (or heat kernel). Later on it was developed into a very effective tool for analyzing the transition probabilities of Markov processes in both the linear and nonlinear cases; see e.g. Nualart [196], Bichteler, Gravereaux and Jacod [36] and Guérin, Méléard and Nualart [91].

5

Unbounded coefficients

So far, apart from in Section 4.2, we have discussed Levy-type generators (4.38) with *bounded coefficients* or, slightly more generally, under assumption (4.39). However, unbounded coefficients arise naturally in many situations, in particular when one is analyzing LLNs for interacting particles. This chapter is devoted to a general approach to the analysis of unbounded coefficients that is based on the so-called Lyapunov or barrier functions. It turns out that the corresponding processes are often not Feller, and an appropriate extension of this notion is needed.

5.1 A growth estimate for Feller processes

In this introductory section we aim at an auxiliary estimate of the growth and continuity of a Feller process using the integral moments of the Lévy measures entering its generator. This estimate plays a crucial role in the extension to unbounded coefficients given later.

Theorem 5.1 *Let an operator L, defined in $C^2_c(\mathbf{R}^d)$ by the usual Lévy–Khintchine form given in* (4.38), *i.e.*

$$Lu(x) = \operatorname{tr}\left[G(x)\frac{\partial^2}{\partial x^2}\right]u(x) + (b(x),\nabla)u(x) + \int \left[u(x+y) - u(x) - \mathbf{1}_{B_1}(y)(y,\nabla)u(x)\right]\nu(x,dy), \tag{5.1}$$

satisfy the boundedness requirement (4.39)*:*

$$\sup_x \left(\frac{\|G(x)\|}{1+|x|^2} + \frac{|b(x)|}{1+|x|} + \frac{1}{(1+|x|)^2}\int_{B_1}|y|^2\nu(x,dy) + \int_{\{|y|>1\}}\nu(x,dy)\right) < \infty. \tag{5.2}$$

Suppose that the additional moment condition

$$\sup_x (1+|x|)^{-p} \int_{\{|y|>1\}} |y|^p \nu(x,dy) < \infty \tag{5.3}$$

holds for $p \in (0,2]$. Let X_t solve the martingale problem when L has domain $C_c^2(\mathbf{R}^d)$ (a solution exists by Theorem 4.20). Then

$$\mathbf{E} \min(|X_t^x - x|^2, |X_t^x - x|^p) \le (e^{ct} - 1)(1 + |x|^2) \tag{5.4}$$

for all t, where the constant c depends on the left-hand sides of (5.3) *and* (5.2)*. Moreover, for any $T > 0$ and a compact set $K \subset \mathbf{R}^d$,*

$$\mathbf{P}\left(\sup_{s\le t} |X_s^x - x| > r\right) \le \frac{t}{r^p} C(T,K) \tag{5.5}$$

for all $t \le T$, $x \in K$ and large enough r and for some constant $C(T,K)$.

Proof Notice first that, from the Cauchy inequality

$$\int_{\{|y|>1\}} |y|^q \nu(x,dy) \\ \le \left(\int_{\{|y|>1\}} |y|^p \nu(x,dy)\right)^{q/p} \left(\int_{\{|y|>1\}} \nu(x,dy)\right)^{(p-q)/p},$$

it follows that (5.3) and (5.2) imply that

$$\sup_x (1+|x|)^{-q} \int_{\{|y|>1\}} |y|^q \nu(x,dy) < \infty$$

for all $q \in (0,p]$. Now, let $f_p(r)$ be an increasing smooth function on $\mathbf{R}_+$ that equals r^2 in a neighborhood of the origin, equals r^p for $r > 1$ and is not less than r^2 for $r < 1$. For instance we can take $f(r) = r^2$ when $p = 2$. Also, let $\chi_q(r)$ be a smooth non-increasing function $[0,\infty) \mapsto [0,1]$ that equals 1 for $r \in [0,1]$ and r^{-q} for $r > 2$. To obtain a bound for the average of the function $f_p^x(y) = f_p(\|y-x\|)$ we approximate it by the increasing sequence of functions $g_n(y) = f_p^x(y)\chi_q(|y-x|/n)$, $n = 1,2,\dots$, $q > p$. The main observation is that

$$|Lg_n(y)| \le c\left[g_n(y) + x^2 + 1\right], \tag{5.6}$$

with a constant c, uniformly for x, y and n. To see this we analyze separately the action of all terms in the expression for L. For instance,

$$\left|\text{tr}\left[G(y)\frac{\partial^2}{\partial y^2}g_n(y)\right]\right|$$
$$\leq c(1+|y|^2)\left[\min(1,|y-x|^{p-2})\chi_q\left(\frac{|y-x|}{n}\right)\right.$$
$$\left.+ f_p(|y-x|)\chi_q''\left(\frac{|y-x|}{n}\right)\frac{1}{n^2} + f_p'(|y-x|)\chi_q'\left(\frac{|y-x|}{n}\right)\frac{1}{n}\right].$$

Taking into account the obvious estimate

$$\chi_q^{(k)}(z) \leq c_k(1+|z|^k)^{-1}\chi_q(z)$$

(which holds for any k, though we need only $k = 1, 2$) and using $|y|^2 \leq 2(y-x)^2 + 2x^2$ yields

$$\left|\text{tr}\left(G(y)\frac{\partial^2}{\partial y^2}g_n(y)\right)\right| \leq c(|g_n(y)| + x^2 + 1),$$

as required. Also, as $g_n(x) = 0$,

$$\begin{aligned}\int_{\{|y|>1\}}\left[g_n(x+y) - g_n(x)\right]\nu(x,dy) &= \int_{\{|y|>1\}} f_p(|y|)\chi_q(|y|/n)\nu(x,dy)\\ &\leq \int_{\{|y|>1\}} |y|^p\nu(x,dy)\\ &\leq c(1+|x|)^p) \leq c(1+|x|^2)\end{aligned}$$

and so on.

Next, as $q > p$ the function $g_n(y)$ belongs to $C_\infty(\mathbf{R}^d)$ and we can establish, by an obvious approximation, that the process

$$M_{g_n}(t) = g_n(X_t^x) - \int_0^t Lg_n(X_s^x)\,ds$$

is a martingale. Recall that $M_f(t)$ is a martingale for any $f \in C_c^2(\mathbf{R}^d)$, because it is supposed that X_t^x solves the martingale problem for L in $C_c^2(\mathbf{R}^d)$. Now using the dominated and monotone convergence theorems and passing to the limit $n \to \infty$ in the equation $\mathbf{E}M_{g_n}(t) = g_n(x) = 0$ (representing the martingale property of M_{g_n}), yields the inequality

$$\mathbf{E}f_p(\|X_t^x - x\|) \leq c\int_0^t [\mathbf{E}f_p(\|X_s^x - x\|) + x^2 + 1]\,ds.$$

This implies that

$$\mathbf{E}f_p(|X_t^x - x|) \leq (e^{ct} - 1)(1+|x|^2)$$

by Gronwall's lemma, and (5.4) follows.

Once the upper bound for $Ef_p(|X_t^x - x|)$ has been obtained it is straightforward to show, by the same approximation as above, that M_f is a martingale for $f = f_p^x$. Moreover, passing to the limit in (5.6) we obtain

$$|Lf_p^x(y)| \leq c\left[f_p^x(y) + x^2 + 1\right]. \tag{5.7}$$

Applying Doob's maximal inequality yields

$$\begin{aligned}\mathbf{P}\left(\sup_{s\leq t}\left|f_p^x(X_s^x) - \int_0^s Lf_p^x(X_\tau^x)\,d\tau\right| \geq r\right) &\leq \frac{1}{r}tc(T)(1+|x|^2)\\ &\leq \frac{1}{r}tc(T,K)).\end{aligned}$$

Hence, with a probability not less than $1 - tc(T,K)/r$, we have

$$\sup_{s\leq t}\left|f_p^x(X_s^x) - \int_0^s Lf_p^x(X_\tau^x)\,d\tau\right| \leq r,$$

implying by Gronwall's lemma and (5.7) that

$$\sup_{t\leq T} f_p^x(X_t^x) \leq c(T)(r + x^2 + 1) \leq 2C(T)r$$

for $x^2 + 1 \leq r$, implying in turn (with a different constant $C(T,K)$) that

$$\mathbf{P}\left(\sup_{s\leq t} f_p(|X_s^x - x|) > r\right) \leq \frac{t}{r}C(T,K).$$

Since $|X_s^x - x| > r$ if and only if $f_p(|X_s^x - x|) > r^p$, the estimate (5.5) follows. □

Exercise 5.1 Show that under the assumptions of Theorem 5.1 the process X_t is conservative in the sense that the dynamics of averages preserves constants, i.e. that $\lim_{n\to\infty}\mathbf{E}\chi(|X_t^x|/n) = 1$ for any $\chi \in C_\infty(\mathbf{R}_+)$ that equals 1 in a neighborhood of the origin. Hint: clearly the limit exists and does not exceed 1. To show the claim use (5.5); choosing r and n large enough, $\mathbf{E}\chi(|X_t^x|/n)$ becomes arbitrarily close to 1.

Exercise 5.2 Show that if the coefficients of L are bounded, i.e. (4.40) holds, then

$$\mathbf{E}\min(|X_t^x - x|^2, |X_t^x - x|^p) \leq (e^{ct} - 1) \tag{5.8}$$

uniformly for all x, and also that (5.5) holds for all x when $C(T,K)$ does not depend on K.

5.2 Extending Feller processes

To formulate our main result on unbounded coefficients, it is convenient to work with weighted spaces of continuous functions. Recall that if $f(x)$ is a continuous positive function, on a locally compact space S, tending to infinity as $x \to \infty$ then we denote by $C_f(S)$ (resp. $C_{f,\infty}(S)$) the space of continuous functions g on S such that $g/f \in C(S)$ (resp. $g/f \in C_\infty(S)$) with norm $\|g\|_{C_f} = \|g/f\|$. Similarly, we define $C^k_f(S)$ (resp. $C^k_{f,\infty}(S)$) as the space of k-times continuously differentiable functions such that $g^{(l)}/f \in C(S)$ (resp. $g^{(l)}/f \in C_\infty(S)$) for all $l \leq k$.

Theorem 5.2, the main result of this section, allows us to construct Markov processes and semigroups from Lévy-type operators with unbounded coefficients, subject to the possibility of localization (i.e. to the assumption that localized problems are well posed) and to the existence of an appropriate Lyapunov function. It is an abstract version of a result from [130] devoted specifically to decomposable generators. The method based on the martingale problem has become standard. However, of importance for us is the identification of a space in which the limiting semigroup is strongly continuous. We shall use the function χ_q defined in the proof of Theorem 5.1 above.

Theorem 5.2 *Let an operator L be defined in $C^2_c(\mathbf{R}^d)$ by* (4.38) *and let $\int_{\{|y|\geq 1\}} |y|^p \nu(x, dy) < \infty$ for a $p \leq 2$ and any x. Assume that a positive function $f_L \in C^2_{1+|x|^p}$ is given (here the subscript L stands for the operator L) such that $f_L(x) \to \infty$ as $x \to \infty$ and*

$$Lf_L \leq c(f_L + 1) \tag{5.9}$$

for a constant c. Set

$$s^p_L(x) = \|G(x)\| + |b(x)| + \int \min(|y|^2, |y|^p)\nu(x, dy).$$

Assume that for a given $q > 1$ the martingale problem for the "normalized" operators $L_n = \chi_q(s^p_L(x)/n)L$, $n = 1, 2, \ldots$, with bounded coefficients is well posed in $C^2_c(\mathbf{R}^d)$ and that the corresponding process is a conservative Feller process (for instance, that one of Theorems 4.13, 4.24, 3.11 or 4.23 applies). Then the martingale problem for L in $C^2_c(\mathbf{R}^d)$ is also well posed, the corresponding process X_t is strong Markov and its contraction semigroup preserves $C(\mathbf{R}^d)$ and extends from $C(\mathbf{R}^d)$ to a strongly continuous semigroup in $C_{f_L,\infty}(\mathbf{R}^d)$ whose domain contains $C^2_c(\mathbf{R}^d)$. Moreover,

$$\mathbf{E} f_L(X^x_t) \leq e^{ct}\left[f_L(x) + c\right] \tag{5.10}$$

and

$$\sup_m \mathbf{P}\Big(\sup_{0\le s\le t} f_L(X_s^x) > r\Big) \le \frac{c(t, f_L(x))}{r}, \tag{5.11}$$

implying in particular that this semigroup in $C_{f_L,\infty}(\mathbf{R}^d)$ is a contraction whenever $c = 0$ in (5.9).

Proof Let $X_{t,m}$ be the Feller processes corresponding to L_m. Approximating f_L by $f_L(y)\chi_p(y/n)$ as in the proof of Theorem 5.1 and using the boundedness of moments (5.4) for processes X_t with bounded generators, it is straightforward to conclude that the processes

$$M_m(t) = f_L(X_{t,m}^x) - \int_0^t L_m f_L(X_{s,m}^x)\, ds$$

are martingales for all m. Moreover, since $\chi_p \le 1$ it follows from our assumptions that $L_m f_L \le c(f_L + 1)$ for all m, implying again by Gronwall's lemma that

$$\mathbf{E} f_L(X_{t,m}^x) \le e^{ct}\left[f_L(x) + c\right]. \tag{5.12}$$

Since by (5.9) and (5.12) the expectation of the negative part of the martingale $M_m(t)$ is uniformly (for $t \le T$) bounded by $c(T)\left[f_L(x) + 1\right]$, we conclude that the expectation of its magnitude is also bounded by $c(T)\left[f_L(x) + 1\right]$ (in fact, for any martingale $M(t)$ one has $M(0) = \mathbf{E}M(t) = \mathbf{E}M^+(t) - \mathbf{E}M^-(t)$, where $M^\pm(t)$ are the positive and negative parts of $M(t)$, implying that $\mathbf{E}M^+(t) = \mathbf{E}M^-(t) + M(0)$). Hence, by the same argument as in the proof of (5.5), one deduces from Doob's inequality for martingales that

$$\sup_m \mathbf{P}\Big(\sup_{0\le s\le t} f_L(X_{s,m}^x) > r\Big) \le \frac{c(t, f_L(x))}{r}$$

uniformly for $t \le T$ with arbitrary T. Since $f_L(x) \to \infty$ as $x \to \infty$, this implies the *compact containment condition* for $X_{t,m}$:

$$\lim_{r\to\infty} \sup_m \mathbf{P}\Big(\sup_{0\le s\le t} |X_{s,m}^x| > r\Big) = 0$$

uniformly for x from any compact set and $t \le T$ with arbitrary T.

Let us estimate the difference between the Feller semigroups of $X_{s,n}$ and $X_{s,m}$. By the compact containment condition, for any $\epsilon > 0$ there exists $r > 0$ such that, for $f \in C(\mathbf{R}^d)$,

$$|\mathbf{E}f(X_{t,m}^x) - \mathbf{E}f(X_{t,n}^x)| \le \left|\mathbf{E}[f(X_{s,m}^x)\mathbf{1}_{t<\tau_r^m}] - \mathbf{E}[f(X_{s,n}^x)\mathbf{1}_{t<\tau_r^n}]\right| + \epsilon\|f\|,$$

where τ_r^m is the exit of $X_{t,m}^x$ from the ball B_r (i.e. it is given by (4.41) with $U = B_r$). Note that for large enough n, m the generators of $X_{t,m}^x$ and $X_{t,n}^x$ coincide in B_r and hence, by Theorem 4.21, the first term on the r.h.s. of the above inequality vanishes. Consequently

$$|\mathbf{E} f(X_{t,m}^x) - \mathbf{E} f(X_{t,n}^x)| \to 0$$

as $n, m \to \infty$ uniformly for x from any compact set. This fact clearly implies that the limit

$$T_t f(x) = \lim_{n \to \infty} \mathbf{E} f(X_{t,n}^x)$$

exists and that T_t is a Markov semigroup preserving $C(\mathbf{R}^d)$ (i.e. it is a C-Feller semigroup) and continuous in the topology of uniform convergence on compact sets, i.e. it is such that $T_t f(x)$ converges to $f(x)$ as $t \to 0$ uniformly for x from any compact set. Clearly this compact containment implies the tightness of the family of transition probabilities for the Markov processes $X_{t,m}^x$. This leads to the conclusion that the limiting semigroup T_t has the form (2.19) for certain transitions p_t and hence specifies a Markov process, which therefore solves the required martingale problem. Uniqueness follows by localization, i.e. by Theorem 4.22. It remains to observe that (5.12) implies (5.10), and this in turn implies (5.11) by the same argument as for the approximations $X_{t,m}$ above. Consequently T_t extends by monotonicity to a semigroup on $C_f(\mathbf{R}^d)$. Since the space $C(\mathbf{R}^d) \subset C_f(\mathbf{R}^d)$ is invariant and T_t is continuous there, in the topology of uniform convergence on compact sets it follows that $T_t f$ converges to f as $t \to 0$ in the topology of $C_f(\mathbf{R}^d)$ for any $f \in C(\mathbf{R}^d)$ and hence (by a standard approximation argument) also for any $f \in C_{f,\infty}(\mathbf{R}^d)$, implying the required strong continuity. □

Let us consider an example of stable-like processes with unbounded coefficients.

Proposition 5.3 *Let L have the form given by (3.55), where ω_p, α_p are as in Proposition 3.18. Let σ, b, a_p be continuously differentiable (a_p as a function of x), a_p be positive, ν depend weakly continuously on x and $\int |y|\nu(x, dy) < \infty$. Then Theorem 5.2 applies, for $f_L(x)$ a twice differentiable function coinciding with $|x|$ for large x, whenever either*

$$\|A(x)\| + \int_P dp \sup_s a_p(x, s) + |x| \int |y|\nu(x, dy) + (b(x), x) \le c|x|^2 \quad (5.13)$$

with $c > 0$, or $(b(x), x)$ is negative and

$$\|A(x)\| + \int_P dp \sup_s a_p(x, s) + |x| \int |y|\nu(x, dy) \le R^{-1}|(b(x), x)| \quad (5.14)$$

for large x with a large enough constant R.

Proof This is straightforward from Theorem 5.2 and Proposition 3.18. □

5.3 Invariant domains

Theorem 5.2 has an important drawback. In the limit $m \to \infty$ we lose all information about the invariant domain of L. Let us describe a method to identify such a domain.

Consider the stochastic equation (3.37):

$$X_t = x + \int_0^t \sigma(X_{s-})dB_s + \int_0^t b(X_{s-})ds + \int_0^t \int F(X_{s-}, y)\tilde{N}(ds\, dy)$$

(taking $G = 1$ for simplicity) under the assumptions of Theorem 3.17(ii) (and with the corresponding generator given by (3.40)). Differentiating it twice with respect to the initial conditions leads to the stochastic equations

$$Z_t = \mathbf{1} + \int_0^t \left(\frac{\partial \sigma}{\partial x}(X_{s-})Z_{s-}dB_s + \frac{\partial b}{\partial x}(X_{s-})Z_{s-}\,ds \right. \\ \left. + \int \frac{\partial F}{\partial x}(X_{s-}, y)Z_{s-}\tilde{N}(ds\, dy) \right), \quad (5.15)$$

and

$$W_t = \int_0^t \left(\left(\frac{\partial^2 \sigma}{\partial x^2}(X_{s-})Z_{s-}, Z_{s-} \right) + \frac{\partial \sigma}{\partial x}(X_{s-})W_{s-} \right) dB_s \\ + \int_0^t \left(\left(\frac{\partial^2 b}{\partial x^2}(X_{s-})Z_{s-}, Z_{s-} \right) + \frac{\partial b}{\partial x}(X_{s-})W_{s-} \right) ds \\ + \int_0^t \int \left(\left(\frac{\partial^2 F}{\partial x^2}(X_{s-}, y)Z_{s-}, Z_{s-} \right) + \frac{\partial F}{\partial x}(X_{s-}, y)W_{s-} \right) \tilde{N}(ds\, dy) \quad (5.16)$$

for

$$Z = \frac{\partial X}{\partial x}, \qquad W = \frac{\partial^2 X}{\partial x^2}.$$

From Proposition 3.13 it follows that the solutions are well defined and can be obtained via the Ito–Euler approximation scheme. Moreover, one sees from Proposition 3.13 that the solutions to equations (3.37) and (5.15) form a Feller process whose generator is given by

$$L_{X,\nabla X} f(x,z) = L^{\mathrm{dif}}_{X,\nabla X} f(x,z) + \int \left[f\left(x + F(x,y), z + \frac{\partial F}{\partial x}(x,y) z\right) - f(x,y) - \left(F(x,y), \frac{\partial f}{\partial x}\right) - \frac{\partial F}{\partial x}(x,y) z \frac{\partial f}{\partial z}\right] \nu(dy), \quad (5.17)$$

where the diffusive part of the generator is given by

$$L^{\mathrm{dif}}_{X,\nabla X} f = \frac{1}{2}\sigma_{il}\sigma_{jl}\frac{\partial^2 f}{\partial x_i \partial x_j} + b_i \frac{\partial f}{\partial x_i} + \frac{\partial b_i}{\partial x_l} z_{lj} \frac{\partial f}{\partial z_{ij}} + \sigma_{il}\frac{\partial \sigma_{pl}}{\partial x_m} z_{mq} \frac{\partial^2 f}{\partial x_i \partial z_{pq}} + \frac{1}{2}\frac{\partial \sigma_{il}}{\partial x_r} z_{rj} \frac{\partial \sigma_{pl}}{\partial x_m} z_{mq} \frac{\partial^2 f}{\partial z_{ij} \partial z_{pq}}$$

(summation over all indices is assumed). The solutions to equations (3.37) and (5.15), (5.16) form a Feller process whose generator is given by

$$\begin{aligned} &L_{X,\nabla X,\nabla^2 X} f(x,z,w) \\ &= L^{\mathrm{dif}}_{X,\nabla X,\nabla^2 X} f(x,z,w) \\ &+ \int \left\{ f\left(x + F(x,y), z + \frac{\partial F}{\partial x}(x,y) z, w + \frac{\partial F}{\partial x}(x,y) w + \left(\frac{\partial^2 F}{\partial x^2}(x,y) z, z\right)\right)\right. \\ &- f(x,z,w) - \left(F(x,y), \frac{\partial f}{\partial x}\right) - \left(\frac{\partial F}{\partial x}(x,y) z, \frac{\partial f}{\partial z}\right) \\ &\left. - \left[\frac{\partial F}{\partial x}(x,y) w + \left(\frac{\partial^2 F}{\partial x^2}(x,y) z, z\right)\right] \frac{\partial f}{\partial w}\right\} \nu(dy), \end{aligned} \quad (5.18)$$

where

$$\begin{aligned} &L^{\mathrm{dif}}_{X,\nabla X,\nabla^2 X} f(x,z,w) \\ &= L^{\mathrm{dif}}_{X,\nabla X} f(x,z) + \left(\frac{\partial^2 b_i}{\partial x_m \partial x_p} z_{mq} z_{pl} + \frac{\partial b_i}{\partial x_m} w^m_{ql}\right) \frac{\partial f}{\partial w^i_{ql}} \\ &+ \sigma_{jk}\left(\frac{\partial^2 \sigma_{ik}}{\partial x_m \partial x_p} z_{mq} z_{pl} + \frac{\partial \sigma_{ik}}{\partial x_m} w^m_{ql}\right) \frac{\partial^2 f}{\partial w^i_{ql} \partial x_j} \\ &+ \frac{\partial \sigma_{jk}}{\partial x_n} z_{nr} \left(\frac{\partial^2 \sigma_{ik}}{\partial x_m \partial x_p} z_{mq} z_{pl} + \frac{\partial \sigma_{ik}}{\partial x_m} w^m_{ql}\right) \frac{\partial^2 f}{\partial w^i_{ql} \partial z_{jr}} \\ &+ \frac{1}{2}\left(\frac{\partial^2 \sigma_{ik}}{\partial x_{m_1} \partial x_{p_1}} z_{m_1 q_1} z_{p_1 l_1} + \frac{\partial \sigma_{ik}}{\partial x_{m_1}} w^{m_1}_{q_1 l_1}\right) \\ &\times \left(\frac{\partial^2 \sigma_{jk}}{\partial x_{m_2} \partial x_{p_2}} z_{m_2 q_2} z_{p_2 l_2} + \frac{\partial \sigma_{jk}}{\partial x_{m_2}} w^{m_2}_{q_2 l_2}\right) \frac{\partial^2 f}{\partial w^i_{q_1 l_1} \partial w^j_{q_2 l_2}}. \end{aligned}$$

The barrier functions for these processes have the useful form

$$f_X^k(x) = |x|^k, \qquad f_Z^k(z) = |z|^k, \qquad f_W^k(w) = |w|^k,$$

where, for the arrays z and w,

$$|z|^2 = \sum_{i,j} z_{ij}^2, \qquad |w|^2 = \sum_{i,j,k} (w_{jk}^i)^2.$$

For these functions the actions of the above generators can be given explicitly as

$$\begin{aligned}
&L_{X,\nabla X,\nabla^2 X} f_X^k(x)\\
&\quad = k|x|^{k-2}\left((b,x) + \frac{k-2}{2}\left\|\sigma^T \frac{x}{|x|}\right\|^2 + \frac{1}{2}\|\sigma\|^2\right)\\
&\qquad + \int \left[\|x+F(x,y)\|^k - \|x\|^k - k\|x\|^{k-2}(x, F(x,y))\right]\nu(dy),\\
&L_{X,\nabla X,\nabla^2 X} f_Z^k(x,z)\\
&\quad = k|z|^{k-2}\left|\operatorname{tr}\left(\frac{\partial b}{\partial x} z z^T\right) + \frac{1}{2}(k-2)\sum_l \left(\sum_{i,r,j} \frac{\partial \sigma_{il}}{\partial x_r}\frac{z_{rj}}{|z|}\right)^2 + \frac{1}{2}\sum_{i,l,j}\left(\sum_r \frac{\partial \sigma_{il}}{\partial x_r} z_{rj}\right)^2\right|\\
&\qquad + \int \left[\|z + F(x,y)z\|^k - \|z\|^k - k\|z\|^{k-2}\left(z, \frac{\partial F}{\partial x}(x,y)z\right)\right]\nu(dy)
\end{aligned}$$

(where we have used indices explicitly to avoid possible ambiguity in the vector notation for arrays) and

$$\begin{aligned}
L_{X,\nabla X,\nabla^2 X} f_W^2(x,z,w) &= \left(\frac{\partial^2 b_i}{\partial x_m \partial x_p} z_{mq} z_{pl} + \frac{\partial b_i}{\partial x_m} w_{ql}^m\right) w_{ql}^i\\
&\quad + \sum_{i,n,l,q}\left[\sum_{m,p}\left(\frac{\partial \sigma_{in}}{\partial x_m \partial x_p} z_{mq} z_{pl} + \frac{\partial \sigma_{in}}{\partial x_m} w_{ql}^m\right)\right]^2\\
&\quad + \int \left[\frac{\partial F}{\partial x}(x,y)w + \left(\frac{\partial^2 F}{\partial x^2}(x,y)z, z\right)\right]^2 \nu(dy).
\end{aligned}$$

As $|z|^4$ is present here, the estimate for the second moment of the second derivatives can be obtained only in conjunction with the estimate to the fourth moment of the first derivative (as one might expect from Theorem 3.17).

Using the Hölder inequality yields the estimate

$$L_{X,\nabla X,\nabla^2 X} f_W^2(x,z,w) \le \left(\frac{\partial b}{\partial x} w, w\right) + c \sup_{i,j,n} \left(\left| \frac{\partial^2 b_i}{\partial x_n \partial x_j} \right| + \left| \frac{\partial \sigma_{ij}}{\partial x_n} \right| \right) \|w\|^2$$
$$+ c \sup_{i,j,n} \left| \frac{\partial \sigma_{ij}}{\partial x_n} \right|^2 \|z\|^4 + \int \left\| \frac{\partial F}{\partial x}(x,y) \right\|^2 \nu(dy) \|w\|^2$$
$$+ \int \left(\frac{\partial^2 F}{\partial x^2}(x,y)z, z \right)^2 \nu(dy). \tag{5.19}$$

Theorem 5.4 *(i) Suppose that the assumptions of Theorem 3.17(iii) hold locally (i.e. for x from any compact domain), and let*

$$L_{X,\nabla X,\nabla^2 X} f_X^k(x) < 0$$

for an even positive k and large enough x. Then the process with generator (3.40), i.e.

$$Lf(x) = \tfrac{1}{2}\left(\sigma(x) G \sigma^T(x) \nabla,\ \nabla\right) f(x) + (b(x), \nabla f(x))$$
$$+ \int \left[f(x + F(x,y)) - f(x) - (F(x,y), \nabla f(x)) \right] \nu(dy),$$

is well defined and its semigroup T_t is strongly continuous in the space $C_{|.|^k,\infty}(\mathbf{R}^d)$. Furthermore, $\mathbf{E}|X_t^x|^k \le |x|^k + ct$ holds true for all times t with a constant c.

(ii) If additionally

$$L_{X,\nabla X,\nabla^2 X} f_Z^4(x,z) < 0$$

for large x (this is essentially a positivity condition for a particular fourth-order form in z with coefficients depending on x), then

$$\mathbf{E} \left\| \frac{\partial X_t^x}{\partial x} \right\|^4 \le c e^{ct} \tag{5.20}$$

for any x, t and the space $C^1(\mathbf{R}^d)$ is invariant under T_t.

(iii) If additionally

$$L_{X,\nabla X,\nabla^2 X} (f_Z^4 + f_W^2)(x,z,w) < 0$$

for large x (owing to (5.19) this is essentially a positivity condition for a particular quadratic form in w and a fourth-order form in z with coefficients depending on x), then

$$\mathbf{E} \left\| \frac{\partial^2 X_t^x}{\partial x^2} \right\|^2 \le c e^{ct} \tag{5.21}$$

for any x, t and the space $C^2(\mathbf{R}^d)$ is invariant under T_t.

(iv) Finally, if additionally the coefficients of the operator L grow more slowly than $|x|^k$, i.e. they belong to $C_{|.|^k,\infty}(\mathbf{R}^d)$, then $C^2(\mathbf{R}^d)$ is an invariant domain and a hence core for the semigroup T_t in $C_{|.|^k,\infty}(\mathbf{R}^d)$.

Proof (i) This follows from Theorem 5.2.
(ii) Working as in Theorem 5.2 with process X_t, Z_t with the generator $L_{X,\nabla X}$ we find that, since

$$L_{X,\nabla X} f_Z^4(x,z) < c\left[1 + f_Z^4(z)\right]$$

for all x, z with a constant c, it follows that

$$\mathbf{E} f_Z^4(X_t^x, Z_t^z) \leq c e^{ct} f_Z^4(z),$$

implying (5.20) since $\partial X_t^x/\partial x = 1$ for $t = 0$. Consequently, for an $f \in C^1(\mathbf{R}^d)$,

$$\left|\frac{\partial}{\partial x}\mathbf{E} f(X_t^x)\right| \leq \mathbf{E}\left\|\frac{\partial f(X_t^x)}{\partial x}\right\| \leq c\mathbf{E}\left\|\frac{\partial X_t^x}{\partial x}\right\| \leq c(t),$$

implying the invariance of the space $C^1(\mathbf{R}^d)$.
(iii) Working as above with the process X_t, Z_t, W_t and the generator $L_{X,\nabla X,\nabla^2 X}$ we find that

$$\mathbf{E}(f_Z^4 + f_W^2)(X_t^x, Z_t^z, W_t)) \leq c e^{ct}(\|z\|^4 + \|w\|^2),$$

implying (5.21). As in (ii), this implies the invariance of functions with bounded second derivatives under the action of T_t.
(iv) Under the assumptions made, the space $C^2(\mathbf{R}^d)$ belongs to the domain. By (iii) it is invariant. □

As an example, let us consider *truncated stable-like generators*.

Proposition 5.5 *Let L have the form given by* (3.55), *where ω_p, α_p are as in Proposition 3.18(ii), and let σ, b, a be thrice continuously differentiable. Suppose that $\nu = 0$, all coefficients belong to $C_{|.|^k,\infty}$ and, for large x and a large enough constant R, the following estimates hold:*

$$(b,x)+\tfrac{1}{2}(k-2)\left\|\sigma^T\frac{x}{|x|}\right\|^2+\tfrac{1}{2}\|\sigma\|^2+R\sup_{p,s,x}a_p(x,s)<0,$$

$$\mathrm{tr}\left(\frac{\partial b}{\partial x}zz^T\right)+\sum_l\left(\sum_{i,r,j}\frac{\partial\sigma_{il}}{\partial x_r}\frac{z_{rj}}{|z|}\right)^2+\tfrac{1}{2}\sum_{i,l,j}\left(\sum_r\frac{\partial\sigma_{il}}{\partial x_r}z_{rj}\right)^2$$
$$+R\sup_{p,s,x}\left|\frac{\partial a_p(x,s)}{\partial x}\right|^2<0,$$

$$\left(\frac{\partial b}{\partial x}w,w\right)+c\sup_{i,j,n}\left(\left|\frac{\partial^2 b_i}{\partial x_n\partial x_j}\right|+\left|\frac{\partial\sigma_{ij}}{\partial x_n}\right|\right)\|w\|^2+c\sup_{i,j,n}\left|\frac{\partial\sigma_{ij}}{\partial x_n}\right|^2\|z\|^4$$
$$+R\sup_{p,s,x}\left(\left|\frac{\partial^2 a_p(x,s)}{\partial x^2}\right|^2+\left|\frac{\partial a_p(x,s)}{\partial x}\right|^4\right)<0$$

(the constant R can be calculated explicitly). Then the process is well defined and its semigroup T_t is strongly continuous in $C_{|.|^k,\infty}(\mathbf{R}^d)$ and has $C^2(\mathbf{R}^d)$ as an invariant core.

Proof This follows from the previous theorem and the obvious estimates for the derivatives of the mapping F from Corollary 3.9. Namely,

$$\frac{\partial F_{x,s}(z)}{\partial x}=\left[K^{-\alpha}+\frac{\alpha}{a}\left(\frac{1}{z}-\frac{1}{K}\right)\right]^{-(1+1/\alpha)}\left(\frac{1}{z}-\frac{1}{K}\right)\nabla\frac{\alpha}{a}(x),$$

so that, for example,

$$\int_0^K\left\|\frac{\partial F_{x,s}(z)}{\partial x}\right\|^2\frac{dz}{z^2}$$
$$=\left\|\nabla\frac{\alpha}{a}\right\|^2\int_0^K\left[K^{-\alpha}+\frac{\alpha}{a}\left(\frac{1}{z}-\frac{1}{K}\right)\right]^{-(2+2/\alpha)}\left(\frac{1}{z}-\frac{1}{K}\right)^2\frac{dz}{z^2}$$
$$=\left\|\nabla\frac{\alpha}{a}(x)\right\|^2\left(\frac{a}{\alpha}\right)^3\int_0^\infty(K^{-\alpha}+r)^{-(2+2/\alpha)}r^2dr,$$

which is of order $\|\nabla a\|^2$ for a bounded below. □

PART II

Nonlinear Markov processes and semigroups

6
Integral generators

This chapter opens the mathematical study of nonlinear Markov semigroups. It is devoted to semigroups with integral generators. This case includes the dynamics described by the spatially homogeneous Smoluchovski and Boltzmann equations, as well as the replicator dynamics of spatially trivial evolutionary games. We start with an introductory section giving a more detailed description of the content of this chapter.

6.1 Overview

For nonlinear extensions of Markov semigroups, the formulation of their duals is very useful. A Feller semigroup Φ_t on $C_\infty(X)$ clearly gives rise to a dual positivity-preserving semigroup $\Phi_t^\star$ on the space $\mathcal{M}(X)$ of bounded Borel measures on X through the duality identity $(\Phi_t f, \mu) = (f, \Phi_t^\star \mu)$, where the pairing (f, μ) is given by integration. If A is the generator of Φ_t then $\mu_t = \Phi_t^\star \mu$ can be characterized by a general equation that in its weak form becomes

$$\frac{d}{dt}(g, \mu_t) = (Ag, \mu_t) = (g, A^\star \mu_t), \tag{6.1}$$

where $A^\star$ is the dual to A. This equation holds for all g from the domain of A.

In this book we are interested in the nonlinear analog of (6.1),

$$\frac{d}{dt}(g, \mu_t) = \Omega(\mu_t) g, \tag{6.2}$$

which holds for g from a certain dense domain D of $C(\mathbf{R}^d)$, where Ω is a nonlinear transformation from a dense subset of $\mathcal{M}(X)$ to the space of linear functionals on $C(X)$ having a common domain containing D.

As we observed in Chapter 1, when describing the LLN limit of a Markov model of interaction, the r.h.s. of (6.2) takes the form $(A_{\mu_t} g, \mu_t)$, for a certain family of conditionally positive linear operators A_μ of Lévy–Khintchine type with variable coefficients depending on μ as a parameter. In Section 1.1, devoted to measures on a finite state space, we called this representation of Ω *stochastic*, as it leads naturally to a stochastic interpretation of an evolution.

A natural question arises (it is posed explicitly in Stroock [227]): can such a representation be deduced, as in the linear case, merely from the assumption that the corresponding measure-valued evolution is positivity preserving? This question was partially answered in [135], where a positive answer was given under the additional assumption that $\Omega(\mu)$ depends polynomially on μ (this assumption can be extended to analytic functionals $\Omega(\mu)$). In order not to interrupt our main theme, we shall discuss this topic in detail in Section 11.5 after developing a nonlinear analog of the notion of conditional positivity, and only the simpler case, for integral operators, will be settled in Section 6.8.

Let us recall that in Chapter 4 we specified three groups of Lévy–Khintchine-type operators Ω (which will appear in a stochastic representation of the r.h.s. of (6.2): integral operators (without smoothness requirements for the domain); operators of order at most one (this domain contains continuously differentiable functions); and the full Lévy–Khintchine operators (this domain contains twice-differentiable functions). These groups differ in the methods of analysis used. In what follows, we shall develop each new step in our investigation separately for the three groups. Well-posedness will be discussed in Chapters 6 and 7, smoothness with respect to initial data in Chapter 8, the LLN for particle approximations in Chapter 9 and the CLT in Chapter 10. To simplify the discussion of the third group (full Lévy–Khintchine operators) we shall reduce our attention to the most natural example, that of possibly degenerate stable-like generators combined with second-order differential operators.

The exposition is given in a form that allows one to read the whole story for each group of operators almost independently of the other groups. In particular, readers who are interested only in pure jump models (including in particular the homogeneous Boltzmann and Smoluchovskii models, as well as the replicator dynamics of evolutionary games) can skip the discussion of the other types of nonlinear stochastic evolution.

This chapter is devoted to the well-posedness of the evolution equations (6.2) for which the r.h.s. has an integral stochastic representation; we start with the case of bounded generators and then discuss unbounded kernels with additive bounds for rates given in terms of a conservation law. This includes the basic spatially trivial models of coagulation and collision. Next we prove a couple of existence results, including unbounded kernels with multiplicative bounds for the rates. Finally we give a characterization of bounded operators generating positivity preserving semigroups, in terms of a nonlinear analog of the notion of conditional positivity.

6.2 Bounded generators

As a warm-up we now consider bounded generators, giving several proofs of the following simple but important result. The methods are instructive and can be applied in various situations.

Theorem 6.1 *Let X be a complete metric space and let*

$$A_\mu f(x) = \int_X f(y)\nu(x, \mu, dy) - a(x, \mu) f(x), \tag{6.3}$$

where $\nu(x, \mu, .)$, $x \in X, \mu \in \mathcal{M}(X)$, is a family of measures from $\mathcal{M}(X)$ depending continuously on x, μ (μ, ν are considered in the weak topology) and where $a(x, \mu)$ is a function continuous in both variables such that

$$\|\nu(x, \mu, .)\| \le a(x, \mu) \le \kappa(\|\mu\|)$$

for a certain positive $\kappa \in C^1(\mathbf{R}_+)$. Finally, let a, ν be locally Lipschitz continuous with respect to the second variable, i.e.

$$\|\nu(x, \xi, .) - \nu(x, \eta, .)\| + |a(x, \xi) - a(x, \eta)| \le c(\lambda_0)\|\xi - \eta\|,$$
$$\xi, \eta \in \lambda_0 \mathcal{P}(X), \tag{6.4}$$

for any $\lambda_0 > 0$ and $\lambda_0 \mathcal{P}(X) = \{\lambda\mu : \lambda \le \lambda_0, \mu \in \mathcal{P}(X)\}$. Then there exists a unique sub-Markov semigroup T_t of (nonlinear) contractions in $\mathcal{M}(S)$ solving globally (for all $t > 0$) the weak nonlinear equation

$$\frac{d}{dt}(g, \mu_t) = (A_{\mu_t} g, \mu_t), \qquad \mu_0 = \mu \in \mathcal{M}(X),\ g \in C(X). \tag{6.5}$$

This solution is in fact strong; that is, the derivative $\dot{\mu}_t$ exists in the norm topology of $\mathcal{M}(S)$ and depends Lipschitz continuously on the initial state, i.e.

$$\sup_{s \le t} \|\mu_s^1 - \mu_s^2\| \le c(t)\|\mu_0^1 - \mu_0^2\|.$$

Finally, if $\|\nu(x, \mu, .)\| = a(x, \mu)$ identically then this semigroup is Markov.

First proof By Theorem 2.33 (more precisely, its obvious time-non-homogeneous extension), for any weakly continuous curve $\mu_t \in \lambda_0 \mathcal{P}(X)$ there exists a backward propagator $U^{t,s}$ in $C(X)$ solving the equation $\dot{g}_t = A_{\mu_t} g$.

Its dual propagator $V^{s,t}$ acts in $\lambda_0\mathcal{P}(X)$. It is important to stress that, by the integral representation of $U^{t,r}$, the dual operators actually act in $\mathcal{M}(X)$ and not just in the dual space $(C(X))^\star$. Owing to assumption (6.4), the existence and uniqueness of a weak solution follows from Theorem 2.12 with $B = D = C(X)$ and $M = \lambda_0\mathcal{P}(X)$. Finally, the existence of the derivative $\dot{T}_t(\mu) = \dot{\mu}_t$ in the norm topology follows clearly from the fact that $A^\star_{\mu_t}\mu_t$ is continuous (even Lipschitz continuous) in t in the norm topology.

Second proof By (6.4), $A^\star_{\mu_t}\mu_t$ is Lipschitz continuous in the norm provided that $\|\mu\|$ remains bounded. Hence, by a standard result for ordinary differential equations in Banach spaces (see Appendix D), the solution $\mu_t = T_t(\mu)$ in the norm topology exists and is unique locally, i.e. as long as it remains bounded, yet it may take values in the whole space of signed measures. As a bound for $\|T_t(\mu)\|$ we can take the solution to the equation $\dot{v}_t = 2\kappa(v_t)$, $v_0 = \|\mu\|$. Hence the solution $T_t(\mu)$ is well defined at least on the interval $t \in [0, t_0]$, where t_0 is defined by the equation $v_{t_0} = 2\|\mu\|$. However, $(1, \mu_t)$ does not increase along the solution. Thus if we can show that the solution remains positive then $\|\mu_t\| = (1, \mu_t)$ does not increase. Consequently, after time t_0 we can iterate the above procedure, producing a unique solution for $t \in [t_0, 2t_0]$ and so on, thus completing the proof. Finally, in order to obtain positivity we can compare μ_t with the solution of the equation $\dot{\xi}_t = -a(x)\xi_t$ (i.e. μ_t is bounded below by ξ_t), which has an obviously positive solution.

Third proof This approach (seemingly first applied in the context of the Boltzmann equation) suggests that we should rewrite our equation as

$$\frac{d}{dt}(g, \mu_t) = -K(g, \mu_t) + [A^\star_{\mu_t}\mu_t + K(g, \mu_t)]$$

and then represent it in integral form (the so-called *mild* or *interaction representation*) using the du Hamel principle (see e.g. (2.35)):

$$\mu_t = e^{-Kt}\mu + \int_0^t e^{-K(t-s)}\left(A^\star_{\mu_s}\mu_s + K\mu_s\right) ds.$$

This equation is a fixed-point equation for a certain nonlinear operator Φ acting in the metric space $C_\mu([0, r], \mathcal{M}(X))$ of continuous functions on $[0, r]$ such that $\mu_0 = \mu$ with values in the Banach space $\mathcal{M}(X)$, μ being equipped with the sup norm $\|\mu(.)\| = \sup_{s\in[0,r]} \|\mu_s\|$. Under the conditions of the theorem we see that ϕ is a contraction for any $K > 0$ and small enough r and, consequently, has (by the fixed-point principle) a unique fixed point that can be approximated by iteration. The crucial observation is the following: if $K > \sup\{a(x, \mu) : \mu \in \lambda_0\mathcal{P}(X)\}$ then Φ preserves positivity so that, starting the iterations from the constant function $\mu_s = \mu$, one obtains necessarily the positive fixed point. Then one can extend the solution to arbitrary times, as in the previous proof.

Fourth proof Finally let us use the method of *T-mappings* or *T-products* (time-ordered or chronological products). Let $V^s[\mu]$ denote the semigroup of contractions in $\lambda_0\mathcal{P}(X)$ solving the Cauchy problem for the equation $\dot{\nu} = A^\star_\mu \nu$ (which is, of course, dual to the semigroup solving $\dot{g} = A_\mu g$). Define the approximations μ^τ to the solution of (6.5) recursively as

$$\mu^\tau_t = V^{t-l\tau}[\mu^\tau_{(l-1)\tau}]\mu^\tau_{(l-1)\tau}, \qquad l\tau < t \leq (l+1)\tau.$$

The limit of such approximations as $\tau \to 0$ is called the *T-mapping* based on the family $V^s[\mu]$. When everything is uniformly bounded (as in our situation), it is easy to see that one can choose a converging subsequence and that the limit satisfies the equation (6.5). Obviously this approach yields positivity-preserving solutions.

Let us stress again that locally (i.e. for small times) the well-posedness of (6.4) is the consequence of a standard result for ODEs in Banach spaces, and the use of positivity is needed to obtain the global solution.

In applications it is often useful to have the following straightforward extension of the previous result.

Theorem 6.2 *Let X, a, ν, A be as in Theorem 6.1 and let $b \in C(X)$ with $\|b\| \leq c$. Then the Cauchy problem*

$$\frac{d}{dt}(g, \mu_t) = (A_{\mu_t} g, \mu_t) + (bg, \mu_t), \qquad \mu_0 = \mu \in \mathcal{M}(X), \; g \in C(X),$$

has a unique global solution $\Phi_t(\mu)$, which is strong. Thus Φ_t forms a semigroup of bounded nonlinear transformations of $\mathcal{M}(X)$ possessing the estimate $\|\Phi_t(\mu)\| \leq e^{ct}\|\mu\|$.

One is often interested in the regularity of the solution semigroup T_t, in particular when it can be defined on functions and not only on measures.

Theorem 6.3 *Under the assumptions of Theorem 6.1 suppose that X is $\mathbf{R}^d$ or its open subset and that for measures μ with densities, say, f_μ with respect to the Lebesgue measure, the continuous dual kernel $\nu'(x, \mu, dy)$ exists such that*

$$\nu(x, \mu, dy)\, dx = \nu'(y, \mu, dx)\, dy$$

and, moreover, the $\|\nu'(x, \mu, .)\|$ are uniformly bounded for bounded μ. Then the semigroup T_t from Theorem 6.1 preserves measures with a density, i.e. it acts in $L_1(X)$.

Proof This is straightforward. Referring to the proofs of Theorems 2.33 or 4.1 one observes that, for a continuous curve $\mu_t \in L_1(X)$, the dual propagator

$V^{s,t}$ would act in $L_1(X)$ owing to the existence of the dual kernel. Hence the required fixed point would also be in $L_1(X)$. □

As a direct application we will derive a simple well-posedness result for the equations of Section 1.5 in the case of bounded coefficients.

Theorem 6.4 *In equation* (1.41) *(which contains as particular cases the spatially homogeneous Smoluchovski and Boltzmann equations), let all kernels P be uniformly bounded and let the P^l be* **1***-subcritical for $l > 1$. Then for any $\mu \in \mathcal{M}(X)$ there exists a unique global solution $T_t(\mu)$ of* (1.41) *in $\mathcal{M}(X)$. This solution is strong (i.e. $\dot{T}_t(\mu)$ exists in the norm topology) and the resolving operators T_t form a semigroup. If the P^l are* **1***-subcritical (resp.* **1***-critical) for all l, this semigroup is sub-Markov (resp. Markov).*

Proof Note that equation (1.41) can be written equivalently as

$$\frac{d}{dt}\int_X g(z)\mu_t(dz) = \sum_{l=1}^{k} \frac{1}{l!}\int_{\mathcal{X}}\int_{X^l} \left[g^{\oplus}(\mathbf{y}) - lg(z_1)\right] P^l(\mathbf{z}, d\mathbf{y})\mu_t^{\otimes l}(d\mathbf{z}), \tag{6.6}$$

which is of the form (6.5), (6.3) with

$$\int g(y)\nu(x,\mu,dy) = \sum_{l=1}^{k} \frac{1}{l!}\int_{\mathcal{X}}\int_{X^{l-1}} g^{\oplus}(\mathbf{y})P^l(x, z_1, \ldots, z_{l-1}, d\mathbf{y}) \times \mu_t(dz_1)\cdots\mu_t(dz_{l-1}),$$

so that in the case of uniformly bounded kernels we have

$$\|\nu(x,\mu,.) - \nu(x,\eta,.)\| \leq c\|\mu - \eta\|$$

uniformly for bounded μ and η. If all P^l are **1**-subcritical, the required result follows directly from Theorem 6.1. Otherwise, if P^1 is allowed not to be subcritical, Theorem 6.2 applies. □

In order to apply Theorem 6.3 one has to know the dual kernel. In the usual models these dual kernels are given explicitly. Consider, for example, the classical Smoluchovski equation, i.e. equation (1.46) with $X = \mathbf{R}_+$, $E(x) = x$ and $K(x_1, x_2, dy) = K(x_1, x_2)\delta(x_1 + x_2 - y)$ for a certain specific symmetric function $K(x_1, x_2)$:

$$\frac{d}{dt}\int_{\mathbf{R}_+} g(z)\mu_t(dz) = \frac{1}{2}\int_{(\mathbf{R}_+)^2} [g(x_1 + x_2) - g(x_1) - g(x_2)]K(x_1, x_2) \times \mu_t(dx_1)\mu_t(dx_2). \tag{6.7}$$

By the symmetry of the coagulation kernel K, this can be also written as

$$\frac{d}{dt}\int_{\mathbf{R}_+} g(z)\mu_t(dz) = \frac{1}{2}\int_{(\mathbf{R}_+)^2} [g(x+z) - 2g(x)]K(x,z)\mu_t(dx)\mu_t(dz). \tag{6.8}$$

This has the form (6.5), (6.3) with

$$\nu(x,\mu,dy) = \int_{z\in\mathbf{R}_+} K(x,z)\delta(y-x-z)\mu(dz)$$

and the dual kernel is

$$\nu'(y,\mu,dx) = \int_0^y \mu(dz)K(y-z,z)\delta(x+z-y),$$

so that

$$\int f(x,y)\nu(x,\mu,dy)dx = \int f(x,y)\nu'(y,\mu,dx)dy$$
$$= \int_0^\infty \int_0^\infty f(x,x+z)K(x,z)\mu(dz)dx$$

for $f \in C_\infty(\mathbf{R}_+^2)$ (i.e. for continuous f vanishing at infinity and on the boundary of $(\mathbf{R}_+)^2$. Hence the strong form of equation (6.7) for the densities f of the measures μ reads as follows:

$$\frac{d}{dt}f_t(x) = \int_0^x f(z)dzK(x-z,z)f(x+z) - f_t(x)\int K(x,z)f(z)dz. \tag{6.9}$$

Theorem 6.3 implies that, if K is continuous and bounded, this equation is well posed and the solutions specify a semigroup in $L_1(\mathbf{R}_+)$.

Similarly, the classical spatially homogeneous Botzmann equation in the weak form (1.52) clearly can be rewritten as

$$\frac{d}{dt}(g,\mu_t) = \frac{1}{2}\int_{S^{d-1}}\iint [g(w_1) - g(v_1)]B(|v_1 - v_2|,\theta)dn\mu_t(dv_1)\mu_t(dv_2), \tag{6.10}$$

where we recall that $B(|v|,\theta) = B(|v|,\pi-\theta)$,

$$w_1 = v_1 - n(v_1 - v_2, n), \qquad w_2 = v_2 + n(v_1 - v_2, n)$$

and θ is the angle between $v_2 - v_1$ and n. This equation has the form (6.5), (6.3) with

$$\int \psi(y)\nu(v_1,\mu,dy) = \iint \psi(v_1 - n(v_1 - v_2, n))B(|v_1 - v_2|,\theta)dn\mu(dv_2).$$

In order to find the dual kernel, observe that when μ has a density f one has

$$\begin{aligned}\int_{\mathbf{R}^{2d}} & g(v_1)\psi(y)\nu(v_1, f, dy)dv_1 \\ &= \int_{S^{d-1}}\int_{\mathbf{R}^{2d}} g(v_1)\psi(w_1)B(|v_1 - v_2|, \theta)dn\, dv_1 f(v_2)dv_2 \\ &= \int_{S^{d-1}}\int_{\mathbf{R}^{2d}} g(w_1)\psi(v_1)B(|v_1 - v_2|, \theta)dnf(w_2)dv_1dv_2 \\ &= \int_{\mathbf{R}^{2d}} g(y)\psi(v_1)\nu'(v_1, f, dy)dv_1 \end{aligned} \tag{6.11}$$

(here we have first relabeled (v_1, v_2) by (w_1, w_2) and then changed dw_1dw_2 to dv_1dv_2, noting that (i) the Jacobian of this (by (1.48)) orthogonal transformation is 1, (ii) $|v_1 - v_2| = |w_1 - w_2|$ and (iii) the angle θ between $v_2 - v_1$ and $w_1 - v_1$ coincides with the angle between $w_2 - w_1$ and $v_1 - w_1$).

Equation (6.11) implies that the dual kernel exists and is given by

$$\int g(y)\nu'(v_1, f, dy) = \int g(v_1 - n(v_1 - v_2, n))B(|v_1 - v_2|, \theta)dnf(w_2)dv_2.$$

Hence the solution operator for the Boltzmann equation preserves the space $L_1(\mathbf{R}^d)$.

It is also useful to know whether the solutions to kinetic equations preserve the space of bounded continuous functions. To answer this question for the Boltzmann equation, we may use the Carleman representation. Namely, from equation (H.1) it follows that for bounded B one has

$$\dot{f}_t(v) = O(1)\int f_t(v)dv\|f_t\|,$$

where as usual $\|f\|$ denotes the sup norm. Consequently, Gronwall's inequality yields the following result.

Proposition 6.5 *If $f_0 \in C(\mathbf{R}^d)$ then the solution f_t to the Boltzmann equation* (G.5) *with initial condition f_0 remains in $C(\mathbf{R}^d)$ for all times and the semigroup $\Phi_t : f_0 \mapsto f_t$ is bounded in $C(\mathbf{R}^d)$.*

6.3 Additive bounds for rates: existence

Application of the fixed-point principle in the spirit of Theorem 2.12 is very effective in solving nonlinear problems. It was used in the first proof of Theorem 6.1 above and will be further demonstrated in the next chapter. However, in some situations the corresponding linear problem is not regular enough for this approach to be applicable, so that other methods must be used. This will be the case for the problems considered from now on in this chapter.

In this section we will consider equation (1.73) for pure jump interactions,

$$\frac{d}{dt}(g, \mu_t) = \sum_{l=1}^{k} \frac{1}{l!} \int_{X^l} \left\{ \int_{\mathcal{X}} [g^{\oplus}(\mathbf{y}) - g^{\oplus}(\mathbf{z})] P(\mu_t, \mathbf{z}; d\mathbf{y}) \right\} \mu_t^{\otimes l}(d\mathbf{z}),$$
$$\mu_0 = \mu, \quad (6.12)$$

and its integral version

$$(g, \mu_t) - (g, \mu) = \int_0^t ds \sum_{l=1}^{k} \frac{1}{l!} \int_{X^l} \left\{ \int_{\mathcal{X}} \left[g^{\oplus}(\mathbf{y}) - g^{\oplus}(\mathbf{z}) \right] P(\mu_s, \mathbf{z}; d\mathbf{y}) \right\} \times \mu_s^{\otimes l}(d\mathbf{z}), \quad (6.13)$$

with unbounded P, where X is a locally compact metric space.

Let us start with some basic definitions concerning the properties of transition rates. Let E be a non-negative function on X. The number $E(x)$ will be called the size of a particle x (E could stand for the mass in mass-exchange models, such as those for coagulation–fragmentation, and for the kinetic energy when modeling Boltzmann-type collisions). We say that the transition kernel $P = P(\mathbf{x}; d\mathbf{y})$ in (1.29) is *E-subcritical* (resp. *E-critical*) if

$$\int \left[E^{\oplus}(\mathbf{y}) - E^{\oplus}(\mathbf{x}) \right] P(\mathbf{x}; d\mathbf{y}) \leq 0 \quad (6.14)$$

for all $\mathbf{x}$ (resp. if equality holds). We say that $P(\mathbf{x}; d\mathbf{y})$ is *E-preserving* (resp. *E-non-increasing*) if the measure $P(\mathbf{x}; d\mathbf{y})$ is supported on the set $\{\mathbf{y} : E^{\oplus}(\mathbf{y}) = E^{\oplus}(\mathbf{x})\}$ (resp. $\{\mathbf{y} : E^{\oplus}(\mathbf{y}) \leq E^{\oplus}(\mathbf{x})\}$). Clearly, if $P(\mathbf{x}; d\mathbf{y})$ is E-preserving (resp. E-non-increasing) then it is also E-critical (resp. E-subcritical). For instance, if $E = \mathbf{1}$ then E-preservation (subcriticality) means that the number of particles remains constant (does not increase on average) during the evolution of the process. As we shall see later, subcriticality enters practically all natural assumptions ensuring the non-explosion of interaction models.

We shall say that our transition kernel P is *multiplicatively E-bounded* or *$E^{\otimes}$-bounded* (resp. *additively E-bounded* or *$E^{\oplus}$-bounded*) whenever $P(\mu; \mathbf{x}) \leq cE^{\otimes}(\mathbf{x})$ (resp. $P(\mu; \mathbf{x}) \leq cE^{\oplus}(\mathbf{x})$) for all μ and $\mathbf{x}$ and some constant $c > 0$, where we have used the notation of (1.27), (1.28).

We shall deal now (and in most further discussion) with additively bounded kernels. However, some existence criteria for multiplicative bounds will be given later.

Lemma 6.6 *The following elementary inequalities hold for all positive a, b, β:*

$$(a+b)^\beta - a^\beta - b^\beta \le 2^\beta (ab^{\beta-1} + ba^{\beta-1}), \tag{6.15}$$

$$(a+b)^\beta - a^\beta \le \beta \max(1, 2^{\beta-1})\, b(b^{\beta-1} + a^{\beta-1}). \tag{6.16}$$

Proof For $\beta \le 1$, inequality (6.15) holds trivially, as the l.h.s. is always non-positive. Hence, by homogeneity, in order to prove (6.15) it is enough to show that

$$(1+x)^\beta - 1 \le 2^\beta x, \qquad \beta \ge 1,\ x \in (0,1). \tag{6.17}$$

Next, the mean value theorem implies that

$$(1+x)^\beta - 1 \le \beta 2^{\beta-1} x, \qquad \beta \ge 1,\ x \in (0,1),$$

yielding (6.17) for $\beta \in [1,2]$. For $\beta \ge 2$ the inequality $g(x) = (1+x)^\beta - 1 - 2^\beta x \le 0$ holds, because $g(0) = 0$, $g(1) = -1$ and $g'(x)$ is increasing. To prove (6.16) observe that the mean value theorem implies that

$$(a+b)^\beta - a^\beta \le \begin{cases} \beta(a+b)^{\beta-1} b, & \beta \ge 1, \\ \beta b a^{\beta-1}, & \beta \in (0,1). \end{cases}$$

□

Theorem 6.7 *Suppose that the transition kernels $P(\mu, \mathbf{x}, .)$ enjoy the following properties.*

(i) $P(\mu, \mathbf{x}, .)$ is a continuous function taking $\mathcal{M}(X) \times \cup_{l=1}^k SX^l$ to $\mathcal{M}(\cup_{l=1}^k SX^l)$ (i.e. not more than k particles can interact or be created simultaneously), where the measures $\mathcal{M}$ are considered in their weak topologies;

(ii) $P(\mu, \mathbf{x}, .)$ is E-non-increasing and $(1+E)^{\oplus}$-bounded for some continuous non-negative function E on X such that $E(x) \to \infty$ as $x \to \infty$;

(iii) the $P(\mu, \mathbf{x}, .)$ are **1**-*subcritical for* $\mathbf{x} \in X^l$, $l \ge 2$.

Suppose that $\int (1+E^\beta)(x)\mu(dx) < \infty$ for the initial condition μ with some $\beta > 1$. Then there exists a global non-negative solution μ_t of (6.13) *that is E-non-increasing, i.e. with $(E, \mu_t) \le (E, \mu)$, $t \ge 0$, such that for an arbitrary T*

$$\sup_{t\in[0,T]} \int (1+E^\beta)(x)\mu_t(dx) \le C(T, \beta, \mu), \tag{6.18}$$

for some constant $C(T, \beta, \mu)$.

Proof Let us first approximate the transition kernel P by cut-off kernels P_n defined by the equation

$$\int g(\mathbf{y})P_n(\mu, \mathbf{z}, d\mathbf{y}) = \int \mathbf{1}_{\{E^{\oplus}(\mathbf{z})\leq n\}}(z)g(\mathbf{y})\mathbf{1}_{\{E^{\oplus}(\mathbf{y})\leq n\}}(y)P(\mu, \mathbf{z}, d\mathbf{y}), \tag{6.19}$$

for arbitrary g. It is easy to see that P_n possesses the same properties (i)–(iii) as P, but at the same time it is bounded and hence solutions μ_t^n to the corresponding kinetic equations with initial condition μ exist by Theorem 6.1. As the evolution defined by P_n clearly does not change measures outside the compact region $\{y : E(y) \leq n\}$, it follows that if $\int(1 + E^{\beta})(x)\mu(dx) < \infty$ then the same holds for μ_t for all t. Our aim now is to obtain a bound for this quantity which is independent of n.

Using (1.75), we can denote by F_g the linear functional on measures $F_g(\mu) = (g, \mu)$ and by $\Lambda F_g(\mu_t)$ the r.h.s. of (6.12). Note first that, by assumption (iii),

$$\Lambda F_{\mathbf{1}}(\mu) \leq ck F_{1+E}(\mu), \tag{6.20}$$

which by Gronwall's lemma implies (6.18) for $\beta = 1$.

Next, for any $\mathbf{y} = (y_1, \dots, y_l)$ in the support of $P(\mu, \mathbf{x}, .)$, we have

$$(E^{\beta})^{\oplus}(\mathbf{y}) \leq [E^{\oplus}(\mathbf{y})]^{\beta} \leq [E^{\oplus}(\mathbf{x})]^{\beta}$$

as P is E-non-increasing and the function $z \mapsto z^{\beta}$ is convex. Consequently,

$$\begin{aligned}\Lambda F_{E^{\beta}}(\mu) &= \sum_{l=1}^{k} \frac{1}{l!}\int_{X^l}[(E^{\beta})^{\oplus}(\mathbf{y}) - (E^{\beta})^{\oplus}(\mathbf{x})]P(\mu, \mathbf{x}, dy)\mu^{\otimes l}(d\mathbf{x}) \\ &\leq \sum_{l=2}^{k} \frac{1}{l!}\int \left\{[E(x_1) + \cdots + E(x_l)]^{\beta} - E(x_1)^{\beta} - \cdots - E^{\beta}(x_l)\right\} \\ &\qquad \times P(\mu, \mathbf{x}, dy)\mu^{\otimes l}(d\mathbf{x}).\end{aligned}$$

Using the symmetry with respect to permutations of $x_1, \dots, x_l$ and the assumption that P is $(1 + E)^{\oplus}$-bounded, one deduces that this expression does not exceed

$$\sum_{l=2}^{k} \frac{1}{(l-1)!}\int \left\{[E(x_1) + \cdots + E(x_l)]^{\beta} - E^{\beta}(x_1) - \cdots - E^{\beta}(x_l)\right\} \times [1 + E(x_1)]\prod_{j=1}^{l}\mu(dx_j).$$

Using (6.15) with $a = E(x_1)$, $b = E(x_2) + \cdots + E(x_l)$ and induction on l yields

$$[E(x_1) + \cdots + E(x_l)]^\beta - E^\beta(x_1) - \cdots - E^\beta(x_l) \le c \sum_{i \ne j} L(x_i) L^{\beta-1}(x_j).$$

Using (6.15) yields

$$\left\{[E(x_1) + \cdots + E(x_l)]^\beta - E^\beta(x_1)\right\} E(x_1)$$
$$\le c \sum_{i \ne 1} [E(x_1)E(x_i)^\beta + E(x_1)^\beta E(x_i)].$$

Again by symmetry, this implies that

$$\Lambda F_{E^\beta}(\mu) \le c \int \sum_{l=2}^{k} \left[E^\beta(x_1) + E^{\beta-1}(x_1)\right] E(x_2) \prod_{j=1}^{l} \mu(dx_j), \qquad (6.21)$$

which, using $E^{\beta-1}(x_1)E(x_2) \le E^\beta(x_1) + E^\beta(x_2)$ can be rewritten as

$$\Lambda F_{E^\beta}(\mu) \le c \int \sum_{l=2}^{k} E^\beta(x_1)\,[1 + E(x_2)] \prod_{j=1}^{l} \mu(dx_j). \qquad (6.22)$$

By (6.20) it follows that

$$\Lambda F_{1+E^\beta}(\mu) \le c \sum_{l=1}^{k} \left[E^\beta(x_1) + 1\right] [E(x_2) + 1] \prod_{j=1}^{l} \mu(dx_j)$$

and, consequently, since $(E, \mu_t^n) \le (E, \mu)$, using (6.18) for $\beta = 1$ the above relation implies that

$$\frac{d}{dt}(1 + E^\beta, \mu_t^n) = \Lambda F_{1+E^\beta}(\mu_t^n) \le c(T, \beta, \mu)(1 + E^\beta, \mu_t^n).$$

Thus by Gronwall's lemma obtain, for arbitrary T,

$$F_{1+E^\beta}(\mu_t^n) < C(T, \beta, \mu)$$

for some $C(T, \beta, \mu)$ and for all $t \in [0, T]$ and all n. This implies that the family μ_t^n is weakly compact for any t and that any limiting point has the estimate (6.18). As the real-valued function $\int g(x)\mu_t^n(dx)$ is absolutely continuous for any $g \in C_c(X)$ (this follows directly from (6.13)), choosing a countable dense sequence of such functions allows us to find a subsequence, denoted μ_t^n also, which converges in the space of continuous functions from $[0, T]$ to $\mathcal{M}(X)$, the latter being taken to have its weak topology. It remains to show that the limit μ_t of this subsequence satisfies (6.13) by passing to the limit in the corresponding equations for μ_t^n. But this is obvious: all integrals outside the domain $\{y : E(y) < K\}$ can be made arbitrarily small by choosing large enough K (because of (6.18)), and inside this domain the result follows from weak convergence. □

Remark 6.8 The assumption $\int(1 + E^\beta)(x)\mu(dx) < \infty$ with $\beta > 1$ is actually not needed to prove existence; see the comments in Section 11.6.

Remark 6.9 A similar existence result holds for the integral version of the equation

$$\frac{d}{dt}(g, \mu_t) = \sum_{l=1}^{k} \frac{1}{l!} \int_{\mathcal{X}} \int_{X^l} \left[g^\oplus(\mathbf{y}) - g^\oplus(\mathbf{z})\right] P^l(\mathbf{z}; d\mathbf{y}) \left(\frac{\mu_t}{\|\mu_t\|}\right)^{\otimes l} (d\mathbf{z})\|\mu_t\|. \tag{6.23}$$

To obtain existence, we have to show that local solutions cannot vanish at any finite time. To this end, observe that any such solution is bounded from below by a solution to the equation

$$\frac{d}{dt}\mu_t(dz) = -\sum_{l=1}^{k} \frac{1}{(l-1)!} \frac{\mu_t(dz)}{\|\mu_t\|^{l-1}} \int_{z_1,\dots,z_{l-1}} \mu_t(dz_1)\cdots\mu_t(dz_{l-1})P(\mu; z, z_1, \dots, z_{l-1})$$

obtained from (6.23) by ignoring the positive part of the r.h.s. In their turn, solutions to this equation are bounded from below by solutions to the equation

$$\frac{d}{dt}\mu_t(dz) = -\sum_{l=1}^{k} \frac{1}{(l-1)!}[1 + E(z)]\,\mu_t(dz) - \sum_{l=2}^{k} \frac{1}{(l-2)!}\mu_t(dz_1)\frac{\int(1 + E(u))\mu(du)}{\|\mu\|}.$$

The solution to this equation can be found explicitly (see the exercise below) and it never vanishes. Hence μ_T does not vanish, as required.

Exercise 6.1 Suppose that X is a locally compact metric space, E is a non-negative continuous function on it and a, b are two positive constants. Then the solution to the measure-valued ordinary differential equation

$$\dot{\mu}(dx) = -aE(x)\mu(dx) - b\,\frac{\int E(u)\mu(du)}{\int \mu(du)}\mu(dx)$$

on (positive) measures μ on an X with initial condition μ_0 equals

$$\mu_t(dx) = \exp[-atE(x)]\mu_0(dx)$$
$$\times \exp\left\{-b\int_0^t \frac{\int E(u)\exp[-asE(u)]\mu_0(du)}{\int \exp[-asE(u)]\mu_0(du)}\,ds\right\}$$

and in particular it does not vanish when $\mu_0 \neq 0$. Hint: This is obtained by simple calculus; the details can be found in [132].

Exercise 6.2 Show (following the proof above) that the estimate (6.18) can be written in a more precise form,

$$\sup_{t\in[0,T]}(L^\beta, \mu_t) \leq c(T, \beta, (1+L, \mu_0))(L^\beta, \mu_0) \tag{6.24}$$

for some constant c.

6.4 Additive bounds for rates: well-posedness

Let us first discuss the regularity of the solutions constructed above.

Theorem 6.10 *Let μ_t be a solution to* (6.13) *satisfying* (6.18) *for some $\beta >$* 1. *Let the transition kernel P be $(1+E^\alpha)^\oplus$-bounded for some $\alpha \in [0, 1]$ and E-non-increasing. Then:*

(i) equation (6.13) *holds for all $g \in C_{1+E^{\beta-\alpha}}(X)$;*
(ii) μ_t is $\star$-weakly continuous in $\mathcal{M}_{1+E^\beta}(X)$;
(iii) μ_t is $\star$-weakly continuously differentiable in $\mathcal{M}_{1+E^{\beta-\alpha}}(X)$;
(iv) μ_t is continuous in the norm topology of $\mathcal{M}_{1+E^{\beta-\epsilon}}(X)$ for any $\epsilon > 0$;
(v) if the kernel P satisfies the additional condition

$$\|P(\mu; z; .) - P(\nu; z; .)\| \leq C(1+E^\alpha)^\oplus(z)\|(1+E)(\mu-\nu)\|, \tag{6.25}$$

for all finite measures μ, ν and some constant C, then the function $t \mapsto \mu_t$ is continuously differentiable in the sense of the total variation norm topology of $\mathcal{M}(X)$, so that the kinetic equation (6.12) *holds in the strong sense.*

Proof (i) Approximating $g \in C_{1+E^{\beta-\alpha}}(X)$ by functions with a compact support one passes to the limit on both sides of (6.13) using the dominated convergence theorem and the estimate

$$\int g^\oplus(\mathbf{y})P(\mu_s; \mathbf{z}; d\mathbf{y})\mu_s^{\tilde{\otimes}}(d\mathbf{z}) \leq c\int (E^{\beta-\alpha})^\oplus(\mathbf{y})P(\mu_s; \mathbf{z}; d\mathbf{y})\mu_s^{\tilde{\otimes}}(d\mathbf{z})$$
$$\leq \int (E^{\beta-\alpha})^\oplus(\mathbf{z})(1+E^\alpha)^\oplus(\mathbf{z})\mu_s^{\tilde{\otimes}}(d\mathbf{z}).$$

(ii) It follows directly from (6.13) that the function $\int g(x)\mu_t(dx)$ is absolutely continuous for any g from (i). The required continuity for more general $g \in$

$C_{1+E^\beta,\infty}(X)$ is obtained, again by approximating such a g by functions with a compact support and using the observation that the integrals of g and its approximations over the set $\{x : E(x) \geq K\}$ can be made uniformly arbitrarily small for all μ_t, $t \in [0, T]$, by choosing K large enough.
(iii) One only needs to show that the r.h.s. of (6.12) is continuous whenever $g \in C_{1+E_\infty^{\beta-\alpha}}(X)$. But this is true, since by (ii) the integral

$$\int \phi(z_1, \ldots, z_l) \prod_{j=1}^{l} \mu_t(dz_j)$$

is a continuous function of t for any function $\phi \in C_{(1+E^\beta)^{\otimes l},\infty}(X^l)$.
(iv) From the same estimates as in (i) it follows that μ_t is Lipschitz continuous in the norm topology of $\mathcal{M}_{1+E^{\beta-\alpha}}(X)$. The required continuity in $\mathcal{M}_{1+E^{\beta-\epsilon}}(X)$ then follows from the uniform boundedness of μ_t in $\mathcal{M}_{1+E^\beta}(X)$.
(v) It is easy to see that if a measure-valued ODE $\dot{\mu}_t = \nu_t$ holds weakly then, in order to conclude that it holds strongly (in the sense of the norm topology), one has to show that the function $t \mapsto \nu_t$ is norm continuous. Hence in our case one has to show that

$$\sup_{|g|\leq 1} \int_{X^l} \int_{\mathcal{X}} [g^\oplus(\mathbf{y}) - g^\oplus(\mathbf{z})] \\ \times \left[P(\mu_t, \mathbf{z}; d\mathbf{y})\mu_t^{\otimes l}(d\mathbf{z}) - P(\mu_0, \mathbf{z}; d\mathbf{y})\mu_0^{\otimes l}(d\mathbf{z})\right] \to 0$$

as $t \to 0$, and this amounts to showing that

$$\|P(\mu_t, \mathbf{z}; d\mathbf{y})\mu_t^{\otimes l}(d\mathbf{z}) - P(\mu_0, \mathbf{z}; d\mathbf{y})\mu_0^{\otimes l}(d\mathbf{z})\| \to 0$$

as $t \to 0$, which follows from (6.25) and the continuity of μ_t. □

The regularity of the solution μ_t (i.e. the number of continuous derivatives) increases with the growth of β in (6.18). As this estimate for μ_t follows from the corresponding estimate for μ_0, the regularity of μ_t depends on the rate of decay of $\mu = \mu_0$ at infinity. For example, the following result is easily deduced.

Proposition 6.11 *If $f_{1+E^\beta}(\mu_0)$ is finite for all positive β and P does not depend explicitly on μ (e.g. no additional mean field interaction is assumed), then the solution μ_t of* (6.12) *obtained above is infinitely differentiable in t (with respect to the norm topology in $\mathcal{M}(X)$).*

We can now prove the main result of this chapter. It relies on a measure-theoretic result, Lemma 6.14 below, which is proved in detail in the next section.

Theorem 6.12 *Suppose that the assumptions of Theorem 6.7 hold, P is $(1+E^\alpha)^\oplus$-bounded for some $\alpha \in [0,1]$ such that $\beta \geq \alpha+1$ and (6.25) holds. Then there exists a unique non-negative solution μ_t to (6.12) satisfying (6.18) and a given initial condition μ_0 such that $\int (1+E^\beta)(x)\mu_0(dx) < \infty$. This μ_t is a strong solution of the corresponding kinetic equation, i.e. the derivative $(d/dt)\mu_t$ exists in the norm topology of $\mathcal{M}(X)$.*

Moreover, the mapping $\mu_0 \mapsto \mu_t$ is Lipschitz continuous in the norm of $\mathcal{M}_{1+E}(X)$, i.e. for any two solutions μ_t and ν_t of (6.12) satisfying (6.18) with initial conditions μ_0 and ν_0 one has

$$\int (1+E)|\mu_t - \nu_t|\,(dx) \leq a e^{at} \int (1+E)|\mu_0 - \nu_0|\,(dx), \tag{6.26}$$

for some constant a, uniformly for all $t \in [0,T]$.

Proof By the previous results we only need to prove (6.26). By Theorem 6.10 (iv), μ_t and ν_t are strong (continuously differentiable) solutions of (6.12), and Lemma 6.14 can be applied to the measure $(1+E)(x)(\mu_t - \nu_t)(dx)$ (see also Remark 6.17). Consequently, denoting by f_t a version of the density of $\mu_t - \nu_t$ with respect to $|\mu_t - \nu_t|$ taken from Lemma 6.14 yields

$$\begin{aligned}
&\int (1+E)(x)|\mu_t - \nu_t|(dx) \\
&= \|(1+E)(\mu_t - \nu_t)\| \\
&= \int (1+E)(x)|\mu_0 - \nu_0|(dx) + \int_0^t ds \int_X f_s(x)(1+E)(x)(\dot{\mu}_s - \dot{\nu}_s)(dx).
\end{aligned}$$

By (6.12) the last integral equals

$$\begin{aligned}
&\int_0^t ds \sum_{l=1}^k \iint \left\{ [f_s(1+E)]^\oplus(\mathbf{y}) - [f_s(1+E)]^\oplus(\mathbf{z}) \right\} \\
&\times \left(P(\mu_s; \mathbf{z}; d\mathbf{y}) \prod_{j=1}^l \mu_s(dz_j) - P(\nu_s; \mathbf{z}; d\mathbf{y}) \prod_{j=1}^l \nu_s(dz_j) \right)
\end{aligned}$$

$$= \int_0^t ds \sum_{l=1}^{k} \iint \left\{ [f_s(1+E)]^{\oplus}(\mathbf{y}) - [f_s(1+E)]^{\oplus}(\mathbf{z}) \right\}$$

$$\times \left\{ P(\mu_s; \mathbf{z}; d\mathbf{y}) \sum_{j=1}^{l} \prod_{i=1}^{j-1} \nu_s(dz_i)(\mu_s - \nu_s)(dz_j) \prod_{i=j+1}^{l} \mu_s(dz_i) \right.$$

$$\left. + \left[P(\mu_s; \mathbf{z}; d\mathbf{y}) - P(\nu_s; \mathbf{z}; d\mathbf{y}) \right] \prod_{j=1}^{l} \nu_s(dz_j) \right\} \tag{6.27}$$

Let us choose arbitrary $l \le k$ and $j \le l$ and estimate the corresponding term of the sum within the large braces of (6.27). We have

$$\iint \left\{ [f_s(1+E)]^{\oplus}(\mathbf{y}) - [f_s(1+E)]^{\oplus}(\mathbf{z}) \right\} P(\mu_s; \mathbf{z}; d\mathbf{y})$$

$$\times (\mu_s - \nu_s)(dz_j) \prod_{i=1}^{j-1} \nu_s(dz_i) \prod_{i=j+1}^{l} \mu_s(dz_i)$$

$$= \iint \left\{ [f_s(1+E)]^{\oplus}(\mathbf{y}) - [f_s(1+E)]^{\oplus}(\mathbf{z}) \right\} P(\mu_s; \mathbf{z}; d\mathbf{y})$$

$$\times f_s(z_j) |\mu_s - \nu_s|(dz_j) \prod_{i=1}^{j-1} \nu_s(dz_i) \prod_{i=j+1}^{l} \mu_s(dz_i). \tag{6.28}$$

As $P(\mu, \mathbf{z}; d\mathbf{y})$ is E-non-increasing,

$$\left\{ [f_s(1+E)]^{\oplus}(\mathbf{y}) - [f_s(1+E)]^{\oplus}(\mathbf{z}) \right\} f_s(z_j)$$

$$\le (1+E)^{\oplus}(\mathbf{y}) - f_s(z_j)[f_s(1+E)]^{\oplus}(\mathbf{z})$$

$$\le 2k + E^{\oplus}(\mathbf{z}) - E(z_j) - \sum_{i \ne j} f_s(z_j) f_s(z_i) E(z_i) \le 2k + 2 \sum_{i \ne j} E(z_i).$$

Hence (6.28) does not exceed

$$\int \left(2k + 2 \sum_{i \ne j} E(z_i) \right) \left(1 + E^{\alpha}(z_j) + \sum_{i \ne j} E^{\alpha}(z_i) \right)$$

$$\times |\mu_s - \nu_s|(dz_j) \prod_{i=1}^{j-1} \nu_s(dz_i) \prod_{i=j+1}^{l} \mu_s(dz_i).$$

Consequently, as $1 + \alpha \le \beta$ and (6.18) holds, and since the second term in the large braces of (6.27) can be estimated by (6.25), it follows that the integral

(6.27) does not exceed

$$c(T)\int_0^t ds \int (1+E)(x)|\mu_t - \nu_t|(dx),$$

for some constant $c(T)$, which implies (6.26) by Gronwall's lemma. □

It can be useful, both for practical calculations and theoretical developments, to know that the approximation μ_t^n solving the cutoff problem used in the proof of Theorem 6.7 actually converges strongly to the solution μ_t. Namely, the following holds.

Theorem 6.13 *Under the assumptions of Theorem 6.12 the approximations μ_t^n introduced in Theorem 6.7 converge to the solution μ_t in the norm topology of $\mathcal{M}_{1+E^\omega}(X)$ for any $\omega \in [1, \beta - \alpha)$ and $\star$-weakly in $\mathcal{M}_{1+E^\beta}(X)$.*

Proof This utilizes the same trick as in the proof of previous theorem. To shorten the formulas, we shall work through only the case $\omega = 1$ and $k = l = 2$ in (6.12). Let σ_t^n denote the sign of the measure $\mu_t^n - \mu_t$ (i.e. the equivalence class of the densities of $\mu_t^n - \mu_t$ with respect to $|\mu_t^n - \mu_t|$ that equals ± 1 respectively in the positive and negative parts of the Hahn decomposition of this measure), so that $|\mu_t^n - \mu_t| = \sigma_t^n(\mu_t^n - \mu_t)$. By Lemma 6.14 below one can choose a representative of the σ_t^n (that we again denote by σ_t^n) in such a way that

$$(1+E, |\mu_t^n - \mu_t|) = \int_0^t \left(\sigma_s^n(1+E), \frac{d}{ds}(\mu_s^n - \mu_s)\right) ds. \tag{6.29}$$

By (6.12) with $k = l = 2$, this implies that

$$\begin{aligned}(1+E, &|\mu_t^n - \mu_t|)\\ &= \frac{1}{2}\int_0^t ds \int \left\{\left[\sigma_s^n(1+E)\right]^{\oplus}(\mathbf{y}) - \left[\sigma_s^n(1+E)\right](x_1) - \left[\sigma_s^n(1+E)\right](x_2)\right\}\\ &\quad \times \left[P_n(x_1,x_2,dy)\mu_s^n(dx_1)\mu_s^n(dx_2) - P(x_1,x_2,dy)\mu_s(dx_1)\mu_s(dx_2)\right].\end{aligned} \tag{6.30}$$

The expression in the last bracket can be rewritten as

$$\begin{aligned}&(P_n - P)(x_1,x_2,dy)\mu_s^n(dx_1)\mu_s^n(dx_2)\\ &\quad + P(x_1,x_2,dy)\Big\{\left[\mu_s^n(dx_1) - \mu_s(dx_1)\right]\mu_s^n(dx_2)\\ &\qquad\qquad + \mu_s(dx_1)\left[\mu_s^n(dx_2) - \mu_s(dx_2)\right]\Big\}.\end{aligned} \tag{6.31}$$

As μ_s^n are uniformly bounded in $\mathcal{M}_{1+E^\beta}$, and as

$$[1+E(x_1)+E(x_2)]\int_X (P_n - P)(x_1,x_2,dy) \le Cn^{-\epsilon}\left[1+E(x_1)+E(x_2)\right]^{2+\epsilon}$$

for $2+\epsilon \le \beta$, the contribution of the first term in (6.31) to the r.h.s. of (6.30) tends to zero as $n \to \infty$. The second and third terms in (6.31) are similar, so let us analyze the second term only. Its contribution to the r.h.s. of (6.30) can be written as

$$\frac{1}{2}\int_0^t ds \int \Big\{ \big[\sigma_s^n(1+E)\big](\mathbf{y}) - \big[\sigma_s^n(1+E)\big](x_1) - \big[\sigma_s^n(1+E)\big](x_2) \Big\}$$
$$\times P(x_1, x_2, dy)\sigma_s^n(x_1)\left|\mu_s^n(dx_1) - \mu_s(dx_1)\right| \mu_s^n(dx_2),$$

which does not exceed

$$\frac{1}{2}\int_0^t ds \int \big[(1+E)(\mathbf{y}) - (1+E)(x_1) + (1+E)(x_2)\big]$$
$$\times P(x_1, x_2, dy)\left|\mu_s^n(dx_1) - \mu_s(dx_1)\right| \mu_s^n(dx_2),$$

because $[\sigma_s^n(x_1)]^2 = 1$ and $|\sigma_s^n(x_j)| \le 1$, $j = 1, 2$. Since P preserves E and is $(1+E)^{\oplus}$- bounded, the latter expression does not exceed

$$c\int_0^t ds \int [1+E(x_2)][1+E(x_1)+E(x_2)]\left|\mu_s^n(dx_1) - \mu_s(dx_1)\right| \mu_s^n(dx_2)$$
$$\le c\int_0^t ds(1+E, |\mu_s^n - \mu_s|)\|\mu_s^n\|_{1+E^2}.$$

Consequently, by Gronwall's lemma one concludes that

$$\|\mu_t^n - \mu_t\|_{1+E} = (1+E, |\mu_t^n - \mu_t|) = O(1)n^{-\epsilon}\exp\left\{t \sup_{s\in[0,t]} \|\mu_s\|_{1+E^2}\right\}.$$

Finally, once convergence in the norm topology of any $\mathcal{M}_{1+E^\gamma}$ with $\gamma \ge 0$ is established, $\star$-weak convergence in $\mathcal{M}_{1+E^\beta}$ follows from the uniform (in n) boundedness of μ_n in $\mathcal{M}_{1+E^\beta}$. □

Exercise 6.3 Fill in the details needed for the proof of the above theorem for $\omega \in (1, \beta - \alpha)$.

Exercise 6.4 Under the assumptions of Theorem 6.12, show that the mapping $\mu_0 \mapsto \mu_t$ is Lipschitz continuous in the norm of $\mathcal{M}_{1+E^\omega}(X)$ for any $\omega \in [1, \beta - \alpha]$ (in the theorem the case $\omega = 1$ was considered). Hint: for this extension the estimates from the proof of Theorem 6.7 are needed.

6.5 A tool for proving uniqueness

The following lemma supplies the main tool for proving uniqueness for kinetic equations with jump-type nonlinearities.

Lemma 6.14 *Let Y be a measurable space, and let the mapping $t \mapsto \mu_t$ from $[0, T]$ to $\mathcal{M}(Y)$ be continuously differentiable with respect to the topology specified by the norm in $\mathcal{M}(Y)$, with (continuous) derivative $\dot{\mu}_t = \nu_t$. Let σ_t denote the density of μ_t with respect to its total variation $|\mu_t|$, i.e. σ_t is in the class of measurable functions taking three values $-1, 0, 1$ and such that $\mu_t = \sigma_t|\mu_t|$ and $|\mu_t| = \sigma_t\mu_t$ a.s. with respect to $|\mu_t|$. Then there exists a measurable function $f_t(x)$ on $[0, T] \times Y$ such that f_t is a representative of class σ_t for any $t \in [0, T]$ and*

$$\|\mu_t\| = \|\mu_0\| + \int_0^t ds \int_Y f_s(y)\nu_s(dy). \tag{6.32}$$

Proof *Step 1.* As μ_t is continuously differentiable, $\|\mu_t - \mu_s\| = O(t - s)$ uniformly for $0 \le s \le t \le T$. Hence $\|\mu_t\|$ is an absolutely continuous real-valued function. Consequently, this function has a derivative, say ω_s, almost everywhere on $[0, T]$ and an integral representation $\|\mu_t\| = \|\mu_0\| + \int_0^t \omega_s\, ds$ valid for all $t \in [0, T]$. It remains to calculate ω_s. To simplify this calculation, we observe that as the right and left derivatives of an absolutely continuous function coincide almost everywhere (according to the Lebesgue theorem), it is enough to calculate only the right derivative of $\|\mu_t\|$. Hence from now on we shall consider only the limit $t \to s$ with $t \ge s$.

Step 2. For an arbitrary measurable $A \subset Y$ and an arbitrary representative σ_t of the density, we have

$$\begin{aligned} O(t - s) &= \int_A |\mu_t|(dy) - \int_A |\mu_s|(dy) = \int_A \sigma_t\mu_t(dy) - \int_A \sigma_s\mu_s(dy) \\ &= \int_A (\sigma_t - \sigma_s)\mu_s(dy) + \int_A \sigma_t(\mu_t - \mu_s)(dy). \end{aligned}$$

As the second term here is also of order $O(t - s)$, we conclude that the first term must be of order $O(t - s)$ uniformly for all A and s, t. Hence $\sigma_t \to \sigma_s$ almost surely with respect to $|\mu_s|$ as $t \to s$. As σ_t takes only three values, $0, 1, -1$, it follows that $\dot{\sigma}_s(x)$ exists and vanishes for almost all x with respect to $|\mu_s|$.

Remark 6.15 Now writing formally

$$\frac{d}{dt}\|\mu_t\| = \frac{d}{dt}\int_Y \sigma_t\mu_t(dy) = \int_Y \dot{\sigma}_t\mu_t(dy) + \int_Y \sigma_t\dot{\mu}_t(dy),$$

and noticing that the first term here vanishes by step 2 of the proof, yields (6.32). However, this formal calculation cannot be justified for an arbitrary choice of σ_t.

Step 3. Let us choose now an appropriate σ_t. For this purpose we perform the Lebesgue decomposition of ν_t into the sum $\nu_t = \nu_t^s + \nu_t^a$ of a singular and an absolutely continuous measure with respect to $|\mu_t|$. Let η_t be the density of ν_t^s with respect to its total variation measure $|\nu_t^s|$, i.e. η_t is in the class of functions taking values $0, 1, -1$ and such that $\nu_t^s = \eta_t |\nu_t^s|$ almost surely with respect to $|\nu_t^s|$. Now let us pick an f_t from the intersection $\sigma_t \cap \eta_t$, i.e. f is a representative of both density classes simultaneously. Such a choice is possible, because μ_t and ν_t^s are mutually singular. From this definition of f_t and from step 2 it follows that $f_t \to f_s$ as $t \to s$ (and $t \geq s$) a.s. with respect to both μ_s and ν_s. In fact, let B be either the positive part of the Hahn decomposition of ν_s^s or any measurable subset of the latter. Then $f_s = 1$ on B. Moreover,

$$\mu_t(B) = (t - s)\nu_s^s(B) + o(t - s), \qquad t - s \to 0,$$

and is positive and hence $f_t = 1$ on B for t close enough to s.

Step 4. By definition,

$$\begin{aligned} \frac{d}{ds}\|\mu_s\| &= \lim_{t\to s} \frac{\|\mu_t\| - \|\mu_s\|}{t - s} \\ &= \lim_{t\to s} \int \frac{f_t - f_s}{t - s}\mu_s(dy) + \lim_{t\to s} \int f_t \frac{\mu_t - \mu_s}{t - s}(dy) \end{aligned} \tag{6.33}$$

(if both limits exist, of course). It is easy to see that the second limit here does always exist and equals $\int f_s \nu_s(dy)$. In fact,

$$\begin{aligned} \int f_t \frac{\mu_t - \mu_s}{t - s}(dy) &= \int f_s \frac{\mu_t - \mu_s}{t - s}(dy) \\ &\quad + \int (f_t - f_s)\left(\frac{\mu_t - \mu_s}{t - s} - \nu_s\right)(dy) + \int (f_t - f_s)\nu_s(dy), \end{aligned}$$

and the limit of the first integral equals $\int f_s \nu_s(dy)$, the second integral is $o(t - s)$ by the definition of the derivative, and hence vanishes in the limit, and the limit of the third integral is zero because (owing to our choice of f_t in Step 3) $f_t - f_s \to 0$ as $t \to s$ $(t \geq s)$ a.s. with respect to ν_s. Consequently, to complete the proof it remains to show that the first term on the r.h.s. of (6.33) vanishes.

Remark 6.16 As we showed in step 2, the function $(f_t - f_s)/(t - s)$ under the integral in this term tends to zero a.s. with respect to μ_s, but unfortunately this is not enough to conclude that the limit of the integral vanishes; an additional argument is required.

Step 5. In order to show that the first term in (6.33) vanishes, it is enough to show that the corresponding limits of the integrals over the sets A^+ and A^- vanish, where $Y = A^+ \cup A^- \cup A^0$ is the Hahn decomposition of Y with respect to the measure μ_s. Let us consider only A^+ (A^- may be treated similarly). As $f_s = 1$ on A^+ a.s., we need to show that

$$\lim_{t \to s} \int_A \frac{f_t - 1}{t - s} \mu_s(dy) = 0, \tag{6.34}$$

where A is a measurable subset of Y such that $(\mu_s)|_A$ is a positive measure. Using now the Lebesgue decomposition of $(\nu_s)|_A$ into the sum of a singular and an absolutely continuous part with respect to μ_s, we can reduce the discussion to the case when ν_s is absolutely continuous with respect to μ_s on A.

Introducing the set $A_t = \{y \in A : f_t(y) \leq 0\}$ one can clearly replace A by A_t in (6.34). Consequently, to obtain (6.34) it is enough to show that $\mu_s(A_t) = o(t - s)$ as $t \to s$. This will be done in the next and final step of the proof.

Step 6. From the definition of A_t it follows that $\mu_t(B) \leq 0$ for any $B \subset A_t$ and hence

$$\mu_s(B) + (t - s)\nu_s(B) + o(t - s) \leq 0, \tag{6.35}$$

where $o(t-s)$ is uniform, i.e. $\|o(t-s)\|/(t-s) \to 0$ as $t \to s$. Notice first that if $A_t = B_t^+ \cup B_t^- \cup B_t^0$ is the Hahn decomposition of A_t on the positive, negative and zero parts of the measure ν_s then $\mu_s(B_t^+ \cup B_t^0) = o(t - s)$ uniformly (as follows directly from (6.35)), and consequently we can reduce our discussion to the case when ν_s is a negative measure on A. In this case (6.35) implies that

$$\mu_s(A_t) \leq (t - s)(-\nu_s)(A_t) + o(t - s),$$

and it remains to show that $\nu_s(A_t) = o(1)_{t \to s}$. To see this we observe that, for any s, Y has the representation $Y = \cup_{n=0}^{\infty} Y_n$, where $|\mu_s|(Y_0) = 0$ and

$$Y_n = \{y \in Y : f_t = f_s \quad \text{for} \quad |t - s| \leq 1/n\}.$$

Clearly $Y_n \subset Y_{n+1}$ for any $n \neq 0$, and $A_t \subset Y \setminus Y_n$ whenever $t - s \leq 1/n$. Hence the A_t are subsets of a decreasing family of sets with intersection of μ_s-measure zero. As ν_s is absolutely continuous with respect to μ_s the same holds for ν_s; hence $\nu_s(A_t) = o(1)_{t \to s}$, which completes the proof of the lemma. □

Remark 6.17 Suppose that the assumptions of Lemma 6.14 hold and that $L(y)$ is a measurable, non-negative and everywhere finite function on Y such that $\|L\mu_s\|$ and $\|L\nu_s\|$ are uniformly bounded for $s \in [0, t]$. Then (6.32) holds for $L\mu_t$ and $L\nu_t$ instead of μ_t and ν_t respectively. In fact, though $s \mapsto L\nu_s$

may be discontinuous with respect to the topology specified by the norm, one can write the required identity first for the space Y_m, instead of Y, where $Y_m = \{y : L(y) \leq m\}$, and then pass to the limit as $m \to \infty$.

The above proof of Lemma 6.14 is based on the original presentation in Kolokoltsov [132]. An elegant and simpler proof, based on discrete approximations, is now available in Bailleul [14]. However, this proof works under rather less general assumptions and does not reveal the structure of σ_t as a common representative of two Radon–Nikodyme derivatives.

6.6 Multiplicative bounds for rates

In many situations one can prove only the existence, not the uniquess, of a global solution to a kinetic equation (possibly in some weak form). In the next two sections we shall discuss two such cases. Here we will be dealing with equation (1.73) for pure jump interactions under the assumption of multiplicatively bounded rates (the definition of such kernels was given in Section 6.3).

We shall work with equation (1.73) in weak integral form, assuming for simplicity that there is no additional mean field dependence;

$$(g, \mu_t - \mu) = \int_0^t ds \sum_{l=1}^k \frac{1}{l!} \int_{X^l} \left\{ \int_{\mathcal{X}} \left[g^{\oplus}(\mathbf{y}) - g^{\oplus}(\mathbf{z}) \right] P(\mathbf{z}, d\mathbf{y}) \right\} \mu_s^{\otimes l}(d\mathbf{z}). \tag{6.36}$$

Assume that X is a locally compact space and E is a continuous function on X.

Theorem 6.18 *Suppose that P is $\mathbf{1}$-non-increasing, E-non-increasing and strongly $(1+E)^{\otimes}$-bounded, meaning that*

$$\|P(z, .)\| = o(1)(1+E)^{\otimes}(z), \qquad z \to \infty. \tag{6.37}$$

Then for any $T > 0$ and $\mu \in \mathcal{M}_{1+E}(X)$ there exists a continuous (in the Banach topology of $\mathcal{M}(X)$) curve $t \mapsto \mu_t \in \mathcal{M}_{1+E}(X)$, $t \in [0, T]$, such that (6.36) *holds for all $g \in C(X)$ and indeed for all $g \in B(X)$.*

Proof As in the proof of Theorem 6.7, we shall use the approximations P_n defined by equation (6.19). As the kernels P_n are bounded, for any $\mu \in \mathcal{M}_{1+E}(X)$ there exists a unique strong solution μ_t^n of the corresponding cut-off equation (6.36) (with P_n instead of P) for an initial condition μ such that $(E, \mu_t^n) \leq (E, \mu)$ and $(1, \mu_t^n) \leq (1, \mu)$ for all t, n, implying that the $\|\dot{\mu}_t^n\|$ are uniformly bounded. Hence the family of curves μ_t^n, $t \in [0, T]$, is uniformly bounded and uniformly Lipschitz continuous in $\mathcal{M}(X)$, implying the existence of a converging subsequence that again we denote by μ_t^n.

Let us denote by $K(\mu)$ the operator under the integral on the r.h.s. of (6.36), so that this equation can be written in the concise form

$$(g, \dot{\mu}_t) = \int_0^t (g, K(\mu_s))\, ds.$$

Let K_n denote the corresponding operators with cut-off kernels P_n instead of P. To prove the theorem it remains to show that

$$\|K_n(\mu_t^n) - K(\mu_t)\| \to 0, \qquad n \to \infty,$$

uniformly in $t \in [0, T]$. By (6.37),

$$\|K_n(\mu) - K(\mu)\| \to 0, \qquad n \to \infty,$$

uniformly for μ with uniformly bounded $(1 + E, \mu)$. Hence one needs only to show that

$$\|K(\mu_t^n) - K(\mu_t)\| \to 0, \qquad n \to \infty.$$

Decomposing the integral defining K into two parts over the set $L^{\oplus}(z) \geq M$ and its complement one sees that the first and second integrals can be made arbitrary small, the first by (6.37) again and the second by the convergence of μ_t^n to μ_t in the Banach topology of $\mathcal{M}(X)$. □

The uniqueness of solutions does not hold generally under the above assumptions. Neither is (E, μ_t) necessarily constant even if P preserves E. The decrease in (E, μ_t) for coagulation processes under the latter condition is interpreted as gelation (the formation of a cluster of infinite size). This is a very interesting effect, attentively studied in the literature (see the comments in Section 11.6).

Exercise 6.5 Show that the statement of the above theorem still holds true if, instead of assuming that P is **1**-non-increasing, one assumes that it is $E^{\otimes}$-bounded.

6.7 Another existence result

Here we apply the method of T-products to provide a rather general existence result for the weak equation

$$\frac{d}{dt}(f, \mu_t) = (A_{\mu_t} f, \mu_t) = \left(\int_X \left[f(y) - f(.) \right] \nu(., \mu_t, dy), \mu_t \right), \qquad (6.38)$$

which is of the type (6.3), (6.5), where ν is a family of transition kernels in a Borel space X depending on $\mu \in \mathcal{M}(X)$. We assume only that the rate function $a(x, \mu) = \|\nu(x, \mu, .)\|$ is locally bounded in the sense that

$$\sup_{\|\mu\| \leq M} a(x, \mu) < \infty \tag{6.39}$$

for a given $M > 0$. Under this condition, the sets

$$M_k = \left\{ x : \sup_{\|\mu\| \leq M} a(x, \mu) < k \right\}$$

exhaust the whole state space. The mild form (compare with (4.16)) of the nonlinear equation (6.38) can be written as

$$\mu_t(dx) = \left[\exp\left(-\int_0^t a(x, \mu_s)\, ds \right) \right] \mu(dx) + \left[\int_0^t \exp\left(-\int_s^t a(x, \mu_\tau)\, d\tau \right) \right] \int_{y \in X} \nu(y, \mu_s, dx) \mu_s(dy). \tag{6.40}$$

The following statement represents a nonlinear counterpart of the results of Section 4.2, though unfortunately the uniqueness (either in terms of minimality or in terms of the solutions to equation (4.17)) is lost.

Theorem 6.19 *Assume that ν is a transition kernel in a Borel space X satisfying* (6.39) *for an $M > 0$ and depending continuously on μ in the norm topology, i.e. $\mu_n \to \mu$ in the Banach topology of $\mathcal{M}(X)$ implies that $\nu(x, \mu_n, .) \to \nu(x, \mu, .)$ in the Banach topology of $\mathcal{M}(X)$. Then for any $\mu \in \mathcal{M}(X)$ with $\|\mu\| \leq M$ and $T > 0$ there exists a curve $\mu_t \in \mathcal{M}(X)$, $t \in [0, T]$, that is continuous in its norm topology, with norm that is non-increasing in t, solving-equation* (6.40).

Proof For a given $\mu \in \mathcal{M}(X)$ with $\|\mu\| \leq M$, $n \in \mathbf{N}$, $\tau = T/n$, let us define, for $0 \leq s \leq r \leq t_1 = \tau$, a minimal backward propagator $U_n^{s,r}$ on $B(X)$ and its dual forward propagator $V_n^{r,s} = (U_n^{s,r})^\star$ on $\mathcal{M}(X)$, associated with the operator A_μ given by

$$A_\mu f = \int_X \left[f(y) - f(x) \right] \nu(x, \mu, dy);$$

this is in accordance with Theorem 4.1. Next, let us define, for $t_1 \leq s \leq r \leq t_2 = 2\tau$, the minimal backward propagator $U_n^{s,r}$ on $B(X)$ and its dual forward propagator $V_n^{r,s} = (U_n^{s,r})^\star$ on $\mathcal{M}(X)$, associated with the operator $A_{\mu_\tau^n}$ given by

$$A_{\mu_\tau^n} f = \int_X [f(y) - f(x)]\, \nu(x, \mu_\tau^n, dy), \qquad \mu_\tau^n = V_n^{t_1,0}\mu,$$

again in accordance with Theorem 4.1. Continuing in the same way and gluing together the constructed propagators yields the complete backward propagator $U_n^{s,r}$ on $B(X)$ and its dual forward propagator $V_n^{r,s} = (U_n^{s,r})^\star$ on $\mathcal{M}(X)$, $0 \le s \le r \le T$. These are associated according to Theorem 4.1 with the time-nonhomogeneous family of operators given by

$$A_s^n f = \int_X [f(y) - f(x)]\, \nu(x, \mu_{[s/\tau]\tau}^n, dy),$$

where

$$\mu_{k\tau}^n = V_n^{k\tau,0}\mu = V_n^{k\tau,(k-1)\tau}\mu_{(k-1)\tau}^n.$$

By Corollary 4.4 the curve $\mu_t^n = V_n^{t,0}\mu$ depends continuously on t in the Banach topology of $\mathcal{M}(X)$, its norm is non-increasing and it satisfies the equation

$$\begin{aligned}\mu_t^n(dx) = {} & \left[\exp\left(-\int_0^t a(x, \mu_{[s/\tau]\tau}^n)\, ds\right)\right]\mu(dx) \\ & + \left[\int_0^t \exp\left(-\int_s^t a(x, \mu_{[s/\tau]\tau}^n)\, d\tau\right)\right]\int_{y\in X} \nu(y, \mu_{[s/\tau]\tau}^n, dx)\mu_s^n(dy).\end{aligned} \tag{6.41}$$

By Theorem 4.1 (or more precisely (4.13)), the family $U_n^{r,s}$ is locally equicontinuous in the sense that

$$\sup_{\|f\|\le 1} \|(U_n^{r,s_1} - U_n^{r,s_2}) f \mathbf{1}_{M_k}\| \le 2k|s_1 - s_2|.$$

Hence, using the Arzela–Ascoli theorem and diagonalization, one can choose a subsequence of backward propagators $U_n^{r,s}$ (which we again denote by U_n) converging to a propagator $U^{r,s}$ on $B(X)$ in the sense that, for any k,

$$\sup_{\|f\|\le 1,\, s\le r\le t} \|(U^{r,s} - U_n^{r,s}) f \mathbf{1}_{M_k}\| \to 0, \qquad n \to \infty.$$

By duality this implies that the sequence $\mu_t^n = V_n^{t,0}\mu$ converges strongly to a strongly continuous curve μ_t. Passing to the limit in equation (6.41) and using the continuous dependence of $\nu(x, \mu, .)$ on μ yields (6.40). □

Corollary 6.20 *Under the assumptions of Theorem 6.19, assume additionally that, for any k,*

$$\sup_{x \in X, \|\mu\| \le M} \nu_t(x, \mu, M_k) < \infty. \tag{6.42}$$

Then a continuous (in the Banach topology of $\mathcal{M}(X)$) solution μ_t to (6.41) *solves equation* (6.38) *strongly on M_k, i.e. for any k we have*

$$\frac{d}{dt}\mathbf{1}_{M_k}(x)\mu_t(dx) = -\mathbf{1}_{M_k}(x)a_t(x, \mu_t)\mu_t(dx)$$
$$+ \int_{z \in X} \mu_t(dz)\mathbf{1}_{M_k}(x)\nu_t(z, \mu_t, dx),$$

where the derivative exists in the Banach topology of $\mathcal{M}(X)$.

Exercise 6.6 Prove the statement in Corollary 6.20, using the argument from Proposition 4.5.

6.8 Conditional positivity

In this section we shall show that bounded generators of measure-valued positivity-preserving evolutions necessarily have the form given by (6.3); in Section 1.1 this was called a stochastic representation, as it leads directly to a probabilistic interpretation of the corresponding evolution. For a Borel space X we shall say that a mapping $\Omega : \mathcal{M}(X) \to \mathcal{M}^{\text{signed}}(X)$ is *conditionally positive* if the negative part $\Omega^-(\mu)$ of the Hahn decomposition of the measure $\Omega(\mu)$ is absolutely continuous with respect to μ for all μ. This is a straightforward extension of the definition of conditional positivity given in Section 1.1, and one easily deduces that continuous generators of positivity-preserving evolutions should be conditionally positive in this sense.

Theorem 6.21 *Let X be a Borel space and $\Omega : \mathcal{M}(X) \to \mathcal{M}^{\text{signed}}(X)$ be a conditionally positive mapping. Then there exists a non-negative function $a(x, \mu)$ and a family of kernels $\nu(x, \mu, \cdot)$ in X such that*

$$\Omega(\mu) = \int_X \mu(dz)\nu(z, \mu, \cdot) - a(\cdot, \mu)\mu. \tag{6.43}$$

If, moreover, $\int \Omega(\mu)(dx) = 0$ for all μ (the condition of conservativity) then this representation can be chosen in such a way that $a(x, \mu) = \|\nu(x, \mu, \cdot)\|$, in which case

$$(g, \Omega(\mu)) = \int_X \left[g(y) - g(x)\right] \nu(x, \mu, dy).$$

Proof One can take $a(x, \mu)$ to be the Radon–Nicodyme derivative of $\Omega^-(\mu)$ with respect to μ, yielding

$$\nu(x, \mu, dy) = \left(\int \Omega^-(\mu)(dz) \right)^{-1} a(x, \mu)\Omega^+(\mu)(dy). \qquad \square$$

Remark 6.22 The choice of $a(x, \mu)$ made in the above proof is in some sense canonical as it is minimal, i.e. it yields the minimum of all possible $a(x, \mu)$ for which a representation of the type (6.43) can be given.

7

Generators of Lévy–Khintchine type

In this chapter we discuss the well-posedness of nonlinear semigroups with generators of Lévy–Khintchine type. Two approaches to this analysis will be developed. One is given in the first two sections and is based on duality and fixed-point arguments in conjunction with the regularity of the corresponding time-nonhomogeneous linear problems. Another approach is a direct SDE construction, which is a nonlinear counterpart of the theory developed in Chapter 3.

7.1 Nonlinear Lévy processes and semigroups

As a warm-up, we will show how the first method mentioned above works in the simplest situation, where the coefficients of the generator do not depend explicitly on position but only on its distribution, i.e. the case of nonlinear Lévy processes introduced in Section 1.4. Referring to Section 1.4 for the analytic definition of a Lévy process, we start here with an obvious extension of this concept. Namely, we define a *time-nonhomogeneous Lévy process* with continuous coefficients as a time-nonhomogeneous Markov process generated by the time-dependent family of operators L_t of Lévy–Khintchine form given by

$$L_t f(x) = \frac{1}{2}(G_t \nabla, \nabla) f(x) + (b_t, \nabla f)(x) + \int [f(x+y) - f(x) - (y, \nabla f(x))\mathbf{1}_{B_1}(y)]\nu_t(dy), \qquad (7.1)$$

where G_t, b_t, ν_t depend continuously on t and ν is taken in its weak topology, i.e. $\int f(y)\nu_t(dy)$ depends continuously on t for any continuous f on $\mathbf{R}^d \setminus \{0\}$ with $|f| \leq c \min(|y|^2, 1)$). More precisely, by the *Lévy process generated by the family* (7.1) we mean a process X_t for which

$$\mathbf{E}(f(X_t)|X_s = x) = (\Phi^{s,t} f)(x), \qquad f \in C(\mathbf{R}^d),$$

where $\Phi^{s,t}$ is the propagator of positive linear contractions in $C_\infty(\mathbf{R}^d)$ depending strongly continuously on $s \leq t$ and such that, for any $f \in (C_\infty \cap C^2)(\mathbf{R}^d)$, the function $f_s = \Phi^{s,t} f$ is the unique solution in $(C_\infty \cap C^2)(\mathbf{R}^d)$ of the inverse-time Cauchy problem

$$\dot{f}_s = -L_s f_s, \qquad s \leq t, \; f_t = f. \tag{7.2}$$

From the theory of Markov processes, the existence of the family $\Phi^{s,t}$ with the required properties implies the existence and uniqueness of the corresponding Markov process. Thus the question of the existence of the process for a given family L_t is reduced to the question of the existence of a strongly continuous family $\Phi^{s,t}$ with the required properties. This is settled in the following statement.

Proposition 7.1 *For a given family L_t of the form* (7.1) *with coefficients continuous in t, there exists a family $\Phi^{s,t}$ of positive linear contractions in $C_\infty(\mathbf{R}^d)$ depending strongly continuously on $s \leq t$ such that, for any $f \in C^2_\infty(\mathbf{R}^d)$, the function $f_s = \Phi^{s,t} f$ is the unique solution in $C^2_\infty(\mathbf{R}^d)$ of the Cauchy problem* (7.2).

Proof Let f belong to the Schwartz space $S(\mathbf{R}^d)$. Then its Fourier transform

$$g(p) = (Ff)(p) = (2\pi)^{-d/2} \int_{\mathbf{R}^d} e^{-ipx} f(x)\, dx$$

also belongs to $S(\mathbf{R}^d)$. As the Fourier transform of equation (7.2) has the form

$$\dot{g}_s(p) = -\left[-\tfrac{1}{2}(G_s p, p) + i(b_s, p) + \int (e^{ipy} - 1 - ipy\mathbf{1}_{B_1})\nu_s(dy) \right] g_s(p),$$

it has the obvious unique solution

$$g_s(p) = \exp\left\{ \int_s^t \left[-\tfrac{1}{2}(G_\tau p, p) + i(b_\tau, p) \right.\right.$$
$$\left.\left. + \int (e^{ipy} - 1 - ipy\mathbf{1}_{B_1})\nu_\tau(dy) \right] d\tau \right\} g(p), \tag{7.3}$$

which belongs to $L^1(\mathbf{R}^d)$, so that $f_s = F^{-1} g_s = \Phi^{s,t} f$ belongs to $C_\infty(\mathbf{R}^d)$. The simplest way to deduce the required property of this propagator is to observe that, for any fixed s, t, the operator $\Phi^{s,t}$ coincides with an operator from the semigroup of a certain homogeneous Lévy process, implying that each $\Phi^{s,t}$ is a positivity-preserving contraction in $C_\infty(\mathbf{R}^d)$ preserving the spaces $(C_\infty \cap C^2)(\mathbf{R}^d)$ and $C^2_\infty(\mathbf{R}^d)$. Strong continuity then follows as in the standard (homogeneous) case. □

Corollary 7.2 *Under the assumptions of Proposition 7.1, the family of dual operators on measures $V^{t,s} = (\Phi^{s,t})'$ depends weakly continuously on s, t and Lipschitz continuously on the norm topology of the Banach dual $(C^2_\infty(\mathbf{R}^d))'$ to $C^2_\infty(\mathbf{R}^d)$. That is,*

$$\|V^{t,s}(\mu) - \mu\|_{(C^2_\infty(\mathbf{R}^d))'} = \sup_{\|f\|_{C^2_\infty(\mathbf{R}^d)} \leq 1} |(f, V^{t,s}(\mu) - \mu)|$$

$$\leq (t-s) \sup_{\tau \in [s,t]} \left(\|A_\tau\| + \|b_\tau\| + \int \min(1, |y|^2)\nu_\tau(dy) \right). \qquad (7.4)$$

Moreover, for any $\mu \in \mathcal{P}(\mathbf{R}^d)$, $V^{t,s}(\mu)$ yields the unique solution of the weak Cauchy problem

$$\frac{d}{dt}(f, \mu_t) = (L_t f, \mu_t), \qquad s \leq t, \ \mu_s = \mu, \qquad (7.5)$$

which holds for any $f \in C^2_\infty(\mathbf{R}^d)$.

Proof The weak continuity of $V^{t,s}$ is straightforward from the strong continuity of $\Phi^{s,t}$ and duality. Next, again by duality,

$$\sup_{\|f\|_{C^2(\mathbf{R}^d)} \leq 1} |(f, V^{t,s}(\mu) - \mu)| = \sup_{\|f\|_{C^2(\mathbf{R}^d)} \leq 1} |(\Phi^{s,t} f - f, \mu)|$$

$$= \sup_{\|f\|_{C^2(\mathbf{R}^d)} \leq 1} \left\| \int_s^t L_\tau \Phi^{\tau,t} f \, d\tau \right\|_{C(\mathbf{R}^d)}$$

$$\leq (t-s) \sup_{\tau \in [s,t]} \sup_{\|f\|_{C^2(\mathbf{R}^d)} \leq 1} \|L_\tau f\|_{C(\mathbf{R}^d)},$$

because $\Phi^{s,t}$ must also be a contraction in $\mathbf{C}^2(\mathbf{R}^d)$ since, as we have noted, the derivatives of f satisfy the same equation as f itself. This implies (7.4). Equation (7.5) is again a direct consequence of duality. Only uniqueness is not obvious here, but it follows from a general duality argument; see Theorem 2.10. □

We now have all the tools we need to analyze nonlinear Lévy processes. First let us recall their definition from Section 1.4. We defined a family of Lévy–Khintchine generators by (see (1.23))

$$A_\mu f(x) = \frac{1}{2}(G(\mu)\nabla, \nabla) f(x) + (b(\mu), \nabla f)(x)$$
$$+ \int [f(x+y) - f(x) - (y, \nabla f(x))\mathbf{1}_{B_1}(y)]\nu(\mu, dy) \qquad (7.6)$$

for $\mu \in \mathcal{P}(\mathbf{R}^d)$. The *nonlinear Lévy semigroup* generated by A_μ was defined as the weakly continuous semigroup V^t of weakly continuous transformations of $\mathcal{P}(\mathbf{R}^d)$ such that, for any $\mu \in \mathcal{P}(\mathbf{R}^d)$ and any $f \in C^2_\infty(\mathbf{R}^d)$, the measure-valued curve $\mu_t = V^t(\mu)$ satisfies

$$\frac{d}{dt}(f, \mu_t) = (A_{\mu_t} f, \mu_t), \qquad t \geq 0, \ \mu_0 = \mu.$$

Once the Lévy semigroup was constructed, we defined the corresponding *nonlinear Lévy process* with initial law μ as the time-nonhomogeneous Lévy process generated by the family

$$L_t f(x) = A_{V^t \mu} f(x) = \frac{1}{2}(G(V^t(\mu))\nabla, \nabla) f(x) + (b(V^t(\mu)), \nabla f)(x)$$
$$+ \int [f(x+y) - f(x) - (y, \nabla f(x))\mathbf{1}_{B_1}(y)]\nu(V^t(\mu), dy).$$

with law μ at $t = 0$.

Theorem 7.3 *Suppose that the coefficients of the family* (7.6) *depend on* μ *Lipschitz continuously in the norm of the Banach space* $(C^2_\infty(\mathbf{R}^d))'$ *dual to* $C^2_\infty(\mathbf{R}^d)$, *i.e.*

$$\|G(\mu) - G(\eta)\| + \|b(\mu) - b(\eta)\| + \int \min(1, |y|^2)|\nu(\mu, dy) - \nu(\eta, dy)|$$
$$\leq \kappa \|\mu - \eta\|_{(C^2_\infty(\mathbf{R}^d))'} = \kappa \sup_{\|f\|_{C^2_\infty(\mathbf{R}^d)} \leq 1} |(f, \mu - \eta)|, \tag{7.7}$$

with constant κ. *Then there exists a unique nonlinear Lévy semigroup generated by* A_μ, *and hence a unique nonlinear Lévy process.*

Proof Let us introduce the distance d on $\mathcal{P}(\mathbf{R}^d)$ induced by its embedding in $(C^2_\infty(\mathbf{R}^d))'$:

$$d(\mu, \eta) = \sup\left\{|(f, \mu - \eta)| : f \in C^2_\infty(\mathbf{R}^d), \ \|f\|_{C^2_\infty(\mathbf{R}^d)} \leq 1\right\}.$$

Observe that $\mathcal{P}(\mathbf{R}^d)$ is a closed subset of $(C^2_\infty(\mathbf{R}^d))'$ with respect to this metric. In fact, as it clearly holds that

$$d(\mu, \eta) = \sup\left\{|(f, \mu - \eta)| : f \in C^2(\mathbf{R}^d), \ \|f\|_{C^2(\mathbf{R}^d)} \leq 1\right\},$$

the convergence $\mu_n \to \mu$, $\mu_n \in \mathcal{P}(\mathbf{R}^d)$, with respect to this metric implies the convergence $(f, \mu_n) \to (f, \mu)$ for all $f \in C^2(\mathbf{R}^d)$ and hence for all $f \in C_\infty(\mathbf{R}^d)$ and for constant f. This implies the tightness of the family μ_n and that the limit $\mu \in \mathcal{P}(\mathbf{R}^d)$. Hence the set $M_\mu(t)$ of curves $s \in [0, t] \mapsto \mathcal{P}(\mathbf{R}^d)$ that are continuous with respect to the distance d, such that $\mu_0 = \mu$, is a complete metric space with respect to the uniform distance

$$d^t_u(\mu[.], \eta[.]) = \sup_{s \in [0,t]} d(\mu_s, \eta_s).$$

By Proposition 7.1 and Corollary 7.2, for any curve $\mu[.] \in M_\mu(t)$ the nonhomogeneous Lévy semigroup $\Phi^{s,t}(\mu[.])$ corresponding to $L(\mu_t)$ and its dual $V^{t,s}(\mu[.])$ are well defined, and the curve $V^{t,s}(\mu[.])\mu$ belongs to $M_\mu(t)$. Clearly, to prove the theorem it is enough to show the existence of the unique fixed point of the mapping of $M_\mu(t)$ to itself given by $\mu[.] \mapsto V^{.,0}(\mu[.])\mu$. By the contraction principle it is enough to show that this mapping is a (strict) contraction. To this end, one can write

$$d_u^t\left(V^{.,0}(\mu[.])\mu,\, V^{.,0}(\eta[.])\mu\right)$$
$$= \sup_{\|f\|_{C^2(\mathbf{R}^d)}\leq 1} \sup_{s\in[0,t]} |(f,\, V^{s,0}(\mu[.])\mu - V^{s,0}(\eta[.])\mu)|$$
$$= \sup_{\|f\|_{C^2(\mathbf{R}^d)}\leq 1} \sup_{s\in[0,t]} \left|(\Phi^{0,s}(\mu[.])f - \Phi^{0,s}(\eta[.])f,\, \mu)\right|.$$

Now we use a well-known trick (applied repeatedly in our exposition) for estimating the difference of two propagators, writing

$$\Phi^{0,s}(\mu[.])f - \Phi^{0,s}(\eta[.])f = \Phi^{0,\tau}(\mu[.])\Phi^{\tau,s}(\eta[.])f\Big|_{\tau=0}^{\tau=s}$$
$$= \int_0^s d\tau \frac{d}{d\tau}\Phi^{0,\tau}(\mu[.])\Phi^{\tau,s}(\eta[.])f$$
$$= \int_0^s \Phi^{0,\tau}(\mu[.])(A_{\mu_\tau} - A_{\eta_\tau})\Phi^{\tau,s}(\eta[.])f,$$

where a differential equation like (7.2) has been used for Φ. Consequently,

$$d_u^t\left(V^{.,0}(\mu[.])\mu,\, V^{.,0}(\eta[.])\mu\right)$$
$$\leq t \sup_{s\in[0,t]} \sup_{\|f\|_{C^2(\mathbf{R}^d)}\leq 1} \left\|\Phi^{0,\tau}(\mu[.])(A_{\mu_\tau} - A_{\eta_\tau})\Phi^{\tau,s}(\eta[.])f\right\|_{C(\mathbf{R}^d)}.$$

As the family of transformations $\Phi^{s,t}$ increases neither the norm in $C(\mathbf{R}^d)$ nor the norm in $C^2(\mathbf{R}^d)$ (because, as mentioned above, the derivatives of f satisfy the same equation as f itself owing to spatial homogeneity), the above expression is bounded by

$$t \sup_{s\in[0,t]} \sup_{\|f\|_{C^2_\infty(\mathbf{R}^d)}\leq 1} \|(A_{\mu_\tau} - A_{\eta_\tau})f\|_{C(\mathbf{R}^d)}$$
$$\leq t \sup_{s\in[0,t]} \Bigg(\|G(\mu_s) - G(\eta_s)\| + \|b(\mu_s) - b(\eta_s)\|$$
$$+ \int \min(1, |y|^2)|\nu(\mu_s, dy) - \nu(\eta_s, dy)| \Bigg),$$

which by (7.7) does not exceed $t\kappa d_u^t(\mu[.], \eta[.])$. Hence for $t < 1/\kappa$ our mapping is a contraction, showing the existence and uniqueness of the fixed point for such t. Of course, this can be extended to arbitrary times by iteration (as usual in the theory of ordinary differential equations), completing the proof of the theorem. □

Remark 7.4 Condition (7.7) is not as strange as it seems. It is satisfied, for instance, when the coefficients G, b, ν depend on μ via certain integrals (possibly multiple) with sufficiently smooth densities, as usually occurs in applications.

7.2 Variable coefficients via fixed-point arguments

By Theorem 2.12 one can get well-posedness for a nonlinear problem of the type

$$\frac{d}{dt}(g, \mu_t) = (A_{\mu_t} g, \mu_t), \qquad \mu_0 = \mu, \tag{7.8}$$

from the regularity of the time-nonhomogeneous problem obtained by fixing μ_t in the expression A_{μ_t}, since this yields natural nonlinear analogs of all results from Part I. Moreover, when A_μ is of Lévy–Khintchine form the dual operator often has a similar expression, allowing one to deduce additional regularity for the dual problem and consequently also for the nonlinear problem. Yet the smoothing properties of a linear semigroup (say, if it has continuous transition densities), which are usually linked with a certain kind of non-degeneracy, have a nonlinear counterpart. The results given in this section exemplify more or less straightforward applications of this approach.

We start with nonlinear evolutions generated by integro-differential *operators of order at most one*, i.e. by the operator equation

$$A_\mu f(x) = (b(x, \mu), \nabla f(x)) + \int_{\mathbf{R}^d \setminus \{0\}} \left[f(x+y) - f(x)\right] \nu(x, \mu, dy). \tag{7.9}$$

The nonlinear evolutions governed by operators of this type include the Vlasov equations, the mollified Boltzmann equation and nonlinear stable-like processes whose index of stability is less than 1. Stable-like processes with higher indices are analyzed at the end of this section.

Theorem 7.5 *Assume that, for any $\mu \in \mathcal{P}(\mathbf{R}^d)$, $b(., \mu) \in C^1(\mathbf{R}^d)$ and that $\nabla \nu(x, \mu, dy)$ (the gradient with respect to x) exists in the weak sense as a signed measure and depends weakly continuously on x. Let the following conditions hold:*

$$\sup_{x,\mu} \int \min(1, |y|) \nu(x, \mu, dy) < \infty, \quad \sup_{x,\mu} \int \min(1, |y|) |\nabla \nu(x, \mu, dy)| < \infty; \tag{7.10}$$

for any $\epsilon > 0$ there exists a $K > 0$ such that

$$\sup_{x,\mu} \int_{\mathbf{R}^d \setminus B_K} \nu(x,\mu,dy) < \epsilon, \quad \sup_{x,\mu} \int_{\mathbf{R}^d \setminus B_K} |\nabla \nu(x,\mu,dy)| < \epsilon \tag{7.11}$$

$$\sup_{x,\mu} \int_{B_{1/K}} |y| \nu(x,\mu,dy) < \epsilon; \tag{7.12}$$

and, finally,

$$\sup_x \int \min(1,|y|) |\nu(x,\mu_1,dy) - \nu(x,\mu_2,dy)| \le c\|\mu_1 - \mu_2\|_{(C^1_\infty(\mathbf{R}^d))^\star}, \tag{7.13}$$

$$\sup_x |b(x,\mu_1) - b(x,\mu_2)| \le c\|\mu_1 - \mu_2\|_{(C^1_\infty(\mathbf{R}^d))^\star} \tag{7.14}$$

uniformly for bounded μ_1, μ_2. Then the weak nonlinear Cauchy problem (7.8) *with A_μ given by* (7.9) *is well posed, i.e. for any $\mu \in \mathcal{M}(\mathbf{R}^d)$ it has a unique solution $T_t(\mu) \in \mathcal{M}(\mathbf{R}^d)$ (so that* (7.8) *holds for all $g \in C^1_\infty(\mathbf{R}^d)$) preserving the norm, and the transformations T_t of $\mathcal{P}(\mathbf{R}^d)$ or, more generally, $\mathcal{M}(\mathbf{R}^d)$, $t \ge 0$, form a semigroup depending Lipschitz continuously on time t and the initial data in the norm of $(C^1_\infty(\mathbf{R}^d))^\star$.*

Proof This is straightforward from Theorems 4.17, and 2.12. Alternatively one can use Theorem 3.12. □

We shall say that a family of functions $f_\alpha(x)$ on a locally compact space S belongs to $C_\infty(\mathbf{R}^d)$ uniformly in α if, for any $\epsilon > 0$, there exists a compact K such that $|f_\alpha(x)| < \epsilon$ for all $x \notin K$ and all α.

Remark 7.6 Clearly (7.13), (7.14) hold whenever b and ν have variational derivatives such that

$$\left|\frac{\delta b(x,\mu)}{\delta \mu(v)}\right| + \left|\frac{\partial}{\partial v}\frac{\delta b(x,\mu)}{\delta \mu(v)}\right| \in C_\infty(\mathbf{R}^d), \tag{7.15}$$

$$\int \min(1,|y|)\left(\left\|\frac{\delta \nu}{\delta \mu(v)}(x,\mu,dy)\right\| + \left\|\frac{\partial}{\partial v}\frac{\delta \nu}{\delta \mu(v)}(x,\mu,dy)\right\|\right) \in C_\infty(\mathbf{R}^d) \tag{7.16}$$

as functions of v uniformly for $x \in \mathbf{R}^d$, $\|\mu\| \le M$. Hint: use (F.4).

We shall discuss now the regularity of the solution to (7.8) and its stability with respect to small perturbations of A, for which we need to write down the action of the operator dual to (7.9) on functions, not just on measures. Since we are not aiming to describe the most general situation, we shall reduce our attention to the case of Lévy measures with densities.

Theorem 7.7 *Let $k \in \mathbf{N}$, $k \geq 2$. Assume that the assumptions of the previous theorem hold and moreover that the measures $\nu(x, \mu, .)$ have densities $\nu(x, \mu, y)$ with respect to Lebesgue measure such that*

$$\int \min(1, |y|) \sup_{x \in \mathbf{R}^d, \|\mu\| \leq M} \left(\nu(x, \mu, y) + \left| \frac{\partial \nu}{\partial x}(x, \mu, y) \right| + \dots + \left| \frac{\partial^k \nu}{\partial x^k}(x, \mu, y) \right| \right) dy < \infty. \tag{7.17}$$

(i) Then the nonlinear semigroup T_t of Theorem 7.5 preserves the space of measures with smooth densities, i.e. the Sobolev spaces $W_1^l = W_1^l(\mathbf{R}^d)$, $l = 0, \dots, k-1$, are invariant under T_t and T_t is a bounded strongly continuous semigroup (of nonlinear transformations) in each of these Banach spaces.

Furthermore, with some abuse of notation we shall identify the measures with their densities, denoting by T_t the action of T_t on these densities (for an $f \in L_1(\mathbf{R}^d)$, $T_t(f)$ is the density of the measure that is the image under T_t of the measure with density f). A similar convention will apply to the notation for the coefficients b and ν.

(ii) If additionally

$$\int \min(1, |y|) \sup_x \left| \frac{\partial^l}{\partial x^l} \nu(x, f_1, y) - \frac{\partial^l}{\partial x^l} \nu(x, f_2, y) \right| dy \leq c \|f_1 - f_2\|_{W_1^l}, \tag{7.18}$$

$$\sup_x \left| \frac{\partial^l}{\partial x^l} b(x, f_1) - \frac{\partial^l}{\partial x^l} b(x, f_2) \right| \leq c \|f_1 - f_2\|_{W_1^l} \tag{7.19}$$

for $l = 1, \dots, k$ then the mapping T_t reduced to any space W_1^l is Lipschitz continuous in the norm of W_1^{l-1}, i.e. uniformly for finite times we have

$$\|T_t(f_1) - T_t(f_2)\|_{W_1^{l-1}} \leq c(\|f_1\|_{W_1^l} + \|f_2\|_{W_1^l}) \|f_1 - f_2\|_{W_1^{l-1}} \tag{7.20}$$

for a continuous function c on $\mathbf{R}_+$. Moreover, for any $f \in W_1^l$ the curve $f_t = T_t(f)$ satisfies equation (7.8) strongly in the sense that

$$\frac{d}{dt} f_t = A_{f_t}^{\star} f_t \tag{7.21}$$

in the norm topology of W_1^{l-1}.

Proof Statement (i) follows from Theorem 4.18. In particular, in order to see that T_t is strongly continuous, i.e. $\|T_t(f) - f\|_{W_1^l} \to 0$ as $t \to 0$, one observes that $T_t(f) = V^{t,0}[f]f$ and that $V^{t,0}[f]$ is strongly continuous by

Theorem 4.18. Statement (ii) is proved similarly to Theorem 2.12. Namely, from (7.18), (7.19) it follows that (see (4.36) for the explicit form of $A_f^\star$)

$$\|(A_f^\star - A_g^\star)\phi\|_{W_1^{l-1}} \leq c\|\phi\|_{W_1^l}\|f-g\|_{W_1^{l-1}}. \tag{7.22}$$

Since

$$\begin{aligned} f_t - g_t = T_t(f) - T_t(g) &= (V^{t,0}[f.] - V^{t,0}[g.])f + V^{t,0}[g.](f-g) \\ &= \int_0^t V^{t,0}[g.](A_{f_s}^\star - A_{g_s}^\star)V^{s,0}[f.]f\, ds + V^{t,0}[g.](f-g), \end{aligned}$$

one has

$$\sup_{s\leq t}\|f_s - g_s\|_{W_1^{l-1}} \leq t\kappa \sup_{s\leq t}\|f_s - g_s\|_{W_1^{l-1}}\|f\|_{W_1^l} + c\|f-g\|_{W_1^{l-1}}.$$

Consequently (7.20) follows, first for small t and then for all finite t by iteration. Finally, in order to see that (7.21) holds in W_1^{l-1} for an $f \in W_1^l$ one needs to show that the rh.s. of (7.21) is continuous in W_1^{l-1}; this is clear because

$$A_{f_t}^\star f_t - A_f^\star f = A_{f_t}^\star(f_t - f) + (A_{f_t}^\star - A_f^\star)f,$$

where the first (resp. the second) term is small in W_1^{l-1} owing to the strong continuity of $f_t = T_t(f)$ in W_1^l (resp. owing to (7.22)). □

Theorem 7.8 *Under the assumptions of Theorem 7.7, suppose that we have in addition a sequence* A_μ^n *of operators, of the form given by (7.9), also satisfying the conditions of Theorem 7.7 and such that*

$$\left\| \left[(A_f^n)^\star - A_f^\star\right]\phi\right\|_{W_1^l} \leq \alpha_n\kappa(\|\phi\|_{W_1^m}) \tag{7.23}$$

for a certain $l \leq m \leq k-1$ *and for a sequence* α_n *tending to zero as* $n \to \infty$ *and a certain continuous function* κ *on* $\mathbf{R}_+$. *Then the corresponding nonlinear semigroups* T_t^n *converge to* T_t *in the sense that*

$$\|T_t^n(f) - T_t(f)\|_{W_1^l} \to 0, \qquad n \to \infty, \tag{7.24}$$

uniformly for f *from bounded subsets of* W_1^m.

Proof Using the same method as in the proof of Theorem 2.15 one obtains the estimate

$$\|T_t^n(f) - T_t(f)\|_{W_1^l} \leq \alpha_n c(\|f\|_{W_1^m}),$$

implying (7.24). □

Further regularity for this problem will be discussed in the next chapter.

The approach to general nonlinear generators of Lévy–Khintchine type will be demonstrated for the model of nonlinear diffusion combined with stable-like processes considered in Proposition 3.17. One can of course formulate more general results based on other linear models analyzed in Part I. Moreover, using the additional regularity of linear models (say, by invoking Theorem 4.25 for the non-degenerate spectral-measure case) yields additional regularity for the linear problem.

Assume that

$$\begin{aligned} A_\mu f(x) &= \tfrac{1}{2}\operatorname{tr}\left[\sigma_\mu(x)\sigma_\mu^T(x)\nabla^2 f(x)\right] + (b_\mu(x), \nabla f(x)) + \int [f(x+y) - f(x)]\,\nu_\mu(x, dy) \\ &\quad + \int_P dp \int_0^K d|y| \int_{S^{d-1}} a_{p,\mu}(x,s)\frac{f(x+y) - f(x) - (y, \nabla f(x))}{|y|^{\alpha_p(x,s)+1}} \\ &\quad \times d|y|\,\omega_{p,\mu}(ds), \end{aligned} \tag{7.25}$$

where $s = y/|y|$, $K > 0$, (P, dp) is a Borel space with finite measure dp and the ω_p are certain finite Borel measures on S^{d-1}.

Theorem 7.9 *Suppose that the assumptions of Proposition 3.18 (ii) hold uniformly for all probability measures μ and, moreover, that*

$$\|A_{\mu_1} - A_{\mu_2}\|_{C^2_\infty(\mathbf{R}^d)\mapsto C_\infty(\mathbf{R}^d)} \le c\|\mu_1 - \mu_2\|_{(C^2_\infty(\mathbf{R}^d))^\star}, \qquad \mu_1, \mu_2 \in \mathcal{P}(\mathbf{R}^d)$$

(this assumption of smooth dependence of the coefficients on μ is easy to check). Then the weak equation (7.8) *with A_μ of the form given by* (7.25) *is well posed, i.e. for any $\mu \in \mathcal{M}(\mathbf{R}^d)$ there exists a unique weakly continuous curve $\mu_t \in \mathcal{M}(\mathbf{R}^d)$ such that* (7.8) *holds for all $g \in (C^2_\infty \cap C^2_{\mathrm{Lip}})(\mathbf{R}^d)$.*

Proof This follows from Theorem 2.12 and an obvious nonhomogeneous extension of Proposition 3.18. □

Another example of the use of linear solutions to construct nonlinear solutions relates to nonlinear curvilinear Ornstein–Uhlenbeck processes, which are discussed in Chapter 11.4.

7.3 Nonlinear SDE construction

Here we describe how the SDE approach can be used to solve weak equations of the form

$$\frac{d}{dt}(f, \mu_t) = (A_{\mu_t} f, \mu_t), \qquad \mu_t \in \mathcal{P}(\mathbf{R}^d),\ \mu_0 = \mu, \tag{7.26}$$

that we expect to hold for, say, all $f \in C_c^2(\mathbf{R}^d)$, with

$$A_\mu f(x) = \frac{1}{2}(G(x,\mu)\nabla,\nabla)f(x) + (b(x,\mu),\nabla f(x)) + \int \big[f(x+y) - f(x) - (\nabla f(x), y)\big]\,\nu(x,\mu,dy) \tag{7.27}$$

and $\nu(x,\mu,.) \in \mathcal{M}_2(\mathbf{R}^d)$, using the SDE approach. This method is a natural extension of that used in Theorem 3.11.

Let $Y_t(z,\mu)$ be a family of Lévy processes parametrized by points z and probability measures μ in $\mathbf{R}^d$ and specified by their generators $L[z,\mu]$, where

$$L[z,\mu]f(x) = \frac{1}{2}(G(z,\mu)\nabla,\nabla)f(x) + (b(z,\mu),\nabla f(x)) + \int \big[f(x+y) - f(x) - (\nabla f(x), y)\big]\,\nu(z,\mu,dy) \tag{7.28}$$

and $\nu(z,\mu,.) \in \mathcal{M}_2(\mathbf{R}^d)$. Our approach to solving (7.26) is to use the solution to the following nonlinear distribution-dependent stochastic equation with nonlinear Lévy-type integrators:

$$X_t = x + \int_0^t dY_s(X_s, \mathcal{L}(X_s)), \qquad \mathcal{L}(x) = \mu, \tag{7.29}$$

with a given initial distribution μ and a random variable x that is independent of $Y_\tau(z,\mu)$. Let us define the solution through the Euler–Ito-type approximation scheme, i.e. by means of the approximations $X_t^{\mu,\tau}$, where

$$X_t^{\mu,\tau} = X_{l\tau}^{\mu,\tau} + Y_{t-l\tau}^l(X_{l\tau}^{\mu,\tau}, \mathcal{L}(X_{l\tau}^{\mu,\tau})), \qquad \mathcal{L}(X_0^{\mu,\tau}) = \mu, \tag{7.30}$$

for $l\tau < t \le (l+1)\tau$, and where $Y_\tau^l(x,\mu)$ is a collection (for $l = 0, 1, 2, \ldots$) of independent families of the Lévy processes $Y_\tau(x,\mu)$ depending measurably on x, μ (which can be constructed via Lemma 3.1 under the conditions of Theorem 7.10 below). We define the approximations $X^{\mu,\tau}$ by

$$X_t^{\mu,\tau} = X_{l\tau}^{\mu,\tau} + Y_{t-l\tau}^l(X_{l\tau}^{\mu,\tau}), \qquad \mathcal{L}(X_0^{\mu,\tau}) = \mu,$$

for $l\tau < t \le (l+1)\tau$, where $\mathcal{L}(X)$ is the probability law of X.

Clearly these approximation processes are càdlàg. Let us stress for clarity that the Y_τ depend on x, μ only via the parameters of the generator, i.e., say, the random variable $\xi = x + Y_\tau(x, \mathcal{L}(x))$ has the characteristic function

$$\mathbf{E}e^{ip\xi} = \int \mathbf{E}e^{ip(x+Y_\tau(x,\mathcal{L}(x))}\mu(dx).$$

For $x \in \mathbf{R}^d$ we shall write for brevity $X_{k\tau}^{x,\tau}$ instead of $X_{k\tau}^{\delta_x,\tau}$.

By the weak solution to (7.29) we shall mean the weak limit of $X_\mu^{\tau_k}$, $\tau_k = 2^{-k}$, $k \to \infty$, in the sense of distributions on the Skorohod space of càdlàg paths (which is of course implied by the convergence of the distributions in the sense of the distance (A.4)). Alternatively, one could define it as the solution to the corresponding nonlinear martingale problem or using more direct notions of a solution, as in Section 3.1.

Theorem 7.10 *Let an operator* A_μ *have the form given by* (7.27) *and let*

$$\|\sqrt{G(x,\mu)} - \sqrt{G(z,\eta)}\| + |b(x,\mu) - b(z,\eta)| + W_2(\nu(x,\mu;.), \nu(z,\eta,.)) \leq \kappa(|x-z| + W_2(\mu,\eta)) \qquad (7.31)$$

hold true for a constant κ. *Then:*

(i) for any $\mu \in \mathcal{P}(\mathbf{R}^d) \cap \mathcal{M}_2(\mathbf{R}^d)$ *there exists a process* $X_\mu(t)$ *solving* (7.29) *such that*

$$\sup_{\mu:\|\mu\|_{\mathcal{M}_2(\mathbf{R}^d)}<M} W^2_{2,t,\text{un}}\left(X^{\mu,\tau_k}, X^\mu\right) \leq c(t)\tau_k; \qquad (7.32)$$

(ii) the distributions $\mu_t = \mathcal{L}(X_t)$ *depend* $1/2$*-Hölder continuously on* t *in the metric* W_2 *and the* X_t^μ *depend Lipschitz continuously on the initial condition in the sense that*

$$W_2^2(X_t^\mu, X_t^\eta) \leq c(t_0) W_2^2(\mu,\eta); \qquad (7.33)$$

(iii) the processes

$$M(t) = f(X_t^\mu) - f(x) - \int_0^t L[X_s^\mu, \mathcal{L}(X_s^\mu)]f(X_s^\mu)\,ds$$

are martingales for any $f \in C^2(\mathbf{R}^d)$*; in other words, the process* X_t^μ *solves the corresponding (nonlinear) martingale problem;*

(iv) the distributions $\mu_t = \mathcal{L}(X_t)$ *satisfy the weak nonlinear equation* (7.26) *(which holds for all* $f \in C^2(\mathbf{R}^d)$*);*

(v) the resolving operators $U_t : \mu \mapsto \mu_t$ *of the Cauchy problem* (7.26) *form a nonlinear Markov semigroup, i.e. they are continuous mappings from* $\mathcal{P}(\mathbf{R}^d) \cap \mathcal{M}_2(\mathbf{R}^d)$*, equipped with the metric* W_2*, to itself such that* U_0 *is the identity mapping and* $U_{t+s} = U_t U_s$ *for all* $s, t \geq 0$.

Proof This is an extension of the corresponding result for Feller processes, Theorem 3.11. The details can be found in [138] and will not be reproduced here. □

7.4 Unbounded coefficients

Processes having unbounded coefficients are usually obtained by a limiting procedure from the corresponding bounded ones. For example, this was the strategy used in Section 6.4 for constructing pure jump interactions with a bound for the rates that is additive with respect to a conservation law. Usually this procedure yields only an existence result (one proves the compactness of bounded approximations and then chooses a converging subsequence) and the uniqueness question is settled by other methods. However, if a corresponding linear problem with unbounded coefficients is sufficiently regular, one can establish well posedness by the same direct procedure as that used in Section 7.2 for the case of bounded coefficients. We give here an example on *nonlinear localized stable-like processes* with unbounded coefficients.

Theorem 7.11 *Let*

$$
\begin{aligned}
A_\mu f(x) &= \tfrac{1}{2}\operatorname{tr}\left[\sigma(x,\mu)\sigma^T(x,\mu)\nabla^2 f(x)\right] + (b(x,\mu),\nabla f(x)) \\
&\quad + \int_P (dp)\int_0^K d|y|\int_{S^{d-1}} a_p(x,s,\mu) \\
&\qquad\qquad \times \frac{f(x+y)-f(x)-(y,\nabla f(x))}{|y|^{\alpha_p(x,s,\mu)+1}}\, d|y|\,\omega_p(ds),
\end{aligned}
\tag{7.34}
$$

where $s = y/|y|$ and the coefficients satisfy the assumptions of Proposition 5.5 uniformly for all probability measures μ on $\mathbf{R}^d$ and furthermore

$$
\|A_{\mu_1} - A_{\mu_2}\|_{C^2_\infty(\mathbf{R}^d)\mapsto C_{|.|^k,\infty}} \le \kappa\|\mu_1 - \mu_2\|_{(C^2_\infty(\mathbf{R}^d))^\star}.
$$

Then the weak equation (7.8) with A_μ of the form (3.55) is well posed, i.e. for any $\mu \in \mathcal{M}(\mathbf{R}^d)$ there exists a unique weakly continuous curve $\mu_t \in \mathcal{M}(\mathbf{R}^d)$ such that (7.8) holds for all $g \in C^2_\infty(\mathbf{R}^d)$.

Proof This is a consequence of Theorem 2.12 and an obvious nonhomogeneous extension of Proposition 5.5. □

8

Smoothness with respect to initial data

In this chapter we analyze the derivatives of the solutions to nonlinear kinetic equations with respect to the initial data. Rather precise estimates of these derivatives are given. As usual we consider sequentially our basic three cases of nonlinear integral generators, generators of order at most one and stable-like generators. This chapter contains possibly the most technical material in the book. Hence some motivation is in order, which is given in the following introductory section.

8.1 Motivation and plan; a warm-up result

The main result on the convergence of semigroups states that a sequence of strongly continuous semigroups T_t^n converges to a strongly continuous semigroup T_t whenever the generators A_n of T_t^n converge to the generator A of T_t on the set of vectors f forming a core for A. If this core is invariant, then precise estimates of the convergence can be obtained; see Theorem 2.11 for a more general statement. Using this fact, a functional central limit theorem can easily be proved as follows. Let Z_t be a continuous-time random walk on $\mathbf{Z}$ moving in each direction with equal probability. This process is specified by the generator L given by

$$Lf(x) = a[f(x+1) + f(x-1) - 2f(x)]$$

with coefficient $a > 0$. By scaling we can define a new random walk Z_t^h on $h\mathbf{Z}$ with generator L_h given by

$$L_h f(x) = \frac{a}{h^2}\left[f(x+h) + f(x-h) - 2f(x)\right].$$

If f is thrice continuously differentiable then

$$L_h f(x) = af''(x) + O(1) \sup_y |f^{(3)}(y)|.$$

Hence, for $f \in C^3_\infty(\mathbf{R}^d)$ the generators L_h converge to the generator L, where $Lf = af''$, for a Brownian motion. Since $C^3_\infty(\mathbf{R}^d)$ is invariant for the semigroup of this Brownian motion, we conclude that the semigroups of Z^h_t converge to the heat semigroup of the Brownian motion. We can also apply Theorem C.5 to deduce convergence in distribution on the Skorohod space of càdlàg paths.

Remark 8.1 The process Z^h_t is a random walk. Namely, when starting at the origin, it equals $h(S_1 + \cdots + S_{N_t})$, where N_t is the number of jumps and the S_i are i.i.d. Bernoulli random variables. By Exercise 2.1 $\mathbf{E}(N_t) = 2at/h^2$, so that h is of order $1/\sqrt{N_t}$.

The above argument has a more or less straightforward extension to quite general classes of (even position-dependent) continuous-time random walks; see e.g. Kolokoltsov [135] for the case of stable-like processes. The main non-trivial technical problem in implementing this approach lies in identifying a core for the limiting generator.

In Part III we will use this method twice in an infinite-dimensional setting, showing first that the approximating interacting-particle systems converge to the deterministic limit described by the kinetic equations and second that the fluctuation process converges to a limiting infinite-dimensional Gaussian process. In both cases we need a core for the limiting semigroup, and to obtain the rates of convergence we need an invariant core.

Looking at the generators in (1.75) we can conclude that, for ΛF to make sense, the function F on measures should have a variational derivative. But the limiting evolution is deterministic, i.e. it has the form $F_t(\mu) = F(\mu_t)$ where μ_t is a solution to the kinetic equation. We therefore have

$$\frac{\delta F_t}{\delta \mu} = \frac{\delta F}{\delta \mu_t} \frac{\delta \mu_t}{\delta \mu},$$

which shows that, in order to apply the general method outlined above, we need first-order variational derivatives of solutions to the kinetic equation that are sufficiently regular with respect to the initial data. Similarly, the generators of fluctuation processes are written in terms of second-order variational derivatives, making it necessary to study the second-order variational derivatives of solutions to the kinetic equation with respect to the initial data.

In this chapter we address this issue by analyzing the smoothness with respect to the initial data of the nonlinear Markov semigroups $T_t(\mu) = \mu_t$

constructed in previous chapters from equations of the type

$$\frac{d}{dt}(g, \mu_t) = (A_{\mu_t} g, \mu_t), \qquad \mu_0 = \mu \in \mathcal{M}(S), \; g \in D \subset C(S), \qquad (8.1)$$

where A_μ is a Lévy–Khintchine type operator depending on μ as on a parameter. Our aim is to identify an invariant core for the usual (linear) Markov semigroup Φ_t, where $\Phi_t F(\mu) = F(\mu_t)$, of the deterministic measure-valued Markov process μ_t. The strong form of equation (8.1) is

$$\frac{d}{dt}\mu_t = A^\star_{\mu_t}\mu_t, \qquad \mu_0 = \mu \in \mathcal{M}(S). \qquad (8.2)$$

We are concerned with derivatives of first and second order only, because they are responsible for the LLN (the first-order stochastic approximation to the limiting kinetic equation) and the CLT (the second-order approximation) for the corresponding interacting-particle systems.

Namely, we will study the signed measures

$$\xi_t(\mu_0, \xi) = D_\xi \mu_t(\mu_0) = \lim_{s \to 0_+} \frac{1}{s}\left[\mu_t(\mu_0 + s\xi) - \mu_t(\mu_0)\right] \qquad (8.3)$$

and

$$\eta_t(\mu_0, \xi, \eta) = D_\eta \xi_t(\mu_0, \xi) = \lim_{s \to 0_+} \frac{1}{s}\left[\xi_t(\mu_0 + s\eta, \xi) - \xi_t(\mu_0, \xi)\right], \qquad (8.4)$$

denoting (with some abuse of notation) the corresponding variational derivatives by

$$\xi_t(\mu_0, x) = D_{\delta_x}\mu_t(\mu_0) = \frac{\delta \mu_t}{\delta \mu_0(x)} = \lim_{s \to 0_+} \frac{1}{s}\left[\mu_t(\mu_0 + s\delta_x) - \mu_t(\mu_0)\right] \qquad (8.5)$$

and

$$\eta_t(\mu_0, x, w) = D_{\delta_w}\xi_t(\mu_0, x) = \frac{\delta \xi_t(\mu_0, x)}{\delta \mu_0(w)} = \lim_{s \to 0_+} \frac{1}{s}\left[\xi_t(\mu_0 + s\delta_w, x) - \xi_t(\mu_0, x)\right] \qquad (8.6)$$

and occasionally omitting the arguments of ξ_t, η_t to shorten the formulas.

In the following sections we shall analyze separately various classes of kinetic equations, aiming at proving the existence of the above derivatives and obtaining effective estimates for them. The most subtle estimates are obtained, in Sections 8.4 and 8.6, for the processes of coagulation and collision.

The following formal calculations will form the basis for this analysis. Differentiating (8.1) with respect to the initial data yields

$$\frac{d}{dt}(g, \xi_t(\mu_0, \xi)) = \int \left(A_{\mu_t} g(v) + \int \frac{\delta A_{\mu_t}}{\delta \mu_t(v)} g(z) \mu_t(dz) \right) \xi_t(\mu_0, \xi, dv) \tag{8.7}$$

and

$$\begin{aligned}\frac{d}{dt}(g, \eta_t(\mu_0, \xi, \eta)) &= \int \left(A_{\mu_t} g(v) + \int \frac{\delta A_{\mu_t}}{\delta \mu_t(v)} g(z) \mu_t(dz) \right) \eta_t(\mu_0, \xi, \eta, dv) \\ &\quad + \iint \left(\frac{\delta A_{\mu_t}}{\delta \mu_t(u)} g(v) + \frac{\delta A_{\mu_t}}{\delta \mu_t(v)} g(u) \right. \\ &\quad \left. + \int \frac{\delta^2 A_{\mu_t}}{\delta \mu_t(v) \delta \mu_t(u)} g(z) \mu_t(dz) \right) \\ &\quad \times \xi_t(\mu_0, \xi, dv) \xi_t(\mu_0, \eta, du), \end{aligned} \tag{8.8}$$

where the variational derivative for the operator A is obtained by variational differentiation of its coefficients (see below for explicit formulas).

In the strong form, if μ_t, ξ_t or η_t have densities with respect to Lebesgue measure, which are also denoted μ_t, ξ_t, η_t by the usual abuse of notation, and if the operator $A^\star$ and its variational derivatives act in $L_1(\mathbf{R}^d)$ (i.e. $A^\star$ preserves the set of measures with densities) then equations (8.7), (8.8) clearly take the form

$$\frac{d}{dt}\xi_t(\mu_0, \xi, x) = A^\star_{\mu_t} \xi_t(\mu_0, \xi, x) + \int \xi_t(\mu_0, \xi, v) dv \frac{\delta A^\star_{\mu_t}}{\delta \mu_t(v)} \mu_t(x) \tag{8.9}$$

and

$$\begin{aligned}\frac{d}{dt}\eta_t(\mu_0, \xi, \eta, x) &= A^\star_{\mu_t} \eta_t(\mu_0, \xi, \eta, x) + \int \eta_t(\mu_0, \xi, \eta, v)\, dv \frac{\delta A^\star_{\mu_t}}{\delta \mu_t(v)} \mu_t(x) \\ &\quad + \int \xi_t(\mu_0, \xi, v)\, dv\, \xi_t(\mu_0, \eta, u)\, du \frac{\delta^2 A^\star_{\mu_t}}{\delta \mu_t(v) \delta \mu_t(u)} \mu_t(x) \\ &\quad + \int \xi_t(\mu_0, \eta, u)\, du \frac{\delta A^\star_{\mu_t}}{\delta \mu_t(u)} \xi_t(\mu_0, \xi, x) \\ &\quad + \int \xi_t(\mu_0, \xi, v)\, dv \frac{\delta A^\star_{\mu_t}}{\delta \mu_t(v)} \xi_t(\mu_0, \eta, x). \end{aligned} \tag{8.10}$$

To conclude this introductory section, let us analyze the simplest case of bounded generators, i.e. the problem (8.1) with A_μ given by (6.3),

$$A_\mu f(x) = \int_X f(y) \nu(x, \mu, dy) - a(x, \mu) f(x),$$

for which the required smoothness with respect to the initial data is a direct consequence of the standard theory of ODEs in Banach spaces (once global existence is proved, of course).

Theorem 8.2 *Under the assumptions of Theorem 6.1, suppose that the variational derivatives*

$$\frac{\delta \nu}{\delta \mu(x)}(z, \mu, dy), \qquad \frac{\delta a}{\delta \mu(x)}(z, \mu)$$

exist, depend weakly continuously on x, z, μ and are bounded on bounded subsets of μ. Then for any $\xi \in \mathcal{M}(\mathcal{X})$ the derivative (8.3) *exists strongly (the limit exists in the norm topology of $\mathcal{M}(X)$) and is a unique solution of the evolution equation*

$$\frac{d}{dt}(g, \xi_t(\mu_0, \xi)) = \int \xi_t(\mu_0, x, dv) \left[A_{\mu_t} g(v) + \int \left(\int g(y) \frac{\delta \nu}{\delta \mu(v)}(z, \mu, dy) \right.\right.$$
$$\left.\left. - \frac{\delta a}{\delta \mu(v)}(z, \mu) g(z) \right) \mu_t(dz) \right] \qquad (8.11)$$

with initial condition $\xi_0(\mu_0, \xi)) = \xi$. If the second variational derivatives of ν and a exist and are continuous and bounded on bounded subsets of μ, then the second derivative (8.4) *exists and is a unique solution of the equation obtained by differentiating* (8.11).

Proof Equation (8.11) is obtained by differentiating (8.1) using the chain rule for variational derivatives (F.7). By Theorem 6.1 this is a linear equation with coefficients that are bounded uniformly in time. The rest follows from the standard theory of ODEs in Banach spaces; see Appendix D. The statement about the second derivative is proved similarly. □

In the rest of this chapter our strategy will be the following: approximate a given equation by another equation with the same structure but with a bounded generator, apply the standard result on smoothness with respect to the initial data to this approximation and then pass to the limit.

8.2 Lévy–Khintchine-type generators

We shall start with the semigroup specified by the operator equation (7.9), with all Lévy measures absolutely continuous with respect to Lebesgue measure.

Equation (7.9) reads

$$A_\mu f(x) = (b(x,\mu), \nabla f(x)) + \int_{\mathbf{R}^d} [f(x+y) - f(x)]\, \nu(x,\mu,y)dy, \quad (8.12)$$

and clearly we have

$$\frac{\delta A}{\delta \mu(v)} g(z) = \left(\frac{\delta b}{\delta \mu(v)}(z,\mu), \nabla g(z) \right) + \int_{\mathbf{R}^d} [g(z+y) - g(z)] \frac{\delta \nu}{\delta \mu(v)}(z,\mu,y)dy$$

with a similar expression for the second variational derivative.

A particular case that is important for applications occurs when b, ν depend on μ as integral linear operators:

$$A_\mu f(x) = \left(b(x) + \int \tilde{b}(x,z)\mu(dz), \nabla f(x) \right) + \int_{\mathbf{R}^d} [f(x+y) - f(x)](\nu(x,y) + \int \tilde{\nu}(x,z,y)\mu(dz)(dy) \quad (8.13)$$

for some functions $b(x)$, $\tilde{b}(x,y)$, $\nu(x,y)$, $\tilde{\nu}(x,y,z)$, where clearly

$$\frac{\delta A}{\delta \mu(v)} g(z) = (\tilde{b}(z,v), \nabla g(z)) + \int [g(z+y) - g(z)] \tilde{\nu}(z,v,y)dy,$$

$$\frac{\delta A^2}{\delta \mu(v) \delta \mu(u)} = 0.$$

This situation is already rich enough to include, say, the Vlasov equation (when ν and $\tilde{\nu}$ vanish) and the LLN limit for a binary interacting α-stable process with $\alpha < 1$.

In the case of operators given by (8.12), the solution to (8.7) generally does not belong to $\mathcal{M}(\mathbf{R}^d)$ but to the larger Banach space $(C^1_\infty(\mathbf{R}^d))^\star$.

Theorem 8.3 *Let $k \in \mathbf{N}, k \geq 2$ and let the coefficient functions b, ν be k times continuously differentiable in x, so that*

$$\sup_{\|\mu\| \leq M, x \in \mathbf{R}^d} \left(b(x,\mu) + \left| \frac{\partial b}{\partial x}(x,\mu) \right| + \cdots + \left| \frac{\partial^k b}{\partial x^k}(x,\mu) \right| \right) < \infty, \quad (8.14)$$

$$\int \min(1,|y|) \sup_{\|\mu\| \leq M, x \in \mathbf{R}^d} \sum_{l=0}^{k} \left| \frac{\partial^l \nu}{\partial x^l}(x,\mu,y) \right| dy < \infty. \quad (8.15)$$

For a $\mu \in \mathcal{M}(\mathbf{R}^d)$ or a $\mu \in W^l_1(\mathbf{R}^d)$, $l = 0, \ldots, k-1$, we denote by μ_t the corresponding unique solution to (8.1), (8.12) *with initial data $\mu_0 = \mu$ given by Theorem 7.5. Moreover, let the variational derivatives of b, ν exist and* (7.15), (7.16) *hold. Then equation* (8.7) *is well posed in $(C^1_\infty(\mathbf{R}^d))^\star$. More* *precisely, for any* $\mu \in \mathcal{M}(\mathbf{R}^d)$ *and* $\xi \in (C^1_\infty(\mathbf{R}^d))^\star$ *there exists a unique* $\star$*-weakly continuous curve* $\xi_t \in (C^1_\infty(\mathbf{R}^d))^\star$ *with* $\xi_0 = \xi$ *solving* (8.7) $\star$*-weakly in* $(C^2_\infty(\mathbf{R}^d))^\star$*. Furthermore* ξ_t *depends continuously on* μ *in the sense that if* $\mu^n \to \mu$ *in the norm topology of* $(C^1_\infty(\mathbf{R}^d))^\star$ *then the corresponding* ξ^n_t *converge to* ξ_t *in the norm topology of* $(C^2_\infty(\mathbf{R}^d))^\star$ *and* $\star$*-weakly in* $(C^1_\infty(\mathbf{R}^d))^\star$*. Finally, if* $\xi \in \mathcal{M}^{\text{signed}}(X)$ *then* ξ_t *represents the Gateaux derivative* (8.3)*, where the limit exists in the norm topology of* $(C^2_\infty(\mathbf{R}^d))^\star$ *and* $\star$*-weakly in* $(C^1_\infty(\mathbf{R}^d))^\star$*.*

Proof The equation

$$\frac{d}{dt}(g, \xi_t(\mu, \xi)) = (A_{\mu_t} g,\ \xi_t(\mu_0, \xi))$$

is well posed, by (8.14), (8.15) and Theorem 4.18. Conditions (7.15), (7.16) ensure that the operator

$$g \mapsto \int \frac{\delta A_{\mu_t}}{\delta \mu_t(.)} g(z) \mu_t(dz) \tag{8.16}$$

appearing on the r.h.s. of (8.7) is a bounded linear mapping $C^1(\mathbf{R}^d) \mapsto C^1_\infty(\mathbf{R}^d)$ that is uniform for bounded μ, so that the stated well-posedness of (8.7) follows from perturbation theory.

Remark 8.4 Notice that this operator is not defined on $C(\mathbf{R}^d)$, hence the necessity to work with the evolution of ξ_t in $(C^1_\infty(\mathbf{R}^d))^\star$ and not in $\mathcal{M}(\mathbf{R}^d)$.

The continuous dependence of ξ_t on μ follows from the continuous dependence of μ_t on μ in the norm topology of $(C^1_\infty(\mathbf{R}^d))^\star$, the standard propagator convergence (Theorem 2.11 with $D = C^2_\infty(\mathbf{R}^d)$, $B = C^1_\infty(\mathbf{R}^d)$) and the observation that

$$\left\| (A_{\mu_1} - A_{\mu_2}) g + \int \frac{\delta A_{\mu_1}}{\delta \mu(.)} g(z) \mu_1(dz) - \int \frac{\delta A_{\mu_2}}{\delta \mu(.)} g(z) \mu_2(dz) \right\|_{C^1(\mathbf{R}^d)}$$
$$\leq c \|g\|_{C^2(\mathbf{R}^d)} \|\mu_1 - \mu_2\|_{(C^1_\infty(\mathbf{R}^d))^\star}, \tag{8.17}$$

which follows from (7.15) and (7.16) (or more precisely from their implications, (7.13) and (7.14)).

In order to prove (8.3) we shall approximate the operator A_μ by a sequence of operators A^n_μ that are bounded in $C_\infty(\mathbf{R}^d)$ and are given by

$$A^n_\mu f(x) = [f(x + b(x, \mu)/n) - f(x)]n + \int_{\mathbf{R}^d \setminus B_{1/n}} [f(x+y) - f(x)] \nu(x, \mu, y) dy \tag{8.18}$$

for $n \in \mathbf{N}$, so that

$$\|(A^n_\mu - A_\mu)f\| \le \frac{1}{n}c\|f\|_{C^2(\mathbf{R}^d)} \tag{8.19}$$

uniformly for bounded μ. Since these operators enjoy the same properties as A_μ, well-posedness holds for them in the same form as for A_μ. Let us denote by $\mu^n_t, \xi^n_t, \eta^n_t$ the objects analogous to μ_t, ξ_t, η_t constructed from A^n_μ.

Since the $(A^n_\mu)^\star$ are bounded linear operators in $\mathcal{M}^{\text{signed}}(\mathbf{R}^d)$ and $(C^1_\infty(\mathbf{R}^d))^\star$, the equations for μ_t and ξ_t are both well posed in the strong sense in $(C^1_\infty(\mathbf{R}^d))^\star$ and $\mathcal{M}^{\text{signed}}(\mathbf{R}^d)$. Hence the standard result on differentiation with respect to the initial data is applicable (see Theorem D.1), leading to the conclusion that the ξ^n_t represent the Gateaux derivatives of μ^n_t with respect to the initial data in the norm of $(C^1_\infty(\mathbf{R}^d))^\star$. Consequently

$$\mu^n_t(\mu + h\xi) - \mu^n_t(\mu) = h\int_0^1 \xi^n_t(\mu + sh\xi)\,ds \tag{8.20}$$

holds as an equation in $(C^1_\infty(\mathbf{R}^d))^\star$ (and in $\mathcal{M}^{\text{signed}}(\mathbf{R}^d)$ whenever $\xi \in \mathcal{M}^{\text{signed}}(\mathbf{R}^d)$).

We aim at passing to the limit $n \to \infty$ in equation (8.20) in the norm topology of $(C^2_\infty(\mathbf{R}^d))^\star$, in the case when $\xi \in \mathcal{M}^{\text{signed}}(\mathbf{R}^d)$. Using Theorem 2.15 with $D = C^2_\infty(\mathbf{R}^d)$ we deduce the convergence of the μ^n_t in the norm of $(C^2_\infty(\mathbf{R}^d))^\star$. Next,

$$\begin{aligned}\|\xi^n_t(\mu) - \xi_t(\mu)\|_{(C^2_\infty(\mathbf{R}^d))^\star} &= \sup_{\|g\|_{C^2_\infty(\mathbf{R}^d)}\le 1} \left|\left(g, \xi^n_t(\mu) - \xi_t(\mu)\right)\right| \\ &= \sup_{\|g\|_{C^2_\infty(\mathbf{R}^d)}\le 1} (U^t g - U^t_n g, \xi) \le \|\xi\|_{\mathcal{M}(\mathbf{R}^d)}\|U^t g - U^t_n g\|_{C(\mathbf{R}^d)},\end{aligned}$$

where $U^t = U^{0,t}$ denotes the backward propagator in $C_\infty(\mathbf{R}^d)$ that is dual to the propagator on ξ given by equation (8.7). The last expression above can now be estimated by the standard convergence of propagators, given in Theorem 2.11, for $D = C^2_\infty(\mathbf{R}^d)$, $B = C_\infty(\mathbf{R}^d)$ together with the observation that combining the estimates (8.17) and (8.19) yields

$$\begin{aligned}&\left\|(A^n_{\mu^n} - A_\mu)g + \int \frac{\delta A^n_{\mu^n}}{\delta \mu^n(.)} g(z)\mu^n(dz) - \int \frac{\delta A_\mu}{\delta \mu(.)} g(z)\mu(dz)\right\|_{C(\mathbf{R}^d)} \\ &\qquad \le c\|g\|_{C^2(\mathbf{R}^d)}\left(\frac{1}{n} + \|\mu^n - \mu\|_{(C^2_\infty(\mathbf{R}^d))^\star}\right).\end{aligned}$$

Consequently, passing to the limit $n \to \infty$ in (8.20) yields the equation

$$\mu_t(\mu + h\xi) - \mu_t(\mu) = h\int_0^1 \xi_t(\mu + sh\xi)\,ds, \tag{8.21}$$

where all objects are well defined in $(C^1_\infty(\mathbf{R}^d))^\star$.

This equation together with the stated continuous dependence of ξ_t on μ implies (8.3) in the sense required. □

In the following exercise the additional regularity of the variational derivatives ξ_t is discussed.

Exercise 8.1 Under the assumptions of the above theorem, let an $m_1 \in [1, k]$ exist such that

$$\sup_{\|\mu\| \le M, x, v \in \mathbf{R}^d} \sum_{l=0}^{m_1} \left| \frac{\partial^l}{\partial x^l} \frac{\delta b}{\delta \mu(v)}(x, \mu) \right| < \infty, \tag{8.22}$$

$$\int \min(1, |y|) \sup_{\|\mu\| \le M, x, v \in \mathbf{R}^d} \sum_{l=0}^{m_1} \left| \frac{\partial^l}{\partial x^l} \frac{\delta \nu}{\delta \mu(v)}(x, \mu, y) \right| dy < \infty \tag{8.23}$$

(of course, all the derivatives in these estimates are supposed to exist). Show that, for any $\mu \in W_1^l$, $\xi \in W_1^m$, $l = 1, \ldots, k-1$, $m \le \min(m_1 - 1, l - 1)$, there exists a unique global solution $\xi_t = \xi_t(\mu_0, \xi) \in W_1^m$ of equation (8.9) considered as an equation for the densities ξ_t of the measures ξ_t (we now identify the measures and their densities both for μ_t and ξ_t) in the norm topology of W_1^{m-1}. Also, show that this solution represents the Gateaux derivative (8.3), where the limit exists in the norm topology of W_1^{m-1}. Hint: Conditions (8.22) and (8.23) imply that the second term on the r.h.s. of equation (8.9) defines the bounded operator $L_1(\mathbf{R}^d) \mapsto W_1^{\min(l-1, m_1-1)}$. Hence the well-posedness of equation (8.9) follows from perturbation theory (Theorem 2.9) and from Theorem 4.18. The other statements follow by the same arguments as in the theorem above.

Theorem 8.5 *Under the assumptions of Theorem 8.3 suppose that the second variational derivatives of b, ν exist and are smooth, so that*

$$\sup_{\|\mu\| \le M, x, v, u \in \mathbf{R}^d} \left(\left| \frac{\partial}{\partial v} \frac{\delta^2 b}{\delta \mu(v) \delta \mu(u)}(x, \mu) \right| + \left| \frac{\partial}{\partial u} \frac{\delta^2 b}{\delta \mu(v) \delta \mu(u)}(x, \mu) \right| \right.$$
$$\left. + \left| \frac{\delta^2 b}{\delta \mu(v) \delta \mu(u)}(x, \mu) \right| + \left| \frac{\partial^2}{\partial v \partial u} \frac{\delta^2 b}{\delta \mu(v) \delta \mu(u)}(x, \mu) \right| \right) < \infty$$

and the function

$$\sup_{\|\mu\| \le M, x, v, u \in \mathbf{R}^d} \left(\left| \frac{\partial}{\partial v} \frac{\delta^2 \nu}{\delta \mu(v) \delta \mu(u)}(x, \mu, y) \right| + \left| \frac{\partial}{\partial u} \frac{\delta^2 \nu}{\delta \mu(v) \delta \mu(u)}(x, \mu, y) \right| \right.$$
$$\left. + \left| \frac{\delta^2 \nu}{\delta \mu(v) \delta \mu(u)}(x, \mu, y) \right| + \left| \frac{\partial^2}{\partial v \partial u} \frac{\delta^2 \nu}{\delta \mu(v) \delta \mu(u)}(x, \mu, y) \right| \right)$$

is integrable with respect to the measure $\min(1, |y|)dy$.

Then, for any $\mu \in \mathcal{M}(\mathbf{R}^d)$, $\xi, \eta \in (C^1_\infty(\mathbf{R}^d))^\star$, *there exists a unique weakly (and vaguely) continuous curve* $\eta_t \in (C^1_\infty(\mathbf{R}^d))^\star$, *with* $\eta_0 = 0$, *solving* (8.8) *weakly in* $(C^2_\infty(\mathbf{R}^d))^\star$. *Moreover,* η_t *depends continuously on* μ *in the sense that if* $\mu^n \to \mu$ *in the norm topology of* $(C^1_\infty(\mathbf{R}^d))^\star$ *then the corresponding* η^n_t *converge to* η_t *in the norm topology of* $(C^2_\infty(\mathbf{R}^d))^\star$ *and weakly in* $(C^1_\infty(\mathbf{R}^d))^\star$. *Finally, if* $\xi, \eta \in \mathcal{M}^{\text{signed}}(X)$ *then* η_t *represents the Gateaux derivative* (8.3), *where the limit exists in the norm topology of* $(C^2_\infty(\mathbf{R}^d))^\star$ *and weakly in* $(C^1_\infty(\mathbf{R}^d))^\star$.

Proof Well-posedness for the nonhomogeneous linear equation (8.10) follows from the well-posedness of the corresponding homogeneous equation, (8.9), obtained above, from the du Hamel principle (see Theorem 2.40) and from the observation that the assumed smoothness of the variational derivatives ensures that the second integral in (8.10) is well defined for $g \in C^1(\mathbf{R}^d)$ and $\xi_t \in (C^1_\infty(\mathbf{R}^d))^\star$. The validity of (8.4) follows, as above, by the approximation argument. □

The smoothness of the solutions μ_t to the kinetic equations is of interest both theoretically and practically (say, for the analysis of numerical solution schemes). Moreover, it plays a crucial role in the analysis of the *Markov semigroup* Φ_t, where $\Phi_t F(\mu) = F(\mu_t(\mu))$, *arising from the deterministic measure-valued Markov process* $\mu_t(\mu)$ and its Feller properties. Clearly, the operators Φ_t form a semigroup of contractions on the space of bounded measurable functionals on measures. For the analysis of this Markov semigroup, naturally arising problems are the identification of a space where it is strongly continuous and the specification of an appropriate core (preferably invariant). The above results allow us to solve these problems easily and effectively.

We shall need the notation introduced in Appendix D and some natural extensions of this. In particular, we denote by $C^{1,k}_{\text{weak}}(\mathcal{M}_M(\mathbf{R}^d))$ (resp. $C^{1,k}_{\text{vague}}(\mathcal{M}_M(\mathbf{R}^d)))$ the space of weakly (resp. vaguely) continuous functionals on $\mathcal{M}_M(X)$ (measures with norms bounded by M) such that the variational derivative $\delta F/\delta Y(x)$ exists for all Y, x, is a bounded continuous function of two variables with Y considered in the weak topology and is k times differentiable in x, so that $(\partial^k/\partial x^k)(\delta F/\delta Y(x))$ is also a weakly continuous bounded function of two variables (resp. if Y is considered in the vague topology and if additionally

$$\frac{\delta F(Y)}{\delta Y(.)} \in C^k_\infty(\mathbf{R}^d)$$

uniformly for $Y \in \mathcal{M}_M(\mathbf{R}^d)$). The spaces $C^{1,k}_{\text{weak}}(\mathcal{M}_M(\mathbf{R}^d))$ and $C^{1,k}_{\text{vague}}(\mathcal{M}_M(\mathbf{R}^d))$ become Banach spaces (note that $C^{1,k}_{\text{vague}}(\mathcal{M}_M(\mathbf{R}^d))$ is actually a closed subspace of $C^{1,k}_{\text{weak}}(\mathcal{M}_M(\mathbf{R}^d)))$ if equipped with the norm

$$\|F\|_{C^{1,k}_{\text{weak}}(\mathcal{M}_M(\mathbf{R}^d))} = \sup_Y \left\| \frac{\delta F}{\delta Y(.)} \right\|_{C^k(\mathbf{R}^d)}.$$

Similarly, we denote by $C^{2,2k}_{\text{weak}}(\mathcal{M}_M(\mathbf{R}^d))$ (resp. $C^{2,2k}_{\text{vague}}(\mathcal{M}_M(\mathbf{R}^d)))$ the subspace of $C^2_{\text{weak}}(\mathcal{M}_M(\mathbf{R}^d)) \cap C^{1,k}_{\text{weak}}(\mathcal{M}_M(\mathbf{R}^d))$ such that the derivatives

$$\frac{\partial^{l+m}}{\partial x^l \partial y^m} \frac{\delta^2 F}{\delta Y(x) \delta Y(y)}(Y)$$

exist for $l, m \le k$ and are continuous bounded functions (resp. the same with "vague" instead of "weak" everywhere and if, additionally, the derivatives in the above expression belong to $C_\infty(X^2)$ uniformly for Y). The Banach norm of these spaces is of course defined as the sum of the sup norms of all its relevant derivatives. As we shall see in the next chapter, the spaces $C^{2,2k}_{\text{vague}}(\mathcal{M}_M(\mathbf{R}^d))$ play a crucial role in the study of LLN approximation to nonlinear Markov processes.

Theorem 8.6 *(i) Under the assumptions of Theorem 8.3 the semigroup Φ_t is a Feller semigroup in $C_{\text{vague}}(\mathcal{M}_M(\mathbf{R}^d))$ and $C^{1,1}_{\text{vague}}(\mathcal{M}_M(\mathbf{R}^d))$ represents its invariant core, where the generator Λ is defined by*

$$\Lambda F(Y) = \left(A_Y \frac{\delta F(Y)}{\delta Y(.)}, Y \right). \tag{8.24}$$

(ii) Moreover, the space of polynomial functions generated by the monomials of the form $\int g(x_1, \dots, x_n) Y(dx_1) \cdots Y(dx_n)$ with $g \in C^l_\infty(\mathbf{R}^{dn})$ is also a core (though not necessarily an invariant one) for any $l \ge 1$.

(iii) Under the assumptions of Theorem 8.5 the space $C^{2,2}_{\text{vague}}(\mathcal{M}_M(\mathbf{R}^d))$ is also an invariant core.

Proof (i) Let $F \in C^{1,1}_{\text{vague}}(\mathcal{M}_M(\mathbf{R}^d))$. Then

$$\frac{\delta \Phi_t F(\mu)}{\delta \mu(x)} = \frac{\delta F(\mu_t(\mu))}{\delta \mu(x)} = \int_X \frac{\delta F(\mu_t)}{\delta \mu_t(y)} \frac{\delta \mu_t(\mu)}{\delta \mu(x)}(dy)$$
$$= \int_X \frac{\delta F(\mu_t)}{\delta \mu_t(y)} \xi_t(\mu, x, dy), \tag{8.25}$$

implying that this derivative is well defined, since $\xi_t \in (C^1_\infty(\mathbf{R}^d))^\star$ and $\delta F(\mu_t)/\delta \mu_t(y) \in C^1_\infty(\mathbf{R}^d)$. Moreover

$$(g, \xi_t(\mu, x, .)) = (U^t g, \delta_x) = U^t g(x),$$

where by U we denote, as above, the backward propagator in $C_\infty(\mathbf{R}^d)$ that is dual to the propagator on ξ given by equation (8.7). Since the operators U propagate in $C^1_\infty(\mathbf{R}^d)$ uniformly for bounded μ, it follows that the function (8.25) belongs to $C^1_\infty(\mathbf{R}^d)$ also uniformly for bounded μ. Hence Φ_t is a bounded semigroup in $C^{1,1}_{\text{vague}}(\mathcal{M}_M(\mathbf{R}^d))$. Differentiating $F(\mu_t)$ yields on the one hand

$$\frac{d}{dt}F(\mu_t) = \left(A_{\mu_t}\frac{\delta F(\mu_t)}{\delta \mu_t(.)}, \ \mu_t\right), \tag{8.26}$$

implying (8.24). On the other hand it follows that, for $\|\mu\| \le M$ and $F \in C^{1,1}_{\text{vague}}(\mathcal{M}_M(\mathbf{R}^d))$,

$$|\Phi_t F(\mu) - F(\mu)| \le tc(t, M)\left\|\frac{\delta F(\mu)}{\delta \mu(.)}\right\|_{C^1(\mathbf{R}^d)},$$

implying that Φ_t is strongly continuous in $C^{1,1}_{\text{vague}}(\mathcal{M}_M(\mathbf{R}^d))$ in the topology of $C_{\text{vague}}(\mathcal{M}_M(\mathbf{R}^d))$ and consequently strongly continuous in $C_{\text{vague}}(\mathcal{M}_M(\mathbf{R}^d))$ by the usual approximation argument, since $C^{1,1}_{\text{vague}}(\mathcal{M}_M(\mathbf{R}^d))$ is dense in $C_{\text{vague}}(\mathcal{M}_M(\mathbf{R}^d))$ by the Stone–Weierstrass theorem. It remains to observe that ΛF from (8.24) is well defined as an element of $C_{\text{vague}}(\mathcal{M}_M(\mathbf{R}^d))$ whenever $F \in C^{1,1}_{\text{vague}}(\mathcal{M}_M(\mathbf{R}^d))$.

(ii) This follows from statement (i) and Proposition I.1.

(iii) If $F \in C^{2,2}_{\text{vague}}(\mathcal{M}_M(\mathbf{R}^d))$ then

$$\begin{aligned}\frac{\delta^2 \Phi_t F(\mu)}{\delta\mu(x)\delta\mu(y)} &= \int_X \frac{\delta F(\mu_t)}{\delta \mu_t(v)}\eta_t(\mu, x, y, dv) \\ &\quad + \int_X \frac{\delta^2 F(\mu_t)}{\delta\mu_t(v)\delta\mu_t(u)}\xi_t(\mu, x, dv)\xi_t(\mu, y, du),\end{aligned} \tag{8.27}$$

everything here being well defined owing to Theorem 8.5. It remains to consider differentiability. Since we have supposed that the second variational derivatives of F have continuous mixed second derivatives in u, v, one only needs to show that

$$\frac{\partial}{\partial x}\xi(\mu, x, y), \quad \frac{\partial}{\partial x}\eta(\mu, x, y), \quad \frac{\partial}{\partial y}\eta(\mu, x, y), \quad \frac{\partial^2}{\partial x \partial y}\eta(\mu, x, y)$$

are well defined as the elements of $(C^1_\infty(\mathbf{R}^d))^\star$. And this in turn follows by differentiating equations (8.7) and (8.8). Equation (8.7) with $\xi = \delta_x$ has the same form after differentiation with respect to x. Consequently, $\partial\xi(\mu, x)/\partial x$ solves the same equation with initial condition $\delta'_x \in (C^1_\infty(\mathbf{R}^d))^\star$ and hence belongs to this space owing to the well-posedness of this equation. Next, on differentiating (8.8) we see that the homogeneous part remains the same after differentiation and that the nonhomogeneous part, after, say, two differentiations with respect to x and y, takes the form

$$\iint \left(\frac{\delta A_{\mu_t}}{\delta \mu_t(u)} g(v) + \frac{\delta A_{\mu_t}}{\delta \mu_t(v)} g(u) + \int \frac{\delta^2 A_{\mu_t}}{\delta \mu_t(v) \delta \mu_t(u)} g(z) \mu_t(dz) \right)$$
$$\times \frac{\partial}{\partial x} \xi_t(\mu_0, x, dv) \frac{\partial}{\partial y} \xi_t(\mu_0, y, du).$$

Here everything is well defined, since

$$\frac{\partial}{\partial x} \xi_t(\mu_0, x) \in (C^1_\infty(\mathbf{R}^d))^\star.$$

Continuity of the derivatives of η with respect to x and y (and of the second mixed derivative) follows from the continuity of the derivatives of ξ. □

We have shown how one can work with an arbitrary smooth dependence of the coefficients on measures. As the results and proofs are rather heavy at the present level of generality, we shall consider now an example of *nonlinear stable-like processes* with the most elementary type of nonlinearity. More general extensions can be obtained if needed using variational derivatives as above.

Thus, let A_μ have the form given in (3.55), with σ, ω, ν independent of μ and with

$$b_\mu(x) = \int b(x, y) \mu(dy), \qquad a_{p,\mu}(x, s) = \int a_p(x, s, y) \mu(dy) \tag{8.28}$$

for some functions b, a. In this case clearly

$$\frac{\delta A}{\delta \mu(v)} g(x) = (b(x, v), \nabla g(x)) + \int_P dp \int_0^K d|y| \int_{S^{d-1}} a_p(x, s, v)$$
$$\times \frac{g(x+y) - g(x) - (y, \nabla g(x))}{|y|^{\alpha_p(x,s)+1}} d|y| \omega_p(ds).$$

Unlike the situation previously discussed the operator (8.16) is defined in $C^2_\infty(\mathbf{R}^d)$ and not just in $C^1_\infty(\mathbf{R}^d)$. The following result is proved in the same way as Theorem 8.6 but taking into account this additional feature.

Theorem 8.7 *Under the assumptions of Theorem 7.9 suppose that the nonlinear dependence is as described above, with* $b(x, .), a(x, s, .) \in C^5_\infty(\mathbf{R}^d)$ *uniformly for* x *and* s *and that other coefficients satisfy the assumptions of Proposition 3.17(ii) for* $k > 4$. *Then the equations* (8.7) *and* (8.8) *for* ξ_t *and* η_t *are well posed in* $(C^2_\infty(\mathbf{R}^d))^\star$ *(they solve* (8.7) *and* (8.8) $\star$*-weakly in* $(C^4_\infty(\mathbf{R}^d))^\star$*) and are given by the limits* (8.3), (8.4) *in the norm topology of* $(C^4_\infty(\mathbf{R}^d))^\star$. *Moreover, the semigroup given by* $\Phi_t F(\mu) = F(\mu_t)$ *is* *a Feller semigroup in* $C_{\text{vague}}(\mathcal{M}_M(\mathbf{R}^d))$ *and the spaces* $C^{1,2}_{\text{vague}}(\mathcal{M}_M(\mathbf{R}^d))$ *and* $C^{2,4}_{\text{vague}}(\mathcal{M}_M(\mathbf{R}^d))$ *represent its invariant cores.*

One can treat the nonlinear counterpart of the unbounded-coefficient evolution from Proposition 5.5 similarly.

Exercise 8.2 State and prove the analog of the regularity statement (ii) from Theorem 8.3 for the second derivative of η.

Exercise 8.3 Under the assumptions of Theorem 8.3 the semigroup Φ_t is also strongly continuous on the closure of $C^{1,1}_{\text{weak}}(\mathcal{M}_M(\mathbf{R}^d))$ in the space $C_{\text{weak}}(\mathcal{M}_M(X))$, the space $C^{1,1}_{\text{weak}}(\mathcal{M}_M(\mathbf{R}^d))$ (and, under the assumptions of Theorem 8.5, $C^{2,2}_{\text{weak}}(\mathcal{M}_M(\mathbf{R}^d))$ also) is invariant, and for $F \in C^{1,1}_{\text{weak}}(\mathcal{M}_M(\mathbf{R}^d))$ equation (8.24) holds for each Y with $\Lambda F \in C_{\text{weak}}(\mathcal{M}_M(\mathbf{R}^d))$. Moreover, to achieve this result one can essentially relax the conditions of Theorem 8.3 by, say, allowing, in conditions (7.15), (7.16), the corresponding functions to belong to $C(\mathbf{R}^d)$ and not necessarily to $C_\infty(\mathbf{R}^d)$. Hint: See Exercise 11.5.

8.3 Jump-type models

In this section we shall prove smoothness with respect to the initial data for the kinetic equations with pure jump interactions, focusing our attention on the most important binary interactions (extension to interactions of other orders being more or less straightforward) for non-increasing particle number. (The latter assumption simplifies calculation but is not essential for the validity of the methods developed.) Namely, we are going to study the signed measures (8.5) and (8.6) for the solutions of the weak kinetic equation

$$\begin{aligned}\frac{d}{dt}(g,\mu_t) &= \tfrac{1}{2}\int_{\mathcal{X}}\int_{X^2}[g^{\oplus}(\mathbf{y}) - g(x_1) - g(x_2)]P(x_1,x_2,d\mathbf{y})\mu_t(dx_1)\mu_t(dx_2)\\ &= \tfrac{1}{2}\int_{X^2}\left\{\int_X [g(y) - g(x_1) - g(x_2)]P(x_1,x_2,dy)\right.\\ &\qquad \left. + \int_{X^2}[g(y_1)+g(y_2)-g(x_1)-g(x_2)]P(x_1,x_2,dy_1dy_2)\right\}\\ &\qquad \times \mu_t(dx_1)\mu_t(dx_2), \end{aligned} \tag{8.29}$$

which includes as particular cases the spatially homogeneous Boltzmann and Smoluchovski models. As neither the number of particles nor the quantity E are increased by the interaction we shall assume that all initial measures (and hence under an appropriate well-posedness condition all solutions) belong to the subset

$$\mathcal{M}_{e_0,e_1}(X) = \{\mu \in \mathcal{M}(X) : (1, \mu) \leq e_0, (E, \mu) \leq e_1\}. \tag{8.30}$$

The generally known results on the derivatives of evolution systems with respect to the initial data are not applicable directly to this equation in the case of unbounded coefficients. Thus we shall adhere to the same strategy as in the previous section, introducing approximations with bounded kernels (Theorem 6.13), applying standard results on variational derivatives to them and then passing to the limit.

Differentiating formally equation (8.29) with respect to the initial measure μ_0 one obtains for ξ_t and η_t the equations

$$\frac{d}{dt}(g, \xi_t) = \int_{X\times X}\int_{\mathcal{X}} [g^{\oplus}(\mathbf{y}) - g(x_1) - g(x_2)]P(x_1, x_2, d\mathbf{y})\xi_t(dx_1)\mu_t(dx_2) \tag{8.31}$$

and

$$\frac{d}{dt}(g, \eta_t(x, w; , \cdot)) = \int_{X\times X}\int_{\mathcal{X}} \left[g^{\oplus}(\mathbf{y}) - g(x_1) - g(x_2)\right] P(x_1, x_2, d\mathbf{y}) \times [\eta_t(x, w; dx_1)\mu_t(dx_2) + \xi_t(x; dx_1)\xi_t(w; dx_2)]. \tag{8.32}$$

The evolution on functions that is dual to the dynamics (8.31) is specified by the equation

$$\frac{d}{dt} g(x) = \int_X \int_{\mathcal{X}} \left[g^{\oplus}(\mathbf{y}) - g(x) - g(z)\right] P(x, z, d\mathbf{y})\mu_t(dz). \tag{8.33}$$

More precisely, the proper dual problem to the Cauchy problem of equation (8.31) is given by the inverse Cauchy problem for equation (8.33):

$$\frac{d}{dt} g(x) = -\int_X \int_{\mathcal{X}} \left[g^{\oplus}(\mathbf{y}) - g(x) - g(z)\right] P(x, z, d\mathbf{y})\mu_t(dz), \quad t \in [0, r], \; g_r = g. \tag{8.34}$$

As mentioned above we shall use the cut-off kernel P_n, which, as in Theorem 6.7, will be defined by equation (6.19). We shall denote by μ_t^n, ξ_t^n, etc. the analogs of μ_t, ξ_t arising from the corresponding cut-off equations (with P_n instead of P, μ^n instead of μ, etc.).

The following result collects together the qualitative properties of the first variational derivatives of the solutions to kinetic equation (8.29) and their cut-off approximations.

Theorem 8.8 *Suppose that $P(\mathbf{x}, .)$ is a continuous function taking SX^2 to $\mathcal{M}(X \cup SX^2)$ (where the measures are considered in the weak topology) and that it is E-non-increasing and $(1 + E)^{\oplus}$-bounded for some continuous non-negative function E on X such that $E(x) \to \infty$ as $x \to \infty$. Suppose that*

$\mu \in \mathcal{M}_{e_0,e_1}(X)$ *and* $(1 + E^\beta, \mu) < \infty$ *for the initial condition* $\mu = \mu_0$ *and for some* $\beta > 3$.

(i) Then the inverse Cauchy problem (8.34) *is well posed in* C_{1+E^ω}, $\omega \leq \beta - 2$. *More precisely: there exists a backward propagator* $U^{t,r} = U^{t,r}_\mu$ *in* $C_{1+E^{\beta-1}}(X)$ *which preserves the spaces* $C_{1+E^\omega}(X)$ *for* $\omega \in [1, \beta-1]$, *specifies a bounded propagator in each of these spaces and is strongly continuous in* $C_{1+E^\omega}(X)$ *for any* $\omega \in (1, \beta - 1]$. *For any* $g \in C_{1+E^\omega}(X)$, $\omega \in [1, \beta - 2]$, $U^{t,r}g = g_t$ *is the unique solution to* (8.34) *in the sense that* $g_t \in C_{1+E^\omega}(X)$ *for all t and* (8.34) *holds uniformly on the sets* $\{x : L(x) \leq l\}$, $l > 0$ *and, for any* $g \in C_{1+E^\omega,\infty}(X)$ *and* $\omega \in (1, \beta - 2]$, $U^{t,r}g = g_t$ *solves* (8.34) *in the norm topology of* $C_{1+E^{\omega+1}}(X)$.

(ii) The dual (forward) propagator $V^{s,t}$ *is well defined in each space* $\mathcal{M}^{\text{signed}}_{1+E^\omega}(X)$, $\omega \in [1, \beta - 1]$, *and, for any* $\xi \in \mathcal{M}^{\text{signed}}_{1+E^\omega}(X)$, $V^{s,t}\xi$ *is the unique weakly continuous curve in* $\mathcal{M}^{\text{signed}}_{1+E^{\beta-1}}(X)$ *that solves the Cauchy problem* (8.31) *(with initial data* ξ *at time s) weakly in* $\mathcal{M}^{\text{signed}}_{1+E^{\beta-2}}(X)$. *In particular,* $\xi_t(\mu, x) = V^{0,t}\delta_x$ *is well defined.*

(iii) $\xi^n_t(\mu, x) \to \xi_t(\mu, x)$ *as* $n \to \infty$ *in the norm topology of* $\mathcal{M}_{1+E^\omega}$ *with* $\omega \in [1, \beta - 2)$ *and* $\star$*-weakly in* $\mathcal{M}_{1+E^{\beta-1}}$.

(iv) ξ_t *depends Lipschitz continuously on* μ_0 *in the sense that, for* $\omega \in [1, \beta - 2]$,

$$\sup_{s \leq t} \|\xi_s(\mu_0^1, x) - \xi_s(\mu_0^2, x)\|_{1+E^\omega} \leq c(t, (1 + E^{2+\omega}, \mu_0^1 + \mu_0^2)) \times [1 + E^{1+\omega}(x)] \|\mu_0^1 - \mu_0^2\|_{1+E^{1+\omega}}.$$

(v) Equation (8.5) *holds for* ξ_t *if the limit exists in the norm topology of* $\mathcal{M}_{1+E^\omega}(X)$ *with* $\omega \in [1, \beta - 1)$ *and weakly in* $\mathcal{M}_{1+E^{\beta-1}}$.

Proof (i) Equation (8.33) can be written in the form

$$\dot{g} = A_t g - B_t g \tag{8.35}$$

with

$$A_t g(x) = \int_X \int_{\mathcal{X}} \left[g^\oplus(\mathbf{y}) - g(x) \right] P(x, z, d\mathbf{y}) \mu_t(dz) \tag{8.36}$$

and

$$B_t g(x) = \int_X g(z) \int_X \int_{\mathcal{X}} P(x, z, d\mathbf{y}) \mu_t(dz). \tag{8.37}$$

Thus it is natural to analyze this equation by perturbation theory, considering B_t as a perturbation. The equation $\dot{g}_t = A_t g_t$ has the form (4.1) with

$$a_t(x) = \int_X \int_{\mathcal{X}} P(x, z, d\mathbf{y}) \mu_t(dz) \leq c\,[1 + E(x)]\, \|\mu_t\|_{1+E},$$

and, for all $\omega \le \beta - 1$,

$$\|B_t g\|_{1+E} = \left\| \frac{B_t g}{1+E} \right\| \le C \sup_x \left(\frac{\int g(z)(1+E(x)+E(z))\mu_t(dz)}{1+E(x)} \right)$$

$$\le C\|g\|_{1+E^\omega} \int \left[1+E^\omega(z)\right][1+E(z)]\,\mu_t(dz) \le 3C\|g\|_{1+E^\omega}\|\mu_t\|_{1+E^{\omega+1}}.$$

Moreover, as $\omega \ge 1$,

$$A_t(1+E^\omega)(x) \le C \int_X \left\{1+[E(x)+E(z)]^\omega - E^\omega(x)\right\}[1+E(z)+E(x)]\,\mu_t(dz)$$
$$\le Cc(\omega) \int_X \left[E^{\omega-1}(x)E(z)+E^\omega(z)\right][1+E(z)+E(x)]\,\mu_t(dz). \tag{8.38}$$

Applying the second inequality in Exercise 6.1, at the end of Section 6.3, allows us to estimate the integrand as

$$c(\omega)\left\{1+E^{\omega+1}(z)+E^\omega(x)\,[1+E(z)]\right\},$$

implying that

$$A_t(1+E^\omega)(x) \le c(\omega,t,e_0,e_1)[1+E^\omega(x)+(E^{\omega+1},\mu)]. \tag{8.39}$$

By Gronwall's lemma this implies that the backward propagator $U_A^{t,r}$ of the equation $\dot g_t = -A_t g_t$ possesses the estimate

$$|U_A^{t,r} g(x)| \le c(\omega,t,e_0,e_1)\|g\|_{1+E^\omega}[1+E^\omega(x)+(E^{\omega+1},\mu)], \tag{8.40}$$

and the required well-posedness of the equation $\dot g = A_t g$ follows from Theorems 4.8 and 4.9 with $\psi_1 = 1+E^s$, $s \in [1,\beta-2]$ and $\psi_2 = 1+E^{\beta-1}$. Well-posedness for equation (8.33) follows from perturbation theory.
(ii) This is a direct consequence of (i) and duality.
(iii) The required convergence in the norm topology can be proved in the same way as Theorem 6.13, and we omit the details. The corresponding $\star$-weak convergence follows from the uniform boundedess of ξ_t^n and ξ_t in $\mathcal{M}_{1+E^{\beta-1}}$.
(iv) This is similar to the corresponding continuity statement (Theorem 6.12) for the solution of the kinetic equation itself. Namely, setting $\xi_t^j = \xi_t(\mu_0^j, x)$, $j = 1, 2$, one writes

$$\|\xi_t^1 - \xi_t^2\|_{1+E^\omega}$$
$$= \int_0^t ds \int [(\sigma_s(1+E^\omega))^\oplus(\mathbf{y}) - (\sigma_s(1+E^\omega))(x_1) - (\sigma_s(1+E^\omega))(x_2)]$$
$$\times P(x_1,x_2,d\mathbf{y})[\xi_s^1(dx_1)\mu_s^1(dx_2) - \xi_s^2(dx_1)\mu_s^2(dx_2)],$$

where σ_s denotes the sign of the measure $\xi_t^1 - \xi_t^2$ (again chosen according to Lemma 6.14). Next, writing

$$\xi_s^1(dx_1)\mu_s^1(dx_2) - \xi_s^2(dx_1)\mu_s^2(dx_2)$$
$$= \sigma_s(x_1)|\xi_s^1 - \xi_s^2|(dx_1)\mu_s^1(dx_2) + \xi_s^2(dx_1)(\mu_s^1 - \mu_s^2)(dx_2),$$

one may given an estimate for the contribution of the first term in the above expression for $\|\xi_t^1 - \xi_t^2\|_{1+E^\omega}$:

$$\int_0^t ds \int \{[E(x_1) + E(x_2)]^\omega - E^\omega(x_1) + E^\omega(x_2) + 1\} P(x_1, x_2, d\mathbf{y})$$
$$\times |\xi_s^1 - \xi_s^2|(dx_1)\mu_s^1(dx_2)$$
$$\leq c \int_0^t ds \int [E^{\omega-1}(x_1)E(x_2) + E^\omega(x_2) + 1][1 + E(x_1) + E(x_2)]$$
$$\times |\xi_s^1 - \xi_s^2\|(dx_1)\mu_s^1(dx_2)$$
$$\leq c(\omega, e_0, e_1) \int_0^t ds \|\xi_s^1 - \xi_s^2\|_{1+E^\omega} \|\mu_s^1\|_{1+E^{\omega+1}}.$$

The contribution of the second term is

$$c(\omega, e_0, e_1) \int_0^t ds \|\mu_s^1 - \mu_s^2\|_{1+E^{\omega+1}} \|\xi_s^2\|_{1+E^{\omega+1}}$$
$$\leq c(\omega, (1 + E^{2+\omega}, \mu_0^1 + \mu_0^2))t \|\mu_0^1 - \mu_0^2\|_{1+E^{\omega+1}} \|\xi_0^2\|_{1+E^{\omega+1}}.$$

Applying Gronwall's lemma completes the proof of statement (iii).
(v) As can be easily seen (adapting the proof of Theorem 6.10 to the simpler situation of bounded kernels), the solution μ_t^n to the cut-off version of the kinetic equation satisfies this equation strongly in the norm topology of $\mathcal{M}_{1+E^{\beta-\epsilon}}$ for any $\epsilon > 0$. Moreover, μ_n^t depends Lipschitz continuously on μ_0 in the same topology and ξ_t^n satisfies the corresponding equation in variational form in the same topology. Hence it follows from Theorem D.1 that

$$\xi_t^n = \xi_t^n(\mu_0; x) = \lim_{s \to 0_+} \frac{1}{s} \left[\mu_t^n(\mu_0, +s\delta_x) - \mu_t(\mu_0)\right]$$

in the norm topology of $\mathcal{M}_{1+E^{\beta-\epsilon}}$, with $\epsilon > 0$. Consequently,

$$(g, \mu_t^n(\mu_0 + h\delta_x)) - (g, \mu_t^n(\mu_0)) = \int_0^h (g, \xi_t^n(\mu_0 + s\delta_x; x; .))\, ds$$

for all $g \in C_{1+E^{\beta-\epsilon}}(X)$ and $\epsilon > 0$. Using statement (ii) and the dominated convergence theorem one deduces that

$$(g, \mu_t(\mu_0 + h\delta_x)) - (g, \mu_t(\mu_0)) = \int_0^h (g, \xi_t(\mu_0 + s\delta_x; x; .))\, ds \qquad (8.41)$$

for all $g \in C_{1+E^\gamma}(X)$, with $\gamma < \beta - 2$. Again using dominated convergence and the fact that the ξ_t are bounded in $\mathcal{M}_{1+E^{\beta-1}}$ (since they are $\star$-weakly continuous there), one deduces that (8.41) holds for $g \in C_{1+E^{\beta-1},\infty}(X)$. Next, for such g the expression under the integral on the r.h.s. of (8.41) depends continuously on s due to (iii), which justifies the limit (8.5) in the $\star$-weak topology of $\mathcal{M}_{1+E^{\beta-1}}$. Moreover, as ξ_t depends Lipshitz continuously on μ_0 in the norm topology of $\mathcal{M}_{1+E^{\beta-2}}$ and since the ξ_t are bounded in $\mathcal{M}_{1+E^{\beta-2}}$, ξ_t must depend continuously on s in the r.h.s. of (8.41) in the norm topology of $\mathcal{M}_{1+E^\gamma}$, with $\gamma < \beta - 1$. Hence (8.41) justifies the limit (8.5) in the norm topology of $\mathcal{M}_{1+E^\gamma}(X)$, with $\gamma < \beta - 1$. □

Theorem 8.9 *Suppose that the assumptions of the previous theorem hold.*

(i) There exists a unique solution $\eta_t = \eta(\mu, x, w)$ to (8.32) *in the sense that $\eta_0 = 0$, η_t is a $*$-weakly continuous function $t \mapsto \mathcal{M}_{1+E^{\beta-2}}$ and* (8.32) *holds for $g \in C(X)$.*

(ii) $\eta_t^n(\mu, x, w) \to \eta_t(\mu, x, w)$ $*$-weakly in $\mathcal{M}_{1+E^{\beta-2}}$.

(iii) $\eta_t(\mu, x, w)$ can be defined by the r.h.s. of (8.6) *in the norm topology of $\mathcal{M}_{1+E^\gamma}$ with $\gamma < \beta - 2$ and in the $*$-weak topology of $\mathcal{M}_{1+E^{\beta-2}}$.*

Proof (i) The linear equation (8.32) differs from equation (8.31) by the addition of a nonhomogeneous term. Hence one deduces from Theorem 8.8 the well-posedness of equation (8.32) and the explicit formula

$$\eta_t(x, w) = \int_0^t V^{t,s}\Omega_s(x, w)\, ds, \tag{8.42}$$

where $V^{t,s}$ is a resolving operator for the Cauchy problem of equation (8.31) that is given by Theorem 8.8 and $\Omega_s(x, w)$ is the measure defined weakly as

$$(g, \Omega_s(x, w)) = \int_{X\times X}\int_X \big[g(y) - g(z_1) - g(z_2)\big]\, P(z_1, z_2; dy) \times \xi_t(x; dz_1)\xi_t(w; dz_2). \tag{8.43}$$

Thus

$$\|\Omega_t(x, w)\|_{1+E^\omega} \le c\int_{X^2}\Big[1 + E^{\omega+1}(z_1) + E^{\omega+1}(z_2)\Big]\,\xi_t(x; dz_1)\xi_t(w; dz_2).$$

(ii) This follows from (8.42) and Theorem 8.8 (ii).
(iii) As in the case of ξ_t, we first prove the formula

$$(g, \xi_t(\mu_0 + h\delta_w; x, .)) - (g, \xi_t(\mu_0; x, .)) = \int_0^h (g, \eta_t(\mu_0 + s\delta_w; x, w; .))\, ds \tag{8.44}$$

for $g \in C_\infty(X)$ by using the approximation η_t^n and dominated convergence. Then the validity of (8.44) is extended to all $g \in C_{1+E^{\beta-2},\infty}$ using dominated convergence and the bounds obtained above for η_t and ξ_t. By continuity of the expression under the integral on the r.h.s. of (8.43) we can justify the limit (8.6) in the $\star$-weak topology of $\mathcal{M}_{1+E^{\beta-2}}(X)$, completing the proof of Theorem 8.9. □

As a consequence of these results we can get additional regularity for the solutions of the kinetic equations and the corresponding semigroup given by $\Phi_t F(\mu) = F(\mu_t)$.

Let $\mathcal{M}^\beta_{e_0,e_1}(X) = \mathcal{M}_{e_0,e_1}(X) \cap \mathcal{M}_{1+E^\beta}(X)$; this is clearly a closed subset of $\mathcal{M}_{1+E^\beta}(X)$. Let $C_{\text{vague}}(\mathcal{M}^\beta_{e_0,e_1}(X))$ be the Banach space of $\star$-weakly (in the sense of $\mathcal{M}_{1+E^\beta}(X)$) continuous functions on $\mathcal{M}^\beta_{e_0,e_1}(X)$. The starting point for the analysis of Φ_t is given by the following theorem.

Theorem 8.10 *Under the assumptions of Theorem 8.8, the mapping* $(t, \mu) \mapsto \mu_t$ *is $\star$-weakly continuous in* $\mathcal{M}^\beta_{e_0,e_1}(X)$ *(if both μ and μ_t are considered in the $\star$-weak topology of $\mathcal{M}_{1+E^\beta}(X)$), and consequently the semigroup Φ_t is a contraction semigroup in* $C_{\text{vague}}(\mathcal{M}^\beta_{e_0,e_1}(X))$.

Proof Suppose that $\mu, \mu_1, \mu_2, \ldots \in \mathcal{M}^\beta_{e_0,e_1}(X)$ and that $\mu^n \to \mu$ $\star$-weakly in $\mathcal{M}_{1+E^\beta}(X)$. By the uniform boundedness of μ_t^n in $\mathcal{M}_{1+E^\beta}(X)$, in order to show that $\mu_t(\mu^n) \to \mu_t(\mu)$ $\star$-weakly in $\mathcal{M}_{1+E^\beta}(X)$, it is enough to show the convergence $(g, \mu_t(\mu^n)) \to (g, \mu_t(\mu))$ for any $g \in C_c(X)$. From Theorem 8.8 it follows that

$$
\begin{aligned}
&(g, \mu_t(\mu^n)) - (g, \mu_t(\mu)) \\
&\quad = \int_0^h ds \iint g(y)\xi_t(\mu + s(\mu^n - \mu), x, dy)(\mu^n - \mu)(dx) \qquad (8.45)
\end{aligned}
$$

for any $g \in C_{1+E^{\beta-1},\infty}(X)$. To show that the r.h.s. tends to zero it is sufficient to prove that the family of functions

$$\int g(y)\xi_t(\mu + s(\mu^n - \mu), ., dy)$$

is relatively compact in $C_{1+E^\beta,\infty}(X)$. As this family is uniformly bounded in $C_{1+E^{\beta-1}}(X)$, again by Theorem 8.8, it remains to show that by the Arzela–Ascoli theorem it is uniformly (with respect to n) continuous in any set where E is bounded. But this follows from the proof of Theorem 4.1 (see equation (4.20)). □

We are now interested in a rather useful invariant subspace of $C_{\text{vague}}(\mathcal{M}^\beta_{e_0,e_1}(X))$, where Φ_t is strongly continuous, and an appropriate invariant core. To begin with we observe that if the first variational derivative of F exists and is uniformly bounded on $\mathcal{M}^{\beta}_{e0,e_1}(X)$ then $\Phi_t F$ is strongly continuous (and even smooth in t). However, it is not at all clear that the space of such functionals is invariant under Φ_t, since all our estimates of the variational derivatives of μ_t depend on the $\mathcal{M}_{1+E^\beta}(X)$-norm of μ. Hence the choice of invariant subspace must be a bit subtler, as is explained in the next result.

Theorem 8.11 *Under the assumptions of Theorem 8.8 the subspaces K_ω, $\omega \in [1, \beta - 1]$, of functions F from $C_{\text{vague}}(\mathcal{M}^{\beta}_{e0,e_1}(X))$ having the properties that*

(i) $\delta F(\mu)/\delta\mu(x)$ exists and is vaguely continuous for $x \in \mathbf{R}^d$, $\mu \in \mathcal{M}^{\beta}_{e0,e_1}(X)$,

(ii) $\delta F(\mu)/\delta\mu(.) \in C_{1+E^\omega}$ uniformly for bounded $\|\mu\|_{1+E^\beta}$,

(iii) $\sup_{\mu\in\mathcal{M}^{\beta}_{e0,e_1}(X)} < \infty$

are invariant under Φ_t, and Φ_t is strongly continuous (in the topology of $C_{\text{vague}}(\mathcal{M}^{\beta}_{e0,e_1}(X))$) in these spaces.

Proof

$$\frac{\delta F(\mu_t(mu))}{\delta\mu(x)} = \int_X \frac{\delta F(\mu_t(mu))}{\delta\mu_t(y)} \xi_t(\mu, x, dy).$$

As $\|\xi_t(\mu, x, .)\|_{1+E^{\beta-1}} \leq c[1 + E^{\beta-1}(x)]$ uniformly for bounded $\|\mu\|_{1+E^\beta}$, it follows that properties (i) and (ii) are invariant (the continuity of the variational derivatives in μ follows from duality). The invariance in (iii) follows from the equation

$$\left(\frac{\delta F(\mu_t(mu))}{\delta\mu(.)}, \Lambda_\mu\right) = \left(\frac{\delta F(\mu_t)}{\delta\mu_t(.)}, \Lambda_{\mu_t}\right).$$

This equation simply expresses the commutativity of the semigroup and its generator, $\Phi_t LF(\mu) = L\Phi_t F(\mu)$. However, as we do not know that our semigroup is strongly continuous, we must justify the above equation by the usual cut-off approximations. □

8.4 Estimates for Smoluchovski's equation

Unfortunately, application of the perturbation series representation to bound the norms of $U^{t,r}$ in $C_{1+E^\omega}(X)$ would yield only the exponential dependence of these norms on $(E^{\beta-1}, \mu)$, and this does not allow for effective estimates of the LLN rate of convergence. To obtain better estimates is not a simple task. We shall derive them now only for pure coagulation processes, imposing further assumptions. Namely, we shall prove two results on the linear dependence of the variational derivatives of the moments of μ, assuming either additional bounds on the rates or some smoothness of the kernel. These estimates will play a crucial role in obtaining precise rates of convergence for the LLN for coagulation processes.

In what follows we shall often use the elementary inequalities

$$(E^l, \nu)(E^k, \nu) \leq 2(E^{k+l-1}, \nu)(E, \nu), \tag{8.46a}$$

$$(E^k, \nu)E(x) \leq (E^{k+1}, \nu) + (E, \nu)E^k(x), \tag{8.46b}$$

which are valid for arbitrary positive ν and $k, l \geq 1$.

Remark 8.12 These inequalities are easy to prove. For example, to get (8.46b), one writes

$$(E^l, \nu)(E^k, \nu) = \iint E^l(x)E^k(y)\nu(dx)\nu(dy)$$

and decomposes this integral into the sum of two integrals over the domain $\{E(x) \geq E(y)\}$ and its complement. Then the first integral, say, is estimated as $\iint E^{l+k-1}(x)E(y)\nu(dx)\nu(dy)$.

Theorem 8.13 *Under the assumptions of Theorem 8.8 assume additionally that we have pure coagulation, i.e that $P(x_1, x_2, dy_1dy_2)$ vanishes identically and the remaining coagulation kernel $P = K$ is $(1+\sqrt{E})^{\otimes}$-bounded, i.e. that*

$$K(x_1, x_2) \leq C\left[1 + \sqrt{E(x_1)}\right]\left[1 + \sqrt{E(x_2)}\right]. \tag{8.47}$$

Then

$$\|U^{t,r}\|_{C_{1+\sqrt{E}}} \leq c(r, e_0, e_1) \tag{8.48}$$

(using notation from Theorem 8.8). Moreover, for any $\omega \in [1/2, \beta - 1/2]$ we have

$$|U^{t,r}g(x)| \leq c(r, e_0, e_1)\|g\|_{1+E^\omega}\left\{1 + E^\omega(x) + \left[1 + \sqrt{E(x)}\right](E^{\omega+1/2}, \mu)\right\}, \tag{8.49}$$

$$\|V^{s,t}\xi\|_{\mathcal{M}_{1+E^\omega}(X)} \leq c(t, e_0, e_1)\left(\|\xi\|_{\mathcal{M}_{1+E^\omega}(X)} + (E^{\omega+1/2}, \mu)\|\xi\|_{1+\sqrt{E}}\right), \tag{8.50}$$

$$\sup_{s\leq t}\|\xi_s(\mu; x)\|_{1+E^\omega} \leq c(t, e_0, e_1)\left\{1 + E^\omega(x) + (E^{\omega+1/2}, \mu)\left[1 + \sqrt{E(x)}\right]\right\}. \tag{8.51}$$

Finally, for $\omega \in [1/2, \beta - 2]$,

$$\sup_{s \le t} \|\eta_s(\mu, x, w; \cdot)\|_{1+E^\omega}$$
$$\le c(t, \omega, e_0, e_1)[1 + E^{\omega+1}(x) + E^{\omega+1}(w) + (E^{\omega+2}, \mu)]. \tag{8.52}$$

Proof To begin, observe that the estimate

$$A_t(1 + E^\omega)(x)$$
$$\le C \int_X \{1 + [E(x) + E(z)]^\omega - E^\omega(x)\} \left[1 + \sqrt{E(z)}\right] \left[1 + \sqrt{E(x)}\right] \mu_t(dz)$$

holds for all $\omega \ge 0$, and not only for $\omega \ge 1$ (here the absence of collisions is taken into account). After simplifying with the help of Exercise 6.6 we obtain

$$A_t(1 + E^\omega)(x) \le C \left\{1 + E^\omega(x) + \left[1 + \sqrt{E(x)}\right] (E^{\omega+1/2}, \mu)\right\}$$

for $\omega \ge 1/2$ and, consequently, by Gronwall's lemma,

$$|U_A^{t,r} g(x)| \le c(r, e_0, e_1) \|g\|_{1+E^\omega} \left\{1 + E^\omega(x) + \left[1 + \sqrt{E(x)}\right] (E^{\omega+1/2}, \mu)\right\} \tag{8.53}$$

for $g \in C_{1+E^\omega}(X)$, $\omega \ge 1/2$. In particular,

$$|U_A^{t,r} g(x)|_{C_{1+\sqrt{E}}} \le c(r, e_0, e_1) \|g\|_{C_{1+\sqrt{E}}}.$$

Moreover, as

$$\|B_t g\|_{1+\sqrt{E}} \le c \int g(z) \left[1 + \sqrt{E(z)}\right] \mu_t(dz)$$
$$\le c \|g\|_{1+E^\omega} \|\mu\|_{1+E^{\omega+1/2}}$$

for $\omega \ge 0$, the operator B is bounded in $C_{1+\sqrt{E}}$ with estimates depending only on e_0, e_1 so that from perturbation theory applied to the equation $\dot{g} = -(A_t - B_t)g$ one obtains (8.48) directly. Moreover, it follows from (8.53) that

$$|B_t U_A^{t,r} g(x)| \le c(r, e_0, e - 1) \|g\|_{1+E^\omega} \left[1 + \sqrt{E(x)}\right] \left[1 + (E^{\omega+1/2}, \mu)\right],$$

and since from perturbation theory

$$U^{t,r} g = U_A^{t,r} g + \int_t^r U^{r,s} B_t U_A^{t,r} g \, ds,$$

one obtains (8.49) from (8.48) and (8.53). Formula (8.50) follows then by duality, and (8.51) is its direct consequence.

Finally, if $\omega \geq 1/2$ it follows from (8.43) that

$$\|\Omega_t(x,w)\|_{1+E^\omega} \leq c \int_{X^2} [1+E^\omega(z_1)+E^\omega(z_2)]\left[1+E^{1/2}(z_1)\right]\left[1+E^{1/2}(z_2)\right] \times \xi_t(x;dz_1)\xi_t(w;dz_2).$$
$$\leq c\left(\|\xi_t(\mu,x)\|_{1+E^{\omega+1/2}}\|\xi_t(\mu,w)\|_{1+E^{1/2}} + \|\xi_t(\mu,w)\|_{1+E^{\omega+1/2}}\|\xi_t(\mu,x)\|_{1+E^{1/2}}\right),$$

which by (8.51) does not exceed

$$c\left\{1+E^{\omega+1/2}(x)+(E^{\omega+1/2},\mu)\left[1+\sqrt{E(x)}\right]\right\}\left[1+\sqrt{E(w)}\right]$$
$$+c\left\{1+E^{\omega+1/2}(w)+(E^{\omega+1/2},\mu)\left[1+\sqrt{E(w)}\right]\right\}\left[1+\sqrt{E(x)}\right]$$
$$\leq c[1+E^{\omega+1}(x)+E^{\omega+1}(w)+(E^{\omega+3/2},\mu)].$$

Hence by (8.50)

$$\sup_{s\leq t}\|\eta_t(x,w;\cdot)\|_{1+E^\omega}$$
$$\leq c(t,e_0,e_1)\sup_{s\leq t}\|\Omega_t(x,w)\|_{1+E^\omega}+(E^{\omega+1/2},\mu)\sup_{s\leq t}\|\Omega_s(x,w)\|_{1+\sqrt{E}}$$
$$\leq c(t,e_0,e_1)\left\{1+E^{\omega+1}(x)+E^{\omega+1}(w)+(E^{\omega+2},\mu)\right.$$
$$\left.+(E^{\omega+1/2},\mu)\left[1+E^{3/2}(x)+E^{3/2}(w)+(E^{5/2},\mu)\right]\right\}.$$

It remains to observe that the last term here can be estimated by the previous ones, owing to (8.46b). □

Our next result deals with the case of a smooth coagulation kernel. For simplicity, we shall consider here only the classical Smoluchovski model, where a particle is characterized exclusively by its mass, i.e. the state space is $\mathbf{R}_+$. Let us introduce now some useful functional spaces.

Let $X = \mathbf{R}_+ = \{x > 0\}$. For positive f we denote by $C_f^{1,0} = C_f^{1,0}(X)$ the Banach space of continuously differentiable functions ϕ on $X = \mathbf{R}_+$ such that $\lim_{x\to 0}\phi(x) = 0$ and the norm

$$\|\phi\|_{C_f^{1,0}(X)} = \|\phi'\|_{C_f(X)}$$

is finite. By $C_f^{2,0} = C_f^{2,0}(X)$ we denote the space of twice continuously differentiable functions such that $\lim_{x\to 0}\phi(x) = 0$ and the norm

$$\|\phi\|_{C_f^{2,0}(X)} = \|\phi'\|_f + \|\phi''\|_f$$

is finite. By $\mathcal{M}^1_f(X)$ and $\mathcal{M}^2_f(X)$ we denote the Banach spaces dual to $C^{1,0}_f$ and $C^{2,0}_f$ respectively; actually, we need only the topology they induce on (signed) measures, so that for $\nu \in \mathcal{M}(X) \cap \mathcal{M}^i_f(X)$, $i = 1, 2$,

$$\|\nu\|_{\mathcal{M}^i_f(X)} = \sup\{(\phi, \nu) : \|\phi\|_{C^{i,0}_f(X)} \le 1\}.$$

Theorem 8.14 *Assume that $X = \mathbf{R}_+$, $K(x_1, x_2, dy) = K(x_1, x_2)\delta(y - x_1 - x_2)$ and $E(x) = x$ and that K is non-decreasing in each argument that is twice continuously differentiable on $(\mathbf{R}_+)^2$ up to the boundary, all the first and second partial derivatives being bounded by a constant C. Then for any $k \ge 0$ the spaces $C^{1,0}_{1+E^k}$ and $C^{2,0}_{1+E^k}$ are invariant under $U^{t,r}$ and*

$$|(U^{t,r}g)'(x)| \le \kappa(C, r, k, e_0, e_1)\|g\|_{C^{1,0}_{1+E^k}}\left[1 + E^k(x) + (E^{k+1}, \mu_0)\right], \tag{8.54a}$$

$$|(U^{t,r}g)''(x)| \le \kappa(C, r, k, e_0, e_1)\|g\|_{C^{2,0}_{1+E^k}}\left[1 + E^k(x) + (E^{k+1}, \mu_0)\right], \tag{8.54b}$$

$$\sup_{s\le t} \|\xi_s(\mu_0; x; \cdot)\|_{\mathcal{M}^1_{1+E^k}} \le \kappa(C, r, k, e_0, e_1)\left\{E(x)\left[1 + (E^{k+1}, \mu_0)\right] + E^{k+1}(x)\right\} \tag{8.55}$$

and

$$\sup_{s\le t} \|\eta_s(\mu_0; x, w; , \cdot)\|_{\mathcal{M}^2_{1+E^k}} \le \kappa(C, t, k, e_0, e_1)\left[1 + (E^{k+1}, \mu_0)\right] \times \left\{\left[E(x)(1 + E^{k+2}, \mu_0) + E^{k+2}(x)\right]E(w) + \left[E(w)(1 + E^{k+2}, \mu_0) + E^{k+2}(w)\right]E(x)\right\}. \tag{8.56}$$

Proof Notice first that if $g_r(0) = 0$ then $g_t = 0$ for all t according to the evolution described by the equation $\dot g = -(A_t - B_t)g$. Hence the space of functions vanishing at the origin is invariant under this evolution.

Recall that $E(x) = x$. Differentiating the equation $\dot g = -(A_t - B_t)g$ with respect to the space variable x leads to the equation

$$\dot g'(x) = -A_t(g')(x) - \int [g(x+z) - g(x) - g(z)]\frac{\partial K}{\partial x}(x, z)\mu_t(dz). \tag{8.57}$$

For functions g vanishing at the origin this can be rewritten as

$$\dot g'(x) = -A_t g' - D_t g'$$

with

$$D_t\phi(x) = \int \left(\int_x^{x+z} \phi(y)\,dy - \int_0^z \phi(y)\,dy \right) \frac{\partial K}{\partial x}(x,z)\mu_t(dz).$$

Since

$$\|D_t\phi\| \le 2C\|\phi\|(E,\mu_t) = 2Ce_1\|\phi\|$$

and $U_A^{t,r}$ is a contraction, it follows from the perturbation series representation, with D_t considered as the perturbation, that

$$\|U^{t,r}\|_{C_1^{1,0}(X)} \le c(r, e_0, e_1),$$

proving (8.54a) for $k = 0$. Next, for $k > 0$,

$$|D_t\phi(x)| \le c\|\phi\|_{1+E^k} \int \left[(x+z)^{k+1} - x^{k+1} + z^{k+1} + z \right] \mu_t(dz)$$
$$\le c(k)\|\phi\|_{1+E^k} \int (x^k z + z^{k+1} + z)\mu_t(dz),$$

which by the propagation of the moments of μ_t does not exceed

$$c(k,e_1)\|\phi\|_{1+E^k}[(1+x^k) + (E^{k+1},\mu_0)].$$

Hence, by equation (8.40),

$$\int_t^r \left| U_A^{t,s} D_s U_A^{s,r} g(x) \right| ds \le (r-t)\kappa(r,k,e_0,e_1)\|g\|_{1+E^k}$$
$$\times\ [1 + E^k(x) + (E^{k+1},\mu_0)],$$

which by induction implies that

$$\int_{t \le s_1 \le \cdots \le s_n \le r} \left| U_A^{t,s_1} D_{s_1} \cdots D_{s_n} U_A^{s_n,r} g(x) \right| ds_1 \cdots ds_n$$
$$\le \frac{(r-t)^n}{n!} \kappa^n(r,k,e_0,e_1)\|g\|_{1+E^k}[1 + E^k(x) + (E^{k+1},\mu_0)].$$

Hence (8.54a) follows from the perturbation series representation to the solution of (8.57).

Differentiating (8.57) leads to the equation

$$\dot{g}''(x) = -A_t(g'')(x) - \psi_t, \tag{8.58}$$

where

$$\psi_t = 2\int [g'(x+z) - g'(x)] \frac{\partial K}{\partial x}(x,z)\mu_t(dz)$$
$$+ \int \left(\int_x^{x+z} g'(y)\,dy - \int_0^z g'(y)\,dy \right) \frac{\partial^2 K}{\partial x^2}(x,z)\mu_t(dz).$$

We know already that for $g_r \in C^2_{1+E^k}$ the function g' belongs to $1+E^k$ with the bound given by (8.54a). Hence by the du Hamel principle the solution to (8.58) can be represented as

$$g''_t = U_A^{t,r} g''_r + \int_t^r U_A^{t,s} \psi_s \, ds.$$

As

$$|\psi_t(x)| \leq \kappa(C, r-t, e_0, e_1) \left[1 + E^k(x) + (E^{k+1}, \mu_0)\right],$$

(8.54b) follows, completing the proof of (8.54), which by duality implies (8.55).

Next, from (8.42), (8.43), duality and (8.54) one obtains

$$\begin{aligned}\sup_{s\leq t} \|\eta_s(\mu_0; x, w; .)\|_{\mathcal{M}^2_{1+E^k}} &\leq t \sup_{s\leq t} \sup\{|(U^{s,t} g, \Omega_s(x, w))| : \|g\|_{C^{2,0}_{1+E^k}} \leq 1\} \\ &\leq \kappa(C, t, e_0, e_1) \sup_{s\leq t} \sup_{g\in\Pi_k} (g, \Omega_s(x, w)),\end{aligned}$$

where

$$\Pi_k = \left\{g : g(0) = 0, \max(|g'(y)|, |g''(y)|) \leq 1 + E^k(y) + (E^{k+1}, \mu_0)\right\}.$$

It is convenient to introduce a twice continuously differentiable function χ on $\mathbf{R}$ such that $\chi(x) \in [0, 1]$ for all x and $\chi(x)$ equals 1 or 0 respectively for $x \geq 1$ and $x \leq -1$. Then one writes $\Omega_s = \Omega_s^1 + \Omega_s^2$, with Ω^1 (resp. Ω^2) obtained from (8.43) with $\chi(x_1 - x_2)K(x_1, x_2)$ (resp. $[1-\chi(x_1-x_2)]K(x_1, x_2)$) instead of $K(x_1, x_2)$. If $g \in \Pi_k$, one has

$$\begin{aligned}(g, \Omega_s^1(x, w)) = \iint & \left[g(x_1 + x_2) - g(x_1) - g(x_2)\right] \chi(x_1 - x_2) \\ & \times K(x_1, x_2)\xi_s(x; dx_1)\xi_s(w; dx_2),\end{aligned}$$

which is bounded in magnitude by

$$\begin{aligned}&\|\xi_s(w, .)\|_{\mathcal{M}^1_1(X)} \sup_{x_2} \left| \frac{\partial}{\partial x_2} \int \left\{ \left[g(x_1 + x_2) - g(x_1) - g(x_2)\right] \right.\right. \\ &\qquad\qquad \left.\left. \times \chi(x_1 - x_2)K(x_1, x_2)\right\} \xi_s(x; dx_1) \right| \\ &\leq \|\xi_s(w, .)\|_{\mathcal{M}^1_1(X)} \|\xi_s(x, .)\|_{\mathcal{M}^1_{1+E^{k+1}}(X)} \\ &\quad \times \sup_{x_1, x_2} \left| \left[1 + E^{k+1}(x_1)\right]^{-1} \frac{\partial^2}{\partial x_2 \partial x_1} \left\{ \left[g(x_1 + x_2) - g(x_1) - g(x_2)\right] \right.\right. \\ &\qquad\qquad \left.\left. \times \chi(x_1 - x_2)K(x_1, x_2)\right\} \right|.\end{aligned}$$

Since

$$\frac{\partial^2}{\partial x_2 \partial x_1}\left\{\left[g(x_1+x_2)-g(x_1)-g(x_2)\right]\chi(x_1-x_2)K(x_1,x_2)\right\}$$
$$= g''(x_1+x_2)(\chi K)(x_1,x_2)+\left[g'(x_1+x_2)-g'(x_2)\right]\frac{\partial(\chi K)(x_1,x_2)}{\partial x_1}$$
$$+\left[g'(x_1+x_2)-g'(x_1)\right]\frac{\partial(\chi K)(x_1,x_2)}{\partial x_2}$$
$$+\left[g(x_1+x_2)-g(x_1)-g(x_2)\right]\frac{\partial^2(\chi K)(x_1,x_2)}{\partial x_1 \partial x_2},$$

this expression does not exceed in magnitude

$$1+E^{k+1}(x_1)+(E^{k+1},\mu_0)(1+E(x_1)$$

(up to a constant multiplier). Consequently,

$$|(g,\Omega_s^1(x,w)|\leq \kappa(C)\|\xi_t(w,.)\|_{\mathcal{M}_1^1(X)}\|\xi_t(x,.)\|_{\mathcal{M}^1_{1+E^{k+1}}(X)}$$
$$\times\left[1+(E^{k+1},\mu_0)\right].$$

The norm of Ω_s^2 may be estimated in the same way. Consequently (8.55) leads to (8.56) and this completes the proof of the theorem. □

We shall prove now the Lipschitz continuity of the solutions to our kinetic equation with respect to the initial data in the norm topology of the space $\mathcal{M}^1_{1+E^k}$.

Proposition 8.15 *Under the assumptions of Theorem 8.14, for $k\geq 0$ and $m=1,2$,*

$$\sup_{s\leq t}\|\mu_s(\mu_0^1)-\mu_s(\mu_0^2)\|_{\mathcal{M}^m_{1+E^k}}\leq\kappa(C,t,k,e_0,e_1)(1+E^{1+k},\mu_0^1+\mu_0^2)$$
$$\times\|\mu_0^1-\mu_0^2\|_{\mathcal{M}^m_{1+E^k}}. \tag{8.59}$$

Proof By Theorem 8.8,

$$(g,\mu_t(\mu_0^1)-\mu_t(\mu_0^2))=\int_0^t ds\iint g(y)\xi_t(\mu_0^2+s(\mu_0^1-\mu_0^2);x;dy)$$
$$\times(\mu_0^1-\mu_0^2)(dx). \tag{8.60}$$

Since

$$(g,\xi_t(Y;x;.))=(U^{0,t}g,\xi_0(Y,x;.))=(U^{0,t}g)(x),$$

it follows from Theorem 8.14 that $(g, \xi_t(Y; x; .))$ belongs to $C^{m,0}_{1+E^k}$ as a function of x whenever g belongs to this space and that

$$\|(g, \xi_t(Y; x; .))\|_{C^{m,0}_{1+E^k}(X)} \leq \kappa(C, t, k, e_0, e_1)\|g\|_{C^{m,0}_{1+E^k}(X)} \left[1 + (E^{k+1}, Y)\right].$$

Consequently (8.59) follows from (8.60). □

Exercise 8.4 By Lemma 6.14 and Theorem 8.8, the norm of the solution to the Cauchy problem (8.31) with initial data $\xi \in \mathcal{M}_{1+E^\omega}(X)$, $\omega \leq \beta - 1$ satisfies the equation

$$\begin{aligned}\|\xi_t\|_{\mathcal{M}_{1+E^\omega}(X)} &\\ = \|\xi\|_{\mathcal{M}_{1+E^\omega}(X)} &+ \int_0^t ds \int P(x, z, d\mathbf{y})\xi_s(dx)\mu_s(dz) \\ &\times [(\sigma_s(1 + E^\omega))^{\oplus}(\mathbf{y}) - (\sigma_s(1 + E^\omega))(x) - (\sigma_s(1 + E^\omega))(z)],\end{aligned}$$

where σ_t is the density of ξ_t with respect to its positive variation $|\xi_t|$. Deduce from it that

$$\begin{aligned}\|\xi_t\|_{\mathcal{M}_{1+E^\omega}(X)} \leq \|\xi\|_{\mathcal{M}_{1+E^\omega}(X)} &+ c(e_0, e_1, t) \\ &\times \int_0^t [\|\xi_s\|_{\mathcal{M}_{1+E^\omega}(X)} + \|\xi_t\|_{1+\sqrt{E}}(E^{\omega+1/2}, \mu)]\, ds,\end{aligned}$$

which implies (8.50) directly, bypassing the use of dual equations on functions.

8.5 Propagation and production of moments for the Boltzmann equation

Here, in preparation for the next section, we prove two well-known special features of Boltzmann-equation moment propagation (in comparison with the general jump-type processes of Section 6.4): first, the bound to the moments can be made uniform in time and, second, finite moments are produced by the evolution, i.e. the moments become finite at any positive time even if for the initial state they were infinite. The exposition follows essentially the paper of Lu and Wennberg [164].

Recall that the Boltzmann equation in its weak form (1.52) or (G.3) can be written down as

$$\begin{aligned}\frac{d}{dt}(g, \mu_t) = \tfrac{1}{4} \int_{S^{d-1}} \int_{R^{2d}} &[g(v') + g(w') - g(v) - g(w)] B(|v - w|, \theta) \\ &\times dn\, \mu_t(dv)\mu_t(dw),\end{aligned} \tag{8.61}$$

where the input and output pairs of velocities are denoted by (v, w) and (v', w'). We shall work with the hard-potential case, using a cut-off; i.e. will assume that

$$B(|v|, \theta) = b(\theta)|v|^\beta, \qquad \beta \in (0, 1], \tag{8.62}$$

where $b(\theta)$ is a bounded, not identically vanishing, function. Notice that we exclude the case of bounded kernels (i.e. kernels with $\beta = 0$).

Remark 8.16 For applications it is important to allow certain non-integrable singularities in $b(\theta)$, for θ near 0 and $\pi/2$. As can be seen from the proof of Proposition 8.17, below, our estimates remain valid only if

$$\int_0^{\pi/2} b(\theta) \cos\theta \, \sin^{d-1}\theta \, d\theta < \infty.$$

Here we shall use the following shortened notation for the moments of positive functions f and measures μ:

$$\|f\|_s = \int f(x)(1 + |v|^2)^{s/2} \, dx, \qquad \|\mu\|_s = \int (1 + |v|^2)^{s/2} \, \mu(dx).$$

In our general notation, $\|\mu\|_s$ is written as $\|\mu\|_{\mathcal{M}_{(1+E)^{s/2}}(\mathbf{R}^d)}$, where $E(v) = v^2$ is the kinetic energy.

Proposition 8.17 *Under condition* (8.62) *let $\mu = \mu_0$ be a finite measure on $\mathbf{R}^d$ with $\|\mu_0\|_s < \infty$ for some $s > 2$. Then*

$$\|\mu_t\|_s \le \left[\frac{a_s}{b_s + (a_s \|\mu\|_s^{-\beta/(s-2)} - b_s) e^{-a_s \beta t/(s-2)}} \right]^{\beta/(s-2)}, \tag{8.63}$$

where

$$a_s = \tilde{a}_s \|\mu\|_2, \qquad b_s = \tilde{b}_s \|\mu\|_0 \|\mu\|_2^{-\beta(s-2)} \tag{8.64}$$

and $\tilde{a}_s, \tilde{b}_s$ are positive constants depending only on s. In particular

$$\|\mu_t\|_{2+\beta} \le \frac{a}{b + (a\|\mu_0\|_{2+\beta}^{-1} - b)e^{-at}}, \tag{8.65}$$

where a, b depend on $\|\mu_0\|_0$ and $\|\mu_0\|_2$.

Proof From the general results in Theorem 6.7 we know that the $\|\mu_t\|_s$ remain finite for all t (in fact one can prove this without referring to Theorem 6.7 by means of the usual approximations combined with the a priori estimates given below). From the Boltzmann equation (1.52), for any non-negative solution μ_t we get

$$\frac{d}{dt}\|\mu_t\|_s = \frac{1}{4}\int_{S^{d-1}}\int_{R^{2d}}\mu_t(dv)\mu_t(dw)|v-w|^\beta b(\theta)dn \times [(1+|v'|^2)^{s/2}+(1+|w'|^2)^{s/2} - (1+|v|^2)^{s/2}-(1+|w|^2)^{s/2}],$$

which by the collision inequality (G.7) does not exceed

$$2^s\kappa_1\int_{R^{2d}}\mu_t(dv)\mu_t(dw)|v-w|^\beta(1+|v|^2)^{(s-1)/2}(1+|w|^2)^{1/2} - \tfrac{1}{2}\min\left(\tfrac{1}{4}s(s-2),2\right)\kappa_2\int_{R^{2d}}\mu_t(dv)\mu_t(dw)|v-w|^\beta(1+|v|^2)^{s/2}, \tag{8.66}$$

where

$$\kappa_1 = \int_{S^{d-1}} b(\theta)|\cos\theta||\sin\theta|\,dn = 2A_{d-2}\int_0^{\pi/2} b(\theta)\cos\theta\sin^{d-1}\theta\,d\theta,$$

$$\kappa_2 = \int_{S^{d-1}} b(\theta)\cos^2\theta\sin^2\theta\,dn = 2A_{d-2}\int_0^{\pi/2} b(\theta)\cos^2\theta\sin^{d}\theta\,d\theta.$$

Here A_d denotes the surface area of the d-dimensional unit sphere. Using the elementary inequality

$$|v-w|^\beta \ge (1+|v|^2)^{\beta/2}-(1+|w|^2)^{\beta/2},$$

one deduces that

$$\int_{R^{2d}}\mu_t(dv)\mu_t(dw)|v-w|^\beta(1+|v|^2)^{s/2} \ge \|\mu_t\|_0\|\mu_t\|_{s+\beta}-\|\mu_t\|_\beta\|\mu_t\|_s \ge \|\mu_t\|_0\|\mu_t\|_s^{1+\beta(s-2)}\|\mu_t\|_2^{-\beta(s-2)}-\|\mu_t\|_\beta\|\mu_t\|_s.$$

The latter inequality follows from the Hölder inequality with $q=1+\beta/(s-2)$, $p=1+(s-2)/\beta$, i.e.

$$\|\mu\|_s^{1+\beta(s-2)} \le \|\mu\|_2^{\beta(s-2)}\|\mu\|_{s+\beta}.$$

Consequently, using $|v-w|^\beta \le |v|^\beta+|w|^\beta$ for the first term in (8.66), it follows from (8.66) that

$$\frac{d}{dt}\|\mu_t\|_s \le a_s\|\mu_t\|_s - b_s\|\mu_t\|_s^{1+\beta(s-2)}, \tag{8.67}$$

with a_s, b_s from (8.64). Applying the result of the easy exercise given below completes the proof. □

Corollary 8.18 *Under the condition* (8.62) *let* $\|\mu_0\|_2 < \infty$. *Then there exists a solution* μ_t, *preserving mass and energy and having all moments finite for all positive times, that possesses the estimate*

$$\|\mu_t\|_s \leq \left[\frac{a_s}{b_s(1+e^{-a\beta t/(s-2)})}\right]^{\beta/(s-2)}. \tag{8.68}$$

Proof For μ_0 with a finite moment $\|\mu_0\|_s$ this follows directly from (8.63). For other μ_0 one obtains the required solution by approximating μ_0 with μ_0^n having a finite moment. □

Remark 8.19 From (8.68) one can also deduce the uniqueness of the solutions preserving mass and energy; see Mishler and Wennberg [188].

Exercise 8.5 Show that the ODE

$$\dot{x} = ax - bx^{1+\omega} \tag{8.69}$$

in $\mathbf{R}_+$, where a, b are positive numbers, has the explicit solution

$$x = \left[\frac{a}{b + (ax_0^{-\omega} - b)e^{-a\omega t}}\right]^{1/\omega} \tag{8.70}$$

and, moreover, that the sign of the expression $ax^{-\omega} - b$ remains constant in the solutions. Hint: check that the function $z = ax^{-\omega} - b$ satisfies the linear equation

$$\dot{z} = -\omega a z.$$

8.6 Estimates for the Boltzmann equation

Theorem 8.8 allows us to estimate the moments of the derivatives ξ_t, with respect to the initial data, of the solutions μ_t to kinetic equations of jump type, in terms of the moments of the μ_t. However, as mentioned already, following the proof of Theorem 8.8 yields only the exponential dependence of such estimates of the moments of μ, which would not be enough to obtain effective estimates for the LLN developed later. With the latter objective in mind, in Section 8.4 we obtained estimates for the norms of ξ_t having a linear dependence on the moments of μ, using specific features of coagulation processes. In the present section we shall use specific features of Boltzmann collisions (developed in the previous section) to get polynomial estimates of the moments of ξ_t in terms of the moments of μ_t. We shall use the notation for derivatives introduced at the beginning of this chapter in (8.5) and (8.6).

Theorem 8.20 *Assume that* (8.62) *holds for a bounded non-vanishing* $b(\theta)$. *Then, for* $s \geq 2$,

$$|U^{t,r}g(v)| \leq c(r,s,e_0,e_1)\|g\|_{(1+E)^{s/2}}(1+|v|^2)^{s/2} \times \left(1+\|\mu_0\|_{s+\beta}\|\mu_0\|_{2+\beta}^{\omega(r,e_0,e_1)}\right), \tag{8.71}$$

$$\|V^{r,t}\xi\|_s \leq c(t,s,e_0,e_1)\left(1+\|\mu_0\|_{s+\beta}\|\mu_0\|_{2+\beta}^{\omega(r,e_0,e_1)}\right)\|\xi\|_s, \tag{8.72}$$

$$\sup_{r\leq t}\|\xi_r(\mu,v)\|_s \leq c(t,e_0,e_1)(1+|v|^2)^{s/2}\left(1+\|\mu_0\|_{s+\beta}\|\mu_0\|_{2+\beta}^{\omega(r,s,e_0,e_1)}\right), \tag{8.73}$$

$$\sup_{s\leq t}\|\eta_s(\mu,x,w;\cdot)\|_s \leq c(t,s,e_0,e_1)\left(1+\|\mu_0\|_{s+3}\|\mu_0\|_3^{\omega(r,s,e_0,e_1)}\right) \times [(1+|v|^2)^{s/2}(1+|w|^2)+(1+|w|^2)^{s/2}(1+|v|^2)]. \tag{8.74}$$

Proof As in Exercise 8.4, we begin by noting that, by Lemma 6.14 and Theorem 8.8, the norm of the solution to the Cauchy problem equation (8.31) with initial data $\xi \in \mathcal{M}_{1+E^\omega}(X)$, $\omega \leq \beta - 1$, satisfies the equation

$$\|\xi_t\|_{\mathcal{M}_{1+E^\omega}(X)} = \|\xi\|_{\mathcal{M}_{1+E^\omega}(X)} + \int_0^t d\tau \int P(x,z,d\mathbf{y})\xi_\tau(dx)\mu_\tau(dz) \times [(\sigma_\tau(1+E^\omega))^\oplus(\mathbf{y}) - (\sigma_\tau(1+E^\omega))(x) - (\sigma_\tau(1+E^\omega))(z)],$$

where σ_t is the density of ξ_t with respect to its positive variation $|\xi_t|$. Under the assumption of the present theorem it follows that

$$\|\xi_t\|_s \leq \|\xi_0\|_s + c\int_0^t\int_{R^{2d}}(|v|^\beta+|w|^\beta)\xi_\tau(dv)\mu_\tau(dw)\,d\tau \times [(1+|v|^2+1+|w|^2)^{s/2} - (1+|v|^2)^{s/2} + (1+|w|^2)^{s/2}],$$

and consequently, by inequality (6.16),

$$\|\xi_t\|_s \leq \|\xi_0\|_s + c\int_0^t\int_{R^{2d}}(|v|^\beta+|w|^\beta)\xi_\tau(dv)\mu_\tau(dw)\,d\tau \times [(1+|w|^2)((1+|v|^2)^{s/2-1}+(1+|w|^2)^{s/2-1})+(1+|w|^2)^{s/2}],$$

implying that

$$\|\xi_t\|_s \leq \|\xi_0\|_s + c\int_0^t d\tau(\|\xi_\tau\|_s\|f_\tau\|_2 + \|\xi_\tau\|_2\|f_\tau\|_{s+\beta}). \tag{8.75}$$

Thus

$$\|\xi_t\|_2 \leq \exp\left(c\int_0^t \|\mu_s\|_{2+\beta}\, ds\right)\|\xi_0\|_2.$$

Hence, if $\|\mu_0\|_{2+\beta} \leq a/b$ (see equation (8.65)) then $\|\mu_s\|_{2+\beta} \leq a/b$ for all s and

$$\|\xi_t\|_2 \leq \exp(cta/b)\,\|\xi_0\|_2.$$

If $\|\mu_0\|_{2+\beta} > a/b$ then, by (8.65),

$$\|\xi_t\|_2 \leq \exp\left(c\int_0^t \frac{a\, ds}{b - (b - a\|\mu_0\|_{2+\beta}^{-1})e^{-as}}\right)\|\xi_0\|_2.$$

Using the elementary integral

$$\int \frac{dy}{1 - ce^{-y}} = \log(e^y - c), \qquad c < 1,\ y \geq 0,$$

one then deduces that

$$\begin{aligned}\int_0^t \frac{a\, ds}{b - (b - a\|\mu_0\|_{2+\beta}^{-1})e^{-as}} &= \frac{1}{b}\log\left(e^{as} - \frac{b - a\|\mu_0\|_{2+\beta}^{-1}}{b}\right)\Bigg|_0^t \\ &= \frac{1}{b}\log\left(1 + \frac{b}{a}(e^{at} - 1)\|\mu_0\|_{2+\beta}\right),\end{aligned}$$

implying that

$$\|\xi_t\|_2 \leq \left(1 + \frac{b}{a}(e^{at} - 1)\|\mu_0\|_{2+\beta}\right)^{1/b}\|\xi_0\|_2 \tag{8.76}$$

and hence that (8.73) holds for $s = 2$. Now one may conclude from (8.75), (8.76) and Gronwall's lemma that (8.73) holds also for $s > 2$. One obtains (8.72) similarly, and (8.71) follows from duality. Finally, (8.74) is obtained from (8.42) as in the coagulation case. □

PART III

Applications to interacting particles

9

The dynamic law of large numbers

In the introduction to this book general kinetic equations were obtained as the law of large numbers (LLN) limit of rather general Markov models of interacting particles. This deduction can be called informal, because the limit was performed (albeit quite rigorously) on the forms of the corresponding equations rather than on their solutions, and only the latter type of limit can make any practical sense. Thus it was noted that, in order to make the theory work properly, one has to complete two tasks: to obtain the well-posedness of the limiting kinetic equations (specifying nonlinear Markov processes) and to prove the convergence of the approximating processes to the solutions of these kinetic equations. The first task was settled in Part II. In this chapter we address the second task by proving the convergence of approximations and also supplying precise estimates for error terms.

We can proceed either analytically using semigroup methods or by working directly with the convergence of stochastic processes. Each method has its advantages, and we shall demonstrate both. To obtain the convergence of semigroups we need to estimate the difference between the approximating and limiting generators on a sufficiently rich class of functionals (forming a core for the limiting generator). In Section 9.1 we calculate this difference explicitly for functionals on measures having well-defined first- and second-order variational derivatives. Section 9.2 is devoted to the case of limiting generators of Lévy–Khintchine type with bounded coefficients. The remaining sections deal with pure jump models having unbounded rates.

9.1 Manipulations with generators

In this section we carry out some instructive manipulations with the generators of approximating Markov chains, leading to an alternative derivation of the limiting equation (1.75). Instead of first deducing (1.73) as the limit of the action of the approximating generators on linear functionals F_g and then lifting the obtained evolution to arbitrary functionals, we will obtain the action of the generators (1.66) on arbitrary (smooth enough) functionals on measures in a closed form. This approach leads not only to (1.75) but also to an explicit expression for the difference between (1.66) and its limiting generator, which is useful when estimating this difference in an appropriate functional space. For arbitrary k, $G_{\leq k}$ from (1.67) and A^k from (1.66), these calculations were carried out in Kolokoltsov [134]. Here we shall simplify the story by considering only binary interactions, which are by far the most important for applications, and also only conditionally positive A^k as only such operators are relevant for (at least classical) interacting particles.

Let us start with interactions preserving the number of particles, i.e. those given in $(SX)^n$ by

$$\sum_{i=1}^{n}(B_\mu^1)_i f(x_1,\ldots,x_n) + \sum_{\{i,j\}\subset\{1,\ldots,n\}} (B_\mu^2)_{ij} f(x_1,\ldots,x_n), \tag{9.1}$$

where $(B_\mu^1)_i$ and $(B_\mu^2)_{ij}$ denote the actions of the operators B_μ^1 and B_μ^2 on the variables x_i and x_i, x_j respectively. Here B_μ^1 and B_μ^2 are Lévy-type operators, in $C(x)$ and $C^{\mathrm{sym}}(X^2)$ respectively, depending on a measure μ as on a parameter and allowing for an additional mean field interaction:

$$\begin{aligned}B_\mu^1 f(x) &= \tfrac{1}{2}(G_\mu(x)\nabla,\nabla)f(x) + (b_\mu(x),\nabla f(x))\\ &\quad + \int \left[f(x+y) - f(x) - (\nabla f(x), y)\mathbf{1}_{B_1}(y)\right]\nu_\mu(x,dy)\end{aligned}$$

and

$$\begin{aligned}&B_\mu^2 f(x,y)\\ &= \left[\frac{1}{2}\left(G_\mu(x,y)\frac{\partial}{\partial x},\frac{\partial}{\partial x}\right) + \frac{1}{2}\left(G_\mu(y,x)\frac{\partial}{\partial y},\frac{\partial}{\partial y}\right) + \left(\gamma_\mu(x,y)\frac{\partial}{\partial x},\frac{\partial}{\partial y}\right)\right]f(x,y)\\ &\quad + \left[\left(b_\mu(x,y),\frac{\partial}{\partial x}\right) + \left(b_\mu(y,x),\frac{\partial}{\partial y}\right)f(x,y)\right]\\ &\quad + \int_{X^2}\nu_\mu(x,y,dv_1dv_2)\left[f(x+v_1,y+v_2) - f(x,y)\right.\\ &\qquad \left. - \left(\frac{\partial f}{\partial x}(x,y),v_1\right)\mathbf{1}_{B_1}(v_1) - \left(\frac{\partial f}{\partial y}(x,y),v_2\right)\mathbf{1}_{B_1}(v_2)\right],\end{aligned}$$

where $G(x,y)$ and $\gamma(x,y)$ are symmetric matrices such that $\gamma(x,y) = \gamma(y,x)$ and the Lévy kernels ν satisfy the relation

$$\nu_\mu(x,y,dv_1dv_2) = \nu_\mu(y,x,dv_2dv_1).$$

Remark 9.1 The symmetry condition imposed on the coefficients of B^2 is necessary and sufficient for this operator to preserve the set of functions $f(x, y)$ that are symmetric with respect to the permutation of x and y.

The procedure leading to the kinetic equations consists of scaling B^2 by a multiplier h and substituting the function $f(\mathbf{x})$ on $S\mathcal{X}$ by the functional $F(h\delta_{\mathbf{x}})$ on measures. This leads to a generator given by

$$(\Lambda_h^1 + \Lambda_h^2)F(h\delta\mathbf{x}) = \sum_{i=1}^{n}(B^1_{h\delta_{\mathbf{x}}})_i F(h\delta_{\mathbf{x}}) + h \sum_{\{i,j\}\subset\{1,\dots,n\}} (B^2_{h\delta_{\mathbf{x}}})_{ij} F(h\delta_{\mathbf{x}}).$$

The calculation of this generator is summarized in the following.

Proposition 9.2 *If F is smooth enough, which means that all the derivatives in the formulas below must be well defined, then*

$$\begin{aligned}
\Lambda_h^1 F(Y) &= \int_X \left[\left(B_Y^1 \frac{\delta F}{\delta Y(.)}\right)(x) + \frac{h}{2}\left(G_Y(x)\frac{\partial}{\partial x}, \frac{\partial}{\partial y}\right) \frac{\delta^2 F}{\delta Y(x)\delta Y(y)}\bigg|_{y=x}\right] Y(dx) \\
&\quad + h\int_0^1 (1-s)\,ds \int_{X^2}\left(\frac{\delta^2 F}{\delta Y(.)\delta Y(.)}(Y + sh(\delta_{x+y}-\delta_x)), (\delta_{x+y}-\delta_x)^{\otimes 2}\right) \\
&\qquad \times \nu_Y(x, dy)Y(dx)
\end{aligned} \tag{9.2}$$

and

$$\begin{aligned}
\Lambda_h^2 F(Y) &= \frac{1}{2}\int_{X^2}\left(B_Y^2\left(\frac{\delta F}{\delta Y(.)}\right)^{\oplus}\right)(x, y)Y(dx)Y(dy) \\
&\quad - \frac{h}{2}\int_X \left(B_Y^2\left(\frac{\delta F}{\delta Y(.)}\right)^{\oplus}\right)(x, x)Y(dx) + h^3 \sum_{\{i,j\}\subset\{1,\dots,n\}} \Omega_Y(x_i, x_j),
\end{aligned} \tag{9.3}$$

where $Y = h\delta_{\mathbf{x}}$, $x = (x_1, \dots, x_n)$ and

$$\begin{aligned}
\Omega_Y(x, y) &= \left(\gamma_Y(x, y)\frac{\partial}{\partial x}, \frac{\partial}{\partial y}\right)\frac{\delta^2 F}{\delta Y(x)\delta Y(y)} \\
&\quad + \left(G_Y(x, y)\frac{\partial}{\partial x}, \frac{\partial}{\partial y}\right)\frac{\delta^2 F}{\delta Y(x)\delta Y(y)}\bigg|_{y=x}
\end{aligned}$$

$$+\int_0^1 (1-s)\,ds \int \nu_Y(x,y,dv_1dv_2)$$
$$\times\left(\frac{\delta^2 F}{\delta Y(.)\delta Y(.)}(Y+sh(\delta_{x+v_1}-\delta_x+\delta_{y+v_2}-\delta_y)),\right.$$
$$\left.(\delta_{x+v_1}-\delta_x+\delta_{y+v_2}-\delta_y)^{\otimes 2}\right).$$

Proof For differentiation we use the rule

$$\frac{\partial}{\partial x_i}F(h\delta_{\mathbf{x}}) = h\frac{\partial}{\partial x_i}\frac{\delta F(Y)}{\delta Y(x_i)}, \qquad Y = h\delta_{\mathbf{x}},\ \ \mathbf{x}=(x_1,\ldots,x_n)\in\mathcal{X}$$

(see Lemma F.4), which implies that

$$\left(G\frac{\partial}{\partial x_i},\frac{\partial}{\partial x_i}\right)F(h\delta_{\mathbf{x}})$$
$$= h\left(G\frac{\partial}{\partial x_i},\frac{\partial}{\partial x_i}\right)\frac{\delta F(Y)}{\delta Y(x_i)} + h^2\left(G\frac{\partial}{\partial y},\frac{\partial}{\partial z}\right)\frac{\delta^2 F(Y)}{\delta Y(y)\delta Y(z)}\bigg|_{y=z=x_i}.$$

Thus

$$\Lambda_h^1 F(Y) = \sum_{i=1}^n (B^1_{h\delta_{\mathbf{x}}})_i F(h\delta_{\mathbf{x}})$$
$$= \sum_i^n \left\{h\left(b_Y(x_i),\frac{\partial}{\partial x_i}\frac{\delta F(Y)}{\delta Y(x_i)}\right) + \frac{h}{2}\left(G_Y(x_i)\frac{\partial}{\partial x_i},\frac{\partial}{\partial x_i}\right)\frac{\delta F(Y)}{\delta Y(x_i)}\right.$$
$$+\frac{h^2}{2}\left(G_Y(x_i)\frac{\partial}{\partial y},\frac{\partial}{\partial z}\right)\frac{\delta^2 F(Y)}{\delta Y(y)\delta Y(z)}\bigg|_{y=z=x_i}$$
$$+\int\Big[F(h\delta_{\mathbf{x}}+h\delta_{x_i+y}-h\delta_{x_i})$$
$$\left.-\ F(h\delta_{\mathbf{x}}) - h\left(\frac{\partial}{\partial x_i}\frac{\delta F(Y)}{\delta Y(x_i)},y\right)\mathbf{1}_{B_1}(y)\Big]\,\nu_Y(x_i,dy)\right\}.$$

By (F.5a),

$$F(h\delta_{\mathbf{x}}+h\delta_{x_i+y}-h\delta_{x_i}) - F(h\delta_{\mathbf{x}})$$
$$= h\left(\frac{\delta F(Y)}{\delta Y(.)},\ \delta_{x_i+y}-\delta_{x_i}\right)$$
$$+h^2\int_0^1(1-s)\left(\frac{\delta^2 F(Y+hs(\delta_{x_i+y}-\delta_{x_i}))}{\delta Y(.)\delta Y(.)},\ (\delta_{x_i+y}-\delta_{x_i})^{\otimes 2}\right)ds,$$

implying (9.2).

Similarly,

$$\Lambda_h^2 F(Y) = h\sum_{i,j}\left\{\frac{h}{2}\left(G_Y(x_i,x_j)\frac{\partial}{\partial x_i},\frac{\partial}{\partial x_i}\right)\frac{\delta F(Y)}{\delta Y(x_i)}\right.$$
$$+\frac{h}{2}\left(G_Y(x_j,x_i)\frac{\partial}{\partial x_j},\frac{\partial}{\partial x_j}\right)\frac{\delta F(Y)}{\delta Y(x_j)}$$
$$+h^2\left(G_Y(x_i,x_j)\frac{\partial}{\partial y},\frac{\partial}{\partial z}\right)\frac{\delta^2 F(Y)}{\delta Y(y)\delta Y(z)}\bigg|_{y=z=x_i}$$
$$+h^2\left(\gamma_Y(x_i,x_j)\frac{\partial}{\partial x_i},\frac{\partial}{\partial x_j}\right)\frac{\delta^2 F(Y)}{\delta Y(x_i)\delta Y(x_j)}$$
$$+h\left(b_Y(x_i,x_j),\frac{\partial}{\partial x_i}\frac{\delta F(Y)}{\delta Y(x_i)}\right)+h\left(b_Y(x_j,x_i),\frac{\partial}{\partial x_j}\frac{\delta F(Y)}{\delta Y(x_j)}\right)$$
$$+\int\Bigg[F(h\delta_{\mathbf{x}}+h\delta_{x_i+v_1}-h\delta_{x_i}+h\delta_{x_j+v_2}-h\delta_{x_j})-F(h\delta_{\mathbf{x}})$$
$$-h\left(\frac{\partial}{\partial x_i}\frac{\delta F(Y)}{\delta Y(x_i)},v_1\right)\mathbf{1}_{B_1}(v_1)-h\left(\frac{\partial}{\partial x_j}\frac{\delta F(Y)}{\delta Y(x_j)},v_2\right)\mathbf{1}_{B_1}(v_2)\Bigg]$$
$$\left.\times\nu_Y(x_i,x_j,dv_1dv_2)\right\}.$$

Applying formula (F.5) to express

$$F(h\delta_{\mathbf{x}}+h\delta_{x_i+v_1}-h\delta_{x_i}+h\delta_{x_j+v_2}-h\delta_{x_j})-F(h\delta_{\mathbf{x}})$$

in terms of variational derivatives yields (9.3). □

Remark 9.3 (i) Formula (9.2) can be used to extend the action of Λ_h^1 to arbitrary, not necessarily discrete, measures Y but only for functionals F that are defined on signed measures, since the measure $Y+sh(\delta_{x+y}-\delta_x)$ may not be positive. This is important to have in mind, as we are often interested in functionals of the type $F(\mu_t(\mu_0))$, where μ_t solves the kinetic equation and hence is defined only for positive measures μ_0.

(ii) By (H.1) one can rewrite the last term in (9.3) as an integral:

$$h^3\sum_{\{i,j\}\subset\{1,\ldots,n\}}\Omega_Y(x,x_j)$$
$$=h\iint\Omega_Y(x,y)Y(dx)Y(dy)-h\int\Omega_Y(x,x)Y(dx),$$

which is more convenient for analysis but can be used only for functionals F that extend beyond the set of positive measures, as $\Omega_Y(x,x)$ is not defined otherwise even for $Y=h\delta_{\mathbf{x}}$.

Similar formulas are valid for interactions that change the number of particles. See, for instance, the following exercises.

Exercise 9.1 Show that if Λ_2^h is given by (1.38) then

$$
\begin{aligned}
&\Lambda_2^h F(Y) \\
&= \frac{1}{2}\int_{X^2}\int_{\mathcal{X}}\left[\left(\frac{\delta F(Y)}{\delta Y(.)}\right)^{\oplus}(\mathbf{y}) - \left(\frac{\delta F(Y)}{\delta Y(.)}\right)^{\oplus}(\mathbf{z})\right] P(\mathbf{z}, d\mathbf{y}) Y^{\otimes 2}(d\mathbf{z}) \\
&\quad - \frac{h}{2}\int_{X}\int_{\mathcal{X}}\left[\left(\frac{\delta F(Y)}{\delta Y(.)}\right)^{\oplus}(\mathbf{y}) - 2\frac{\delta F(Y)}{\delta Y(z)}\right] P(z, z, d\mathbf{y}) Y(dz) \\
&\quad + h^3 \sum_{\{i,j\}\subset\{1,\ldots,n\}} \int_0^1 (1-s)\, ds \int_{\mathcal{X}} P(x_i, x_j, d\mathbf{y}) \\
&\quad \times \left(\frac{\delta^2 F}{\delta Y(.)\delta Y(.)}(Y + sh(\delta_{\mathbf{y}} - \delta_{x_i} - \delta_{x_j})), (\delta_{\mathbf{y}} - \delta_{x_i} - \delta_{x_j})^{\otimes 2}\right) \qquad (9.4)
\end{aligned}
$$

Exercise 9.2 Extend formulas (9.2)–(9.4) to the case of k-ary interactions, giving the answer in terms of the operators $\Phi_l^k[f]$ from equation (H.8). Hint: consult [134], where the full expansion in h is obtained for the generators Λ_h^k on analytic functionals $F(Y)$.

The main conclusion to be drawn from the calculations above is the following. To approximate the generator Λ of a nonlinear process by the generators of approximating interacting-particle systems we must work with functionals F whose second-order variational derivative has the regularity (with respect to the spatial variable) needed for it to belong to the domain of the generator of the corresponding approximating system. For instance, this second-order derivative should be at least twice continuously differentiable for full Lévy–Khintchine generators – hence the the spaces $C^{2,k}_{\text{vague}}(\mathcal{M}(\mathbf{R}^d))$ appear in a natural way – or it should satisfy certain bounds for unbounded generators of integral type.

To conclude this section, we consider the model of binary interaction described by the generator given by (1.38) or (9.4) with limiting equation (8.29) and under the assumptions of Theorem 8.8 and identify natural classes of strong continuity for a semigroup T_t^h, specified below. The natural state space to work with turns out to be the set

$$\mathcal{M}_{h\delta}^{e_0,e_1} = \mathcal{M}_{e_0,e_1}(X) \cap \mathcal{M}_{h\delta}^{+}(X),$$

where $\mathcal{M}_{e_0,e_1}(X)$ is given by (8.30). As usual, we denote by $\mu_t = \mu_t(\mu_0)$ the solution to the Cauchy problem (8.29).

Proposition 9.4 *For any positive e_0, e_1 and $1 \le l \le m$, the operator Λ_h^2 is bounded in the space $C_{(1+E^l,\cdot)^m}(\mathcal{M}_{h\delta}^{e_0,e_1})$ and defines a strongly continuous semigroup T_t^h there such that*

$$\|T_t^h\|_{C_{(1+E^l,\cdot)^m}(\mathcal{M}_{h\delta}^{e_0,e_1})} \le \exp\left[c(C,m,l)e_1 t\right]. \tag{9.5}$$

Proof Let us show that

$$\Lambda_h^2 F(Y) \le c(C,m,l)e_1 F(Y) \tag{9.6}$$

for $Y = h\delta_{\mathbf{x}}$ and $F(Y) = (1+E^l, Y)^m$. Then (9.5) will follow by Theorem 4.1 or Gronwall's lemma.

We have

$$\Lambda_h^2 F(Y) = h \sum_{I\subset\{1,\dots,n\}:|I|=2} \int \left[(1+E^l, Y + h(\delta_y - \delta_{\mathbf{x}_I}))^m - (1+E^l, Y)^m\right] K(\mathbf{x}_I; dy).$$

Because

$$\begin{aligned}(1+E^l, h(\delta_y - \delta_{x_i} - \delta_{x_j})) &\le h\left\{\left[E(x_i)+E(x_j)\right]^l - E^l(x_i) - E^l(x_j)\right\}\\ &\le hc(l)[E^{l-1}(x_i)E(x_j) + E(x_i)E^{l-1}(x_j)],\end{aligned}$$

and using the inequality $(a+b)^m - a^m \le c(m)(a^{m-1}b + b^m)$, we obtain

$$\begin{aligned}\Lambda_h^2 F(Y) \le hc(m,l) \sum_{I\subset\{1,\dots,n\}:|I|=2} &\Big\{(1+E^l, Y)^{m-1} h\left[E^{l-1}(x_i)E(x_j) + E(x_i)E^{l-1}(x_j)\right]\\ &+ h^m\left[E^{l-1}(x_i)E(x_j) + E(x_i)E^{l-1}(x_j)\right]^m\Big\} K(\mathbf{x}_I; dy)\\ \le c(C,m,l) \iint &\Big\{(1+E^l, Y)^{m-1}\left[E^{l-1}(z_1)E(z_2) + E(z_1)E^{l-1}(z_2)\right]\\ &+ h^{m-1}\left[E^{l-1}(z_1)E(z_2) + E(z_1)E^{l-1}(z_2)\right]^m\Big\}\\ &\times [1+E(z_1)+E(z_2)]\, Y(dz_1)Y(dz_2).\end{aligned}$$

By symmetry it is enough to estimate the integral over the set where $E(z_1) \ge E(z_2)$. Consequently $\Lambda_h^2 F(Y)$ does not exceed

$$\begin{aligned}&c\int \left\{(1+E^l, Y)^{m-1}E^{l-1}(z_1)E(z_2) + h^{m-1}\left[E^{l-1}(z_1)E(z_2)\right]^m\right\}\\ &\quad\times [1+E(z_1)]\, Y(dz_1)Y(dz_2)\\ &\quad\le c(1+E^l, Y)^m(E, Y) + h^{m-1}c\int E^{m(l-1)+1}(z_1)E^m(z_2)Y(dz_1)Y(dz_2).\end{aligned}$$

To prove (9.6) it remains to show that the second term on the r.h.s. of that expression can be estimated by its first term. This follows from the estimates

$$(E^m, Y) = h \sum E^m(x_i) \le h \left(\sum E^l(x_i)\right)^{m/l} = h^{1-m/l}(E^l, Y)^{m/l},$$
$$(E^{m(l-1)+1}, Y) \le h^{-1}(E^{m(l-1)}, Y)(E, Y)$$
$$\le h^{-m(1-1/l)}(E^l, Y)^{m(1-1/l)}(E, Y). \qquad \square$$

The following statement is a straightforward extension of Proposition 9.4 and its proof is omitted.

Proposition 9.5 *The statement of Proposition 9.4 remains true if the space* $C_{(1+E^l,\cdot)^m}$ *is replaced by the more general space*

$$C_{(1+E^{l_1},\cdot)^{m_1}\dots(1+E^{l_j},\cdot)^{m_j}}.$$

9.2 Interacting diffusions, stable-like and Vlasov processes

As already mentioned, we can roughly distinguish two approaches to the proof of the LLN: analytical (via semigroups and generators) and probabilistic (using the tightness of the approximating probability laws and choosing a converging subsequence). The first leads to rather precise estimates for the error term of the approximating averages and finite-dimensional approximations, and the second yields a stronger convergence in distribution of the trajectories. The second, probabilistic, approach can also be used to obtain the existence of solutions to the kinetic equations. We shall pay attention to both these approaches, starting with mean-field-interaction Markov processes of a rather general form (with Lévy–Khintchine-type generators) and then moving to popular and practically important models of binary jump-type interactions possibly subject to a spatial motion.

We begin with mean-field-interaction Markov processes having *generator of order at most one*, i.e. with the Markov process in $(\mathbf{R}^d)^n$ generated by an operator G of the form (9.1) with vanishing B^2 and $B^1_\mu = A_\mu$ of the form (8.12), i.e.

$$A_\mu f(x) = (b(x,\mu), \nabla f(x)) + \int_{\mathbf{R}^d} \left[f(x+y) - f(x)\right] \nu(x,\mu,y)dy.$$

Then

$$Gf(x_1,\dots,x_n) = \sum_{i=1}^n (A_{\delta_{\mathbf{x}}})_i f(x_1,\dots,x_n), \tag{9.7}$$

where $\delta_{\mathbf{x}} = \delta_{x_1} + \cdots + \delta_{x_n}$ and $(A_{\delta_{\mathbf{x}}})_i$ denotes the action of the operator $A_{\delta_{\mathbf{x}}}$ on the variables x_i. According to Proposition 9.2 the corresponding scaled operator acting on smooth functionals $F(h\delta_{\mathbf{x}})$ is defined by the formula

$$\Lambda_h F(Y) = \Lambda F(Y) + h \int_0^1 (1-s)\, ds \int_{X^2} \nu_Y(x, dy) Y(dx)$$
$$\times \left(\frac{\delta^2 F}{\delta Y(.) \delta Y(.)} (Y + sh(\delta_{x+y} - \delta_x)), (\delta_{x+y} - \delta_x)^{\otimes 2} \right) \qquad (9.8)$$

with

$$\Lambda F(Y) = \left(A_Y \frac{\delta F}{\delta Y(.)}, Y \right).$$

Clearly $\Lambda f \in C_{\text{vague}}(\mathcal{M}_M(\mathbf{R}^d))$ whenever $F \in C^{1,1}_{\text{vague}}(\mathcal{M}_M(\mathbf{R}^d))$. Notice that the contents of the large parentheses in (9.8) equal

$$\left(\frac{\delta^2 F}{\delta Y(x+y) \delta Y(x+y)} - 2 \frac{\delta^2 F}{\delta Y(x+y) \delta Y(x)} + \frac{\delta^2 F}{\delta Y(x) \delta Y(x)} \right) (Y + sh(\delta_{x+y} - \delta_x));$$

this expression is integrable with respect to a Lévy measure $\nu_Y(x, dy)$.

Recall now that under the assumptions of Theorem 8.6 the solutions to the Cauchy problem

$$\frac{d}{dt} \mu_t = A^\star_{\mu_t} \mu_t, \qquad \mu_0 = \mu, \qquad (9.9)$$

are well defined and specify the Feller semigroup $\Phi_t F(\mu) = F(\mu_t)$ in $C_{\text{vague}}(\mathcal{M}_M(\mathbf{R}^d))$ generated by Λ.

The following result states the *dynamic LLN for mean-field-interaction Markov processes with generators of order at most one* in analytic form.

Theorem 9.6 *Under the assumptions of Theorem 8.6, the operator given by (9.7) generates a uniquely defined Feller process in $\mathbf{R}^{dn}$ with invariant domain $C^1_\infty(\mathbf{R}^{dn})$. The corresponding Feller semigroup U^h_s on $C(\mathcal{M}^+_{h\delta}(\mathbf{R}^d))$ of the scaled process Z^h_t in $\mathcal{M}^+_{h\delta}(\mathbf{R}^d)$ generated by (9.8) converges strongly, for any $M > 0$, to the Feller semigroup Φ_t in the sense that*

$$\sup_{s \le t} \sup_{\|h\delta_{\mathbf{x}_n}\| \le M} |U^h_s F(h\delta_{\mathbf{x}_n}) - \Phi_s F(h\delta_{\mathbf{x}_n})| \to 0, \qquad n \to \infty,$$

for any $t > 0$ and $F \in C_{\text{vague}}(\mathcal{M}_M(\mathbf{R}^d))$. Moreover, in the case $F \in C^{2,2}_{\text{vague}}(\mathcal{M}_M(\mathbf{R}^d))$ (using the notation introduced before Theorem 8.6) one has the estimate

$$\sup_{s\le t}\sup_{\|h\delta_{\mathbf{x}_n}\|\le M}|U_s^h F(h\delta_{\mathbf{x}_n})-\Phi_s F(h\delta_{\mathbf{x}_n})|\le c(t,M)h\|F\|_{C^{2,2}_{\text{vague}}(\mathcal{M}_M(\mathbf{R}^d))}. \tag{9.10}$$

Proof The hard work needed for the proof has been carried out already, in previous chapters. The first statement follows directly from Theorem 4.13, and the second from Theorem 8.6, equation (9.8) and Theorem 2.11 (or, better, an obvious modification in which the approximating propagators act in different spaces). □

Theorem 9.7 *Under the assumptions of Theorem 9.6, suppose that the initial measures $h\delta_{\mathbf{x}(h)}$ of the approximating processes Z_t^h generated by Λ_h converge weakly to a measure μ. Then the distributions of the approximating processes in $D(\mathbf{R}_+,\mathcal{M}^+_{h\delta}(\mathbf{R}^d))$ are tight, with measures considered in their vague topology, and converge weakly to the distributions of the deterministic process μ_t solving equation* (9.9).

First proof This follows directly from Theorems 9.6 and C.5.

Remark 9.8 We have avoided any discussion of the compact containment condition that applies when one is working in the large compact space of positive bounded measures equipped with the vague topology. The price paid is, of course, a weaker convergence result (vague but not weak convergence) than one would expect. Proving this compact containment condition for the approximations in the weak rather than vague topology, which is not very difficult, would allow us to strengthen the result obtained. This is also relevant to the second proof of the theorem, which now follows.

Second proof Here we give a proof which is independent of the analysis of Chapter 8 and even of most of Chapter 6. As we shall see, the straightforward probabilistic argument yields the tightness of Z_t^h and that the limit of any converging subsequence solves the corresponding kinetic equation. Thus the only part of the previous analysis needed to carry out this proof is the existence of the Feller processes generated by (9.7) and the uniqueness of the solutions to the Cauchy problem of the kinetic equation (the latter being needed, of course, to conclude that not only does a converging subsequence exist but the whole family of approximations converges). This method, however, does not supply the error estimates (9.10).

Independently of the convergence of the semigroup, the tightness of the approximations follows directly via Theorem C.6 from equation (9.8). Let us show that a converging subsequence, which we shall still denote by Z_t^h, converges to a solution of the kinetic equation.

Recall now that on the linear functionals $F_g(Y) = (g, Y)$ the scaled operator Λ_h acts according to

$$\Lambda_h F_g(h\delta_{\mathbf{x}}) = h \sum_{i=1}^{n} (A_{h\delta_{\mathbf{x}}})_i g^{\oplus}(\mathbf{x}) = (A_{h\delta_{\mathbf{x}}} g, h\delta_{\mathbf{x}}),$$

as follows from (9.8) but also directly from the definition of the mean field scaling. Hence, by Dynkin's formula and the fact that (9.7) generates a Feller semigroup (by Theorem 4.13) it follows that

$$M_g^h(t) = (g, h\nu_t) - (g, h\nu) - \int_0^t (A_{h\nu_s} g, h\nu_s)\, ds \tag{9.11}$$

is a martingale, for any initial measure $\nu = \delta_{\mathbf{x}}$, with respect to the natural filtration of the process Z_t^h. Here $\nu_t = \delta_{\mathbf{x}_t^h}$ where $\mathbf{x}_t^h$ is the process generated by

$$\sum_{i=1}^{n} (A_{h\delta_{\mathbf{x}}})_i f(x_1, \ldots, x_n)$$

in $\mathbf{R}^{dn}$. Assuming that $h\nu_t$ converges in distribution to a measure-valued process μ_t it can be seen directly that the r.h.s. of (9.11) converges to

$$(g, \mu_t) - (g, \mu) - \int_0^t (A_{\mu_s} g, \mu_s)\, ds$$

for any $g \in C^1_\infty(\mathbf{R}^d)$.

Hence, in order to demonstrate that the limiting measure μ_t solves the kinetic equation (9.9), it remains to show that the martingale (9.11) converges to zero for any $g \in C^1_\infty(\mathbf{R}^d)$. Taking into account that $\mathbf{E}(M_g^h)^2 = \mathbf{E}[M_g^h]$ (see e.g. Appendix C on the basic properties of quadratic variations, in particular Proposition C.7), it is enough to show that the expectation of the quadratic variation $\mathbf{E}[M_g^h]$ tends to zero as $h \to 0$. But $[M_g^h]$ equals the quadratic variation $[(g, h\nu_t)]$ of $(g, h\nu_t)$, and

$$[(g, h\nu_t)] = h^2 [g^{\oplus}(\mathbf{x}_t^h)],$$

so that

$$\begin{aligned} \mathbf{E}[M_g^h]_t &= h^2\, \mathbf{E}[g^{\oplus}(\mathbf{x}_t^h)] \le nh^2 \max_i \mathbf{E}[g((x_t^h)_i)] \\ &\le ch^2 n \le ch \end{aligned}$$

with constant c, because hn is bounded (since the measures $h\nu_t$ are bounded) and we know that $\mathbf{E}[g((x_t^h)_i)]$ is also bounded, because it can be estimated by the expectation of the square of the corresponding Dynkin's martingale. This is in turn bounded by the assumption $g \in C^1_\infty(\mathbf{R}^d)$. The proof is complete.

Remark 9.9 Juggling the various criteria of tightness listed in Appendix C yields various modifications of the above proof. For instance, one can avoid Proposition 9.2 and equation (9.8) altogether. Namely it is enough to establish, say by Theorem C.2, tightness for polynomials of linear functionals F_g for which the action of Λ_h is straightforward. To do this one can use either Theorem C.9 (estimating the quadratic variation as above) or Theorem C.10, thus working with only predictable quadratic variations, which are easy to calculate using (C.7).

The method of obtaining the LLN for mean field interactions, discussed above in detail for generators of order at most one, works similarly in other situations once all the ingredients (well-posedness for the approximations and regularity for the kinetic equation) are in place. Let us consider, for instance, interacting diffusions and stable-like processes.

Theorem 9.10 *Under the assumptions of Theorem 8.7, suppose that the initial measures $h\delta_{\mathbf{x}(h)}$ of the approximating processes Z_t^h generated by Λ_h converge weakly to a measure μ. Then the distributions of the approximating processes in $D(\mathbf{R}_+, \mathcal{M}_{h\delta}^+(\mathbf{R}^d))$ are tight with measures considered in the vague topology and converge weakly to the distributions of the deterministic process μ_t solving equation* (9.9). *The estimates of convergence* (9.10) *also hold for $F \in C_{\text{vague}}^{2,k}(\mathcal{M}(\mathbf{R}^d))$.*

Proof It is the same as that of Theorem 9.7 and so is omitted. □

The LLN can be obtained in a similar way for models with unbounded coefficients or for the stochastic systems on manifolds discussed in Sections 7.4 and 11.4.

9.3 Pure jump models: probabilistic approach

In this section we take a probabilistic approach to the analysis of kinetic equations and the LLN for pure jump models of interaction. These results will not be used in what follows.

We shall exploit the properties of and notation for transition kernels introduced in Section 6.3. Recall, in particular, that the transition kernel P is called *multiplicatively E-bounded* or *$E^{\otimes}$-bounded* (resp. *additively E-bounded* or *$E^{\oplus}$-bounded*) whenever $P(\mu; \mathbf{x}) \leq cE^{\otimes}(\mathbf{x})$ (resp. $P(\mu; \mathbf{x}) \leq cE^{\oplus}(\mathbf{x})$) for all μ and $\mathbf{x}$ and some constant $c > 0$. In future we shall often take $c = 1$ for brevity.

We start with multiplicatively bounded kernels and then turn to additively bounded ones. Notice that if a kernel P is $(1+E)^{\otimes}$-bounded then it is also $(1+E)^{\oplus}$-bounded, so the next result includes $(1+E)^{\oplus}$-bounded kernels.

Theorem 9.11 *Suppose that the transition kernel $P(\mu, \mathbf{x}, .)$ enjoys the following properties.*

(a) $P(\mu, \mathbf{x}, .)$ is a continuous function

$$\mathcal{M}(X) \times \cup_{l=1}^{k} SX^l \mapsto \mathcal{M}(\cup_{l=1}^{k} SX^l),$$

where the measures are considered in their weak topologies.

(b) $P(\mu, \mathbf{x}, .)$ is E-subcritical and $(1+E)^{\otimes}$-bounded for some continuous non-negative function E on X such that $E(x) \to \infty$ as $x \to \infty$;

*(c) The $P(\mu, \mathbf{x}, .)$ are **1**-subcritical for $\mathbf{x} \in X^l$, $l \geq 2$. Then one has the following.*

(i) The Markov processes $Z_t^h(h\nu)$ ($h\nu$ denotes the starting point) in $\mathcal{M}_{h\delta}^+(X)$ are correctly and uniquely defined by generators of type (1.34) *or* (1.36), *and the processes $(E, Z_t^h(h\nu))$ are non-negative supermartingales.*

(ii) Given arbitrary $b > 0$, $T > 0$, $h \in [0,1]$ and $\nu = \delta_{\mathbf{x}}$ with $h(1+E, \nu) \leq b$, one has

$$\mathbf{P}\left(\sup_{t\in[0,T]} (1+E, Z_t^h(h\nu)) > r\right) \leq \frac{c(T,b)}{r} \tag{9.12}$$

for all $r > 0$, where the constant $c(T,b)$ does not depend on h; moreover, $Z_t^h(h\nu)$ enjoys the compact containment condition, i.e. for arbitrary $\eta > 0$, $T > 0$ there exists a compact subset $\Gamma_{\eta,T} \subset \mathcal{M}(X)$ for which

$$\inf_{h\nu} \mathbf{P}(Z^{h\nu}(t) \in \Gamma_{\eta,T} \; for \; 0 \leq t \leq T) \geq 1 - \eta. \tag{9.13}$$

(iii) If the family of measures $\nu = \nu(h)$ is such that $(1+E, h\nu)$ is uniformly bounded then the family of processes $Z_t^h(h\nu)$ defined by (1.34) *is tight as a family of processes with sample paths in $D(\mathbf{R}_+, \mathcal{M}(X))$; the same holds for processes defined by* (1.36) *whenever P is $(1+E)^{\oplus}$- bounded.*

(iv) If additionally P is L-non-increasing and interactions of order $l > 1$ do not increase the number of particles then the moment measures of $Z_t^h(h\nu)$ are uniformly bounded, i.e. the $\mathbf{E}(\|Z^{h\nu}(t)\|^r)$ are uniformly bounded for $t \in [0,T]$, for arbitrary $r \geq 1$ and $T > 0$.

Proof (i) Recall that we denote by F_g the linear functionals on measures: $F_g(\mu) = (g, \mu)$. By E-subcriticality,

$$\Lambda^h F_E(h\nu) \leq 0, \qquad \nu \in \mathcal{M}_\delta^+(X).$$

By the **1**-subcriticality of interactions of order $l \geq 2$,

$$\Lambda^h F_{\mathbf{1}}(h\nu) \leq h \sum_{i=1}^{n} \int \left[\mathbf{1}^{\oplus}(\mathbf{y}) - 1\right] P(h\nu, x_i, d\mathbf{y})$$

for $\nu = \delta_{x_1} + \cdots + \delta_{x_n}$, and hence

$$\Lambda^h F_{\mathbf{1}}(h\nu) \leq hk \sum_{i=1}^{n} (\mathbf{1} + E)(x_i) = kF_{\mathbf{1}+E}(h\nu),$$

as the number of particles created at one time is bounded by k. Moreover, the intensity of jumps corresponding to the generator (1.34) equals

$$q(h\nu) = \frac{1}{h} \sum_{l=1}^{k} h^l \sum_{I \subset \{x_1, \ldots, x_n\}, |I|=l} \int P(h\nu; \mathbf{x}_I; d\mathbf{y}) \leq \frac{1}{h} \sum_{l=1}^{k} \frac{1}{l!} \left[f_{1+E}(h\nu)\right]^l \tag{9.14}$$

for $\nu = \delta_{x_1} + \cdots + \delta_{x_n}$. In the case of (1.36) the same estimate holds with an additional multiplier $h^{-(l-1)}$. Hence the conditions of Theorem 4.10 are met if F_{1+E} plays the role of the barrier function.

(ii) The estimate (9.12) also follows from Theorem 4.10 and estimate (9.13) is its direct consequence.

(iii) By the Jakubovski criterion (see Theorem C.2), when the compact containment condition (9.13) has been proved, in order to get tightness it is enough to show the tightness of the family of real-valued processes $f(Z_t^h(h\nu))$ (as a family of processes with sample paths in $D(\mathbf{R}_+, \mathbf{R})$ for any f from a dense subset, in the topology of uniform convergence on compact sets, of $C(\mathcal{M}(X))$. By the Weierstrass theorem, it is thus enough to verify the tightness of $f(Z_t^h(h\nu))$ for an f from the algebra generated by F_g with $g \in C_c(X)$. Let us show this for $f = F_g$ (as will be seen from the proof, it is straightforward to generalize this to the sums and products of these functions). By Dynkin's formula, the process

$$M_g^h(t) = F_g(Z_t^h(h\nu)) - F_g(h\nu) - \int_0^t \Lambda^h F_g(Z_s^h(h\nu))\, ds \tag{9.15}$$

is a martingale for any $g \in C_c(X)$. Consequently, by Theorems C.3 and C.9, in order to prove tightness for $F_g(Z_t^h(h\nu))$ one needs to show that

$$V_g^h(t) = \int_0^t \Lambda^h F_g(Z_s^h(h\nu))\, ds$$

and that the quadratic variation $[M_g^h(t)]$ of $M_g^h(t)$ satisfies the Aldous condition, i.e. that, given a sequence $h_n \to 0$ as $n \to \infty$ and a sequence of stopping times τ_n bounded by a constant T and an arbitrary $\epsilon > 0$, there exist $\delta > 0$ and $n_0 > 0$ such that

$$\sup_{n \geq n_0} \sup_{\theta \in [0,\delta]} P\left[\left|V_g^h(\tau_n + \theta) - V_g^h(\tau_n)\right| > \epsilon\right] \leq \epsilon,$$

and

$$\sup_{n \geq n_0} \sup_{\theta \in [0,\delta]} P\left[\left|[M_g^h](\tau_n + \theta) - [M_g^h](\tau_n)\right| > \epsilon\right] \leq \epsilon.$$

For V_g^h this fact is clear. Let us prove it for $[M_g^h(t)]$.

Since the process $Z_t^h(h\nu)$ is a pure jump process,

$$[M_g^h(t)] = \sum_{s \leq t} \left\{\Delta f_g(Z^{h\nu}(s))\right\}^2,$$

where $\Delta Z(s) = Z(s) - Z(s_-)$ denotes the jump of a process $Z(s)$ (see (C.6)). Since the number of particles created at one time is bounded by k, it follows that

$$|\Delta f_g(Z^{h\mu}(s))|^2 \leq 4k^2 \|g\|^2 h^2 \tag{9.16}$$

for any s, so that

$$[M_g^h(t)] - [M_g^h(s)] \leq ch^2(N_t - N_s),$$

where N_t denotes the number of jumps in $[0, t]$. By the Lévy formula for Markov chains (see Exercise 2.1) the process $N_t - \int_0^t q(Z_s^h)\,ds$ is a martingale where $q(Y)$ denotes the intensity of jumps at Y; see (9.14). Hence, as $t - s \leq \theta$,

$$\mathbf{E}(N_t - N_s) \leq \theta\, \mathbf{E} \sup_{s \leq t} q(Z_s^h)\,ds. \tag{9.17}$$

Consequently, by the compact containment condition and Chebyshev's inequality, conditioned to a subset of probability arbitrarily close to 1, $N_t - N_s$ is of order $(t-s)/h$ uniformly for all $s \leq t$ and, consequently, $[M_g^h(t) - M_g^h(s)]$ is uniformly of order $(t - s)h$ with probability arbitrarily close to 1, implying the Aldous condition for $[M_g^h(t)]$. This completes the proof of the theorem in the case (1.34). In the case (1.36), assuming the $(1 + E)^{\oplus}$-boundedness of P one obtains the intensity estimate

$$q(h\nu) \leq \frac{1}{h} \sum_{l=1}^{k} \frac{1}{l!} \left[F_{1+E}(h\nu)\right],$$

which again implies that $q(h\nu)$ is of order h^{-1} and thus completes the proof in the case of the generator given by (1.36).

(iv) Observe that

$$\Lambda^h (F_1)^r (h\nu) = \frac{1}{h}\sum_{l=1}^{k} h^l \sum_{I\subset\{1,\dots,n\},|I|=l} h^r \sum_{m=0}^{\infty} \left[(n+m-l)^r - n^r\right] P(\mathbf{x}_I; dy_1 \dots dy_n),$$

and since the interactions of order $l > 1$ do not increase the number of particles, this does not exceed

$$\sum_{i=1}^{n} h^r \left[(n+k-1)^r - n^r\right] P(x_i; d\mathbf{y}) \le c F_{1+E}(h\nu) h^{r-1}(n+k)^{r-1}$$
$$\le c F_{1+E}(h\nu)\left[F_1(h\nu)\right]^{r-1}.$$

As P is E-non-increasing, $F_E(Z_t^h(h\nu))$ is a.s. uniformly bounded, and using induction in r we can conclude that

$$\mathbf{E}\left[\Lambda^h (F_1)^r (Z_t^h(h\nu))\right] \le c\left[1 + \mathbf{E}(F_1)^r (Z_t^h(h\nu))\right].$$

This implies statement (iv) by, as usual, Gronwall's lemma and Dynkin's martingale applied to $(F_1)^r$. □

We shall turn now to the convergence of many-particle approximations, i.e. to the dynamic LLN, focusing our attention on equation (1.34) and referring to [132] for a similar treatment of the processes defined by (1.36). Notice that in the next theorem no uniqueness is claimed.

We shall say that P is *strongly multiplicatively E-bounded* whenever $P(\mu; \mathbf{x}) = o(1)_{\mathbf{x}\to\infty} E^{\otimes}(\mathbf{x})$, i.e. if

$$P(\mu; x_1, \dots, x_l) = o(1)_{(x_1,\dots,x_l)\to\infty} \prod_{j=1}^{l} E(x_j)$$

for $l = 1, \dots, k$.

Theorem 9.12 *Under the assumptions of Theorem 9.11 (including condition(iv)) suppose that the family of initial measures $h\nu$ converges (weakly) to a measure $\mu \in \mathcal{M}(X)$ and that the kernel P is strongly $(1 + E)^{\otimes}$-bounded. Then there is a subsequence of the family of processes $Z_t^h(h\nu)$ defined by the generator* (1.34) *that weakly converges to a global non-negative solution μ_t of the integral version of the corresponding weak kinetic equation, i.e.*

$$\int g(z)\mu_t(dz) - \int g(z)\mu_0(dz) - \int_0^t ds \int \left[g^{\oplus}(\mathbf{y}) - g^{\oplus}(\mathbf{z})\right] P(\mu, \mathbf{z}, d\mathbf{y}) \mu_s^{\tilde{\otimes}}(d\mathbf{z}) = 0 \quad (9.18)$$

for all t and all $g \in C_c(X)$.

Proof By Theorem 9.11 one can choose a sequence of positive numbers h tending to zero such that the family $Z_t^h(h\nu)$ is weakly converging as $h \to 0$ along this sequence. Let us denote the limit by μ_t and prove that it satisfies (9.18). By Skorohod's theorem, we can assume that $Z_t^h(h\nu)$ converges to μ_t a.s. The idea now is to pass to the limit in equation (9.15), as in the second proof of Theorem 9.7, though additional care is needed to take into account unbounded coefficients. From the estimates of the quadratic variation $[M_g^h](t)$ from Theorem 9.11 it follows that it tends to zero a.s. Moreover, statement (iv) of Theorem 9.11 implies the uniform bound $O(1)\theta/h$ for the r.h.s. of (9.17), and consequently the convergence $\mathbf{E}[M_g^h](t) \to 0$ as $h \to 0$. This implies that the martingale on the l.h.s. of (9.15) tends to zero a.s. and also in L_2. Moreover, owing to Proposition B.3 and the estimate (9.16), any limiting process μ_t has continuous sample paths a.s. The positivity of μ_t is obvious as it is the limit of positive measures.

Clearly, the first two terms on the r.h.s. of (9.15) tend to the first two terms on the l.h.s. of (9.18). So, we need to show that the integral on the r.h.s. of (9.15) tends to the last integral on the l.h.s. of (9.18). Let us show first that a.s.

$$|\Lambda^h f_g(Z_s^h(h\nu)) - \Lambda f_g(Z_s^h(h\nu))| \to 0, \qquad h \to 0, \tag{9.19}$$

uniformly for $s \in t$. Since, for $\eta = \delta_{v_1} + \cdots + \delta_{v_m}$,

$$\Lambda^h F_g(h\eta) = \sum_{l=1}^k h^l \sum_{I \subset \{1,\dots,m\}, |I|=l} \int \left(g^{\oplus}(\mathbf{y}) - \sum_{i \in I} g(v_i) \right) P(h\eta; \mathbf{v}_I; d\mathbf{y})$$

we may conclude by (H.5) that in order to prove (9.19) we need to show that

$$h \int_{X^{k-1} \times \mathcal{X}} P(h\eta, z_1, z_1, z_2, \dots, z_{k-1}; d\mathbf{y}) \prod_{j=1}^{k-1} (h\eta(dz_j)) \to 0, \qquad h \to 0, \tag{9.20}$$

uniformly for all $\eta \in \mathcal{M}_\delta^+(X)$ with uniformly bounded $F_{1+E}(h\eta)$.

Owing to the assumed strong multiplicative boundedness of P, it is enough to show that

$$h \int_{X^{k-1}} \left[1 + E(z_1) \right]^2 \left[1 + E(z_2) \right] \cdots \left[1 + E(z_{k-1}) \right] \times o(1)_{z_1 \to \infty} \prod_{j=1}^{k-1} (h\eta(dz_j)) \to 0, \qquad h \to 0,$$

with uniformly bounded $F_{1+E}(h\eta)$, and consequently that

$$h \int_X [1 + E(z)]^2 o(1)_{z \to \infty} h\eta(dz) \to 0, \qquad h \to 0$$

with uniformly bounded $F_{1+E}(h\eta)$. As the boundedness of $F_E(h\eta)$ implies, in particular, that the support of the η is contained in a set for which E is bounded by c/h with some constant c, it is now enough to show that

$$h \int_{\{z: E(z) \leq c/h\}} [1 + E(z)]^2 \, o(1)_{z\to\infty} h\eta(dz) \to 0, \qquad h \to 0.$$

To this end, let us write the integral here as the sum of two integrals over sets where $E(z) \leq K$ and $E(z) > K$ respectively, for some $K > 0$. Then the first integral clearly tends to zero, as $E(z)$ is bounded. The second does not exceed

$$\int_{z: E(z) \geq K} [1 + E(z)] \, o(1)_{K\to\infty}(h\eta)(dz)$$

and tends to zero as $K \to \infty$.

Once the convergence (9.19) is proved, it remains to show that

$$\int_0^t |\Lambda F_g(Z_s^h(h\nu)) - \Lambda F_g(\mu_s)| \to 0$$

or, more explicitly, that the integral

$$\int_0^t ds \int \left[g^\oplus(\mathbf{y}) - g^\oplus(\mathbf{z})\right] [P(\mu_s; \mathbf{z}, d\mathbf{y})\mu_s^{\tilde{\otimes}}(d\mathbf{z})$$
$$- P(Z_s^h(h\nu), \mathbf{z}, d\mathbf{y})(Z_s^h(h\nu))^{\tilde{\otimes}}(d\mathbf{z})] \qquad (9.21)$$

tends to zero as $h \to 0$ for bounded g. But from weak convergence, and the fact that μ_t has continuous sample paths, it follows (see Proposition B.3(ii)) that $Z_s^h(h\nu)$ converges to μ_s for all $s \in [0, t]$. Hence we need to show that

$$\int_0^t ds \, P(\mu, \mathbf{z}, d\mathbf{y})[\mu_s^{\tilde{\otimes}}(d\mathbf{z}) - (Z_s^h(h\nu))^{\tilde{\otimes}}(d\mathbf{z})]$$
$$+ \int_0^t ds[P(\mu, \mathbf{z}, d\mathbf{y}) - P(Z_s^h(h\nu), \mathbf{z}, d\mathbf{y})](Z_s^h(h\nu))^{\tilde{\otimes}}(d\mathbf{z}) \qquad (9.22)$$

tends to zero as $h \to 0$ under the condition that $Z_s^h(h\nu)$ weakly converges to μ_t. Moreover, as in the proof of (9.20), one can consider $(1 + E, \mu_s)$ and $(1 + L, Z_s^h(h\nu))$ to be uniformly bounded, as this holds a.s. by the compact containment condition.

We now decompose the integrals in (9.22) into the sum of two integrals, by decomposing the domain of integration into the domain $\{\mathbf{z} = (z_1, \ldots, z_m) : \max E(z_j) \geq K\}$ and its complement. By strong $(1 + E)^\otimes$-boundedness, both integrals over the first domain tend to zero as $K \to \infty$. On the second domain the integrand is uniformly bounded and hence the weak convergence of $Z_s^h(h\nu)$ to μ_t ensures the smallness of the l.h.s. of (9.22). Theorem 9.12 is proved. □

Let us return now to our main focus, the case of additively bounded kernels, concentrating for simplicity on the case of binary interactions ($k \leq 2$) (see [132] for k-ary interactions).

Theorem 9.13 *Under assumptions of Theorem 9.11 suppose additionally that the family of initial measures $h\nu$ converges (weakly) to a measure $\mu \in \mathcal{M}(X)$ and that these initial conditions $h\nu$ are such that $\int (E^\beta)(x)h\nu(dx) \leq C$ for all $h\nu$ and some $C > 0$, $\beta > 1$, the transition kernel P is $(1+E)^\oplus$-bounded and E-non-increasing and $k \leq 2$.*

Then there is a subsequence of the family of processes $Z_t^h(h\nu)$ defined by (1.34) *that weakly converges to a global non-negative solution μ_t of* (9.18) *such that*

$$\sup_{t\in[0,T]} \int (\mathbf{1} + E^\beta)(x)\mu_t(dx) \leq C(T), \tag{9.23}$$

for some constant $C(T)$ and for arbitrary T.

Moreover, if P is not only E-non-increasing but also E-preserving then the obtained solution μ_t is also E-preserving, i.e. for all t

$$\int E(x)\mu_t(dx) = F_E(\mu_t) = F_E(\mu_0) = \int E(x)\mu_0(dx).$$

Proof Let us first show (9.23). To this end, let us prove that

$$\mathbf{E}F_{E^\beta}(Z_s^h(h\nu)) \leq c\left[1 + F_{E^\beta}(h\nu)\right] \tag{9.24}$$

uniformly for all $s \in [0, t]$ with arbitrary t and some constant c depending on t and β but not on h. As the process is E-non-increasing, the process $Z_s^h(h\nu)$ lives on measures with support on a compact space $\{x \in X : E(x) \leq c/h\}$ for some constant c. Hence $F_{E^\beta}(Z_s^h(h\nu))$ is uniformly bounded (but not necessarily uniform in h), and one can apply Dynkin's formula for $g = E^\beta$. Hence

$$\mathbf{E}F_{E^\beta}(Z_s^h(h\nu)) \leq F_{E^\beta}(h\nu) + \int_0^t \mathbf{E}\Lambda^h(F_{E^\beta}(Z_s^h(h\nu)))\,ds. \tag{9.25}$$

From (6.21) we know that

$$\Lambda F_{E^\beta}(\mu) \leq c\int\int \left[E^\beta(x_1) + E^{\beta-1}(x_1)\right] E(x_2)\mu(dx_1)\mu(dx_2)$$

(here we assume that $k \leq 2$). Consequently, since P is E-non-increasing, so that $(E, h\nu_t) \leq (E, h\nu)$, one concludes using induction in β that

$$\mathbf{E}\Lambda^h F_{E^\beta}(h\nu_t) \leq c\left[1 + \mathbf{E}F_{E^\beta}(h\nu_t)\right],$$

implying (9.24), by Gronwall's lemma, and consequently (9.23).

Now following the same strategy as in the proof of Theorem 9.12, we need to show the convergence (9.19). But this is implied by (9.20), which is obvious for $(1+E)^{\oplus}$-bounded kernels. It remains to show (9.21). Again one decomposes the domain of integration into the domain $\{\mathbf{z} = (z_1, \ldots, z_m) : \max E(z_j) \geq K\}$ and its complement. The second part is dealt with precisely as in Theorem 9.12. Hence we need only to show that, for any l, by choosing K arbitrarily large one can make

$$\int_{L(z_1)\geq K} [1+E(z_1)] \left(\prod_{j=1}^{2} \mu_s(dz_j) + \prod_{j=1}^{2} (Z_s^h(h\nu))(dz_j) \right)$$

arbitrary small. But this integral does not exceed

$$K^{-(\beta-1)} \int [1+E(z_1)][E(z_1)]^{\beta-1} \left(\prod_{j=1}^{2} \mu_s(dz_j) + \prod_{j=1}^{2} (Z_s^h(h\nu))(dz_j) \right),$$

which is finite and tends to zero as $K \to \infty$, in view of (9.23).

It remains to show that μ_t is E-preserving whenever the transition kernel $P(\mathbf{x}; d\mathbf{y})$ is E-preserving. As the approximations $Z_s^h(h\nu)$ are then E-preserving, we need to show only that

$$\lim_{h\to 0} \int E(x)\big[\mu_t - Z_t^h(h\nu)\big](dx) = 0.$$

This is done as above. Decomposing the domain of integration into two parts, $\{x : E(x) \leq K\}$ and its complement, we observe that the required limit for the first integral is zero owing to the weak convergence of $Z_s^h(h\nu)$ to μ_t, and in the second part the integral can be made arbitrarily small by choosing K large enough. □

As a consequence, we may obtain now a version of the *propagation of chaos* property for approximating interacting-particle systems. In general, this property means that the moment measures for some random measures tend to their product measures when passing to a certain limit. The moment measures μ_t^m of the jump processes $Z_t^h(h\nu)$ (see Section 1.9) are defined as

$$\mu_{t,h}^m(dx_1 \cdots dx_m) = \mathbf{E}\left[Z_t^h(h\nu)(dx_1) \cdots Z_t^h(h\nu)(dx_m) \right].$$

Theorem 9.14 *Under the conditions of Theorem 9.13 suppose that P is $(1+E^\alpha)^{\oplus}$-bounded for some $\alpha \in [0,1]$, where $\beta \geq \alpha + 1$, and that* (6.25) *holds. Let the family of initial measures $h\nu = h\nu(h)$ converge weakly to a certain measure μ_0 as $h \to 0$. Then, for any $m = 1, 2, \ldots,$ the moment measures $\mu_{t,h}^m$ converge weakly to the product measure $\mu_t^{\otimes m}$.*

Proof By Theorems 9.13 and 6.12, for any $g \in C_c(SX^m)$ the random variables

$$\eta_h = \int g(x_1, \dots, x_m) Z_t^h(h\nu)(dx_1) \cdots Z_t^h(h\nu)(dx_m)$$

converge a.s. to the integrals $\int g(x_1, \dots, x_m) \prod_{j=1}^m \mu_t(dx_j)$. To conclude that the expectations of the random variables η_h converge to a pointwise deterministic limit, one only need show that the variances of the η_h are uniformly bounded. Consequently, to complete the proof one needs to show that $E(\|Z_t^h(h\nu)\|^r)$ are bounded, for any positive integer r, uniformly for $t \in [0, T]$ with arbitrary $T > 0$. But this was done in Theorem 9.11(iv). □

9.4 Rates of convergence for Smoluchovski coagulation

Theorem 9.15 *Under the assumptions of Theorem 8.13 let g be a continuous symmetric function on X^m and let $F(Y) = (g, Y^{\otimes m})$; assume that $Y = h\delta_{\mathbf{x}}$ belongs to $\mathcal{M}_{h\delta}^{e_0, e_1}$, where $\mathbf{x} = (x_1, \dots, x_n)$. Then*

$$\sup_{s \le t} |T_t^h F(Y) - T_t F(Y)| \le hc(m, k, t, e_0, e_1) \|g\|_{(1+E^k)^{\otimes m}} (1 + E^{2k+1/2}, Y)(1 + E^k, Y)^{m-1} \tag{9.26}$$

for any $2k \in [1, \beta - 1]$.

Remark 9.16 Theorem 9.15 gives the hierarchy of estimates for the error term, making precise an intuitively clear fact that the power of growth of the polynomial functions on measures for which the LLN can be established depends on the order of the finite moments of the initial measure. One can extend the estimate (9.26) to functionals F that are not necessarily polynomials.

Proof For brevity we shall write $Y_t = \mu_t(Y)$, so that $T_t F(Y) = F(\mu_t(Y)) = F(Y_t)$. For a function $F(Y) = (g, Y^{\otimes m})$ with $g \in C^{\mathrm{sym}}_{(1+E)^{\otimes m}, \infty}(X^m)$, $m \ge 1$ and $Y = h\delta_{\mathbf{x}}$ one has

$$T_t F(Y) - T_t^h F(Y) = \int_0^t T_{t-s}^h (\Lambda_h^2 - \Lambda_2) T_s F(Y)\, ds. \tag{9.27}$$

As $T_t F(Y) = (g, Y_t^{\otimes m})$, Theorem 8.8 yields

$$\frac{\delta T_t F(Y)}{\delta Y(x)} = m \int_{X^m} g(y_1, y_2, \dots, y_m) \xi_t(Y; x; dy_1) Y_t^{\otimes (m-1)}(dy_2 \dots dy_m)$$

and

$$\frac{\delta^2 T_t F(Y)}{\delta Y(x)\delta Y(w)} = m\int_{X^m} g(y_1, y_2, \dots, y_m) \times \eta_t(Y; x, w; dy_1) Y_t^{\otimes(m-1)}(dy_2 \dots dy_m) + m(m-1)\int_{X^m} g(y_1, y_2, \dots, y_m)\xi_t(Y; x; dy_1) \times \xi_t(Y; w; dy_2) Y_t^{\otimes(m-2)}(dy_3 \dots dy_m). \tag{9.28}$$

Let us estimate the difference $(\Lambda_2^h - \Lambda_2)T_t F(Y)$ using (9.4), i.e.

$$\Lambda_2^h F(Y) - \Lambda_2 F(Y) = -\frac{h}{2}\int_X\int_X \left(\frac{\delta F(Y)}{\delta Y(y)} - 2\frac{\delta F(Y)}{\delta Y(z)}\right) P(z, z, dy) Y(dz) + h^3 \sum_{\{i,j\}\subset\{1,\dots,n\}} \int_0^1 (1-s)\,ds \int_X P(x_i, x_j, dy) \times \left(\frac{\delta^2 F}{\delta Y(.)\delta Y(.)}(Y + sh(\delta_y - \delta_{x_i} - \delta_{x_j})), (\delta_y - \delta_{x_i} - \delta_{x_j})^{\otimes 2}\right). \tag{9.29}$$

Let us analyze only the rather strange-looking second term, as the first is analyzed similarly but much more simply. The difficulty lies in the need to assess the second variational derivative at a shifted measure Y. To this end the estimates of Section 8.4 are needed. We will estimate separately the contributions to the second term in (9.29) of the first and second terms of (9.28).

Notice that the norm and the first moment $(E, \cdot)$ of $Y + sh(\delta_y - \delta_{x_i} - \delta_{x_j})$ do not exceed respectively the norm and the first moment of Y. Moreover, for $s \in [0, 1]$, $h > 0$ and $x_i, x_j, y \in X$ with $E(y) = E(x_i) + E(x_j)$, one has, using Exercise 6.1,

$$\begin{aligned}(E^k, Y + sh(\delta_y - \delta_{x_i} - \delta_{x_j})) &= (E^k, Y) + sh\left[E(x_i) + E(x_j)\right]^k - hE^k(x_i) - hE^k(x_j) \\ &\le (E^k, Y) + hc(k)\left[E^{k-1}(x_i)E(x_j) + E(x_i)E^{k-1}(x_j)\right];\end{aligned}$$

$c(k)$ depends only on k. Consequently, by Theorem 8.13, one has

$$\begin{aligned}\|\eta_t(Y + sh(\delta_y - \delta_{x_i} - \delta_{x_j}); x, w; \cdot)\|_{1+E^k} \le c(k, t, e_0, e_1)\Big\{1 + E^{k+1}(x) + E^{k+1}(w) + (E^{k+2}, Y) \\ + hc\left[E^{k+1}(x_i)E(x_j) + E^{k+1}(x_j)E(x_i)\right]\Big\},\end{aligned}$$

Thus the contribution of the first term in (9.28) to the second term in (9.29) does not exceed

$$c(t,k,m,e_0,e_1)\|g\|_{(1+E^k)^{\otimes m}}$$
$$\times h^3\sum_{i\neq j}\int (1+E^k, Y+sh(\delta_y-\delta_{x_i}-\delta_{x_j}))^{m-1}P(x_i,x_j,dy)$$
$$\times \|\eta_t(Y+sh(\delta_y-\delta_{x_i}-\delta_{x_j});x,w;\cdot)\|_{1+E^k}$$
$$\leq c(t,k,m,e_0,e_1)\|g\|_{(1+E^k)^{\otimes m}}h^3$$
$$\times\sum_{i\neq j}\left\{(1+E^k,Y)+h\left[E^k(x_i)+E^k(x_j)\right]\right\}^{m-1}\left[1+E(x_i)+E(x_j)\right]$$
$$\times\left\{1+E^{k+1}(x_i)+E^{k+1}(x_j)+(E^{k+2},Y)\right.$$
$$\left.+h\left[E^{k+1}(x_i)E(x_j)+E^{k+1}(x_j)E(x_i)\right]\right\}.$$

Dividing this sum into two parts, in the first of which $E(x_i)\geq E(x_j)$ and in the second of which $E(x_i)<E(x_j)$, and noting that by symmetry it is enough to estimate only the first part, we can estimate the last expression as

$$c(t,k,m,e_0,e_1)\|g\|_{(1+E^k)^{\otimes m}}h^3\sum_{i\neq j}\left\{(1+E^k,Y)^{(m-1)}+\left[hE^k(x_i)\right]^{m-1}\right\}$$
$$\times\left\{1+E^{k+2}(x_i)+(E^{k+2},Y)\left[1+E(x_i)\right]+hE^{k+2}(x_i)E(x_j)\right\}.$$

Consequently, taking into account that $h\sum_i 1\leq(1,Y)\leq e_0$ and that $\sum_i a_ib_i\leq\sum_i a_i\sum_j b_j$ for any collection of positive numbers a_i, b_i, so that, say,

$$\sum_i\left[hE^k(x_i)\right]^{m-1}\leq\left(\sum_i hE^k(x_i)\right)^{m-1},$$

we conclude that the contribution of the first term in (9.28) to the second term in (9.29) does not exceed

$$h\kappa(C,t,k,m,e_0,e_1)\|g\|_{(1+E^k)^{\otimes m}}$$
$$\times(1+E^k,Y)^{m-1}(1+E^{k+2},Y). \tag{9.30}$$

Turning to the contribution of the second term in (9.28) observe that, again by Theorem 8.13,

$$\begin{aligned}
&\|\xi_t(Y + sh(\delta_y - \delta_{x_i} - \delta_{x_j}); x; \cdot)\|_{1+E^k} \\
&\quad \le c(k, t, e_0, e_1) \\
&\qquad \Big(1 + E^k(x) + \Big[1 + E^{1/2}(x)\Big] \\
&\qquad\quad \times \Big\{(E^{k+1/2}, Y) + hc(k)\Big[E^{k-1/2}(x_i)E^{1/2}(x_j) + E^{k-1/2}(x_j)E^{1/2}(x_i)\Big]\Big\}\Big),
\end{aligned}$$

so that the contribution of the second term in (9.28) does not exceed (where again we take symmetry into account)

$$\begin{aligned}
&c(t, k, m, e_0, e_1)\|g\|_{(1+E^k)^{\otimes m}}(1 + E^k, Y)^{m-2} \\
&\quad \times h^3 \sum_{i \ne j} \Big[1 + E^{1/2}(x_i)\Big]\Big[1 + E^{1/2}(x_j)\Big] \\
&\times \Big\{1 + E^k(x_i) + \Big[1 + E^{1/2}(x_i)\Big]\Big[(E^{k+1/2}, Y) + hE^{k-1/2}(x_i)E^{1/2}(x_j)\Big]\Big\}^2.
\end{aligned}$$

This is estimated by the r.h.s. of (9.26), completing the proof. □

Our second result in this section deals with the case of smooth rates. If f is a positive function on $X^m = \mathbf{R}_+^m$, we denote by $C_f^{1,\mathrm{sym}}(X^m)$ (resp. $C_f^{2,\mathrm{sym}}(X^m)$) the space of symmetric continuous differentiable (resp. twice continuously differentiable) functions g on X^m vanishing whenever at least one argument vanishes, with norm

$$\|g\|_{C_f^{1,\mathrm{sym}}(X^m)} = \left\| \frac{\partial g}{\partial x_1} \right\|_{C_f(X^m)} = \sup_{x,j} \left(\left| \frac{\partial g}{\partial x_j} \right| (f^{-1}) \right)(x)$$

(resp. $\|g\|_{C_f^{2,\mathrm{sym}}(X^m)} = \|\partial g/\partial x_1\|_{C_f(X^m)} + \|\partial^2 g/\partial x_1^2\|_{C_f(X^m)} + \|\partial^2 g/\partial x_1 \partial x_2\|_{C_f(X^m)}$).

Theorem 9.17 *Let the assumptions of Theorem 8.14 hold, i.e.* $X = \mathbf{R}_+$, $K(x_1, x_2, dy) = K(x_1, x_2)\delta(y - x_1 - x_2)$, $E(x) = x$ *and* K *is non-decreasing in each argument function on* $(\mathbf{R}_+)^2$ *that is twice continuously differentiable up to the boundary with bounded first and second derivatives. Let* g *be a twice continuously differentiable symmetric function on* $(\mathbf{R}_+)^m$ *and let* $F(Y) = (g, Y^{\otimes m})$*; assume that* $Y = h\delta_{\mathbf{x}}$ *belongs to* $\mathcal{M}_{h\delta}^{e_0, e_1}$, *where* $\mathbf{x} = (x_1, \dots, x_n)$. *Then*

$$\sup_{s\le t}|T_t^h F(Y) - T_t F(Y)|$$
$$\le hc(m,k,t,e_0,e_1)\|g\|_{C^{2,\mathrm{sym}}_{(1+E^k)^{\otimes m}}(X^m)}(1+E^{2k+3},Y)(1+E^k,Y)^{m-1} \tag{9.31}$$

for any $k \in [0,(\beta-3)/2]$.

Proof This is similar to the proof of Theorem 9.16. Let us indicate the basic technical steps. We again use (9.27), (9.28), (9.29) and assess only the contribution of the second term in (9.29). From our experience with the previous proof, the worst bound comes from the second term in (9.28), so let us concentrate on this term only. Finally, to shorten the formulas let us take $m=2$. Thus we would like to show that

$$h^3 \sum_{\{i,j\}\subset\{1,\dots,n\}} \int_0^1 (1-s)\,ds \int_X P(x_i,x_j,dy) \int_{X^2} g(y_1,y_2)$$
$$\times \Big(\xi_t(Y+sh(\delta_y-\delta_{x_i}-\delta_{x_j}),.,dy_1)$$
$$\times\, \xi_t(Y+sh(\delta_y-\delta_{x_i}-\delta_{x_j}),.,dy_2),\ (\delta_y-\delta_{x_i}-\delta_{x_j})^{\otimes 2}\Big) \tag{9.32}$$

is bounded by the r.h.s. of (9.31) with $m=2$.

Notice that, by the definition of the norm in $C_f^{2,\mathrm{sym}}(X^m)$, one has

$$\left|\int_{X^2} g(y_1,y_2)\xi_t(Y,x,dy_1)\xi_t(Y,z,dy_2)\right|$$
$$\le c\|\xi_t(Y,x)\|_{\mathcal{M}^1_{1+E^k}}\|\xi_t(Y,z)\|_{\mathcal{M}^1_{1+E^k}}\|g\|_{C_f^{2,\mathrm{sym}}(X^2)}.$$

Consequently the expression (9.32) does not exceed

$$c\|g\|_{C_f^{2,\mathrm{sym}}(X^2)}h^3 \sum_{\{i,j\}\subset\{1,\dots,n\}} \int_0^1 (1-s)\,ds \int_X P(x_i,x_j,dy)$$
$$\times \Big(\|\xi_t(Y+sh(\delta_y-\delta_{x_i}-\delta_{x_j}),.)\|_{\mathcal{M}^1_{1+E^k}}$$
$$\times\, \|\xi_t(Y+sh(\delta_y-\delta_{x_i}-\delta_{x_j}),.)\|_{\mathcal{M}^1_{1+E^k}},\ (\delta_y-\delta_{x_i}-\delta_{x_j})^{\otimes 2}\Big),$$

which, by Theorem 8.14, is bounded by

$$c\|g\|_{C_f^{2,\mathrm{sym}}(X^2)}h^3 \sum_{\{i,j\}\subset\{1,\dots,n\}} \int_0^1 (1-s)\,ds \int_X P(x_i,x_j,dy)$$
$$\times \Big(\{E(.)[1+(E^{k+1},Y+sh(\delta_y-\delta_{x_i}-\delta_{x_j}))]+E^{k+1}(.)\}$$
$$\times \{E(.)[1+(E^{k+1},Y+sh(\delta_y-\delta_{x_i}-\delta_{x_j}))]+E^{k+1}(.)\},\ (\delta_y-\delta_{x_i}-\delta_{x_j})^{\otimes 2}\Big).$$

Again by symmetry, choosing $E(x_i) \geq E(x_j)$ (without loss of generality) allows us to estimate this expression as

$$c\|g\|_{C_f^{2,\mathrm{sym}}(X^2)} h^3 \sum_{\{i,j\}\subset\{1,\dots,n\}} [1+E(x_i)]$$

$$\times \left\{E(x_i)\left[1+(E^{k+1},Y)+hE^k(x_i)E(x_j)\right]+E^{k+1}(x_i)\right\}^2$$

$$\leq hc(k,m,t,s_0,s_1)\|g\|_{C_f^{2,\mathrm{sym}}(X^2)}(1+E^{2k+3},Y),$$

as required. □

9.5 Rates of convergence for Boltzmann collisions

Theorem 9.18 *Under the assumptions of Theorem 8.20 let* g *be a continuous symmetric function on* $(\mathbf{R}^d)^m$ *and let* $F(Y) = (g, Y^{\otimes m})$*; assume that* $Y = h\delta_{\mathbf{v}}$ *belongs to* $\mathcal{M}_{h\delta}^{e_0,e_1}$*, where* $\mathbf{v} = (v_1,\dots,v_n)$*. Then*

$$\sup_{s\leq t}|T_t^h F(Y) - T_t F(Y)| \leq hc(m,k,t,e_0,e_1)\|g\|_{(1+E^k)^{\otimes m}}\Omega(Y) \tag{9.33}$$

for any $k \geq 1$*, where* $\Omega(Y)$ *is a polynomial of the moments* $(1+E^l,Y)$*, with* $l \leq 2k+1/2$*, of degree depending on the mass and energy of* Y *(this polynomial can be calculated explicitly from the proof below).*

Proof The proof is basically the same as that of Theorem 9.15 (though based on the estimates of Theorem 8.20), and we shall merely sketch it. One again uses (9.27), (9.28) and (9.4), the latter taking the form (since equal velocities cannot collide)

$$\Lambda_2^h F(Y) - \Lambda_2 F(Y)$$

$$= h^3 \sum_{\{i,j\}\subset\{1,\dots,n\}} \int_0^1 (1-s)\,ds \int_{S^{d-1}} |v_i - v_j|^\beta b(\theta)dn$$

$$\times \left(\frac{\delta^2 F}{\delta Y(.)\delta Y(.)}(Y+sh(\delta_{v_i'}+\delta_{v_j'}-\delta_{v_i}-\delta_{v_j})), (\delta_{v_i'}+\delta_{v_j'}-\delta_{v_i}-\delta_{v_j})^{\otimes 2}\right). \tag{9.34}$$

We shall estimate only the contribution to (9.34) of the second term in (9.28) (the contribution of the first term is estimated similarly). By Theorem 8.20,

$$\|\xi_t(Y+sh(\delta_{v_i'}+\delta_{v_j'}-\delta_{v_i}-\delta_{v_j});v)\|_{1+E^k}$$

$$\leq c(k,t,e_0,e_1)\left[1+E^k(v)\right]$$

$$\times \Big(1 + \{\|Y\|_{2k+\beta} + h[(2 + |v_i|^2 + |v_j|^2)^{k+\beta/2} - (1 + |v_i|^2)^{k+\beta/2} - (1 + |v_j|^2)^{k+\beta/2}]\} \times \{\|Y\|_{2+\beta} + h[(2 + |v_i|^2 + |v_j|^2)^{1+\beta/2} - (1 + |v_i|^2)^{1+\beta/2} - (1 + |v_j|^2)^{1+\beta/2}]\}^{\omega}\Big),$$

which does not exceed

$$c(k, t, e_0, e_1)\Big[1 + E^k(v)\Big] \times \Big(1 + \Big[\|Y\|_{2k+\beta} + h(1 + |v_i|^2)^{k-1+\beta/2}(1 + |v_j|^2) + h(1 + |v_j|^2)^{k-1+\beta/2}(1 + |v_i|^2)\Big] \times \Big\{\|Y\|_{2+\beta} + h[(1 + |v_i|^2)^{\beta/2}(1 + |v_j|^2) + (1 + |v_j|^2)^{\beta/2}(1 + |v_i|^2)]\Big\}^{\omega}\Big),$$

so that, again taking symmetry into account, as in the previous section, the contribution to (9.34) of the second term in (9.28) does not exceed

$$c(t, k, m, e_0, e_1)\|g\|_{(1+E^k)^{\otimes m}}(1 + E^k, Y)^{m-2}h^3 \sum_{i \neq j}(1 + |v_i|^\beta) \times \Big(1 + E^k(v_i) + \Big\{\|Y\|_{2k+\beta} + h\Big[(1 + |v_i|^2)^{k-1+\beta/2}(1 + |v_j|^2) + (1 + |v_j|^2)^{k-1+\beta/2}(1 + |v_i|^2)\Big]\Big\} \times \Big[\|Y\|_{2+\beta} + h(1 + |v_i|^2)(1 + |v_j|^2)^{\beta/2}\Big]^{\omega}\Big)^2,$$

and (9.33) follows as in Theorem 9.15. □

10

The dynamic central limit theorem

Our program of interpreting a nonlinear Markov process as the LLN limit of an approximating Markov interacting-particle system was fulfilled in Chapter 9 for a wide class of interactions. In this chapter we address the natural next step in the analysis of approximating systems of interacting particles. Namely, we deal with processes involving fluctuations around the dynamic LLN limit. The objective is to show that in many cases the limiting behavior of a fluctuation process is described by an infinite-dimensional Gaussian process of Ornstein–Uhlenbeck type. This statement can be called a *dynamic central limit theorem (CLT)*. As in Chapter 9 we start with a formal calculation of the generator for the fluctuation process in order to be able to compare it with the limiting second-order Ornstein–Uhlenbeck generator. Then we deduce a weak form of the CLT, though with precise convergence rates. Finally we sketch the proof of the full result (i.e. the convergence of fluctuation processes in a certain Skorohod space of càdlàg paths with values in weighted Sobolev spaces) for a basic coagulation model, referring for details to the original paper.

10.1 Generators for fluctuation processes

In this section we calculate generators for fluctuation processes of approximating Markov interacting-particle systems around their LLNs, which are given by solutions to kinetic equations. Here we undertake only general, formal, calculations without paying much attention to the precise conditions under which the various manipulations actually make sense. We postpone to later sections justifying the validity of these calculations for concrete models in various strong or weak topologies under differing assumptions. The calculations are lengthy but straightforward.

Suppose that S is a closed subset of a linear topological space Y and that Z_t is a Markov process on S specified by its C-Feller Markov semigroup Ψ_t

on $C(S)$. Assume that D is an invariant subspace of $C(S)$ and A is a linear operator taking D to $C(S)$ such that $\dot{\Psi}_t f(x) = A\Psi_t f(x)$ for any $f \in D$ and for each $x \in S$. Let $\Omega_t(z) = (z - \xi_t)/a$ be a family of linear transformations on Y, where a is a positive constant and ξ_t, $t \geq 0$, is a differentiable curve in Y. The transformation $Y_t = \Omega_t(Z_t^x)$ is again a Markov process, though in a slightly generalized sense: not only is it time nonhomogeneous but also its state space $\Omega_t(S)$ is time dependent. In many situations of interest the sets $\Omega_t(S)$ could be pairwise disjoint for different t. The corresponding averaging operators

$$U^{s,t} f(y) = \mathbf{E}(f(Y_t)|Y_s = y), \qquad s \leq t,$$

form a backward Markov propagator, each $U^{s,t}$ being a conservative contraction taking $C(\Omega_t(S))$ to $C(\Omega_s(S))$ that can be expressed in terms of Φ_t as

$$U^{s,t} f(y) = \mathbf{E}\left(f(\Omega_t(Z_t^x))|\Omega_s(Z_s^x) = y\right) = \Psi_{t-s}[f \circ \Omega_t](\Omega_s^{-1}(y)).$$

Lifting the transformations Ω_t to operators on functions, i.e. writing

$$\tilde{\Omega}_t f(y) = (f \circ \Omega_t)(y) = f(\Omega_t(y)),$$

we obtain the operator equation

$$U^{s,t} f = \tilde{\Omega}_s^{-1} \Psi_{t-s} \tilde{\Omega}_t f, \qquad f \in C(\Omega_t(S)),\ s \leq t. \tag{10.1}$$

Differentiating by the chain rule (assuming that all derivatives are defined and sufficiently regular) yields

$$\frac{d}{dt}\tilde{\Omega}_t f(y) = \frac{d}{dt} f(\Omega_t(y)) = -\frac{1}{a}(D_{\dot{\xi}_t} f)(\Omega_t(y)) = -\frac{1}{a}\tilde{\Omega}_t D_{\dot{\xi}_t} f(y),$$

where D_η denotes the Gateaux derivative in the direction η. Consequently,

$$\frac{d}{dt} U^{s,t} f = \tilde{\Omega}_s^{-1} \Psi_{t-s} A \tilde{\Omega}_t f - \frac{1}{a}\tilde{\Omega}_s^{-1} \Psi_{t-s} \tilde{\Omega}_t D_{\dot{\xi}_t} f,$$

implying that

$$\frac{d}{dt} U^{s,t} f = U^{s,t} L_t f, \qquad s \leq t \tag{10.2}$$

where

$$L_t f = \tilde{\Omega}_t^{-1} A \tilde{\Omega}_t f - \frac{1}{a} D_{\dot{\xi}_t} f. \tag{10.3}$$

Similarly, since

$$\frac{d}{ds} U^{s,t} f = -\tilde{\Omega}_s^{-1} A \Psi_{t-s} \tilde{\Omega}_t f + \tilde{\Omega}_s^{-1} D_{\dot{\xi}_s} \Psi_{t-s} \tilde{\Omega}_t f,$$

using the identity

$$\tilde{\Omega}_s^{-1} D_{\dot{\xi}_s} \tilde{\Omega}_s = a^{-1} D_{\dot{\xi}_s}$$

we obtain

$$\frac{d}{ds} U^{s,t} f = -L_s U^{s,t} f, \qquad s \leq t. \tag{10.4}$$

We are going to apply these formulas to the backward propagator

$$U_{fl}^{h;s,r} : C(\Omega_r^h(\mathcal{M}_{h\delta}^+)) \mapsto C(\Omega_s^h(\mathcal{M}_{h\delta}^+))$$

of the fluctuation process F_t^h obtained from Z_t^h by the deterministic linear transformation $\Omega_t^h(Y) = h^{-1/2}(Y - \mu_t)$. According to (10.1) this propagator is given by the formula

$$U_{fl}^{h;s,r} F = (\tilde{\Omega}_s^h)^{-1} T_{r-s}^h \tilde{\Omega}_r^h F, \tag{10.5}$$

where $\tilde{\Omega}_t^h F(Y) = F(\Omega_t^h Y)$, and under appropriate regularity assumptions it satisfies the equation

$$\frac{d}{dt} U_{fl}^{h;s,t} F = U_{fl}^{h;s,t} O_t^h F, \qquad s < t < T, \tag{10.6}$$

where

$$O_t^h \psi = (\tilde{\Omega}_t^h)^{-1} \Lambda_h \tilde{\Omega}_t^h \psi - h^{-1/2} \left(\frac{\delta \psi}{\delta Y}, \dot{\mu}_t \right). \tag{10.7}$$

The aim of this section is to calculate expression (10.7) explicitly in terms of the variational derivatives of F, identifying both the main and the error terms in its asymptotic expansion in small h.

Let us start with the general mean field approximation.

Proposition 10.1 *Assume that the approximating Markov process on $\mathcal{M}_{h\delta}^+(\mathbf{R}^d)$ is specified by the generator Λ_h^1 from (9.2), with B_μ^1 of Lévy–Khintchine form:*

$$\begin{aligned} B_\mu^1 f(x) &= \tfrac{1}{2}(G_\mu(x)\nabla, \nabla) f(x) + (b_\mu(x), \nabla f(x)) \\ &\quad + \int \left[f(x+y) - f(x) - (\nabla f(x), y)\mathbf{1}_{B_1}(y) \right] \nu_\mu(x, dy), \end{aligned} \tag{10.8}$$

so that μ_t solves the corresponding mean field kinetic equation

$$\frac{d}{dt}(g, \mu_t) = (B_{\mu_t}^1 g, \mu_t).$$

Then the operator O_t^h from (10.6) *describing the fluctuation process is written down as*

$$O_t^h F = O_t F + O(\sqrt{h}) \tag{10.9}$$

where

$$O_t F(Y) = \left(\frac{\delta}{\delta \mu_t} \left(B^1_{\mu_t} \frac{\delta F}{\delta Y}, \mu_t \right), Y \right) \\ + \int_{\mathbf{R}^d} B^1_{\mu_t} \left(\frac{1}{2} \frac{\delta^2 F}{\delta Y^2(.)} - \frac{\delta^2 F}{\delta Y(y) \delta Y(.)} \right) (x) \bigg|_{y=x} \mu(dx). \quad (10.10)$$

Here

$$\left(\frac{\delta}{\delta \mu_t} \left(B^1_{\mu_t} \frac{\delta F}{\delta Y}, \mu_t \right), Y \right) \\ = \int \left(B^1_{\mu_t} \frac{\delta F}{\delta Y(.)} \right) (x) Y(dx) \\ + \int_{X^2} \mu_t(dx) Y(dz) \left\{ \frac{1}{2} \left(\frac{\delta G_{\mu_t}(x)}{\delta \mu_t(z)} \frac{\partial}{\partial x}, \frac{\partial}{\partial x} \right) \frac{\delta F(Y)}{\delta Y(x)} + \frac{\delta b_{\mu_t}(x)}{\delta \mu_t(z)} \frac{\partial}{\partial x} \frac{\delta F(Y)}{\delta Y(x)} \right. \\ \left. + \int \left[\frac{\delta F(Y)}{\delta Y(x+y)} - \frac{\delta F(Y)}{\delta Y(x)} - \left(\frac{\partial}{\partial x} \frac{\delta F(Y)}{\delta Y(x)}, y \right) \mathbf{1}_{B_1}(y) \right] \frac{\delta \nu_{\mu_t}}{\delta \mu_t(z)}(x, dy) \right\}$$

and

$$\int_{\mathbf{R}^d} B^1_{\mu_t} \left(\frac{1}{2} \frac{\delta^2 F}{\delta Y^2(.)} - \frac{\delta^2 F}{\delta Y(y) \delta Y(.)} \right) (x) \bigg|_{y=x} \mu(dx) \\ = \frac{1}{2} \int \left(G_{\mu_t}(x) \frac{\partial}{\partial x}, \frac{\partial}{\partial y} \right) \frac{\delta^2 F}{\delta Y(x) \delta Y(y)} \bigg|_{y=x} \mu_t(dx) \\ + \frac{1}{2} \iint \left(\frac{\delta^2 F}{\delta Y^2(x+y)} - 2 \frac{\delta^2 F}{\delta Y(x+y) \delta Y(x)} + \frac{\delta^2 F}{\delta Y^2(x)} \right) \nu_{\mu_t}(x, dy) \mu_t(dx). \quad (10.11)$$

The error term $O(\sqrt{h})$ *in* (10.9) *equals*

$$(O^h_t F - O_t F)(Y) \\ = \sqrt{h} \int_0^1 ds \left(\frac{\delta}{\delta \mu_t(.)} B^1_{\mu_t + s\sqrt{h}Y}, Y \right) \frac{\delta F}{\delta Y}(x) Y(dx) \\ + \sqrt{h} \int_0^1 (1-s) ds \left(\frac{\delta^2}{\delta \mu_t^2(.)} B^1_{\mu_t + s\sqrt{h}Y}, Y^{\otimes 2} \right) \frac{\delta F}{\delta Y}(x) \mu_t(dx)) \\ + \frac{\sqrt{h}}{2} \int \left(G_{\mu_t + \sqrt{h}Y}(x) \frac{\partial}{\partial x}, \frac{\partial}{\partial y} \right) \frac{\delta^2 F}{\delta Y(y) \delta Y(x)} \bigg|_{y=x} Y(dx) \\ + \frac{\sqrt{h}}{2} \iint_0^1 ds \left(\left(\frac{\delta G_{\mu_t + s\sqrt{h}Y}(x)}{\delta \mu_t(.)}, Y \right) \frac{\partial}{\partial x}, \frac{\partial}{\partial y} \right) \frac{\delta^2 F}{\delta Y(y) \delta Y(x)} \bigg|_{y=x} \mu_t(dx)$$

$$+\int_0^1 (1-s)ds \int \left(\frac{\delta^2 F}{\delta Y^2(x+y)} \right.$$
$$\left. -2\frac{\delta^2 F}{\delta Y(x+y)\delta Y(x)} + \frac{\delta^2 F}{\delta Y^2(x)} \right) (Y + sh(\delta_{x+y} - \delta_x))$$
$$\times \left[\nu_{\mu_t + \sqrt{h}Y}(x, dy)(\mu_t + \sqrt{h}Y)(dx) - \nu_t(x, dy)\mu_t(dx) \right]$$
$$+ \int \nu_t(x, dy)\mu_t(dx) \int_0^1 (1-s)ds$$
$$\times \left(\frac{\delta^2 F(Y + sh(\delta_{x+y} - \delta_x))}{\delta Y(.)\delta Y(.)} - \frac{\delta^2 F(Y)}{\delta Y(.)\delta Y(.)}, (\delta_{x+y} - \delta_x)^{\otimes 2} \right).$$

Explicit expressions for the variational derivatives $\delta/\delta\mu_t$ are contained in the proof below.

Remark 10.2 Notice that the last term in the expression for $(O_t^h F - O_t F)(Y) = O(\sqrt{h})$ is of order $O(h)$ if the third variational derivative of F is well defined and finite. For this term to be $O(\sqrt{h})$, like the other terms, it is of course sufficient that the second variational derivative of $F(Y)$ is 1/2-Hölder continuous with respect to Y. The main conclusion to be drawn from these lengthy formulas is the following. To approximate the limiting $O_t F$ by $O_t^h F$ one has to work with functionals F having a sufficiently regular second variational derivative (i.e. *twice differentiable in space variables for differential generators and at least Hölder continuous in Y for jump-type generators).*

Proof By linearity one can do the calculations separately for a B_μ^1 having one part that is only differential (involving G and b) and the other part only integral. Thus, assume first that $\nu_\mu = 0$. Then, taking into account that

$$\frac{\delta}{\delta Y(.)} \tilde{\Omega}_t^h F(Y) = \frac{1}{\sqrt{h}} \tilde{\Omega}_t^h \frac{\delta F(Y)}{\delta Y(.)},$$
$$\frac{\delta^2}{\delta Y(.)\delta Y(.)} \tilde{\Omega}_t^h F(Y) = \frac{1}{h} \tilde{\Omega}_t^h \frac{\delta^2 F(Y)}{\delta Y(.)\delta Y(.)},$$

one can write

$$(\Lambda_h^1 \tilde{\Omega}_t^h F)(Y) = \frac{1}{\sqrt{h}} \int b_Y(x) \frac{\partial}{\partial x} \left(\tilde{\Omega}_t^h \frac{\delta F(Y)}{\delta Y(x)} \right) Y(dx)$$
$$+ \frac{1}{2\sqrt{h}} \int \left(G_Y(x) \frac{\partial}{\partial x}, \frac{\partial}{\partial x} \right) \tilde{\Omega}_t^h \frac{\delta F(Y)}{\delta Y(x)} Y(dx)$$
$$+ \frac{1}{2} \int \left(G_Y(x) \frac{\partial}{\partial x}, \frac{\partial}{\partial y} \right) \tilde{\Omega}_t^h \left. \frac{\delta^2 F(Y)}{\delta Y(x)\delta Y(y)} \right|_{y=x} Y(dx),$$

and so $(\tilde{\Omega}_t^h)^{-1}\Lambda_h^1\tilde{\Omega}_t^h F(Y)$ is given by

$$= \frac{1}{\sqrt{h}} \int b_{\mu_t+\sqrt{h}Y}(x) \frac{\partial}{\partial x} \frac{\delta F(Y)}{\delta Y(x)} (\mu_t + \sqrt{h}Y)(dx)$$
$$+ \frac{1}{2\sqrt{h}} \int \left(G_{\mu_t+\sqrt{h}Y}(x) \frac{\partial}{\partial x}, \frac{\partial}{\partial x} \right) \frac{\delta F(Y)}{\delta Y(x)} (\mu_t + \sqrt{h}Y)(dx)$$
$$+ \frac{1}{2} \int \left(G_{\mu_t+\sqrt{h}Y}(x) \frac{\partial}{\partial x}, \frac{\partial}{\partial y} \right) \frac{\delta^2 F(Y)}{\delta Y(x)\delta Y(y)} \bigg|_{y=x} (\mu_t + \sqrt{h}Y)(dx).$$

Taking into account that

$$\left(\frac{\delta F(Y)}{\delta Y}, \dot{\mu}_t \right)$$
$$= \int b_{\mu_t}(x) \frac{\partial}{\partial x} \frac{\delta F(Y)}{\delta Y(x)} \mu_t(dx) + \frac{1}{2\sqrt{h}} \int \left(G_{\mu_t}(x) \frac{\partial}{\partial x}, \frac{\partial}{\partial x} \right) \frac{\delta F(Y)}{\delta Y(x)} \mu_t(dx),$$

one deduces from (10.6) that

$$O_t^h F(Y) = \int \frac{\partial}{\partial x} \frac{\delta F(Y)}{\delta Y(x)} \left[b_{\mu_t}(x) Y(dx) + \left(\frac{\delta b_{\mu_t}(x)}{\delta \mu_t(.)}, Y \right) \mu_t(dx) \right]$$
$$+ \sqrt{h} \int \frac{\partial}{\partial x} \frac{\delta F(Y)}{\delta Y(x)} \int_0^1 ds \left[\left(\frac{\delta b_{\mu_t+s\sqrt{h}Y}(x)}{\delta \mu_t(.)}, Y \right) \right.$$
$$\left. + (1-s) \left(\frac{\delta^2 b_{\mu_t+s\sqrt{h}Y}(x)}{\delta \mu_t^2(.)}, Y^{\otimes 2} \right) \right] Y(dx)$$
$$+ \frac{1}{2} \int \left(G_{\mu_t}(x) \frac{\partial}{\partial x}, \frac{\partial}{\partial x} \right) \frac{\delta F(Y)}{\delta Y(x)} Y(dx)$$
$$+ \frac{1}{2} \int \left(\left(\frac{\delta G_{\mu_t}(x)}{\delta \mu_t(.)}, Y \right) \frac{\partial}{\partial x}, \frac{\partial}{\partial x} \right) \frac{\delta F(Y)}{\delta Y(x)} \mu_t(dx)$$
$$+ \frac{1}{2} \int \left(G_{\mu_t}(x) \frac{\partial}{\partial x}, \frac{\partial}{\partial y} \right) \frac{\delta^2 F(Y)}{\delta Y(x)\delta Y(y)} \bigg|_{y=x} \mu_t(dx)$$
$$+ \frac{\sqrt{h}}{2} \int \left(\left(\frac{\delta G_{\mu_t+s\sqrt{h}Y}(x)}{\delta \mu_t(.)}, Y \right) \frac{\partial}{\partial x}, \frac{\partial}{\partial x} \right) \frac{\delta F(Y)}{\delta Y(x)} Y(dx)$$
$$+ \frac{\sqrt{h}}{2} \int \left(\left(\frac{\delta^2 G_{\mu_t+s\sqrt{h}Y}(x)}{\delta \mu_t^2(.)}, Y^{\otimes 2} \right) \frac{\partial}{\partial x}, \frac{\partial}{\partial x} \right) \frac{\delta F(Y)}{\delta Y(x)} \mu_t(dx)$$
$$+ \frac{\sqrt{h}}{2} \int \left(G_{\mu_t+\sqrt{h}Y}(x) \frac{\partial}{\partial x}, \frac{\partial}{\partial y} \right) \frac{\delta^2 F}{\delta Y(y)\delta Y(x)} \bigg|_{y=x} Y(dx)$$
$$+ \frac{\sqrt{h}}{2} \iint_0^1 ds \left(\left(\frac{\delta G_{\mu_t+s\sqrt{h}Y}(x)}{\delta \mu_t(.)}, Y \right) \frac{\partial}{\partial x}, \frac{\partial}{\partial y} \right) \frac{\delta^2 F}{\delta Y(y)\delta Y(x)} \bigg|_{y=x} \mu_t(dx),$$

proving Proposition 10.1 for the differential operator B^1_μ.

However, for a B^1_μ with vanishing G and b one has

$$\Lambda^1_h F(Y)$$
$$= \int_{X^2} \left[\frac{\delta F}{\delta Y(x+y)} - \frac{\delta F}{\delta Y(x)} - \left(\frac{\partial}{\partial x} \frac{\delta F}{\delta Y(x)} \mathbf{1}_{B_1}(y), y \right) \right] \nu_Y(x, dy) Y(dx)$$
$$+ h \int_0^1 (1-s)\, ds \int_{X^2} \left(\frac{\delta^2 F}{\delta Y(.) \delta Y(.)} (Y + sh(\delta_{x+y} - \delta_x)),\ (\delta_{x+y} - \delta_x)^{\otimes 2} \right)$$
$$\times \nu_Y(x, dy) Y(dx)$$

and consequently

$$(\tilde{\Omega}^h_t)^{-1} \Lambda^1_h \tilde{\Omega}^h_t F(Y)$$
$$= \int_{X^2} \nu_{\mu_t + \sqrt{h} Y}(x, dy)(\mu_t + \sqrt{h} Y)(dx)$$
$$\times \left\{ \frac{1}{\sqrt{h}} \left[\frac{\delta F}{\delta Y(x+y)} - \frac{\delta F}{\delta Y(x)} - \left(\frac{\partial}{\partial x} \frac{\delta F}{\delta Y(x)} \mathbf{1}_{B_1}(y), y \right) \right] \right.$$
$$\left. + \int_0^1 (1-s)\, ds \left(\frac{\delta^2 F}{\delta Y(.) \delta Y(.)} (Y + sh(\delta_{x+y} - \delta_x)), (\delta_{x+y} - \delta_x)^{\otimes 2} \right) \right\},$$

which, on subtracting

$$\frac{1}{\sqrt{h}} \int_{X^2} \nu_{\mu_t}(x, dy) \mu_t(dx) \left[\frac{\delta F}{\delta Y(x+y)} - \frac{\delta F}{\delta Y(x)} - \left(\frac{\partial}{\partial x} \frac{\delta F}{\delta Y(x)} \mathbf{1}_{B_1}(y), y \right) \right]$$

and expanding in a Taylor series with respect to $\sqrt{h}$, yields again the required formulas. □

One deals similarly with binary and higher-order interactions. Namely, the following holds.

Proposition 10.3 *Let B^k be the conditionally positive operator, taking $C(S\mathcal{X})$ to $C(SX^k)$, given by* (1.76), *i.e.*

$$B^k f(x_1, \ldots, x_k) = A^k f(x_1, \ldots, x_k)$$
$$+ \int_{X^k} [f(y_1, \ldots, y_k) - f(x_1, \ldots, x_k)]$$
$$\times P^l(x_1, \ldots, x_k, dy_1 \cdots dy_k),$$

and specifying the scaled generator of the k-ary interaction (compare (1.69)):

$$\Lambda_h^k F(h\delta_{\mathbf{x}}) = h^{k-1} \sum_{I\subset\{1,\dots,n\},|I|=k} B_I^k F(h\delta_{\mathbf{x}}), \qquad \mathbf{x} = (x_1, \dots, x_n)$$

(for simplicity, we assume no additional mean field dependence). Then the operator O_t^h from (10.7) describing the fluctuation process is written down as (10.9), with

$$\begin{aligned} O_t F(Y) = \Bigg(& B^k \left(\frac{\delta F}{\delta Y}\right)^{\oplus}, \, Y \otimes \mu_t^{\tilde{\otimes}(k-1)} \Bigg) \\ & + \Bigg(\frac{1}{2} B^k \sum_{i,j=1}^{k} \frac{\delta^2 F}{\delta Y(y_i)\delta Y(y_j)} \\ & \quad - \left(B^k_{y_1,\dots,y_k} \sum_{i,j=1}^{k} \frac{\delta^2 F}{\delta Y(z_i)\delta Y(y_j)} \right)\Bigg|_{\forall i \; z_i = y_i} , \, \mu_t^{\tilde{\otimes} k} \Bigg), \end{aligned} \tag{10.12}$$

where $B^k_{y_1,\dots,y_k}$ denotes the action of B^k on the variables $y_1, \dots, y_k$. In particular, suppose that $k = 2$ and B^2 preserves the number of particles and hence can be written as

$$\begin{aligned} & B^2 f(x, y) \\ & = \left[\frac{1}{2}\left(G(x, y)\frac{\partial}{\partial x}, \frac{\partial}{\partial x}\right) + \frac{1}{2}\left(G(y, x)\frac{\partial}{\partial y}, \frac{\partial}{\partial y}\right) + \left(\gamma(x, y)\frac{\partial}{\partial x}, \frac{\partial}{\partial y}\right)\right] f(x, y) \\ & \quad + \left[\left(b(x, y), \frac{\partial}{\partial x}\right) + \left(b(y, x), \frac{\partial}{\partial y}\right) f(x, y)\right] + \int_{X^2} \nu(x, y, dv_1 dv_2) \\ & \quad \times \Bigg[f(x + v_1, y + v_2) - f(x, y) \\ & \qquad - \left(\frac{\partial f}{\partial x}(x, y), v_1\right) \mathbf{1}_{B_1}(v_1) - \left(\frac{\partial f}{\partial y}(x, y), v_2\right) \mathbf{1}_{B_1}(v_2) \Bigg], \end{aligned} \tag{10.13}$$

where $G(x, y)$ and $\gamma(x, y)$ are symmetric matrices with $\gamma(x, y) = \gamma(y, x)$ and $\nu(x, y, dv_1 \, dv_2) = \nu(y, x, dv_2 \, dv_1)$. Then

$$\begin{aligned} O_t F(Y) = & \left(B^2 \left(\frac{\delta F}{\delta Y}\right)^{\oplus}, Y \otimes \mu_t \right) \\ & + \int \frac{1}{2} \left(G(x, y)\frac{\partial}{\partial x}, \frac{\partial}{\partial z} \right) \frac{\delta^2 F}{\delta Y(z)\delta Y(x)}\Bigg|_{z=x} \mu_t(dx)\mu_t(dy) \end{aligned}$$

$$+\int \left(\gamma(x,y)\frac{\partial}{\partial x}, \frac{\partial}{\partial y}\right) \frac{\delta^2 F}{\delta Y(y)\delta Y(x)} \mu_t(dx)\mu_t(dy)$$
$$+\frac{1}{4}\int_{X^4} \left(\frac{\delta^2 F}{\delta Y(.)\delta Y(.)}, (\delta_{z_1+v_1}+\delta_{z_2+v_2}-\delta_{z_1}-\delta_{z_2})^{\otimes 2}\right)$$
$$\times\, \nu(z_1,z_2,dv_1dv_2)\mu_t(dz_1)\mu_t(dz_2). \qquad (10.14)$$

Proof We leave these calculations (which are quite similar to those given above) as an exercise. The details can be found in Kolokoltsov [134]. □

Again the term $O(\sqrt{h})$ depends on the third variational derivatives of F evaluated at some shifted measures. We shall give below the precise expression for binary jump interactions to cover the most important models, involving Boltzmann or Smoluchovski interactions.

Before this, a further remark is in order. As was noted at the end of Section 1.7, quite different interacting-particle systems can be described by the same limiting kinetic equation, leading to a natural equivalence relation between interaction models. For instance, the binary interactions specified by the operator (10.13) above lead to the kinetic equation

$$\frac{d}{dt}(g,\mu_t)$$
$$= \int_{X^2} \mu_t(dx)\mu_t(dz) \left\{ \frac{1}{2}\left(G(x,y)\frac{\partial}{\partial x}, \frac{\partial}{\partial x}\right) g(x) + \left(B(x,y), \frac{\partial}{\partial x} g(x)\right)\right.$$
$$\left. + \int_{X^2} [g(x+v_1) - g(x) - (\nabla g(x), v_1)\mathbf{1}_{B_1}(v_1)]\nu(x,y,dv_1dv_2)\right\},$$

which also describes the LLN limit for the mean field interaction specified by the operator B^1_μ from (10.8). The above calculations show, however, that the limiting fluctuation processes may be different; note that the coefficient γ has dropped out of the kinetic equation but is present in (10.14).

Proposition 10.4 *If Λ_2^h is given by* (1.38), *i.e.*

$$\Lambda_2^h F(h\delta_{\mathbf{x}}) = -\frac{1}{2}\int_{\mathcal{X}}\int_X [f(h\delta_{\mathbf{x}} - 2h\delta_z + h\delta_{\mathbf{y}}) - f(h\delta_{\mathbf{x}})] P(z,z;d\mathbf{y})(h\delta_{\mathbf{x}})(dz)$$
$$+\frac{1}{2h}\int_{\mathcal{X}}\int_{X^2} [f(h\delta_{\mathbf{x}} - h\delta_{z_1} - h\delta_{z_2} + h\delta_{\mathbf{y}}) - f(h\delta_{\mathbf{x}})]$$
$$\times\, P(z_1,z_2;d\mathbf{y})(h\delta_{\mathbf{x}})(dz_1)(h\delta_{\mathbf{x}})(dz_2),$$

then we have

$$
\begin{aligned}
&O_t^h F(Y)\\
&= O_t F(Y) + \frac{\sqrt{h}}{2}\int_{X^2}\int_{\mathcal{X}}\left(\frac{\delta F}{\delta Y}, \delta_{\mathbf{y}} - \delta_{z_1} - \delta_{z_2}\right) P(z_1, z_2; d\mathbf{y})Y(dz_1)Y(dz_2)\\
&\quad - \frac{\sqrt{h}}{2}\int_X\int_{\mathcal{X}}\left(\frac{\delta F}{\delta Y}, \delta_{\mathbf{y}} - 2\delta_z\right) P(z, z; d\mathbf{y})(\mu_t + \sqrt{h}Y)(dz)\\
&\quad + \frac{\sqrt{h}}{4}\int_{X^2}\int_{\mathcal{X}}\left(\frac{\delta^2 F}{\delta Y^2}, (\delta_{\mathbf{y}} - \delta_{z_1} - \delta_{z_2})^{\otimes 2}\right) P(z_1, z_2; d\mathbf{y})\\
&\qquad\qquad \times [Y(dz_1)\mu_t(dz_2) + Y(dz_2)\mu_t(dz_1)]\\
&\quad - \frac{h}{4}\iint\left(\frac{\delta^2 F}{\delta Y^2}, (\delta_{\mathbf{y}} - 2\delta_z)^{\otimes 2}\right) P(z, z; d\mathbf{y})(\mu_t + \sqrt{h}Y)(dz)\\
&\quad + \frac{h}{4}\iint\left(\frac{\delta^2 F}{\delta Y^2}, (\delta_{\mathbf{y}} - \delta_{z_1} - \delta_{z_2})^{\otimes 2}\right) P(z_1, z_2; d\mathbf{y})Y(dz_1)Y(dz_2)\\
&\quad + \frac{\sqrt{h}}{4}\int_0^1 (1-s)^2\,ds \iint\left(\frac{\delta^3 F}{\delta Y^3}(Y + s\sqrt{h}(\delta_{\mathbf{y}} - \delta_{z_1} - \delta_{z_2})),\right.\\
&\qquad\qquad\qquad\qquad\qquad\qquad \left.(\delta_{\mathbf{y}} - \delta_{z_1} - \delta_{z_2})^{\otimes 3}\right)\\
&\qquad\qquad \times P(z_1, z_2; d\mathbf{y})(\mu_t + \sqrt{h}Y)(dz_1)(\mu_t + \sqrt{h}Y)(dz_2)\\
&\quad - \frac{h^{3/2}}{4}\int_0^1 (1-s)^2\,ds \iint\left(\frac{\delta^3 F}{\delta Y^3}(Y + s\sqrt{h}(\delta_{\mathbf{y}} - 2\delta_z)), (\delta_y - 2\delta_z)^{\otimes 3}\right)\\
&\qquad\qquad \times P(z, z; d\mathbf{y})(\mu_t + \sqrt{h}Y)(dz), \qquad (10.15)
\end{aligned}
$$

where

$$
\begin{aligned}
&O_t F(Y)\\
&= \int_{X^2}\int_{\mathcal{X}}\left(\frac{\delta F}{\delta Y(.)}, \delta_{\mathbf{y}} - \delta_{z_1} - \delta_{z_2}\right) P(z_1, z_2; d\mathbf{y})Y(dz_1)\mu_t(dz_2)\\
&\quad + \frac{1}{4}\int_{X^2}\int_{\mathcal{X}}\left(\frac{\delta^2 F}{\delta Y(.)\delta Y(.)}, (\delta_{\mathbf{y}} - \delta_{z_1} - \delta_{z_2})^{\otimes 2}\right) P(z_1, z_2; d\mathbf{y})\mu_t(dz_1)\mu_t(dz_2).
\end{aligned} \tag{10.16}
$$

Proof Here one has

$$
\begin{aligned}
&\Lambda_h^2 \tilde{\Omega}_t^h F(Y)\\
&\quad = \frac{1}{2h}\iiint\left[F\left(\frac{Y + h(\delta_{\mathbf{y}} - \delta_{z_1} - \delta_{z_2}) - \mu_t}{\sqrt{h}}\right) - F\left(\frac{Y - \mu_t}{\sqrt{h}}\right)\right]\\
&\qquad\qquad \times P(z_1, z_2; ; d\mathbf{y})Y(dz_1)Y(dz_2)
\end{aligned}
$$

$$- \frac{1}{2} \iint \left[F\left(\frac{Y + h(\delta_{\mathbf{y}} - 2\delta_s) - \mu_t}{\sqrt{h}} \right) - F\left(\frac{Y - \mu_t}{\sqrt{h}} \right) \right]$$
$$\times P(z, z; d\mathbf{y}) Y(dz)$$

and consequently

$$(\Omega_t^h)^{-1} \Lambda_h^2 \Omega_t^h F(Y) = \frac{1}{2h} \iiint \left[(F(Y + \sqrt{h}(\delta_{\mathbf{y}} - \delta_{z_1} - \delta_{z_2}) - F(Y) \right]$$
$$\times P(z_1, z_2; d\mathbf{y})(\sqrt{h}Y + \mu_t)(dz_1)(\sqrt{h}Y + \mu_t)(dz_2)$$
$$- \frac{1}{2} \iint [F(Y + \sqrt{h}(\delta_{\mathbf{y}} - 2\delta_z)) - F(Y)]$$
$$\times P(z, z; d\mathbf{y})(\sqrt{h}Y + \mu_t)(dz).$$

Expanding the r.h.s. in h and using

$$F(Y + \sqrt{h}(\delta_{\mathbf{y}} - \delta_{z_1} - \delta_{z_2})) - F(Y)$$
$$= \sqrt{h} \left(\frac{\delta F}{\delta Y}, \delta_{\mathbf{y}} - \delta_{z_1} - \delta_{z_2} \right) + \frac{h}{2} \left(\frac{\delta^2 F}{\delta Y^2}, (\delta_{\mathbf{y}} - \delta_{z_1} - \delta_{z_2})^{\otimes 2} \right)$$
$$+ \frac{h^{3/2}}{2} \int_0^1 (1-s)^2$$
$$\times \left(\frac{\delta^3 F}{\delta Y^3}(Y + s\sqrt{h}(\delta_{\mathbf{y}} - \delta_{z_1} - \delta_{z_2})), (\delta_{\mathbf{y}} - \delta_{z_1} - \delta_{z_2})^{\otimes 3} \right) ds$$

yields (10.15). □

The infinite-dimensional Gaussian process generated by (10.16) will be constructed later. Here we conclude with an obvious observation that polynomial functionals of the form $F(Y) = (g, Y^{\otimes m})$, $g \in C^{\mathrm{sym}}(X^m)$, on measures are invariant under O_t. In particular, for a linear functional $F(Y) = (g, Y)$ (i.e. for $m = 1$),

$$O_t F(Y) = \int_{X^2} \int_{\mathcal{X}} \left[g(\mathbf{y}) - g(z_1) - g(z_2) \right] P(z_1, z_2; d\mathbf{y}) Y(dz_1) \mu_t(dz_2). \tag{10.17}$$

Hence the evolution (in inverse time) of the linear functionals specified by $\dot{F}_t = -O_t F_t$, $F_t(Y) = (g_t, Y)$, can be described by the equation

$$\dot{g}(z) = -O_t g(z) = -\int_X \int_{\mathcal{X}} \left[g(\mathbf{y}) - g(x) - g(z) \right] P(x, z; d\mathbf{y}) \mu_t(dx) \tag{10.18}$$

for the coefficient functions g_t (with some abuse of notation, we have denoted the action of O_t on the coefficient functions by O_t also). This equation was analyzed in detail in previous chapters.

10.2 Weak CLT with error rates: the Smoluchovski and Boltzmann models, mean field limits and evolutionary games

We can proceed now to the weak form of the CLT, when there is, so to say, just a "trace" of the CLT, for linear functionals. Though this is a sort of reduced CLT, since it "does not feel" the quadratic part of the generator of the limiting Gaussian process, technically it is the major ingredient for proving further-advanced versions of the CLT. For definiteness, we shall carry out the program for the cases of Smoluchovski coagulation and Boltzmann collisions with unbounded rates (under the standard assumptions), relegating to exercises the straightforward modifications needed for Lévy–Khintchine-type generators with bounded coefficients.

Let us start with the coagulation process. Recall that we denote by $U^{t,r}$ the backward propagator of equation (10.18), i.e. the resolving operator of the Cauchy problem $\dot{g} = -\Lambda_t g$ for $t \leq r$ with a given g_r, and that

$$F_t^h(Z_0^h, \mu_0) = h^{-1/2}\left[Z_t^h(Z_0^h) - \mu_t(\mu_0)\right]$$

is a normalized fluctuation process.

We shall start with the weak CLT for the Smoluchovski coagulation model. The main technical ingredient in the proof of the weak form of CLT is given by the following estimate for the second moment of the fluctuation process.

Proposition 10.5 *Let g_2 be a symmetric continuous function on X^2. Suppose that the coagulation kernel $K(\mathbf{x}, .)$ is a continuous function taking SX^2 to $\mathcal{M}(X)$ (where the measures are considered in the weak topology), which is E-non-increasing for a continuous non-negative function E on X such that $E(x) \to \infty$ as $x \to \infty$. Suppose that $\mu \in \mathcal{M}_{e_0,e_1}(X)$ and $(1 + E^\beta, \mu) < \infty$ for the initial condition $\mu = \mu_0$ and for some $\beta > 3$.*

(i) Let the coagulation kernel K be $(1 + \sqrt{E})^{\otimes}$-bounded:

$$K(x_1, x_2) \leq C\left[1 + \sqrt{E(x_1)}\right]\left[1 + \sqrt{E(x_2)}\right]$$

(i.e. we are under the assumptions of Theorem 8.13). Then

$$\sup_{s \leq t}\left|\mathbf{E}\left(g_2, (F_s^h(Z_0^h, \mu_0))^{\otimes 2}\right)\right| = \sup_{s \leq t}\left|\left(U_{fl}^{h;0,s}(g_2, .)\right)(F_0^h)\right| \qquad (10.19)$$

does not exceed

$$\kappa(C,t,k,e_0,e_1)\|g_2\|_{(1+E^k)^{\otimes 2}(X^2)}\big[1+(E^{2k+1/2},Z_0^h+\mu_0)\big]$$
$$\times\big[1+(E^k,Z_0^h+\mu_0)\big]\left(1+\|F_0^h\|^2_{\mathcal{M}_{1+E^k}}\right)$$

for any $k\geq 1$.

(ii) Assume that $X=\mathbf{R}_+$, $K(x_1,x_2,dy)=K(x_1,x_2)\delta(y-x_1-x_2)$, $E(x)=x$, and that K is non-decreasing in each argument that is twice continuously differentiable on $(\mathbf{R}_+)^2$ up to the boundary, all the first and second partial derivatives being bounded by a constant C (i.e. we are under the assumptions of Theorem 8.14). Then expression (10.19) *does not exceed*

$$\kappa(C,t,k,e_0,e_1)\|g_2\|_{C^{2,\mathrm{sym}}_{(1+E^k)^{\otimes 2}(X^2)}}[1+(E^{2k+3},Z_0^h+\mu_0)]$$
$$\times[1+(E^k,Z_0^h+\mu_0)]\left(1+\|F_0^h\|^2_{\mathcal{M}^1_{1+E^k}}\right)$$

for any $k\geq 0$.

Proof To derive the estimate in part(i) of the theorem, one writes

$$\mathbf{E}\left(g_2,\left(\frac{Z_t^h(Z_0^h)-\mu_t(\mu_0)}{\sqrt{h}}\right)^{\otimes 2}\right)$$
$$=\mathbf{E}\left(g_2,\left(\frac{Z_t^h(Z_0^h)-\mu_t(Z_0^h)}{\sqrt{h}}\right)^{\otimes 2}\right)+\left(g_2,\left(\frac{\mu_t(Z_0^h)-\mu_t(\mu_0)}{\sqrt{h}}\right)^{\otimes 2}\right)$$
$$+2\mathbf{E}\left(g_2,\frac{Z_t^h(Z_0^h)-\mu_t(Z_0^h)}{\sqrt{h}}\otimes\frac{\mu_t(Z_0^h)-\mu_t(\mu_0)}{\sqrt{h}}\right).\qquad(10.20)$$

The first term can be rewritten as

$$\frac{1}{h}\mathbf{E}\left(g_2,(Z_t^h(Z_0^h))^{\otimes 2}-(\mu_t(Z_0^h))^{\otimes 2}+\mu_t(Z_0^h)\otimes\big[\mu_t(Z_0^h)-Z_t^h(Z_0^h)\big]\right.$$
$$\left.+\big[\mu_t(Z_0^h)-Z_t^h(Z_0^h)\big]\otimes\mu_t(Z_0^h)\right).$$

Under the assumptions of Theorem 8.15 it can be estimated as

$$\kappa(C,r,e_0,e_1)\|g_2\|_{(1+E^k)^{\otimes 2}}\big[1+(E^{2k+1/2},Z_0^h)\big]\big[1+(E^k,Z_0^h)\big],$$

owing to Theorem 9.15. The second term is estimated as

$$\|g_2\|_{(1+E^k)^{\otimes 2}}\big[1+(E^{k+1},\mu_0+Z_0^h)\big]\left\|\frac{Z_0^h-\mu_0}{\sqrt{h}}\right\|^2_{\mathcal{M}_{1+E^k}}$$

using the Lipschitz continuity of the solutions to Smoluchovski's equation, and the third term is estimated by an obvious combination of these two estimates.

The estimate in part(ii), under the assumptions of Theorem 8.14, is proved analogously. Namely, the first term in the representation (10.20) is again estimated by Theorem 9.15, and the second term is estimated using (8.59) and the observation that

$$
\begin{aligned}
&|(g_2, \nu^{\otimes 2})| \\
&\quad \leq \sup_{x_1} \left| [1 + E^k(x_1)]^{-1} \int \frac{\partial g_2}{\partial x_1}(x_1, x_2)\nu(dx_2) \right| \|\nu\|_{\mathcal{M}^1_{1+E^k}} \\
&\quad \leq \sup_{x_1, x_2} \left| [1 + E^k(x_1)]^{-1} [1 + E^k(x_2)]^{-1} \frac{\partial^2 g_2}{\partial x_1 \partial x_2}(x_1, x_2) \right| \|\nu\|^2_{\mathcal{M}^1_{1+E^k}} \\
&\quad \leq \|g_2\|_{C^{2,sym}_{(1+E^k)^{\otimes 2}}} \|\nu\|^2_{\mathcal{M}^1_{1+E^k}}. \qquad \square
\end{aligned}
$$

Theorem 10.6 (Weak CLT for Smoluchovski coagulation) *Under the assumptions of Proposition 10.5(i)*

$$
\begin{aligned}
&\sup_{s \leq t} \left| \mathbf{E}(g, F^h_s(Z^h_0, \mu_0)) - (U^{0,s} g, F^h_0(Z^h_0, \mu_0)) \right| \\
&\quad \leq \kappa(C, t, k, e_0, e_1)\sqrt{h} \|g\|_{1+E^k} \\
&\quad \times (1 + E^{2k+1/2}, Z^h_0 + \mu_0)^2 \left(1 + \left\| F^h_0(Z^h_0, \mu_0) \right\|^2_{\mathcal{M}_{1+E^{k+1}}(X)} \right)
\end{aligned}
\tag{10.21}
$$

for all $k \geq 1$, $g \in C_{1+E^k}(X)$, *and, under the assumptions of Proposition 10.5(ii),*

$$
\begin{aligned}
&\sup_{s \leq t} \left| \mathbf{E}(g, F^h_s(Z^h_0, \mu_0)) - (U^{0,s} g, F^h_0) \right| \\
&\quad \leq \kappa(C, t, k, e_0, e_1)\sqrt{h} \|g\|_{C^{2,0}_{1+E^k}} \\
&\quad \times (1 + E^{2k+3}, Z^h_0 + \mu_0)^2 \left(1 + \left\| F^h_0(Z^h_0, \mu_0) \right\|^2_{\mathcal{M}^1_{1+E^{k+1}}(X)} \right)
\end{aligned}
\tag{10.22}
$$

for all $k \geq 0$, $g \in C^{2,0}_{1+E^k}(X)$, *where* $\mathbf{E}$ *denotes the expectation with respect to the process* Z^h_t.

Proof Recall that we denoted by $U^{h;t,r}_{fl}$ the backward propagator corresponding to the process $F^h_t = (Z^h_t - \mu_t)/\sqrt{h}$. By (10.6), the l.h.s. of (10.21) can be written as

$$\sup_{s\le t}\left|\left(U_{fl}^{h;0,s}(g,.)\right)(F_0^h)-(U^{0,s}g,F_0^h)\right|$$

$$=\sup_{s\le t}\left|\int_0^s\left[U_{fl}^{h;0,\tau}(O_\tau^h-O_\tau)U^{\tau,s}(g,.)\right]d\tau\,(F_0^h)\right|.$$

Since, by (10.15),

$$(O_\tau^h-O_\tau)(U^{\tau,s}g,\cdot)(Y)=\frac{\sqrt{h}}{2}\iiint\left[U^{\tau,s}g(y)-U^{\tau,s}g(z_1)-U^{\tau,s}g(z_2)\right]$$
$$\times K(z_1,z_2;dy)Y(dz_1)Y(dz_2)$$
$$-\frac{\sqrt{h}}{2}\iint\left[U^{\tau,s}g(y)-2U^{\tau,s}g(z)\right]$$
$$\times K(z,z;dy)(\mu_t+\sqrt{h}Y)(dz)$$

(note that the terms containing the second and third variational derivatives in (10.15) vanish here, as we are applying the operator it to a linear function), the required estimate follows from Proposition 10.5. □

A similar result holds for the Boltzmann equation (8.61):

$$\frac{d}{dt}(g,\mu_t)=\frac{1}{4}\int_{S^{d-1}}\int_{R^{2d}}[g(v')+g(w')-g(v)-g(w)]$$
$$\times B(|v-w|,\theta)\,dn\,\mu_t(dv)\mu_t(dw).$$

Recall that in this situation $E(v)=v^2$.

Theorem 10.7 (Weak CLT for Boltzmann collisions) *Let*

$$B(|v|,\theta)=b(\theta)|v|^\beta,\qquad \beta\in(0,1],$$

where $b(\theta)$ is a bounded not identically vanishing function (i.e. we are under the assumptions of Theorem 8.20). Then the estimate

$$\sup_{s\le t}\left|\mathbf{E}(g,F_s^h(Z_0^h,\mu_0))-(U^{0,s}g,F_0^h(Z_0^h,\mu_0))\right|$$
$$\le\kappa(C,t,k,e_0,e_1)\sqrt{h}\|g\|_{1+E^k}$$
$$\times\Omega(\mu_0,Z_0^h)\left(1+\left\|F_0^h(Z_0^h,\mu_0)\right\|^2_{\mathcal{M}_{1+E^{k+1}}(X)}\right)\tag{10.23}$$

holds true for all $k\ge 1$, $g\in C_{1+E^k}(X)$; the coefficient $\Omega(\mu_0,Z_0^h)$ depends on the moments of μ_0, Z_0^h and can be calculated explicitly.

Proof This is the same as the proof of Theorem 10.6 except that it is based, of course, on the estimates from Theorem 8.20. □

In the same way as above one may obtain the weak CLT with error estimates for other models where the rates of convergence in LLN can be obtained (including processes on manifolds and evolutionary games). For example, the first of the results in Exercises 10.1–10.3 below represents the weak CLT for mean field models with generators of order at most one and the second represents the weak CLT for interacting diffusions combined with stable-like processes.

Exercise 10.1 Show that under the assumption of Theorems 8.3 or 9.6, one has

$$\sup_{t\le T}\left|\mathbf{E}(g, F_s^h(Z_0^h,\mu_0)) - (U^{0,s}g, F_0^h(Z_0^h,\mu_0))\right|$$
$$\le C(T, \|Z_B^h(0)\|)\sqrt{h}\|g\|_{C_\infty^2(X)}\left(1+\left\|F_0^h(Z_0^h,\mu_0)\right\|^2_{(C_\infty^1(X))^*}\right) \tag{10.24}$$

for $g \in C_\infty^2(\mathbf{R}^d)$.

Exercise 10.2 Show that under the assumptions of Theorem 9.10 one has the same estimates as in (10.24) but with $\|g\|_{C_\infty^4(X)}$ and $\|F_0^h(Z_0^h,\mu_0)\|^2_{(C_\infty^2(X))^*}$ instead of $\|g\|_{C_\infty^2(X)}$ and $\|F_0^h(Z_0^h,\mu_0)\|^2_{(C_\infty^1(X))^*}$.

Exercise 10.3 Formulate and prove the corresponding weak CLT for the evolutionary games discussed in Section 1.6. Hint: This is simpler than above as all generators are bounded, implying the required estimates almost automatically. The reduced CLT for general bounded generators was obtained in [134].

10.3 Summarizing the strategy followed

In the previous sections we have developed a method of proving the weak (or reduced) CLT for fluctuations around the LLN limit of a large class of Markov models of interacting particles. Let us summarize the milestones of the path followed. Once the corresponding linear problem and its regularity has been understood, one obtains well-posedness for the given nonlinear Markov process and semigroup. Then the analysis of the derivatives with respect to the initial data is carried out, aiming at the invariance and regularity (smoothness and/or bounds) of functionals on measures having a well-defined second-order variational derivative. The next step is to prove error estimates for the LLN that imply the boundedness of the second moment of the fluctuation process (Proposition 10.5). Finally one deduces error estimates for the CLT. The realization of this program relies heavily on the structure of the model in question.

We have had to use quite different techniques in various situations: the particulars of coagulation processes for Smoluchovski's model, the subtle properties of moment propagation for the Boltzmann equation, the regularity of SDEs for stable-like processes and specific analytical estimates for the generators "of order at most one" including the Vlasov equations, mollified Boltzmann processes and stable-like interacting processes with index $\alpha < 1$.

Though the reduced CLT obtained is useful (for example the convergence rates can be applied for assessing numerical schemes based on a particular interacting-particle approximation), it is not of course the full story since one is naturally interested in the full limit of the fluctuation process. Here again one can identify the general strategy, but its realization for concrete models is rarely straightforward (even when the weak CLT has been proved already). We shall conclude this chapter with a sketch of this strategy in the case of the basic coagulation model, referring for the full story to the original papers.

10.4 Infinite-dimensional Ornstein–Uhlenbeck processes

Infinite-dimensional Ornstein–Uhlenbeck (OU) processes obtained as the fluctuation limits for mean field and kth-order interactions are specified by (10.10) and (10.12) respectively. It is clear how to write down (at least formally) an infinite-dimensional Ito-type stochastic equation driven by infinite-dimensional Gaussian noise with prescribed correlations. However, these correlations are rather singular and degenerate, which makes the corresponding rigorous analysis less obvious. We shall discuss two alternative approaches to the study of the corresponding processes. Namely, in the first approach one constructs the appropriate semigroup on the appropriate space of generalized functions, via finite-dimensional approximations or by means of an infinite-dimensional Riccati equation. The existence of the process itself may be then obtained as a byproduct from the analysis of scaled fluctuation processes for approximating interacting-particle systems, if one can show that these processes form a tight family. Though the appropriate generalized functions can be different for specific models (they should reflect the growth and smoothness properties of the coefficients of B^k), the procedure itself is very much the same. We shall demonstrate this program for Smoluchovski coagulation models with smooth unbounded rates, additively bounded by the mass.

Let us construct the propagator of the equation $\dot{F} = -O_t F$ on the set of cylinder functions $C_k^n = C_k^n(\mathcal{M}^m_{1+E^k})$, $m = 1, 2$, on measures that have the form

$$\Phi_f^{\phi_1,\ldots,\phi_n}(Y) = f((\phi_1, Y), \ldots, (\phi_n, Y)) \qquad (10.25)$$

with $f \in C(\mathbf{R}^n)$ and $\phi_1, \ldots, \phi_n \in C^{m,0}_{1+E^k}$. We shall denote by $\mathcal{C}_k$ the union of the $\mathcal{C}_k^n$ for all $n = 0, 1, \ldots$ (of course, functions from $\mathcal{C}_k^0$ are just constants).

The Banach space of k-times continuously differentiable functions on $\mathbf{R}^d$ (whose norm is the maximum of the sup norms of a function and all its partial derivatives up to and including order k) will be denoted, as usual, by $C^k(\mathbf{R}^d)$.

Theorem 10.8 *Under the assumptions of Proposition 10.5(ii) (or equivalently Theorems 8.14 and 10.5(ii)), for any $k \geq 0$ and a μ_0 such that $(1 + E^{k+1}, \mu_0) < \infty$ there exists a propagator $OU^{t,r}$ of contractions on $\mathcal{C}_k$ preserving the subspaces $\mathcal{C}_k^n$, $n = 0, 1, 2, \ldots$, such that $OU^{t,r}F$, $F \in \mathcal{C}_k$, depends continuously on t in the topology of uniform convergence on bounded subsets of $\mathcal{M}^m_{1+E^k}$, $m = 1, 2$, and solves the equation $\dot{F} = -O_t F$ in the sense that if $f \in C^2(\mathbf{R}^d)$ in (10.25) then*

$$\frac{d}{dt} OU^{t,r} \Phi_f^{\phi_1, \ldots, \phi_n}(Y) = -\Lambda_t OU^{t,r} \Phi_f^{\phi_1, \ldots, \phi_n}(Y), \qquad 0 \leq t \leq r, \quad (10.26)$$

uniformly for Y from bounded subsets of $\mathcal{M}^m_{1+E^k}$.

Proof Substituting a function $\Phi_{f_t}^{\phi_1^t, \ldots, \phi_n^t}$ of the form (10.25) (having twice continuously differentiable f_t) with a given initial condition $\Phi_r(Y) = \Phi_{f_r}^{\phi_1^r, \ldots, \phi_n^r}(Y)$ at $t = r$ into the equation $\dot{F}_t = -O_t F_t$ and taking into account that

$$\frac{\delta \Phi}{\delta Y(.)} = \sum_i \frac{\partial f_t}{\partial x_i} \phi_i^t(.), \qquad \frac{\delta^2 \Phi}{\delta Y(.) \delta Y(.)} = \sum_{i,j} \frac{\partial^2 f_t}{\partial x_i \partial x_j} \phi_i^t(.) \phi_j^t(.),$$

yields the equation

$$\begin{aligned}
&\frac{\partial f_t}{\partial t} + \frac{\partial f_t}{\partial x_1}(\dot{\phi}_1^t, Y) + \cdots + \frac{\partial f_t}{\partial x_n}(\dot{\phi}_n^t, Y) \\
&\quad = -\iiint \sum_{j=1}^n \frac{\partial f_t}{\partial x_j}(\phi_j^t, \delta_y - \delta_{z_1} - \delta_{z_2}) K(z_1, z_2; dy) \\
&\qquad \times Y(dz_1)\mu_t(dz_2) \\
&\quad - \frac{1}{4} \iiint \sum_{j,l=1}^n \frac{\partial^2 f_t}{\partial x_j \partial x_l}(\phi_j^t \otimes \phi_l^t, (\delta_y - \delta_{z_1} - \delta_{z_2})^{\otimes 2}) K(z_1, z_2; dy) \\
&\qquad \times \mu_t(dz_1)\mu_t(dz_2),
\end{aligned}$$

where $f_t(x_1, \ldots, x_n)$ and all its derivatives are evaluated at the points $x_j = (\phi_j^t, Y)$ (here and in what follows we denote by a dot the derivative d/dt with respect to time). This equation is clearly satisfied whenever

$$\dot{f}_t(x_1, \dots, x_n) = -\sum_{j,k=1}^{n} \Pi(t, \phi_j^t, \phi_k^t) \frac{\partial^2 f_t}{\partial x_j \partial x_k}(x_1, \dots, x_n) \tag{10.27}$$

and

$$\dot{\phi}_j^t(z) = -\iint \left[\phi_j^t(y) - \phi_j^t(z) - \phi_j^t(w)\right] K(z, w; dy)\mu_t(dw) = -O_t\phi_j^t(z);$$

in (10.27) Π is given by

$$\Pi(t, \phi, \psi) = \tfrac{1}{4} \iiint \left(\phi \otimes \psi, (\delta_y - \delta_{z_1} - \delta_{z_2})^{\otimes 2}\right) K(z_1, z_2; dy)\mu_t(dz_1) \times \mu_t(dz_2). \tag{10.28}$$

Consequently

$$OU^{t,r}\Phi_r(Y) = \Phi_t(Y) = (\mathcal{U}^{t,r} f_r)\left((U^{t,r}\phi_1^r, Y), \dots, (U^{t,r}\phi_n^r, Y)\right), \tag{10.29}$$

where $\mathcal{U}^{t,r} f_r = \mathcal{U}_\Pi^{t,r} f_r$ is defined as the resolving operator for the (inverse-time) Cauchy problem of equation (10.27) (it is well defined, as (10.27) is just a spatially invariant second-order evolution). The resolving operator $U^{t,r}$ was constructed in Section 8.4, and

$$\Pi(t, \phi_j^t, \phi_k^t) = \Pi(t, U^{t,r}\phi_j^r, U^{t,r}\phi_k^r).$$

All the statements of Theorem 10.8 follow from the explicit formula (10.29), the semigroup property of the solution to the finite-dimensional equation (10.27) and the estimates from Section 8.4. □

In the second approach the propagator is constructed by means of the generalized Riccati equation discussed in Appendix K.

10.5 Full CLT for coagulation processes (a sketch)

Once the weak CLT is obtained, it is natural to ask whether the fluctuation process corresponding to a particular model converges to the limiting OU process. To do this one has to choose an appropriate infinite-dimensional space (usually a Hilbert space is the most convenient choice) where this OU process is well defined and then use some compactness argument to get convergence. Though this is clear in principle, the technical details can be non-trivial for any particular model (especially for unbounded generators). For the case of Smoluchovski coagulation this program was carried out in Kolokoltsov [136]. Referring to this paper for the full story with its rather lengthy manipulations (see also the bibliographical comments in [136] for related papers), we shall simply sketch the main steps to be taken and the tools to be used.

Theorem 10.9 (CLT for coagulation: convergence of semigroups) *Suppose that $k \geq 0$ and $h_0 > 0$ are given such that*

$$\sup_{h \leq h_0} (1 + E^{2k+5}, Z_0^h + \mu_0) < \infty. \tag{10.30}$$

(i) Let $\Phi \in C_k^n(\mathcal{M}_{1+E^k}^2)$ be given by (10.25) with $f \in C^3(\mathbf{R}^n)$ and all $\phi_j \in C_{1+E^k}^{2,0}(X)$. Then

$$\sup_{s \leq t} \left| \mathbf{E}\Phi(F_t^h(Z_0^h, \mu_0)) - OU^{0,t}\Phi(F_0^h) \right|$$
$$\leq \kappa(C, t, k, e_0, e_1)\sqrt{h} \max_j \|\phi_j\|_{C_{1+E^k}^{2,0}} \|f\|_{C^3(\mathbf{R}^n)}$$
$$\times (1 + E^{2k+5}, Z_0^h + \mu_0)^2 \left(1 + \|F_0^h\|^2_{\mathcal{M}^1_{1+E^{k+1}}(X)} \right). \tag{10.31}$$

(ii) If $\Phi \in C_k^n(\mathcal{M}_{1+E^k}^2)$ (with not necessarily smooth f in the representation (10.25)) and F_0^h converges to some F_0 as $h \to 0$ in the $\star$-weak topology of $\mathcal{M}^1_{1+E^{k+1}}$ then

$$\lim_{h \to 0} \left| \mathbf{E}\Phi(F_t^h(Z_0^h, \mu_0)) - OU^{0,t}\Phi(F_0) \right| = 0 \tag{10.32}$$

uniformly for F_0^h from a bounded subset of $\mathcal{M}^1_{1+E^{k+1}}$ and t from a compact interval.

This result is obtained by extending the arguments from the proof of Theorem 10.6, where now one needs the full expansion of the generator of fluctuations, not only its linear or quadratic part.

To formulate the next result we have to introduce the space of weighted smooth L_p-functions or weighted Sobolev spaces (as we mentioned above, to get effective estimates, Hilbert space methods are more appropriate). Namely, define $L_{p,f} = L_{p,f}(T)$ as the space of measurable functions g on a measurable space T having finite norm $\|g\|_{L_{p,f}} = \|g/f\|_{L_p}$. The spaces $L_{p,f}^{1,0}$ and $L_{p,f}^{2,0}$, $p \geq 1$, are defined as the spaces of absolutely continuous functions ϕ on $X = \mathbf{R}_+$ such that $\lim_{x \to 0} \phi(x) = 0$, with respective norms

$$\|\phi\|_{L_{p,f}^{1,0}(X)} = \|\phi'\|_{L_{p,f}(X)} = \|\phi'/f\|_{L_p(X)},$$
$$\|\phi\|_{L_{p,f}^{2,0}(X)} = \|\phi'/f\|_{L_p(X)} + \|(\phi'/f)'\|_{L_p(X)}$$

and duals $(L^{1,0}_{p,f})'$ and $(L^{2,0}_{p,f})'$. By the Sobolev embedding lemma one has the inclusion $L^{2,0}_2 \subset C^{1,0}$ and hence also $L^{2,0}_{2,f_k} \subset C^{1,0}_{f_k}$ for arbitrary $k > 0$, implying by duality the inclusion $\mathcal{M}^1_{f_k} \subset (L^{2,0}_{2,f_k})'$.

In order to transfer the main estimates for the fluctuation processes to these weighted L_2-spaces, one has to obtain L_2 analogs of the estimates for derivatives (of the solutions to the kinetic equations from Section 8.4), which is rather technical, and also a corresponding extension of Theorem 10.8. Then one obtains the following main ingredient in the proof of the functional CLT.

Proposition 10.10 *Under the assumptions of Theorem 8.14, for any $k > 1/2$,*

$$\sup_{s\le t} \mathbf{E}\| F^h_s(Z^h_0, \mu_0)\|^2_{(L^{2,0}_{2,f_k})'}$$
$$\le \kappa(C, t, k, e_0, e_1)\left[1 + (E^{2k+3}, Z^h_0 + \mu_0)\right]^2 \left(1 + \|F^h_0\|^2_{(L^{2,0}_{2,f_k})'}\right). \tag{10.33}$$

This fact allows us to obtain key estimates when extending the above Theorem 10.8 to the following convergence of finite-dimensional distributions.

Theorem 10.11 (CLT for coagulation: finite-dimensional distributions) *Suppose that* (10.30) *holds, $\phi_1, \dots, \phi_n \in C^{2,0}_{1+E^k}(\mathbf{R}_+)$ and $F^h_0 \in (L^{2,0}_{2,1+E^{k+2}})'$ converges to some F_0 in $(L^{2,0}_{2,1+E^{k+2}})'$ as $h \to 0$. Then the $\mathbf{R}^n$-valued random variables*

$$\Phi^h_{t_1,\dots,t_n} = ((\phi_1, F^h_{t_1}(Z^h_0, \mu_0)), \dots, (\phi_n, F^h_{t_n}(Z^h_0, \mu_0))),$$
$$0 < t_1 \le \dots \le t_n,$$

converge in distribution, as $h \to 0$, to a Gaussian random variable with characteristic function

$$g_{t_1,\dots,t_n}(p_1, \dots, p_n)$$
$$= \exp\left(i \sum_{j=1}^n p_j(U^{0,t_j}\phi_j, F_0) - \sum_{j=1}^n \int_{t_{j-1}}^{t_j} \sum_{l,k=j}^n p_l p_k \Pi(s, U^{s,t_l}\phi_l, U^{s,t_k}\phi_k)\, ds\right), \tag{10.34}$$

where $t_0 = 0$ and Π is given by (10.28). *In particular, for $t = t_1 = \dots = t_n$ this implies that*

$$\lim_{h\to 0} \mathbf{E} \exp\left(i \sum_{j=1}^{n} (\phi_j, F_t^h)\right)$$

$$= \exp\left(i \sum_{j=1}^{n} (U^{0,t}\phi_j, F_0) - \sum_{j,k=1}^{n} \int_0^t \Pi(s, U^{s,t}\phi_j, U^{s,t}\phi_k)\, ds\right).$$

Note that passing from Theorem 10.9 to Theorem 10.11 would be automatic for finite-dimensional Feller processes (owing to Theorem C.5), but in our infinite-dimensional setting this is not at all straightforward and requires in addition the use of Hilbert space methods (yielding an appropriate compact containment condition).

Theorem 10.12 (Full CLT for coagulation) *Suppose that the conditions of Theorems 10.9, 10.11 hold.*

(i) For any $\phi \in C^{2,0}_{1+E^k}(\mathbf{R}_+)$, the real-valued processes $(\phi, F_t^h(Z_0^h, \mu_0))$ converge in distribution in the Skorohod space of càdlàg functions (equipped with the standard J_1-topology) to the Gaussian process with finite-dimensional distributions specified by Theorem 10.11.

(ii) The fluctuation process $F_t^h(Z_0^h, \mu_0)$ converges in distribution on the Skorohod space of càdlàg functions $D([0,T]; (L^{2,0}_{2,1+E^{k+2}}(\mathbf{R}_+))')$ (with J_1-topology), where $(L^{2,0}_{2,1+E^{k+2}}(\mathbf{R}_+))'$ is considered in its weak topology, to the Gaussian process with finite-dimensional distributions specified by Theorem 10.11.

Notice that, once the previous results have been obtained, all that remains to be proved in Theorem 10.12 is the tightness of the approximations, i.e. the existence of a limiting point, because the finite-dimensional distributions of such a point have already been uniquely specified, by Theorems 10.9 and 10.11.

The proof of Theorem 10.12 is carried out in the following steps.

By Dynkin's formula applied to the Markov process Z_t^h one finds that, for a $\phi \in C_{1+E^k}(X)$,

$$M_\phi^h(t) = (\phi, Z_t^h) - (\phi, Z_0^h) - \int_0^t (L_h(\phi, .))(Z_s^h) ds$$

is a martingale and that

$$(\phi, F_t^h) = \frac{M_\phi^h}{\sqrt{h}} + \frac{1}{\sqrt{h}}\left((\phi, Z_0^h) + \int_0^t (L_h(\phi, .))(Z_s^h) ds - (\phi, \mu_t)\right).$$

To show the compactness of this family one uses Theorem C.9 in conjunction with the result in Exercise 2.1 to estimate the quadratic variation. This implies (i). In order to prove (ii), one needs only to prove the tightness of the family of normalized fluctuations F_t^h, as by Theorem 10.11 the limiting process is uniquely defined whenever it exists. By Theorem C.2, to prove the tightness it remains to establish the following compact containment condition: for every $\epsilon > 0$ and $T > 0$ there exists a $K > 0$ such that, for any h,

$$P\left(\sup_{t\in[0,T]} \|F_t^h\|_{(L^{2,0}_{2,f_k})'} > K\right) \leq \epsilon,$$

and this can be obtained from Proposition 10.10.

11

Developments and comments

In this chapter we discuss briefly several possible developments of the theory. Each section can be read independently.

11.1 Measure-valued processes as stochastic dynamic LLNs for interacting particles; duality of one-dimensional processes

We have introduced nonlinear Markov processes as deterministic dynamic LLNs for interacting particles. As pointed out in Chapter 1, nonlinear Markov processes in a metric space S are just deterministic Markov processes in the space of Borel measures $\mathcal{M}(S)$. Also of interest are nondeterministic (stochastic) LLNs, which are described by more general nondeterministic measure-valued Markov processes. Here we give a short introduction to stochastic LLNs on a simple model of completely identical particles, discussing in passing the duality of Markov processes on $\mathbf{R}_+$.

Suppose that the state of a system is characterized by the number $n \in \mathbf{Z}_+ = \{0, 1, 2, \ldots\}$ of identical indistinguishable particles that it contains at a particular time. If any single particle, independently of the others, can die after a lifetime of random length in which a random number $l \in \mathbf{Z}_+$ of offspring are produced, the generator of this process will be given by

$$(G_1 f)(n) = n \sum_{m=-1}^{\infty} g_m^1 \left[f(n+m) - f(n) \right]$$

for some non-negative constants g_m^1. More generally, if any group of k particles chosen randomly (with uniform distribution, say) from a given group of n particles can be transformed at some random time into a group of $l \in \mathbf{Z}_+$ particles through some process of birth, death or coagulation etc. then the generator of this process will be given by

$$(G_k f)(n) = C_n^k \sum_{m=-k}^{\infty} g_m^k \left[f(n+m) - f(n) \right]$$

for some non-negative constants g_m^k, where the C_n^k are the usual binomial coefficients and where it is understood that these coefficients vanish whenever $n < k$. Finally, if the spontaneous birth (or input) of a random number of particles is allowed to occur, this will contribute a term $\sum_{m=0}^{\infty} g_m^0 \left[f(n+m) - f(n) \right]$ to the generator equation for our process. A generator of the type $\sum_{k=0}^{K} G_k$ describes all k-ary interactions, with $k \leq K$. The usual scaling of the state space $n \mapsto nh$, where h is a positive parameter, combined with the scaling of the interaction $G_k \mapsto h^k G_k$, leads to a Markov chain on $h\mathbf{Z}_+$ with generator

$$G^h = \sum_{k=0}^{K} G_k^h, \qquad (G_k^h f)(hn) = h^k C_n^k \sum_{m=-k}^{\infty} g_m^k \left[f(hn + hm) - f(hn) \right]. \tag{11.1}$$

We are interested in the limits $n \to \infty$, $h \to 0$, with $nh \to x \in \mathbf{R}_+$ and where $g_m^k = g_m^k(h)$ may also depend on h. To analyze this limiting procedure, we will consider the operator (11.1) as a restriction to $h\mathbf{Z}_+$ of the operator, again denoted G^h in an abuse of notation, defined on functions on $(0, \infty)$ by $G^h = \sum_{k=0}^{K} G_k^h$, where

$$(G_k^h f)(x) = \frac{x(x-h)\ldots[x-(k-1)h]}{k!} \sum_{m=\max(-k,-x/h)}^{\infty} g_m^k(h) \left[f(x+hm) - f(x) \right] \tag{11.2}$$

for $x \geq h(k-1)$; the r.h.s. vanishes otherwise. Clearly $x(x-h)\cdots[x-(k-1)h]$ tends to x^k as $h \to 0$, and we expect that, with an appropriate choice of $g_m^k(h)$, the sum of the k-ary interaction generators in (11.2) will tend to the generator of a stochastic process on $\mathbf{R}_+$ of the form $\sum_{k=0}^{K} x^k N_k$, where each N_k is the generator of a spatially homogeneous process with i.i.d. increments (i.e. a Lévy process) on $\mathbf{R}_+$. Hence each N_k is given by the Lévy–Khintchine formula with a Lévy measure that has support in $\mathbf{R}_+$.

Let us formulate a precise result in this simple one-dimensional setting, referring to Kolokoltsov [127] for its proof.

We denote by $C[0, \infty]$ the Banach space of continuous bounded functions on $(0, \infty)$ having limits as $x \to 0$ and as $x \to \infty$ (with the usual sup norm).

We shall also use the closed subspaces $C_0[0,\infty]$ or $C_\infty[0,\infty]$ of $C[0,\infty]$, consisting of functions such that $f(0)=0$ or $f(\infty)=0$ respectively. Consider an operator L in $C[0,\infty]$ given by the formula

$$(Lf)(x) = \sum_{k=1}^K x^k \left\{ a_k f''(x) - b_k f'(x) + \int_0^\infty \left[f(x+y) - f(x) - f'(x)y \right] \nu_k(dy) \right\}, \tag{11.3}$$

where K is a natural number, a_k and b_k are real constants, $k=1,\ldots,K$, the a_k are non-negative and the ν_k are Borel measures on $(0,\infty)$ satisfying

$$\int \min(\xi,\xi^2)\nu_k(d\xi) < \infty.$$

As a natural domain $D(L)$ of L we take the space of twice continuously differentiable functions $f \in C[0,\infty]$ such that $Lf \in C[0,\infty]$.

Let $k_1 \le k_2$ (resp. $l_1 \le l_2$) denote the bounds of those indexes k for which the a_k (resp. b_k) do not vanish, i.e. $a_{k_1} > 0$, $a_{k_2} > 0$ and $a_k = 0$ for $k > k_2$ and $k < k_1$ (resp. $b_{l_1} \ne 0$, $b_{l_2} \ne 0$ and $b_k = 0$ for $k > l_2$ and $k < l_1$).

Theorem 11.1 *Suppose that*

(a) the ν_k vanish for $k < k_1$ and $k > k_2$,
(b) if $l_2 < k_2$ then $\nu_{k_2} = 0$,
(c) if $l_1 = k_1 - 1$ and $b_{l_1} = -l_1 a_{l_1}$ then $\nu_{k_1} = 0$,
(d) $b_{l_2} > 0$ whenever $l_2 \ge k_2 - 1$,
(e) if $l_2 = k_2$ then there exists $\delta > 0$ such that

$$\frac{1}{a_{l_2}} \int_0^\delta \xi^2 \nu_{l_2}(d\xi) + \frac{1}{|b_{l_2}|} \int_\delta^\infty \xi \nu_{l_2}(d\xi) < \frac{1}{4}.$$

Then the following hold.

(i) If $k_1 > 1$ (resp. $k_1 = 1$), L generates a strongly continuous conservative semigroup on $C[0,\infty]$ (resp. non-conservative semigroup on $C_0[0,\infty]$)).

(ii) The corresponding process $X_x(t)$, where x denotes the starting point, is stochastically monotone: the probability $\mathbf{P}(X_x(t) \ge y)$ is a non-decreasing function of x for any y.

(iii) There exists a dual process $\tilde{X}(t)$, with generator given explicitly, whose distribution is connected with the distribution of $X(t)$ by the duality formula

$$\mathbf{P}(\tilde{X}_x(t) \le y) = \mathbf{P}(X_y(t) \ge x).$$

Remark 11.2 The long list of hypotheses in Theorem 11.1 covers the most likely situations, where either the diffusion (the second-order) term or the drift (the first-order) term of L dominates the jumps and where consequently perturbation theory can be used for the analysis of L, with the jump part considered as a perturbation. A simple example with all conditions satisfied is the operator given in (11.3), for $a_1 > 0, a_K > 0, b_K > 0$ and $\nu_1 = \nu_K = 0$.

We shall now describe the approximation of the Markov process $X(t)$ by systems of interacting particles with k-ary interactions, i.e. by Markov chains with generators of the type (11.1), (11.2). This is the simplest and most natural approximation.

Let finite measures $\tilde{\nu}_k$ be defined by $\tilde{\nu}_k(dy) = \min(y, y^2)\nu_k(dy)$. Let β_k^1, β_k^2 be arbitrary positive numbers such that $\beta_k^1 - \beta_k^2 = b_k$ and let ω be an arbitrary constant in (0, 1). Consider the operator $G^h = \sum_{k=1}^K G_k^h$ with

$$
\begin{aligned}
(G_k^h f)(hn) = h^k C_n^k \Bigg[& \frac{a_k}{h^2}(f(hn+h) + f(hn-h) - 2f(hn)) \\
& + \frac{\beta_k^1}{h}(f(hn+h) - f(hn)) + \frac{\beta_k^2}{h}(f(hn-h) - f(hn)) \\
& + \sum_{l=[h^{-\omega}]}^{\infty} \left(f(nh+lh) - f(nh) + lh\frac{f(nh-h) - f(nh)}{h} \right) \\
& \times \nu_k(l, h) \Bigg],
\end{aligned}
$$

where the summation index $\ell = [h^{-\omega}]$ denotes the integer part of $h^{-\omega}$ and where

$$\nu_k(l, h) = \max\left(\frac{1}{hl}, \frac{1}{h^2 l^2} \right) \tilde{\nu}_k[lh, lh + h).$$

Theorem 11.3 *For any $h > 0$, under the assumptions of Theorem 11.1 there exists a unique (and hence non-exploding) Markov chain $X^h(t)$ on $h\mathbf{Z}_+$ with generator G^h as given above. If the initial point nh of this chain tends to a point $x \in \mathbf{R}_+$ as $h \to 0$ then the process $X_{nh}^h(t)$ converges in distribution, as $h \to 0$, to the process $X_x(t)$ from Theorem 11.1.*

An approximating interacting-particle system for a given process on $\mathbf{R}_+$ is by no means unique. The essential features of approximations are the following: (i) a k-ary interaction corresponds to pseudodifferential generators $L(x, \partial/\partial x)$ that are polynomials of degree k in x and requires a common scaling of order h^k; (ii) the acceleration of small jumps (the $g_m^k(h)$ in (11.2), of order h^{-2} for small $|m|$) gives rise to a diffusion term; (iii) the slowing down of large jumps gives rise to non-local terms in the limiting generator.

We conclude that the study of measure-valued limits of processes with k-ary interactions leads to the study of Feller processes having non-local pseudo-differential generators with coefficients growing at least polynomially, if K is finite (more precisely, with polynomially growing symbols).

The extension of the above results beyond the methods of perturbation theory and to the case of a finite number of types of interacting particles, leading in the limit to processes on $\mathbf{R}^d$ with polynomially growing symbols, is developed in Kolokoltsov [129]. No serious study, however, seems to exist concerning extension to an infinite number of types or to the corresponding infinite-dimensional (actually measure-valued) limits. An exception is the well-established development of the theory of superprocesses. These appear as measure-valued limits of multiple branching processes that in the general scheme above correspond to the $k = 1$, non-interacting, case in which the coefficients of the limiting infinite-dimensional pseudo-differential generators depend linearly on position.

11.2 Discrete nonlinear Markov games and controlled processes; the modeling of deception

The theory of controlled stochastic Markov processes has a sound place in the literature, owing to its wide applicability in practice; see e.g. Hernandez-Lerma [95], [94]. Here we shall touch upon the corresponding nonlinear extension just to indicate the possible directions of analysis. A serious presentation would require a book in itself.

Nonlinear Markov games can be considered as a systematic tool for the modeling of deception. In particular, in a game of pursuit – evasion – an evading object can create false objectives or hide in order to deceive the pursuer. Thus, observing this object leads not to its precise location but to a distribution of possible locations, implying that it is necessary to build competitive control on the basis of the distribution of the present state. Moreover, by observing the action of the evading object one can draw conclusions about those of its dynamic characteristics that make the predicted transition probabilities depend on the observed distribution. This is precisely the type of situation modeled by nonlinear Markov games.

The starting point for the analysis is the observation that a nonlinear Markov semigroup is just a deterministic dynamic system (though on a rather strange state space of measures). Thus, just as stochastic control theory is a natural extension of deterministic control, we are going to extend stochastic control by turning back to deterministic control, but of measures, thus exemplifying the usual spiral development of science. The next "turn of the screw" would lead to stochastic measure-valued games, forming a stochastic control counterpart for the class of processes discussed in the previous section.

We shall work directly in the competitive control setting (game theory), which of course includes the usual non-competitive optimization as a particular case, but for simplicity only in discrete time and in a finite original state space $\{1, \ldots, n\}$. The full state space is then chosen as a set of probability measures Σ_n on $\{1, \ldots, n\}$.

Suppose that we are given two metric spaces U, V for the control parameters of two players, a continuous transition cost function $g(u, v, \mu)$, $u \in U, v \in V$, $\mu \in \Sigma_n$, and a transition law $\nu(u, v, \mu)$ prescribing the new state $\nu \in \Sigma_n$ obtained from μ once the players have chosen their strategies $u \in U, v \in V$. The problem corresponding to the one-step game (with sequential moves) consists in calculating the Bellman operator B, where

$$(BS)(\mu) = \min_u \max_v [g(u, v, \mu) + S(\nu(u, v, \mu))] \tag{11.4}$$

for a given final cost function S on Σ_n. According to the dynamic programming principle (see e.g. Bellman [27] or Kolokoltsov, and Malafeyev [139]), the dynamic multi-step game solution is given by the iterations $B^k S$. Often of interest is the behavior of this optimal cost $B^k S(\mu)$ as the number of steps k goes to infinity.

Remark 11.4 In game theory one often (but not always) assumes that min, max in (11.4) are interchangeable, leading to the possibility of making simultaneous moves, but we shall not make this assumption.

The function $\nu(u, v, \mu)$ can be interpreted as a controlled version of the mapping Φ, specifying a nonlinear discrete-time Markov semigroup, discussed in Section 1.1. Assume that a stochastic representation for this mapping is chosen, i.e. that

$$\nu_j(u, v, \mu) = \sum_{i=1}^n \mu_i P_{ij}(u, v, \mu)$$

for a given family of (controlled) stochastic matrices P_{ij}. Then it is natural to assume that the cost function g describes the average over random transitions, so that

$$g(u, v, \mu) = \sum_{i,j=1}^n \mu_i P_{ij}(u, v, \mu) g_{ij}$$

for certain real coefficients g_{ij}. Under this assumption the Bellman operator equation (11.4) takes the form

$$(BS)(\mu) = \min_u \max_v \left[\sum_{i,j=1}^n \mu_i P_{ij}(u, v, \mu) g_{ij} + S\Big(\sum_{i=1}^n \mu_i P_{i.}(u, v, \mu)\Big)\right]. \tag{11.5}$$

We can now identify the (not so obvious) place of the usual stochastic control theory in this nonlinear setting. Namely, assume that the matrices P_{ij} do not depend on μ. Even then, the set of linear functions $S(\mu) = \sum_{i=1}^n s_i \mu^i$ on measures (identified with the set of vectors $S = (s_1, \ldots, s_n)$) is not invariant under B. Hence we are automatically reduced not to the usual stochastic control setting but to a game with incomplete information, where the states are probability laws on $\{1, \ldots, n\}$. Thus, when choosing a move the players do not know the present position of the game precisely, but only its distribution. Only when Dirac measures μ form the state space (i.e. there is no uncertainty in the state) will the Bellman operator be reduced to the usual one for stochastic game theory,

$$(\bar{B}S)_i = \min_u \max_v \sum_{j=1}^n P_{ij}(u, v)(g_{ij} + S_j). \tag{11.6}$$

As an example of a nonlinear result, we will obtain now an analog of the result on the existence of an average income for long-lasting games.

Proposition 11.5 *If the mapping ν is a contraction uniformly in u, v, i.e. if*

$$\|\nu(u, v, \mu^1) - \nu(u, v, \mu^2)\| \le \delta \|\mu^1 - \mu^2\| \tag{11.7}$$

for a $\delta \in (0, 1)$, where $\|\nu\| = \sum_{i=1}^n |\nu_i|$, and if g is Lipschitz continuous, i.e.

$$\|g(u, v, \mu^1) - g(u, v, \mu^2)\| \le C \|\mu^1 - \mu^2\| \tag{11.8}$$

for a constant $C > 0$, then there exist a unique $\lambda \in \mathbf{R}$ and a Lipschitz-continuous function S on Σ_n such that

$$B(S) = \lambda + S \tag{11.9}$$

and, for all $g \in C(\Sigma_n)$, we have

$$\|B^m g - m\lambda\| \le \|S\| + \|S - g\|, \tag{11.10}$$

$$\lim_{m \to \infty} \frac{B^m g}{m} = \lambda. \tag{11.11}$$

Proof Clearly for any constant h and a function S one has $B(h+S) = h + B(S)$. Hence one can project B onto an operator $\tilde{B}$ on the factor space $\tilde{C}(\Sigma_n)$

of $C(\Sigma_n)$ with respect to constant functions. Clearly, in the image $\tilde{C}_{\mathrm{Lip}}(\Sigma_n)$ of the set of Lipschitz-continuous functions $C_{\mathrm{Lip}}(\Sigma_n)$, the Lipschitz constant

$$L(f) = \sup_{\mu^1 \neq \mu^2} \frac{|f(\mu^1) - f(\mu^2)|}{\|\mu^1 - \mu^2\|}$$

is well defined (it does not depend on which representative of the equivalence class is chosen). Moreover, from (11.7) and (11.8) it follows that

$$L(BS) \leq 2C + \delta L(S),$$

implying that the set

$$\Omega_R = \{f \in \tilde{C}_{\mathrm{Lip}}(\Sigma_n) : L(f) \leq R\}$$

is invariant under $\tilde{B}$ whenever $R > C/(1 - \delta)$. Since, by the Arzela–Ascoli theorem, Ω_R is convex and compact one can conclude by the Shauder fixed-point principle that $\tilde{B}$ has a fixed point in Ω_R. Consequently there exists a $\lambda \in \mathbf{R}$ and a Lipschitz-continuous function $\tilde{S}$ such that (11.9) holds.

Notice now that B is non-expansive in the usual sup norm, i.e.

$$\begin{aligned}\|B(S_1) - B(S_2)\| &= \sup_{\mu \in \Sigma_n} |(BS_1)(\mu) - (BS_2)(\mu)| \\ &\leq \sup_{\mu \in \Sigma_n} |S_1(\mu) - S_2(\mu)| = \|S_1 - S_2\|.\end{aligned}$$

Consequently, for any $g \in C(\Sigma_n)$,

$$\|B^m g - B^m S\| = \|B^m(g) - m\lambda - S\| \leq \|g - S\|,$$

implying the first formula in (11.10). The second is a straightforward corollary. This second formula also implies the uniqueness of λ (as well as its interpretation as the average income). The proof is complete.

One can extend other results for stochastic multi-step games to this nonlinear setting, say, the turnpike theorems from Kolokoltsov [123] (see also Kolokoltsov, and Malafeyev [139] and Kolokoltsov, and Maslov [141]), and then study nonlinear Markov analogs of differential games but, as we said, we shall not pursue this theme here.

11.3 Nonlinear quantum dynamic semigroups and the nonlinear Schrödinger equation

The Schrödinger and Heisenberg evolutions are given by the semigroups of unitary operators in a Hilbert space. They describe closed quantum systems not interacting with the outer world. This is of course an idealization, as only the whole universe is closed, strictly speaking. The search for more realistic quantum models leads to the theory of open systems, where an evolution is general rather than unitary, but it still has to be "positive" (in some not quite obvious way). Developing a probabilistic interpretation of the processes underlying this kind of positivity has evolved into the field of quantum probability, where functions (classical observables) are replaced by linear operators in Hilbert spaces and measures (mixed states) are replaced by trace-class operators; see e.g. Meyer [187] for a general introduction to the subject. The LLN for interacting quantum particles leads in this setting to nonlinear quantum semigroups and nonlinear quantum Markov processes.

The aim of this section is to sketch the application of duality to the well-posedness problem for nonlinear quantum evolutions. In particular, we aim to demonstrate that the methods of classical Markov processes often suggest natural approaches to quantum counterparts and that aiming at nonlinear extensions leads to useful insights even for linear problems.

In order for the discussion to be reasonably self-contained (we will not assume that the reader is familiar with quantum physics), we shall start with an informal discussion of the Schrödinger and Heisenberg dual formulations of quantum mechanics and then give rigorous definitions of complete positivity and quantum dynamic semigroups. Finally we shall move to the main point, proving a well-posedness result (that is actually rather simple) for nonlinear quantum dynamics. Trivial particular cases of this result are, for example, existence of the solutions to certain nonlinear Schrödinger equations.

We shall fix a separable complex Hilbert space $\mathcal{H}$ with a scalar product that is linear (resp. anti-linear) with respect to the second (resp. first) variable. The standard norm of both vectors and bounded operators in $\mathcal{H}$ will be denoted by $\|.\|$. Let $\mathcal{B}(\mathcal{H})$ (resp. $\mathcal{B}_c(\mathcal{H})$) denote the Banach space of bounded linear operators in $\mathcal{H}$ (resp. its closed subspace of compact operators), and let $\mathcal{B}_1(\mathcal{H})$ denote the space of trace-class operators considered as a Banach space with respect to the trace norm $\|A\|_1 = \text{tr}|A|$, where $|A| = \sqrt{A^*A}$. We shall denote by $\mathcal{B}^{\text{sa}}(\mathcal{H})$, $\mathcal{B}_1^{\text{sa}}(\mathcal{H})$ and $\mathcal{B}_c^{\text{sa}}(\mathcal{H})$ the subspaces of self-adjoint elements of $\mathcal{B}(\mathcal{H})$, $\mathcal{B}_1(\mathcal{H})$ and $\mathcal{B}_c(\mathcal{H})$ respectively.

As we are interested mostly in self adjoint operators, it is instructive to recall (from the spectral theory for compact operators) that any $Y \in \mathcal{B}_c^{sa}(\mathcal{H})$ has a discrete spectral representation given by $Yv = \sum \lambda_j(e_j, v)e_j$, where the e_j form an orthonormal basis in $\mathcal{H}$ and the λ_j are real numbers. Moreover, such an operator belongs to $\mathcal{B}_1^{\text{sa}}(\mathcal{H})$ if and only if the sum $\sum_{j=1}^{\infty} \lambda_j$ is absolutely convergent, in which case

$$\|Y\|_1 = \mathrm{tr}|Y| = \sum_{j=1}^{\infty} |\lambda_j|, \qquad \mathrm{tr}\, Y = \sum_{j=1}^{\infty} \lambda_j.$$

We shall use basic facts about the space $\mathcal{B}_1(\mathcal{H})$, which can be found e.g. in Meyer [187] or Reed and Simon [205]. The most important is that the space $\mathcal{B}(\mathcal{H})$ is the Banach dual to the space $\mathcal{B}_1(\mathcal{H})$, the duality being given explicitly by $\mathrm{tr}(AB)$, $A \in \mathcal{B}(\mathcal{H})$, $B \in \mathcal{B}_1(\mathcal{H})$. In its turn the space $\mathcal{B}_1(\mathcal{H})$ is the Banach dual to the space $\mathcal{B}_c(\mathcal{H})$ (equipped with the usual operator norm).

Recall that an operator $A \in \mathcal{H}$ is called *positive* if it is self-adjoint and satisfies the inequality $(v, Av) \geq 0$ for any $v \in \mathcal{H}$. This notion specifies the *order relation* in $\mathcal{B}(\mathcal{H})$, i.e. $A \leq B$ means that $B - A$ is positive. Of crucial importance for the theory of open systems is the so called *ultraweak* or *normal topology* in $\mathcal{B}(\mathcal{H})$, which is actually the $*$-weak topology of $\mathcal{B}(\mathcal{H})$ as the Banach dual to $\mathcal{B}_1(\mathcal{H})$. In other words, a sequence $X_n \in \mathcal{B}(\mathcal{H})$ converges to X in ultraweak topology whenever $\mathrm{tr}(X_n Y) \to \mathrm{tr}(XY)$ as $n \to \infty$ for any $Y \in \mathcal{B}_1(\mathcal{H})$. We shall denote the unit operator by I.

Let us turn to quantum mechanics. In the Schrödinger picture it is supposed that a quantum system evolves according to the solution to the Cauchy problem

$$i\dot{\psi}_t = H\psi_t, \qquad \psi_0 = \psi, \tag{11.12}$$

where H is a given (possibly unbounded) self-adjoint operator in $\mathcal{H}$ called the *Hamiltonian* or *energy operator*. The unknown vectors ψ_t are called *wave functions* and describe the *states of the quantum system.* Solutions to (11.12) are given by the exponential formula (rigorously defined via operator calculus):

$$\psi_t = \exp(-iHt)\psi. \tag{11.13}$$

Alternatively the operators in $\mathcal{H}$, which represent *observables*, are considered in the Heisenberg picture to evolve according to the *Heisenberg equation*

$$\dot{X}_t = i[H, X_t], \qquad X_0 = X, \tag{11.14}$$

where $[H, X] = HX - XH$ is the commutator. The solution to this problem is easily seen to be given by writing down the "dressing" of X:

$$X_t = \exp(iHt)\, X\, \exp(-iHt). \tag{11.15}$$

The connection between the Schrödinger and Heisenberg pictures is given by the relation

$$(\psi_t, X\psi_t) = (\psi, X_t\psi), \tag{11.16}$$

which is a direct consequence of (11.13) and (11.15). The physical interpretation of (11.16) signifies the equivalence of the Schrödinger and Heisenberg pictures, as the physically measured quantity is not the value of the wave function itself but precisely the quadratic combination $(\psi, A\psi)$ (where ψ is a wave function and A an operator) that represent the expected value of the observable in the state ψ. In order to have a clear probabilistic interpretation one assumes that the wave functions ψ are normalized, i.e. $\|\psi\| = 1$. This normalization is preserved by the evolution (11.13). Hence the state space in the Schrödinger picture is actually the projective space of complex lines (one-dimensional complex subspaces) in $\mathcal{H}$.

Mathematically equation (11.16) is a *duality relation* between states and obervables, and is the quantum counterpart of the usual linear duality between measures and functions in Markov processes. However, in (11.16) this analogy is in disguise, as the expression on the l.h.s. is quadratic with respect to the state. This apparent difference disappears in a more fundamental formulation of the Schrödinger picture. In this new formulation the states of a quantum system are represented not by vectors but rather by positive normalized trace-class operators that in the physics literature are often referred to as *density matrices*, i.e. they form the convex set

$$\{A \in \mathcal{B}_1(\mathcal{H}) : A \geq 0, \ \|A\|_1 = \operatorname{tr} A = 1\}. \tag{11.17}$$

The evolution of a state Y is now given by the equation

$$\dot{Y}_t = -i[H, Y_t], \qquad Y_0 = Y. \tag{11.18}$$

It is clear that equation (11.14) is dual to equation (11.18), with the respect to the usual duality given by the trace, implying the relation

$$\operatorname{tr}(Y_t X) = \operatorname{tr}(Y X_t) \tag{11.19}$$

between their solutions.

Notice now that for any vectors $\psi, \phi \in \mathcal{H}$ the operator $\psi \otimes \phi$, which by definition acts in $\mathcal{H}$ as $(\psi \otimes \phi)v = (\phi, v)\psi$, is a one-dimensional trace-class operator. In particular, operators of the form $\psi \otimes \bar{\psi}$ belong to the set (11.17) for any normalized ψ. To distinguish this class of density matrices, states of the type $\psi \otimes \bar{\psi}$ are called *pure states* while those not of this type are called *mixed states* (in classical probability a pure state is given by a Dirac point mass). As in classical probability, pure states are (the only) extremal points for the state space (11.17). From the spectral representation for self-adjoint elements of $\mathcal{B}_c(\mathcal{H})$ (and the possibility of representing any element as a linear combination of two self-adjoint elements) it follows that linear combinations of operators of the type $\psi \otimes \bar{\psi}$ are dense both in $\mathcal{B}_1(\mathcal{H})$ and in $\mathcal{B}_c(\mathcal{H})$ (taken with their respective Banach topologies). In particular, convex linear combinations of pure states are dense in the set (11.17) of all density matrices.

It is straightforward to see that if ψ_t satisfies (11.12) then the family of one-dimensional operators $\psi_t \otimes \bar{\psi}_t$ satisfies equation (11.18). Thus the duality (11.16) is a particular case of (11.19) for pure states.

The following extends the correspondence between the normalization of vectors and of density matrices.

In the theory of open quantum systems, the evolution of states and observables is not unitary, but it does have to satisfy a rather strong positivity condition: not only has it to be *positive*, i.e. to take positive operators to positive operators, it also has to remain positive when coupled to another system. Without going into the details of the physical interpretation (see, however, the remark below), let us give only the most transparent mathematical definition. A bounded linear map $\Phi : \mathcal{B}(\mathcal{H}) \to \mathcal{B}(\mathcal{H})$ is called *completely positive* if, for all $n \in \mathbf{N}$ and all sequences $(X_j)_{j=1}^n$, $(Y_j)_{j=1}^n$ of the elements of $\mathcal{B}(\mathcal{H})$, one has

$$\sum_{i,j=1}^{n} Y_i^* \Phi(X_i^* X_j) Y_j \geq 0. \tag{11.20}$$

Remark 11.6 It is not difficult to show that a bounded linear map $\Phi : \mathcal{B}(\mathcal{H}) \to \mathcal{B}(\mathcal{H})$ is completely positive if and only if it is positive (i.e. it takes positive operators to positive operators) and if for any $n \in \mathbf{N}$ the linear mapping of the space $M_n(\mathcal{B}(\mathcal{H}))$ of $\mathcal{B}(\mathcal{H})$-valued $n \times n$ matrices given by

$$A = (a_{ij}) \mapsto \Phi_n(A) = (\Phi a_{ij})$$

is a positive mapping in $M_n(\mathcal{B}(\mathcal{H})) = \mathcal{B}(\mathcal{H} \otimes \mathbf{C}^n)$. On the one hand, this property formalizes the underlying physical intuition indicated above and makes it obvious that the composition of any two completely positive maps is also completely positive. On the other hand, definition (11.20) yields a connection with the notion of positive definite functions and kernels; see e.g. Holevo [99] for the full story.

Remark 11.7 We are working in the simplest noncommutative case for evolutions in $\mathcal{B}(\mathcal{H})$. Generally quantum evolutions act in C^*-algebras, which can be inserted into $\mathcal{B}(\mathcal{H})$ as closed subalgebras. Definition (11.20) is straightforwardly extendable to this setting. The classical probability setting fits into this model when the corresponding C^*-algebra is commutative. Any such algebra can be realized as the algebra $C(K)$ of bounded continuous functions on a compact space K, which clearly can be considered as the algebra of multiplication operators in $L_2(K, \mu)$, where the Borel probability measure μ on K is strictly positive on any open subset of K. In this case complete positivity coincides with the usual notion of positivity.

The well-known *Stinespring theorem* gives a full description of completely positive maps between C^*-algebras. When applied to $\mathcal{B}(\mathcal{H})$ it states that any completely positive bounded linear map $\Phi : \mathcal{B}(\mathcal{H}) \mapsto \mathcal{B}(\mathcal{H})$ has the form

$$\Phi(X) = \sum_{i=1}^{\infty} V_i^* X V_i, \tag{11.21}$$

where the V_i and the sum $\sum_{i=1}^{\infty} V_i^* V_i$ belong to $\mathcal{B}(\mathcal{H})$; see e.g. Davies [56] for a proof.

The following definition is fundamental. A *quantum dynamical semigroup* in $\mathcal{B}(\mathcal{H})$ is a semigroup of completely positive linear contractions Φ_t, $t \geq 0$, in $\mathcal{B}(\mathcal{H})$ such that:

(i) $\Phi_t(I) \leq I$;
(ii) all Φ_t are normal (i.e. are ultraweakly continuous mappings);
(iii) $\Phi_t(X) \to X$ ultraweakly as $t \to 0$ for any $X \in \mathcal{B}(\mathcal{H})$.

Such a semigroup is called *conservative* if $\Phi_t(I) = I$ for all t.

Remark 11.8 One can show that the last condition (together with the positivity assumption) implies an apparently stronger condition that $\Phi_t(X) \to X$ strongly as $t \to 0$ for any $X \in \mathcal{B}(\mathcal{H})$. This is, however, still not the strong continuity of the semigroup Φ_t (familiar from the theory of Feller processes), which would mean that $\Phi_t(X) \to X$ in the norm topology as $t \to 0$ for any $X \in \mathcal{B}(\mathcal{H})$.

The crucial *Lindblad theorem* (see [163]) states that if a dynamic semigroup is norm continuous and hence has a bounded generator L then such an L has the form

$$L(X) = \Psi(X) - \tfrac{1}{2}(\Psi(I)X + X\Psi(I)) + i[H, X], \tag{11.22}$$

where H is a self-adjoint element of $\mathcal{B}(\mathcal{H})$ and Ψ is a completely positive mapping in $\mathcal{B}(\mathcal{H})$. Vice versa, a bounded operator of the type (11.22) generates a norm-continuous dynamic semigroup. In view of the Stinespring theorem one can further specify (11.22) writing it in the form

$$L(X) = \sum_{j=1}^{\infty} \left(V_j^* X V_j - \tfrac{1}{2}\left(V_j^* V_j X + X V_j^* V_j\right)\right) + i[H, X], \tag{11.23}$$

where $V_i, \sum_{i=1}^{\infty} V_i^* V_i \in \mathcal{B}(\mathcal{H})$. A straightforward manipulation shows that the corresponding dual evolution on $\mathcal{B}_1(\mathcal{H})$ (in the Schrödinger picture) is the semigroup with generator

$$L'(Y) = \Psi'(X) - \tfrac{1}{2}(\Psi(I)Y + Y\Psi(I)) - i[H, Y], \qquad (11.24)$$

or, using (11.23),

$$L'(Y) = \tfrac{1}{2}\sum_{j=1}^{\infty}\Big([V_jY, V_j^*] + [V_j, YV_j^*]\Big) - i[H, Y], \qquad (11.25)$$

where L' here (and in what follows) denotes the dual operator with respect to the usual pairing given by the trace.

Concrete physically motivated generators often have the form given by (11.22) or (11.23) but with unbounded Ψ, V_i or H. In this case the existence of a corresponding semigroup is not at all obvious. There is an extensive literature on the construction and properties (say, the conservativity) of such dynamic semigroups from a given formal unbounded generator of the type (11.22); see e.g. [48] or [50] and the references therein. However, from the papers [21] and [22], which consider the quantum analogs of the procedures described in Section 1.3, it becomes clear how naturally the nonlinear counterparts of dynamic semigroups appear as the LLN, i.e. the mean field limit for quantum interacting particles. Namely, one is led to a nonlinear equation of the form

$$\dot{Y}_t = L'_Y(Y) = \tfrac{1}{2}\sum_{j=1}^{\infty}\Big([V_j(Y)Y, V_j^*(Y)] + [V_j(Y), YV_j^*(Y)]\Big) - i[H(Y), Y] \qquad (11.26)$$

in $\mathcal{B}_1(\mathcal{H})$, where the operators V_i and H depend additionally on the current state Y.

The aim of this section is to give a well-posedness result for such an equation, which would yield automatically the existence of the corresponding nonlinear quantum dynamic semigroup in $\mathcal{B}_1(\mathcal{H})$. For simplicity we consider the unbounded part of the generator to be fixed, so that it does not depend on the state (see, however, the remarks below). Our strategy will be the same as in the above analysis of nonlinear Lévy processes. Namely, we shall work out a sufficiently regular class of time-nonhomogeneous problems that can be used in the fixed-point approach to nonlinear situations. But first let us fix a class of simple nonhomogeneous models with an unbounded generator, for which the resulting semigroup is strongly continuous. A fruitful additional idea (suggested by analogy with the classical case) is to concentrate on the analysis of the semigroup generated by (11.22) in the space $\mathcal{B}_c(\mathcal{H})$, where finite-dimensional projectors are dense. Then the evolution of states in $\mathcal{B}_1(\mathcal{H})$ generated by (11.24) acts in the dual space.

Let us fix a (possibly unbounded) self-adjoint operator H_0 in $\mathcal{H}$ and a dense subspace D of $\mathcal{H}$ that is contained in the domain of H_0. Assume further that

D itself is a Banach space with respect to the norm $\|.\|_D \geq \|.\|$, that H_0 has a bounded norm $\|H_0\|_{D\to\mathcal{H}}$ as an operator $D \to \mathcal{H}$ and that the exponents $\exp(iH_0t)$ are bounded in D. Let D' denote the Banach dual of D, so that by duality $\mathcal{H}$ is naturally embedded in D' and H_0 is a bounded operator $\mathcal{H} \mapsto D'$ with norm

$$\|H_0\|_{\mathcal{H}\to D'} \leq \|H_0\|_{D\to\mathcal{H}}.$$

The basic example to have in mind is that where H_0 is the Laplacian Δ in $\mathcal{H} = L^2(\mathbf{R}^d)$, the space D being the Sobolev space of functions from $L^2(\mathbf{R}^d)$ with generalized second derivative also from $L^2(\mathbf{R}^d)$ and corresponding dual Sobolev space D'.

Let $\mathcal{D}_c^{\mathrm{sa}}(\mathcal{H})$ be the subspace of $\mathcal{B}_c(\mathcal{H})$ consisting of self-adjoint operators P whose image is contained in D and having finite norm

$$\|P\|_{\mathcal{D}_c(\mathcal{H})} = \|P\|_{H\to D} = \sup\left(\|Pv\|_D : \|v\| = 1\right).$$

On the one hand, when equipped with this norm $\mathcal{D}_c^{\mathrm{sa}}(\mathcal{H})$ clearly becomes a real Banach space. On the other hand, with respect to the usual operator topology, the space $\mathcal{D}_c^{\mathrm{sa}}(\mathcal{H})$ is dense in the subspace of self-adjoint elements of $\mathcal{B}_c(\mathcal{H})$ (linear combinations of one-dimensional operators of type $\psi \otimes \bar{\psi}$ with $\psi \in D$ are already dense). Again by duality, any element of $\mathcal{D}_c^{\mathrm{sa}}(\mathcal{H})$ extends to a bounded operator that takes D' to H (which we shall consistently denote by P also) having norm

$$\|P\|_{D'\to H} \leq \|P\|_{\mathcal{D}_c(\mathcal{H})}.$$

Proposition 11.9 *(i) Referring to* (11.23), *let*

$$H = H_0 + W,$$

where W and the V_i belong to $\mathcal{B}(\mathcal{H})$, W is self-adjoint and $\sum_{i=1}^{\infty} \|V_i\|^2 < \infty$. Then the operator (11.23) *generates a strongly continuous semigroup Φ_t of completely positive contractions in $\mathcal{B}_c^{\mathrm{as}}(\mathcal{H})$ and the operator* (11.25) *generates the strongly continuous semigroup Φ_t' of completely positive contractions in $\mathcal{B}_1^{\mathrm{as}}(\mathcal{H})$.*

(ii) Assume additionally that W and the V_i are also bounded operators in the Banach space D with $\sum_{i=1}^{\infty} \|V_i\|_D^2 < \infty$. Then the space $\mathcal{D}_c^{\mathrm{as}}(\mathcal{H})$ is an invariant core for Φ_t in $\mathcal{B}_c^{\mathrm{as}}(\mathcal{H})$, where Φ_t is a bounded semigroup of linear operators in the space $\mathcal{D}_c^{\mathrm{as}}(\mathcal{H})$ equipped with the Banach topology.

Proof (i) As we are aiming to apply perturbation theory, let us rewrite the operators (11.23) and (11.25) in the forms

$$L(X) = i[H_0, X] + \tilde{L}(X),$$
$$L'(Y) = -i[H_0, X] + \tilde{L}'(Y).$$

As is easily seen, under the assumptions in part (i) of the theorem the operators $\tilde{L}$ and $\tilde{L}'$ are bounded in $\mathcal{B}_c^{\mathrm{as}}(\mathcal{H})$ and $\mathcal{B}_1^{\mathrm{as}}(\mathcal{H})$ (equipped with their respective Banach norms), and hence by the perturbation theory theorem (see Theorem 2.7) it is enough to prove (i) for vanishing V_i, W. Thus one needs to show strong continuity for the contraction semigroups specified by the equations $\dot{X} = i[H, X]$ and $\dot{Y} = -i[H, Y]$ in $\mathcal{B}_c^{\mathrm{as}}(\mathcal{H})$ and $\mathcal{B}_1^{\mathrm{as}}(\mathcal{H})$ respectively. In both cases it is enough to prove this strong continuity for the one-dimensional operators $\phi \otimes \bar{\phi}$ only, as their combinations are dense. In the case of the second evolution above, i.e. the expression for $\dot{Y}$, one has

$$\|e^{-iH_0t}(\phi \otimes \bar{\phi})e^{iH_0t} - \phi \otimes \bar{\phi}\|_1$$
$$= (e^{-iH_0t}\phi - \phi) \otimes e^{iH_0t}\bar{\phi} + \phi \otimes (e^{iH_0t}\bar{\phi} - \bar{\phi}) \leq 2\|e^{-iH_0t}\phi - \phi\|,$$

where we have used Exercise 11.1 below. A similar argument can be given for the first evolution.

(ii) For vanishing V_i, W the statement follows from the explicit formula (11.15), with $H = H_0$, for the solutions and the assumption that the operators $\exp(iH_0t)$ are bounded in D and hence, by duality, also in D'. In order to obtain the required statement in the general case from perturbation theory, one needs to show that $\tilde{L}$ is bounded as an operator in $\mathcal{D}_c^{\mathrm{as}}(\mathcal{H})$. But this holds true, because the multiplication of $X \in \mathcal{D}_c^{\mathrm{as}}(\mathcal{H})$ from the right or left by an operator that is bounded both in D and $\mathcal{H}$ is a bounded operator in $\mathcal{D}_c^{\mathrm{as}}(\mathcal{H})$. □

The following extension of Proposition 11.9 to the nonhomogeneous case is straightforward and we shall omit the proof, which requires the use of Theorem 2.9 instead of Theorem 2.7.

Proposition 11.10 *Under the same assumptions on H_0 as in Proposition 11.9 assume that*

$$H_t = H_0 + W(t),$$

where $W(t)$ and the $V_i(t)$ and $V_i'(t)$ belong to both $\mathcal{B}(\mathcal{H})$ and $\mathcal{B}(D)$ (the latter equipped with the Banach topology) and depend on t strongly continuously in the operator topologies of $\mathcal{B}(\mathcal{H})$ and $\mathcal{B}(D)$; also, $W(t) \in \mathcal{B}^{\mathrm{sa}}(\mathcal{H})$ and

$$\sum_{i=1}^{\infty} \|V_i(t)\|^2_{D \to D} < \infty$$

uniformly for t from compact intervals. Then the family of operators

$$L_t(X) = i[H_t, X] + \sum_{j=1}^{\infty} \left[V_j^*(t) X V_j(t) - \tfrac{1}{2} \left(V_j^*(t) V_j(t) X + X V_j^*(t) V_j(t) \right) \right] \tag{11.27}$$

generates a strongly continuous backward propagator $U^{s,t}$ of completely positive contractions in $\mathcal{B}_c^{\text{as}}(\mathcal{H})$ such that the space $\mathcal{D}_c^{\text{as}}(\mathcal{H})$ is invariant, the operators $U^{s,t}$ reduced to $\mathcal{D}_c^{\text{as}}(\mathcal{H})$ are bounded and, for any $t \geq 0$ and $X \in \mathcal{D}_c^{\text{as}}(\mathcal{H})$, the function $U^{s,t}Y$ is the unique solution in $\mathcal{D}_c^{\text{as}}(\mathcal{H})$ of the inverse Cauchy problem

$$\frac{d}{ds} U^{s,t} Y = L_s(U^{s,t} Y), \qquad s \leq t, \quad U^{t,t} = Y,$$

where the derivative is understood in the sense of the norm topology of $\mathcal{B}(\mathcal{H})$.

We can now obtain the main result of this section.

Theorem 11.11 *Let H_0 be as in Proposition 11.9. Assume that to any density matrix Y, i.e. an operator Y from the set* (11.17), *there correspond linear operators $W(Y)$ and $V_i(Y)$, $i = 1, 2, \ldots,$ that belong to both $\mathcal{B}(\mathcal{H})$ and $\mathcal{B}(D)$ (the latter equipped with the Banach topology) in such a way that*

$$\|Z(Y_1) - Z(Y_2)\| + \|Z(Y_1) - Z(Y_2)\|_D \leq c \sup_{\|P\|_{\mathcal{D}_c(\mathcal{H})} \leq 1} |\text{tr}\,[(Y_1 - Y_2)P]\,|,$$

where Z stands for W, any V_i or $\sum V_i^* V_i$. If

$$H_t = H_0 + W(Y)$$

then the Cauchy problem for equation (11.26) *is well posed in the sense that for an arbitrary $Y \in \mathcal{B}_1^{\text{sa}}(\mathcal{H})$ there exists a unique weak solution $Y_t = T_t(Y) \in \mathcal{B}_1^{\text{sa}}(\mathcal{H})$, with $\|T_t(Y)\| \leq \|Y\|$, to* (11.26), *i.e.*

$$\frac{d}{dt} \text{tr}(PY_t) = \text{tr}\left[L_{Y_t}(P) Y_t \right], \qquad P \in \mathcal{D}_c^{\text{sa}}(\mathcal{H}).$$

The function $T_t(Y)$ depends continuously on t and Y in the norm topology of the dual Banach space $(\mathcal{D}_c^{\text{sa}}(\mathcal{H}))'$.

Proof This is similar to the proof of Theorem 7.3. In fact, it is a direct consequence of the abstract well-posedness discussed in Theorem 2.12 and of Proposition 11.10. □

The solution $T_t(Y)$ to (11.26) mentioned in the above theorem specifies a *nonlinear quantum dynamic semigroup*.

If the $V_i(Y)$ vanish in (11.26), reducing the situation to one of unitary evolutions, Theorem 11.11 is just an abstract version of the well-known well-posedness results for nonlinear Schrödinger equations where the potential depends continuously on the average of some bounded operator in the current state; see e.g. Maslov [173], Hughes, Kato and Marsden [101] or Kato [116]. The case of bounded $H(Y)$, $V_i(Y)$ depending analytically on Y was analyzed in Belavkin [21], [22].

Remark 11.12 Theorem 11.11 can be extended in many directions. For instance, instead of a fixed H_0 one might well consider the family $H_0(Y)$, provided that the subspace D remains fixed. One could possibly extend the general condition for the existence and conservativity of linear dynamic semigroups from Chebotarev and Fagnola [49], [50] to the nonlinear setting. Once the well-posedness of the nonlinear dynamic semigroup is proved, one can undertake the analysis of the corresponding quantum mean field limits for quantum interacting particles, which is similar to the corresponding analysis in the setting of classical statistical physics developed further in this book.

The useful fact obtained in the following exercise establishes a correspondence between the normalization of vectors and of density matrices.

Exercise 11.1 Show that $\|\psi \otimes \phi\|_1 = \|\psi\| \|\phi\|$ for any $\psi, \phi \in \mathcal{H}$. Hint: this result is a two-dimensional fact, where the calculation of traces is explicit.

The next exercise stresses the similarity of quantum generators and classical Markov generators.

Exercise 11.2 Show that the operators L and L' in (11.23) and (11.25) are conditionally positive in the sense that if X (resp. Y) is positive and $(Xv, v) = 0$ (resp. $(Yv, v) = 0$) for a vector $v \in \mathcal{H}$, then $(L(X)v, v) \geq 0$ (resp. $(L'(Y)v, v) \geq 0$). Hint: use the fact (which follows e.g. from the spectral representation) that for a positive X one has $Xv = 0$ if and only if $(Xv, v) = 0$.

The final exercise in this section shows the crucial difference from the analysis of Φ_t in the whole space $\mathcal{B}(\mathcal{H})$.

Exercise 11.3 Show that the semigroup of the equation $\dot{Y} = i[H, Y]$ with self-adjoint H may not be strongly continuous in $\mathcal{B}(\mathcal{H})$. Hint: choose H with the discrete spectrum $\{n^2\}_{n=1}^{\infty}$ and the initial Y to be the shift operator on the corresponding orthonormal basis, i.e. Y takes e_i to e_{i+1}.

11.4 Curvilinear Ornstein–Uhlenbeck processes (linear and nonlinear) and stochastic geodesic flows on manifolds

In this section we discuss examples of stochastic flows on manifolds. It is meant for readers who have no objection to a stroll on curvilinear ground.

Remark 11.13 The differential geometry used here is minimal. In fact, one only needs to have in mind that a compact d-dimensional Riemannian manifold (M, g) is a compact topological space M such that a neighborhood of any point has a coordinate system, i.e. is homeomorphic to an open subset of $\mathbf{R}^d$, and the Riemannian metric g in local coordinates x is described by a positive $d \times d$ matrix-valued function $g(x) = (g_{ij}(x))_{i,j=1}^d$, which under the change of coordinates $x \mapsto \tilde{x}$ transforms as

$$g(x) \mapsto \tilde{g}(\tilde{x}) = \left(\frac{\partial x}{\partial \tilde{x}}\right)^T g(x(\tilde{x}))\frac{\partial x}{\partial \tilde{x}}. \tag{11.28}$$

Matrices with this transformation rule are called $(0, 2)$-tensors. The tangent space to M at a point x is defined as the d-dimensional linear space of the velocity vectors $\dot{x}$ at x of all smooth curves $x(t)$ in M passing through x. Their inner (or scalar) product is defined by g, i.e. it is given by $(\dot{x}_1, g(x)\dot{x}_2)$. The cotangent space $T_x^\star M$ at x is defined as the space of linear forms on $T_x M$. The vectors from $T_x^\star M$ are often called co-vectors. The mapping

$$\dot{x} \mapsto p = g(x)\dot{x}$$

defines the canonical isometry between the tangent and cotangent spaces, the corresponding distance on $T_x^\star M$ being defined as $(G(x)p_1, p_2)$, where $G(x) = g^{-1}(x)$. It is clear (from the differentiation chain rule for tangent vectors and also from the above isomorphism for co-vectors) that under the change of coordinates $x \mapsto \tilde{x}$ the tangent and cotangent vectors transform as

$$\dot{x} \mapsto \dot{\tilde{x}} = \left(\frac{\partial \tilde{x}}{\partial x}\right)\dot{x}, \qquad p \mapsto \tilde{p} = \left(\frac{\partial x}{\partial \tilde{x}}\right)^T p.$$

This implies in particular that the products $(\dot{x}_1, g(x)\dot{x}_2)$ and $(p_1, G(x)p_2)$ are invariant under this change of coordinates and that the form $(\det G(x))^{1/2}dp$ is the invariant Lebesgue volume in $T_x^\star M$. The *tangent (resp. cotangent) bundle* TM (resp. $T^\star M$) is defined as the union of all tangent (cotangent) spaces.

Finally, it is important to have in mind that any Riemannian manifold (M, g) can be isometrically embedded into an Euclidean space $\mathbf{R}^n$ (of possibly higher dimension) via a smooth mapping $r : M \mapsto \mathbf{R}^n$ (in older definitions the manifolds were defined simply as the images of such an embedding). In this case the vectors $\partial r/\partial x^i$, $i = 1, \dots, d$, form a basis in the tangent space $T_x M$ such that, for all $i, j = 1, \dots, d$,

$$\left(\frac{\partial r}{\partial x^i}, \frac{\partial r}{\partial x^j} \right) = g_{ij}. \tag{11.29}$$

Recall that for a smooth function H on $\mathbf{R}^{2d}$ the system of ODEs

$$\begin{cases} \dot{x} = \dfrac{\partial H}{\partial p}, \\ \dot{p} = -\dfrac{\partial H}{\partial x} \end{cases} \tag{11.30}$$

is called a *Hamiltonian system* with *Hamiltonian function* (or just *Hamiltonian*) H. In particular, the Newton system of classical mechanics

$$\begin{cases} \dot{x} = p, \\ \dot{p} = -\dfrac{\partial V}{\partial x}, \end{cases} \tag{11.31}$$

is a Hamiltonian system with $H(x, p) = p^2/2 + V(x)$, where the function V is called the *potential* or the *potential energy* and $p^2/2$ is interpreted as the *kinetic energy*. Free motion corresponds, of course, to the case of constant V.

More general Hamiltonians H appear in many situations, in particular when one is considering mechanical systems on a non-flat space, i.e. on a manifold. For instance, the analog of the free motion $\dot{x} = p$, $\dot{p} = 0$ on a Riemannian manifold (M, g) is called the *geodesic flow* on M and is defined as the Hamiltonian system on the cotangent bundle $T^\star M$ specified by the Hamiltonian

$$H(x, p) = \frac{1}{2}(G(x)p, p), \qquad G(x) = g^{-1}(x) \tag{11.32}$$

(H again describes the kinetic energy, but in a curvilinear space). The geodesic flow equations are then

$$\begin{cases} \dot{x} = G(x)p, \\ \dot{p} = -\dfrac{1}{2}\left(\dfrac{\partial G}{\partial x} p, p \right). \end{cases} \tag{11.33}$$

The solutions to the system (11.33) (or more often their projections on M) are called *geodesics* on M.

In physics, a heat bath is often used to provide a stochastic input to the system of interest. Mathematically this is expressed by adding a homogeneous noise to the second equation of the Hamiltonian system, i.e. by changing (11.30) to

$$\begin{cases} \dot{x} = \dfrac{\partial H}{\partial p}, \\ dp = -\dfrac{\partial H}{\partial x} dt + dY_t, \end{cases} \tag{11.34}$$

where Y_t is a Lévy process. In the most-studied models, Y_t stands for Brownian motion (BM), with variance proportional to the square root of the temperature. To balance the energy pumped into the system by the noise, one often adds friction to the system, i.e. a non-conservative force proportional to the velocity. In the case of initial free motion this yields the system

$$\begin{cases} \dot{x} = p, \\ dp = -\alpha p\, dt + dY_t \end{cases} \tag{11.35}$$

with non-negative matrix α, which is an Ornstein–Uhlenbeck (OU) system driven by the Lévy noise Y_t. Especially well studied are the cases when Y_t is BM or a stable process (see e.g. Samorodnitski and Taqqu [217]). We aim to construct their curvilinear analogs.

If a random force is not homogeneous, as would be the case on a manifold or in a nonhomogeneous medium, one is led to consider Y_t in (11.34) to be a process depending on the position x, and this leads naturally to the Lévy processes depending on a parameter studied in this chapter. In particular the curvilinear analog of the OU system (11.35) is the process in $T^\star M$ specified by the equations

$$\begin{cases} \dot{x} = \dfrac{\partial H}{\partial p}, \\ dp = -\dfrac{\partial H}{\partial x} dt - \alpha(x) p\, dt + dY_t(x), \end{cases} \tag{11.36}$$

where H is given by (11.32). Assuming for simplicity that the Y_t are zero-mean Lévy processes with Lévy measure absolutely continuous with respect to the invariant Lebesgue measure on $T_x^\star M$ and having finite outer first moment ($\int_{|y|>1} |y|\nu(dy) < \infty$), the generator of Y_t is given by

$$L_Y^x f(p)$$
$$= \tfrac{1}{2}(A(x)\nabla, \nabla) f(p) + \int [f(p+q) - f(p) - \nabla f(p) q] \frac{[\det G(x)]^{1/2}\, dq}{\omega(x, q)}$$

for a certain positive $\omega(x,q)$. Hence the corresponding full generator of the process given by (11.36) has the form

$$Lf(x,p) = \frac{\partial H}{\partial p}\frac{\partial f}{\partial x} - \frac{\partial H}{\partial x}\frac{\partial f}{\partial p} - \left(\alpha(x)p, \frac{\partial f}{\partial p}\right) + \tfrac{1}{2}\operatorname{tr}\left(A(x)\frac{\partial^2 f(x,p)}{\partial p^2}\right)$$
$$+ \int \left(f(x,p+q) - f(x,p) - \frac{\partial f(x,p)}{\partial p}q\right)\frac{[\det G(x)]^{1/2}\,dp}{\omega(x,p)}. \tag{11.37}$$

Of course, in order to have a correctly defined system on a manifold, this expression should be invariant under a change of coordinates, which requires certain transformation rules for the coefficients α, A, ω. This is settled in the next statement.

Proposition 11.14 *The operator* (11.37) *is invariant under the change of coordinates*

$$x \mapsto \tilde{x}, \qquad p \mapsto \tilde{p} = \left(\frac{\partial x}{\partial \tilde{x}}\right)^T p$$

if and only if ω is a function on $T^\star M$, α is a $(1,1)$ *tensor and A is a* $(0,2)$ *tensor, i.e.*

$$\begin{aligned}
\tilde{\omega}(\tilde{x},\tilde{p}) &= \omega(x(\tilde{x}), p(\tilde{x},\tilde{p})),\\
\tilde{\alpha}(\tilde{x}) &= \left(\frac{\partial x}{\partial \tilde{x}}\right)^T \alpha(x(\tilde{x}))\left(\frac{\partial \tilde{x}}{\partial x}\right)^T,\\
\tilde{A}(\tilde{x}) &= \left(\frac{\partial x}{\partial \tilde{x}}\right)^T A(x(\tilde{x}))\frac{\partial x}{\partial \tilde{x}}.
\end{aligned} \tag{11.38}$$

Proof Let U denote the functional transformation

$$Uf(x,p) = f(\tilde{x}(x), \tilde{p}(x,p)) = f\left(\tilde{x}(x), \left(\frac{\partial x}{\partial \tilde{x}}\right)^T p\right).$$

We have to show that $U^{-1}LU = \tilde{L}$, where

$$\tilde{L}f(\tilde{x},\tilde{p}) = \frac{\partial H}{\partial \tilde{p}}\frac{\partial f}{\partial \tilde{x}} - \frac{\partial H}{\partial \tilde{x}}\frac{\partial f}{\partial \tilde{p}} - \left(\tilde{\alpha}(\tilde{x})\tilde{p}, \frac{\partial f}{\partial \tilde{p}}\right) + \tfrac{1}{2}\operatorname{tr}\left(\tilde{A}(\tilde{x})\frac{\partial^2 f(\tilde{x},\tilde{p})}{\partial \tilde{p}^2}\right)$$
$$+ \int \left(f(\tilde{x},\tilde{p}+\tilde{q}) - f(\tilde{x},\tilde{p}) - \frac{\partial f(\tilde{x},\tilde{p})}{\partial \tilde{p}}\tilde{q}\right)\frac{[\det \tilde{G}(\tilde{x})]^{1/2}d\tilde{q}}{\tilde{\omega}(\tilde{x},\tilde{q})}.$$

The invariance of the part containing H is known (since geodesic flow is well defined). Thus it is sufficient to analyze the case $H = 0$. Consider the integral term L_{int}. One has

$$L_{\text{int}}Uf(x,p) = \int \left[f\left(\tilde{x}(x), \left(\frac{\partial x}{\partial \tilde{x}}\right)^T (p+q)\right) - f\left(\tilde{x}, \left(\frac{\partial x}{\partial \tilde{x}}\right)^T p\right) \right.$$
$$\left. - \frac{\partial f(\tilde{x}(x), \tilde{p}(x,p))}{\partial \tilde{p}} \left(\frac{\partial x}{\partial \tilde{x}}\right)^T q \right] \frac{[\det G(x)]^{1/2}\, dq}{\omega(x,q)},$$

and

$$U^{-1}L_{\text{int}}Uf(\tilde{x},\tilde{p})$$
$$= \int \left[f\left(\tilde{x}, \tilde{p} + \left(\frac{\partial x}{\partial \tilde{x}}\right)^T q\right) - f(\tilde{x},\tilde{p}) - \frac{\partial f(\tilde{x},\tilde{p})}{\partial \tilde{p}} \left(\frac{\partial x}{\partial \tilde{x}}\right)^T q \right]$$
$$\times \frac{\left[\det G(x(\tilde{x}))\right]^{1/2}\, dq}{\omega(x(\tilde{x}), q)}$$
$$= \int \left[f(\tilde{x}, \tilde{p}+z) - f(\tilde{x},\tilde{p}) - \frac{\partial f(\tilde{x},\tilde{p})}{\partial \tilde{p}} z \right] \frac{[\det \tilde{G}(\tilde{x})]^{1/2}\, dz}{\omega(x(\tilde{x}), (\partial\tilde{x}/\partial x)^T z)},$$

since, due to (11.28),

$$[\det \tilde{G}(\tilde{x})]^{1/2} = \det\left(\frac{\partial \tilde{x}}{\partial x}\right) [\det G(x(\tilde{x}))]^{1/2}.$$

This implies the first equation in (11.38). The other two formulas are obtained similarly. □

Exercise 11.4 Prove the last two formulas in (11.38).

In particular, if one is interested in processes depending only on the Riemannian structure it is natural to take the metric tensor g to be the tensor A and the function ω to be a function of the energy H; the tensor α can be taken as the product $g(x)G(x)$. For instance, (11.36) defines a *curvilinear OU process* (or *stochastic geodesic flow* in the case $\alpha = 0$), of *diffusion type* if the generator of Y_t is given by

$$L_Y^x f(p) = \tfrac{1}{2}(g(x)\nabla, \nabla) f(p)$$

and is of the *β-stable type*, $\beta \in (0,2)$, if the generator of Y_t is given by

$$L_Y^x f(p) = \int [f(p+q) - f(p) - \nabla f(p) q] \frac{[\det G(x)]^{1/2}\, dq}{(q, G(x)q)^{(\beta+1)/2}}.$$

Theorem 11.15 *If the Riemannian metric g is twice continuously differentiable, the stochastic system* (11.36)*, with either a diffusive or a β-stable Y_t and with either $\alpha = 0$ or $\alpha = gG$, has a unique solution specifying a Markov process in $T^\star M$.*

Proof By localization it is enough to show well-posedness for a stopped process in a cylinder $U \times \mathbf{R}^d$, for any coordinate domain $U \subset M$. By Proposition 11.14 the system is invariant under the change in coordinates. Finally, by Theorem 3.11 the process is well defined in $U \times \mathbf{R}^d$. Namely, by perturbation theory one reduces the discussion to the case of Lévy measures with a bounded support and in this case the coupling is given explicitly by Corollary 3.9. □

An alternative way to extend OU processes to a manifold is by embedding the manifold in a Euclidean space. Namely, observe that one can write $dY_t = (\partial/\partial x)x\,dY_t$ in $\mathbf{R}^n$, meaning that adding a Lévy-noise force is equivalent to adding the singular nonhomogeneous potential $-x\dot{Y}_t$ (the position multiplied by the noise) to the Hamiltonian function. Assume now that a Riemannian manifold (M, g) is embedded in the Euclidean space R^n via a smooth mapping $r : M \mapsto \mathbf{R}^n$ and that the random environment in $\mathbf{R}^n$ is modeled by the Lévy process Y_t. The position of a point x in $\mathbf{R}^n$ is now $r(x)$, so that the analog of xY_t is the product $r(x)Y_t$ and the term Y_t from (11.34) has as curvilinear modification the term

$$\left(\frac{\partial r}{\partial x}\right)^T dY_t = \left\{\sum_{j=1}^n \frac{\partial r^j}{\partial x^i} dY_t^j\right\}_{i=1}^d,$$

which yields the projection of the "free noise" Y_t onto the cotangent bundle of M at x (by (11.29)). In particular, the *stochastic (or stochastically perturbed) geodesic flow induced by the embedding* r can be defined by the stochastic system

$$\begin{cases} \dot{x} = G(x)p, \\ dp = -\dfrac{1}{2}\left(\dfrac{\partial G}{\partial x}p, p\right)dt + \left(\dfrac{\partial r}{\partial x}\right)^T dY_t, \end{cases} \tag{11.39}$$

which represents simultaneously the natural stochastic perturbation of the geodesic flow (11.33) and the curvilinear analog of the stochastically perturbed free motion $\dot{x} = p, dp = dY_t$.

Proposition 11.16 *Let $g(x)$ and $r(x)$ be twice continuously differentiable mappings and Y_t a Lévy process in $\mathbf{R}^n$ specified by the generator equation*

$$\begin{aligned} L_Y f(x) &= \tfrac{1}{2}(A\nabla, \nabla)f(x) + (b, \nabla f(x)) \\ &\quad + \int \left[f(x+y) - f(x) - (\nabla f(x), y)\right]\nu(dy), \end{aligned}$$

with $\int \min(|y|, |y|^2)\nu(dy) < \infty$ (the latter assumption is made for technical simplification; extension to arbitrary Lévy processes is not difficult). Then the *stochastic geodesic process is well defined by the system* (11.39) *and represents a Markov process in* $T^\star M$.

Proof The generator L of the process specified by (11.39) has the form $L = L_1 + L_2 + L_3$ with

$$L_1 f(x,p) = \left(G(x)p, \frac{\partial f}{\partial x}\right) - \frac{1}{2}\left(\frac{\partial G}{\partial x}(x)p, p\right)\frac{\partial f}{\partial p},$$

$$L_2 f(x,p) = \frac{1}{2}\operatorname{tr}\left(\left(\frac{\partial r}{\partial x}\right)^T A\frac{\partial r}{\partial x}\nabla^2 f\right) + \left(\left(\frac{\partial r}{\partial x}\right)^T b, \frac{\partial f}{\partial p}\right)$$

$$L_3 f(x,p) = \int\left[f\left(x, p + \left(\frac{\partial r}{\partial x}\right)^T q\right) - f(x,p) - \frac{\partial f}{\partial p}(x,p)\left(\frac{\partial r}{\partial x}\right)^T q\right] \times \nu(dq).$$

Invariance is now checked as in Proposition 11.14. In particular, for an integral operator A given by

$$Af(x,p) = \int\left(f(x, p+\omega(x)q) - f(x,p) - \frac{\partial f}{\partial p}(x,p)\omega(x)q\right)\nu(dy),$$

one shows that $U^{-1}AU = \tilde{A}$ if and only if

$$\tilde{\omega}(\tilde{x}) = \left(\frac{\partial x}{\partial \tilde{x}}\right)^T \omega(x(\tilde{x})),$$

i.e. the columns of the matrix ω are (co-)vectors in $T_x^\star M$. The rest of the proof is the same as for Theorem 11.15; we will omit the details. □

Notice that the process defined by system (11.39) depends on the embedding. However, if the free process Y_t is a Brownian motion, i.e. if $\nu = 0$, $b = 0$ and A is the unit matrix, the generator L is now given by

$$Lf(x,p) = \left(G(x)p, \frac{\partial f}{\partial x}\right) - \frac{1}{2}\left(\frac{\partial G}{\partial x}(x)p, p\right)\frac{\partial f}{\partial p} + \frac{1}{2}\operatorname{tr}\left[g(x)\nabla^2 f\right];$$

it does not depend on the embedding and coincides with the diffusive geodesic flow analyzed in Theorem 11.15.

One can now obtain the well-posedness of nonlinear curvilinear Ornstein–Uhlenbeck processes or geodesic flows. Consider, say, a nonlinear system that arises from the potentially interacting flows from Theorem 11.15, namely the system

$$\begin{cases} \dot{X}_t = \dfrac{\partial H}{\partial p}(X_t, P_t), \\ dP_t = -\dfrac{\partial H}{\partial x}(X_t, P_t)\,dt - \displaystyle\int \dfrac{\partial V(X_t, y)}{\partial x}\mathcal{L}_{X_t}(dy) \\ \qquad\qquad - \alpha(X_t)P_t\,dt + dY_t(X_t), \end{cases} \tag{11.40}$$

where $H(x, p) = (G(x)p, p)/2$, $V(x, y)$ is a smooth function on the manifold M and $\mathcal{L}_\xi$ means the law of ξ.

Theorem 11.17 *Suppose that the "interaction potential" V is smooth and that the other coefficients of this stochastic system satisfy the assumptions of Theorem 11.15 and are also thrice continuously differentiable. Then the solution X_t, P_t is well defined for any initial distribution X_0, P_0.*

Proof This again follows from Theorem 2.12 and a straightforward nonhomogeneous extension of Theorem 11.15, taking in account the regularity theorem, Theorem 3.17. □

Exercise 11.5 Obtain the corresponding nonlinear version of Theorem 4.14.

11.5 The structure of generators

Now we consider the infinitesimal generators of positivity-preserving nonlinear evolutions on measures. In particular, it will be shown that the general class of evolutions obtained above as the LLN limit for interacting particles, at least for polynomial generators, also arises naturally just from the assumption of positivity of the evolution (a nonlinear version of Courrège's theorem).

Here we shall deal with a nonlinear analog of (6.1), namely

$$\frac{d}{dt}(g, \mu_t) = \Omega(\mu_t)g, \tag{11.41}$$

which holds for g from a certain dense domain D of $C(\mathbf{R}^d)$ if Ω is a nonlinear transformation from a dense subset of $\mathcal{M}(X)$ to the space of linear functionals on $C(X)$ with a common domain containing D.

In Section 6.8 the case of bounded generators was discussed. Here we deal with the general situation. We shall start with an appropriate extension of the notion of conditional positivity.

Suppose that D is a dense subspace of $C_\infty(X)$ and K is a closed subset of X. We shall say that a linear form $A : D \mapsto \mathbf{R}$ is *K-conditionally positive* if

$A(g) \geq 0$ whenever $g \in D$ is non-negative and vanishes in K. The following obvious remark yields a connection with the standard notion of conditional positivity. A linear operator $A : D \mapsto B(X)$ is conditionally positive if the linear form $Ag(x)$ is $\{x\}$-conditionally positive for any $x \in X$. The motivation for introducing K-conditional positivity is given by the following simple but important fact.

Proposition 11.18 *If the solutions to the Cauchy problem of* (6.2) *are defined at least locally for initial measures μ from a subset $M \subset \mathcal{M}(X)$ (so that* (6.2) *holds for all $g \in D$) and preserve positivity, then $\Omega(\mu)$ is* $\mathrm{supp}(\mu)$-*conditionally positive for any $\mu \in M$. In this case we shall say for brevity that $\Omega(\mu)$ is conditionally positive in the class M.*

Proof Let a non-negative $g \in D$ be such that $g|_K = 0$ for $K = \mathrm{supp}(\mu)$. Then $(g, \mu) = 0$, and consequently the condition of positivity-preservation implies that $(d/dt)(g, \mu_t)|_{t=0} = \Omega(\mu)g \geq 0$. □

Of special interest are evolutions with classes of initial measures containing the set $\mathcal{M}_\delta^+(X)$ of finite positive linear combinations of Dirac measures (in a probabilistic setting this allows one to start a process at any fixed point). As a consequence of Courrège's theorem, one obtains the following characterization of conditional positivity for any fixed measure from $\mathcal{M}_\delta^+(X)$.

Proposition 11.19 *Suppose that $X = \mathbf{R}^d$ and the space D from the definition above contains $C_c^2(X)$. If a linear operator A is $\mathbf{x} = \{x_1, \ldots, x_m\}$-conditionally positive, then*

$$A(g) = \sum_{j=1}^{m} \Big\{ c^j g(x_j) + (b^j, \nabla) g(x_j) + \tfrac{1}{2}(G^j \nabla, \nabla) g(x_j) + \int \big[g(x_j + y) - g(x_j) - \mathbf{1}_{B_1}(y)(y, \nabla) g(x_j)\big] \nu^j(dy) \Big\} \tag{11.42}$$

for $g \in C_c^2(X)$, where the G^j are positive definite matrices and the ν^j are Lévy measures.

Proof Let us choose as a partition of unity a family of n non-negative functions $\chi_i \in C_c^2(X)$, $i = 1, \ldots, n$, such that $\sum_{i=1}^m \chi_i = 1$ and each χ_i equals 1 in a neighborhood of x_i (and consequently vanishes in a neighborhood of any other point x_l for which $l \neq i$). By linearity $A = \sum_{i=1}^m A_i$, where $A_j g = A(\chi_j g)$. Clearly each functional $A_i g$ is x_i-conditionally positive. Hence, applying a "fixed-point version" (see Remark 2.27) of Courrège's theorem to each Ω_i, one obtains the representation (11.42). □

Now let Ω be a mapping from $\mathcal{M}(\mathbf{R}^d)$ to linear forms (possibly unbounded) in $C(\mathbf{R}^d)$, with a common domain D containing $C_c^2(\mathbf{R}^d)$, such that $\Omega(\mu)$ is conditionally positive in $\mathcal{M}_\delta^+(\mathbf{R}^d)$. Assume that $\Omega(0) = 0$ and that $\Omega(\mu)g$ is continuously differentiable in μ for $g \in D$ in the sense that the variational derivative of $\Omega(\mu; x)g$ is well defined, continuous in x and weakly continuous in μ (see Lemma F.1 for the variational derivatives). Then

$$\Omega(\mu)g = \left(\int_0^1 \frac{\delta \Omega}{\delta \mu}(s\mu; \cdot)g \, ds, \mu \right),$$

i.e. equation (6.2) can be rewritten as

$$\frac{d}{dt}(g, \mu_t) = (A(\mu_t)g, \mu_t) = (g, A^*(\mu_t)\mu_t), \tag{11.43}$$

for some linear operator $A(\mu)$ depending on μ. As we saw in the previous chapter, this form of nonlinear equation arises from the mean field approximation to interacting-particle systems, in which

$$\begin{aligned} A(\mu)g(x) &= c(x, \mu)g(x) + (b(x, \mu), \nabla)g(x) + \tfrac{1}{2}(G(x, \mu)\nabla, \nabla)g(x) \\ &\quad + \int \left[g(x + y) - g(x) - \mathbf{1}_{B_1}(y)(y, \nabla)g(x) \right] \nu(x, \mu; dy); \end{aligned} \tag{11.44}$$

here c, b, G, and ν depend continuously on μ and x, each G is a non-negative matrix and each ν is a Lévy measure.

We shall show that conditional positivity forces $A(\mu)$ from (6.1) to have the form (11.44), assuming additionally that the linearity mapping Ω is polynomial, i.e. the equation has the form

$$\frac{d}{dt}(g, \mu_t) = \sum_{k=1}^{K} \int \cdots \int (A_k g)(x_1, \ldots, x_k)\mu_t(dx_1) \cdots \mu_t(dx_k), \tag{11.45}$$

where each A_k is a (possibly unbounded) operator from a dense subspace D of $C_\infty(X)$ to the space $C^{\text{sym}}(X^k)$ of symmetric continuous bounded functions of k variables from X.

Specifying the definition of conditional positivity to this case, we shall say that a linear map

$$A = (A_1, \ldots, A_K) : D \quad \mapsto \quad (C(X), C^{\text{sym}}(X^2), \ldots, C^{\text{sym}}(X^K)) \tag{11.46}$$

is *conditionally positive* in $\mathcal{M}_\delta^+(\mathbf{R}^d)$ if, for any collection of different points $x_1, \ldots, x_m$ of X, for any non-negative function $g \in D$ such that $g(x_j) = 0$ for all $j = 1, \ldots, m$ and for any collection of positive numbers ω_j one has

$$\sum_{k=1}^{K}\sum_{i_1=1}^{m}\cdots\sum_{i_k=1}^{m}\omega_{i_1}\cdots\omega_{i_k}A_k g(x_{i_1},\dots x_{i_k})\geq 0. \tag{11.47}$$

In particular, a linear operator $A_k : D \mapsto C^{\text{sym}}(X^k)$ is *conditionally positive* if (11.47) holds for a family $A = (0, \dots, 0, A_k)$ of the type (11.46).

The following result yields the structure of conditionally positive polynomial nonlinearities.

Theorem 11.20 *Suppose again that $X = \mathbf{R}^d$ and D contains $C_c^2(X)$. A linear mapping* (11.46) *is conditionally positive if and only if*

$$\begin{aligned}
&A_k g(x_1,\dots,x_k)\\
&=\sum_{j=1}^{k}\Big\{(c_k(x_j,\mathbf{x}\setminus x_j)g(x_j)\\
&\qquad+(b_k(x_j,\mathbf{x}\setminus x_j),\nabla)g(x_j)+\tfrac{1}{2}(G_k(x_j,\mathbf{x}\setminus x_j)\nabla,\nabla)g(x_j)\\
&\qquad+\int\big[g(x_j+y)-g(x_j)-\mathbf{1}_{B_1}(y)(y,\nabla)g(x_j)\big]\nu_k(x_j,\mathbf{x}\setminus x_j;dy)\Big\},
\end{aligned} \tag{11.48}$$

for $\mathbf{x} = (x_1, \dots, x_k)$, where each G_k is a symmetric matrix and each ν_k is a possibly signed measure on $\mathbf{R}^d \setminus \{0\}$, with $\int \min(1, |y|^2)\,|\nu_k|(\mathbf{x}, dy) < \infty$, such that

$$\sum_{k=1}^{K}k\sum_{i_1=1}^{m}\cdots\sum_{i_{k-1}=1}^{m}\omega_{i_1}\cdots\omega_{i_{k-1}}G_k(x,x_{i_1},\dots,x_{i_{k-1}}) \tag{11.49}$$

is positive definite and the measure

$$\sum_{k=1}^{K}k\sum_{i_1=1}^{m}\cdots\sum_{i_{k-1}=1}^{m}\omega_{i_1}\cdots\omega_{i_{k-1}}\nu_k(x,x_{i_1},\dots,x_{i_{k-1}}) \tag{11.50}$$

is positive for any m and any collection of positive numbers $\omega_1, \dots, \omega_m$ and points $x, x_1, \dots, x_m$. Moreover, c_k, b_k, G_k and ν_k are symmetric with respect to permutations of all arguments apart from the first and depend continuously on $x_1, \dots, x_k$ (the measures ν_k are considered in the weak topology).

Proof For arbitrary fixed $x_1, \dots, x_m$, the functional of g given by

$$\sum_{k=1}^{K}\sum_{i_1=1}^{m}\cdots\sum_{i_k=1}^{m}\omega_{i_1}\cdots\omega_{i_k}A_k g(x_{i_1},\dots,x_{i_k})$$

is ($x = (x_1, \ldots, x_m)$)-conditionally positive and consequently has the form (11.42), as follows from Proposition 11.19. Using $\epsilon\omega_j$ instead of ω_j, dividing by ϵ and then letting $\epsilon \to 0$ one obtains that $\sum_{i=1}^m A_1 g(x_i)$ also has the same form. As m is arbitrary, this implies on the one hand that $A_1 g(x)$ has the same form for arbitrary x (thus giving the required structural result for A_1) and on the other hand that

$$\sum_{k=2}^{K} \epsilon^{k-2} \sum_{i_1=1}^{m} \cdots \sum_{i_k=1}^{m} \omega_{i_1} \cdots \omega_{i_k} A_k g(x_{i_1}, \ldots, x_{i_k})$$

has the required form. Note, however, that by using subtraction we may destroy the positivity of the matrices G and measures ν. Again letting $\epsilon \to 0$ yields the same representation for the functional

$$\sum_{i_1=1}^{m} \sum_{i_2=1}^{m} \omega_{i_1} \omega_{i_2} A_2 g(x_{i_1}, x_{i_2}).$$

As above this implies on the one hand that

$$\omega_1^2 A_2 g(x_1, x_1) + 2\omega_1\omega_2 A_2 g(x_1, x_2) + \omega_2^2 A_2 g(x_2, x_2),$$

and hence also $A_2 g(x_1, x_1)$, has the required form for arbitrary x_1 (to see this put $\omega_2 = 0$ in the previous expression). Therefore $A_2 g(x_1, x_2)$ has this form for arbitrary x_1, x_2 (thus giving the required structural result for A_2). On the other hand the same reasoning implies that

$$\sum_{k=3}^{K} \epsilon^{k-3} \sum_{i_1=1}^{m} \cdots \sum_{i_k=1}^{m} \omega_{i_1} \cdots \omega_{i_k} A_k g(x_{i_1}, \ldots, x_{i_k})$$

has the required form. Following this procedure inductively yields the same representation for all the A_k.

Consequently,

$$\sum_{k=1}^{K} \sum_{i_1=1}^{m} \cdots \sum_{i_k=1}^{m} \omega_{i_1} \cdots \omega_{i_k} A_k g(x_{i_1}, \ldots, x_{i_k})$$
$$= \sum_{k=1}^{K} k \sum_{l=1}^{m} \sum_{i_1=1}^{m} \cdots \sum_{i_{k-1}=1}^{m} \omega_l \omega_{i_1} \cdots \omega_{i_{k-1}}$$

$$\times \Bigg\{ c_k(x_l, x_{i_1}, \ldots, x_{i_{k-1}}) g(x_l) + (b_k(x_l, x_{i_1}, \ldots, x_{i_{k-1}}), \nabla) g(x_l)$$
$$+ \tfrac{1}{2}(G_k(x_l, x_{i_1}, \ldots, x_{i_{k-1}})\nabla, \nabla) g(x_l)$$
$$+ \int \big[g(x_j + y) - g(x_j) - \mathbf{1}_{B_1}(y)(y, \nabla) g(x_j) \big]$$
$$\times \nu_k(x_l, x_{i_1}, \ldots, x_{i_{k-1}}; dy) \Bigg\}.$$

As this functional has to be strongly conditionally positive, the required positivity property of G and ν follows from Proposition 11.19. The required continuity follows from the assumption that A maps continuous functions into continuous functions. □

Taking into account explicitly the symmetry of the generators in (11.45) (note also that in (11.45) the use of nonsymmetric generators or their symmetrizations would specify the same equation) allows us to obtain useful equivalent representation for this equation. This form is also convenient if one is interested in the strong form of the equation in terms of densities. Namely, the following statement is obvious.

Corollary 11.21 *Under the assumptions of Theorem 11.20, equation* (11.45) *can be written equivalently in the form*

$$\frac{d}{dt}(g, \mu_t) = \sum_{k=1}^{K} k \int (A_k^1 g)(x, y_1, \ldots, y_{k-1}) \mu_t(dx) \mu_t(dy_1) \cdots \mu_t(dy_{k-1}), \tag{11.51}$$

where $A_k^1 : C_\infty(X) \mapsto C^{\mathrm{sym}}(X^k)$ *is by*

$$\begin{aligned} A_k^1 g(x, y_1, \ldots, y_{k-1}) &= a_k(x, y_1, \ldots, y_{k-1}) g(x) + (b_k(x, y_1, \ldots, y_{k-1}), \nabla g(x)) \\ &\quad + \tfrac{1}{2}(G_k(x, y_1, \ldots, y_{k-1})\nabla, \nabla) g(x) + \Gamma_k(y_1, \ldots, y_{k-1}) g(x), \end{aligned} \tag{11.52}$$

with

$$\begin{aligned} \Gamma_k(y_1, \ldots, y_{k-1}) g(x) &= \int \big[g(x+z) - g(x) - \chi(z)(z, \nabla) g(x) \big] \nu_k(x, y_1, \ldots, y_{k-1}; dz). \end{aligned} \tag{11.53}$$

As we saw in the previous chapter, in models of interacting particles equation (11.45) often appears in the form

$$\frac{d}{dt}(g, \mu_t) = \sum_{k=1}^{K} \frac{1}{k!} \int (B_k g^{\oplus})(x_1, \ldots, x_k) \mu_t(dx_1) \cdots \mu_t(dx_k), \tag{11.54}$$

where the linear operators B_k act in $C_\infty(X^k)$ and

$$g^\oplus(x_1, \dots, x_k) = g(x_1) + \cdots + g(x_n).$$

Clearly, if the linear operators B_k in $C_\infty(X^k)$, $k = 1, \dots, K$, are conditionally positive in the usual sense then the forms $g \mapsto B_k g^\otimes(x_1, \dots, x_n)$ on $C_\infty(X)$ (or a dense subspace of $C_\infty(X)$) are $\{x_1, \dots, x_n\}$- conditionally positive. The inverse also holds.

Corollary 11.22 *Let X and D be the same as in Proposition 11.19. A linear operator $A_k : D \mapsto C^{\mathrm{sym}}(X^k)$ specifies $\{x_1, \dots, x_k\}$-conditionally positive forms $A_k(x_1, \dots, x_k)$ if and only if A_k has the form given by* (11.48), *in which each G_k (resp. ν_k) is a positive definite matrix (resp. a Lévy measure), and the functions a_k, b_k, c_k, ν_k of variables $x_1, \dots, x_k$ are symmetric with respect to permutations not affecting x_1. Equivalently, A_k is given by*

$$A_k g(x_1, \dots, x_k) = \sum_{j=1}^k c_k(x_j, \mathbf{x} \setminus x_j) g(x_j) + B_k g^\oplus(x_1, \dots, x_k), \quad (11.55)$$

where B_k is a conservative operator in $C_\infty^{\mathrm{sym}}(X^k)$ that is conditionally positive (in the usual sense). If $A_k g = (1/k!) B_k g^\oplus$ then $A_k^1 = (1/k!) B_k \pi$ in (11.51), *where the lifting operator π is given by $\pi g(x_1, \dots, x_k) = g(x_1)$.*

Proof The first statement is obvious. By Courrège's theorem applied to X^k, a conditionally positive B_k in $C_\infty(X^k)$ is given by

$$\begin{aligned} B_k f(x_1, \dots, x_k) &= \tilde{a}(x_1, \dots, x_k) f(x_1, \dots, x_k) \\ &\quad + \sum_{j=1}^k (\tilde{b}^j(x_1, \dots, x_k), \nabla_{x_j}) f(x_1, \dots, x_k)) \\ &\quad + \tfrac{1}{2} (\tilde{c}(x_1, \dots, x_k) \nabla, \nabla) f(x_1, \dots, x_k) \\ &\quad + \int \Big[f(x_1 + y_1, \dots, x_k + y_k) - f(x_1, \dots, x_k) \\ &\qquad - \sum_{i=1}^k \mathbf{1}_{B_1}(y_i)(y_i, \nabla_{x_i}) f(x_1, \dots, x_k) \Big] \\ &\qquad \times \tilde{\nu}(x_1, \dots, x_k; dy_1 \cdots dy_k). \end{aligned} \quad (11.56)$$

Applying this to $f = g^\oplus$ and comparing with (11.48) yields the required result. □

Corollary 11.23 *In the case $K = 2$ the family (A_1, A_2) is conditionally positive in $\mathcal{M}_\delta^+(\mathbf{R}^d)$ if and only if $A_1(x)$ is x-conditionally positive for all x,* *and $A_2(x_1, x_2)$ is (x_1, x_2)-conditionally positive for all x_1, x_2, i.e. if A_1 and A_2 have the form* (11.48), *where G_1 and G_2 (resp. ν_1 and ν_2) are positive definite matrices (resp. Lévy measures). In particular, for $K = 2$ equation* (11.45) *can always be written in the form* (11.54). *In physical language this means that a quadratic mean field dependence can always be realized by a certain binary interaction (this is not the case for discrete X; see Exercise 11.7 below).*

Proof In the case $K = 2$ the positivity of (11.49), say, reads as the positivity of the matrix

$$c_1(x) + 2\sum_{i=1}^{m} \omega_i c_2(x, x_i) \tag{11.57}$$

for all natural m, positive numbers ω_j and points $x, x_j, j = 1, \ldots, m$. Hence c_1 is always positive (to see this put $\omega_j = 0$ for all j). To prove the claim, one has to show that $c(x, y)$ is positive definite for all x, y. But if there exist x, y such that $c(x, y)$ is not positive definite then, by choosing a large enough number of points $x_1, \ldots, x_m$ near y, one would get a matrix of the form (11.57), which is not positive definite (even when all $\omega_j = 1$). The positivity of measures ν is analyzed similarly. This contradiction completes the proof. □

Some remarks about and examples of these results are in order. As in the linear case we shall say that $A(\mu)g$ is conservative if $A(\mu)\phi_n(x)$ tends to zero as $n \to \infty$, where $\phi_n(x) = \phi(x/n)$ and ϕ is an arbitrary function from $C_c^2(\mathbf{R}^d)$ that equals 1 in a neighborhood of the origin and has values in $[0, 1]$. For operators given by (11.44) this is of course equivalent to the condition that c vanishes.

Remark 11.24 It is clear that, in the mapping (11.46), if only two components, say A_i and A_j, do not vanish, and A is conditionally positive then both non-vanishing components A_i and A_j are also conditionally positive (take $\epsilon\omega_j$ instead of ω_j in the definition and then pass to the limits $\epsilon \to 0$ and $\epsilon \to \infty$). In the case where there are more than two non-vanishing components in the family A, the analogous statement is false. Namely, if A is conditionally positive, the "boundary" operators A_1 and A_K are conditionally positive as well (using the same argument as before), but the intermediate operators A_k need not be, as is shown by the following simple example.

Exercise 11.6 Let $A = (A_1, A_2, A_3)$, where

$$A_i g(x_1, \ldots, x_i) = a_i \left[\Delta g(x_1) + \cdots + \Delta g(x_i)\right], \qquad i = 1, 2, 3,$$

with $a_1 = a_3 = 1$. Show that if a_2 is a small enough negative number then A is conditionally positive but its component A_2 is not. Write down an explicit solution to equation (11.45) in this case. Hint: the positivity of (11.49) follows from

$$1 + 2a_2 \sum_{i=1}^{m} \omega_i + 3\Big(\sum_{i=1}^{m} \omega_i\Big)^2 \geq 0.$$

Remark 11.25 We have given results for $X = \mathbf{R}^d$, but using localization arguments (like those for the linear case, see [43]) the same results can be easily extended to closed manifolds. It is seemingly possible to characterize the corresponding boundary conditions in the same way (again generalizing the linear case from [43]), though this is not so straightforward.

Remark 11.26 The basic conditions for the positivity of (11.49), (11.50) can be written in an alternative, integral, form. For example, the positivity of the matrices in (11.49) is equivalent (at least for bounded continuous functions c) to the positivity of the matrices c_k given by

$$\sum_{k=1}^{K} k \int c_k(x, y_1, \ldots, y_{k-1})\mu(dy_1)\cdots\mu(dy_{k-1}) \tag{11.58}$$

for all non-negative Borel measures $\mu(dy)$. (This can be obtained either from the limit of integral sums of the form (11.49) or directly from the conditional positivity of the r.h.s. of (11.51).) Hence the conditions (11.49), (11.50) actually represent modified multi-dimensional matrix-valued or measure-valued versions of the usual notion of positive definite functions. For instance, in the case $k = K = 3$ and $d = 1$, (11.58) means that $\int c_3(x, y, z)\omega(y)\omega(z)\,dydz$ is a non-negative number for any non-negative integrable function ω. The usual notion of a positive definite function c_3 (as a function of the variables y, z) would require the same positivity for arbitrary (not necessarily positive) integrable ω.

Remark 11.27 Corollary 11.23 cannot be extended to $K > 2$. Of course, if each $A_k g(x_1, \ldots, x_m)$ is $\{x_1, \ldots, x_m)$-conditionally positive then the mapping (11.46) is conditionally positive in $\mathcal{M}_\delta^+$, but not vice versa. In a simple example of a conditionally positive operator, in the case $k = K = 3$ without conditionally positive components, the operator is defined by

$$\begin{aligned} A_3 g(x_1, x_2, x_3) &= \cos(x_2 - x_3)\Delta g(x_1) + \cos(x_1 - x_3)\Delta g(x_2) \\ &\quad + \cos(x_1 - x_2)\Delta g(x_3), \end{aligned} \tag{11.59}$$

where Δ is the Laplacian operator. Clearly $A_3 g(x_1, x_2, x_3)$ is not $\{x_1, x_2, x_3\}$-conditionally positive, but the positivity of (11.49) holds. The strong form of equation (11.51) in this case is

$$\frac{d}{dt} f_t(x) = 3\Delta f_t(x) \int f_t(y) f_t(z) \cos(y-z)\, dy\, dz. \tag{11.60}$$

The "nonlinear diffusion coefficient" given by the integral is not strictly positive here. Namely, since

$$\int f(y) f(z) \cos(y-z)\, dydz = \tfrac{1}{2}(|\hat{f}(1)|^2 + |\hat{f}(-1)|^2),$$

where $\hat{f}(p)$ is the Fourier transform of f, this expression does not have to be strictly positive for all non-vanishing non-negative f. However, one can find an explicit solution to the Cauchy problem of (11.60):

$$f_t = \frac{1}{\sqrt{2\pi \omega_t}} \int \exp\left(-\frac{(x-y)^2}{2\omega_t}\right) f_0(y)\, dy,$$

where

$$\omega_t = \ln\left[1 + t(|\hat{f}_0(1)|^2 + |\hat{f}_0(-1)|^2\right].$$

This is easily obtained by passing to the Fourier transform of equation (11.60), which has the form

$$\frac{d}{dt}\hat{f}_t(p) = -\tfrac{1}{2}p^2(|\hat{f}_t(1)|^2 + |\hat{f}_t(-1)|^2)\hat{f}_t(p)$$

and is solved by observing that $\xi_t = |\hat{f}_t(1)|^2 + |\hat{f}_t(-1)|^2$ solves the equation $\dot{\xi}_t = -\xi_t^2$ and consequently equals $(t + \xi_0^{-1})^{-1}$.

Remark 11.28 There is a natural "decomposable" class of operators for which (11.45) reduces straightforwardly to a linear problem. Namely, suppose that $k!A_k g = B_k g^+ = (\tilde{B}_k g)^+$ for all $k = 1, \ldots, K$, for some conservative conditionally positive $\tilde{B}_k$ in $C_\infty(X)$ (in particular $\tilde{B}_k 1 = 0$). Then (11.51) takes the form

$$\frac{d}{dt}(g, \mu_t) = \sum_{k=1}^{K} \frac{1}{(k-1)!}(\tilde{B}_k g, \mu_t)\|\mu_t\|^{(k-1)},$$

which is a linear equation depending on $\|\mu_t\| = \|\mu_0\|$ as parameter.

The following example illustrates the idea of the conditional positivity of the generator of a Markov evolution in the simplest nonlinear situation.

Exercise 11.7 Of special interest for applications is the case of infinitesimal generators depending quadratically on μ (see the next section), which leads to the system of quadratic equations

$$\dot{x}_j = (A^j x, x), \qquad j = 1, 2, \ldots, N, \tag{11.61}$$

where N is a natural number (or more generally $N = \infty$), the unknown $x = (x_1, x_2, \ldots)$ is an N-dimensional vector and the A^j are given square $N \times N$ matrices. Suppose that $\sum_{j=1}^N A^j = 0$. Show that the system (11.61) defines a positivity-preserving semigroup (i.e. if all the coordinates of the initial vector x^0 are non-negative then the solution $x(t)$ is globally uniquely defined and all coordinates of this solution are non-negative for all times), if and only if for each j the matrix $\tilde{A}^j$ obtained from A^j by deleting its jth column and jth row is such that $(\tilde{A}^j v, v) \geq 0$ for any $v \in \mathbf{R}^{N-1}$ with non-negative coordinates. Hint: this condition expresses the fact that if $x_j = 0$ and the other x_i, $i \neq j$, are non-negative, then $\dot{x}_j \geq 0$.

11.6 Bibliographical comments

The first chapter introduced Markov models of interacting particles and their LLN limit, avoiding technical details. A probability model for the deduction of kinetic equations was first suggested by Leontovich [160], who analyzed the case of a discrete state space. For some cases where the state space is continuous, the same approach was developed by Kac [113], McKean [181], [182] and Tanaka [238], [239]; the last-mentioned paper contained the first models of not only binary but also kth-order interactions. For coagulation processes the use of such a method for the deduction of Smoluchovski's equation was suggested by Marcus [169] and Lushnikov [165]. For general Hamiltonian systems the corresponding procedure was carried out by Belavkin and Maslov [26]. The deduction of kinetic equations of Boltzmann type using Bogolyubov chains (see Appendix Appendix J) was suggested in Bogolyubov [42]. The creation–annihilation operator formalism for the analysis of Bogolyubov chains was proposed by Petrina and Vidibida [199] and further developed in Maslov and Tariverdiev [177]. The well-posedness problem for Bogolyubov chains was studied by several authors; see e.g. Sinai and Suchov [221] and references therein. The deduction of the general kth-order interaction equations (1.70) using the Bogolyubov chain approach was carried out in Belavkin and Kolokoltsov [25]. Belavkin [21], [22] developed quantum analogs of the kinetic equations describing a dynamic law of large numbers for particle systems; the evolution of these equations is described by quantum dynamic semigroups and leads to the type of nonlinear equations discussed in Section 11.3. Belavkin [21] also suggested an elegant Hamiltonian approach to the deduction of kinetic equations. Namely, for an evolution of the type (1.68) let us define the Hamiltonian function

$$H(Q, Y) = \sum_{l=0}^{k} \frac{1}{l!} \left((I_l[P^l, A^l]Q^{\otimes})^l, Y^{\otimes l} \right) \tag{11.62}$$

and consider the corresponding infinite-dimensional Hamiltonian system

$$\dot{Y} = \frac{\delta H}{\delta Q}, \qquad \dot{Q} = -\frac{\delta H}{\delta Y}.$$

One can easily see that since $H(\mathbf{1}, Y) = 0$ (by the conservativity of the generator), the above Hamiltonian system has solutions with $Q = 1$ identically. Under this constraint the first Hamiltonian equation, $\dot{Y} = \delta H / \delta Q$, coincides with the kinetic equation (1.70). This yields also a natural link with the semi-classical analysis. Namely, one can easily show (for the details see Belavkin and Kolokoltsov [25]) that in terms of generating functionals the evolution (J.7) is given by the following equation in terms of functional derivatives:

$$h \frac{\partial}{\partial t} \tilde{\Phi}_{\rho_t}(Q) = H\left(Q, h \frac{\delta}{\delta Q} \right) \tilde{\Phi}_{\rho_t}(Q). \tag{11.63}$$

Thus the above Hamiltonian system possesses quasi-classical asymptotics. For a general and accessible introduction to the mathematical theory of interacting particles we refer to the books of Liggett [162], Kipnis and Landim [120], and de Masi and Presutti [61], and the volume edited by Accardi and Fagnola [1] on quantum interactions.

Chapter 2 collected some mostly well-known results of the modern theory of Markov processes, with the emphasis on the connections with analysis (semi-groups, evolution equations etc). Special examples were given (for example in Section 2.4) and more general formulations of some facts were presented (for example at the end of Section 2.1 and in Section 2.4).

Chapter 3 started with a new development in SDEs driven by nonlinear Lévy noise extending the author's paper [137]. The idea here was to solve SDEs with noise depending on a parameter linked with the position of the solution itself, that is to say SDEs driven by noise deriving from feedback from the evolution of the process. Expressed in this general form, this idea had been used already in Kunita [150] and Carmona and Nualart [46]. However, in our setting the dependence of the noise on a parameter is expressed through the generator, so that whether the construction of the process itself varies sufficiently regularly with the evolution of this parameter becomes a problem. Settling this issue properly leads to a spectacular application to the theory of Markov processes, reconciling the analytical theory of general Markov semigroups with the SDE approach. The resulting construction of Feller processes specified by general Lévy–Khintchine-type operators with Lipschitz-continuous coefficients actually represents a probabilistic proof of the convergence of a certain T-product (a chronological product). This T-product has a natural link with the evolution equation, which can be shown using the method of frozen coefficients, well known in the analysis of Ψ DO. In Section 3.3 we explained the connection with the usual stochastic calculus, sketching Ito's approach to the construction of classical SDEs driven by Lévy and/or Poisson noise, as explained in detail in Stroock's monograph [227]. It would seem that Proposition 3.16, on an approximation scheme based on nonlinear functions of increments, is new.

In Chapter 4 we introduced some analytical approaches to the study of Markov processes, the most relevant for the following exposition. Sections 4.2 and 4.3 were based on the Appendix to [136] and developed for arbitrary state spaces an old idea on the construction of countable Markov chains that goes back to Feller and Kolmogorov; see e.g. Anderson [6] for a modern treatment. The results of Sections 4.4 and 4.5 appear to be new. The last section reviewed other developments in the theory. For a more substantial treatment of the subject the reader is referred to the fundamental monograph of Jacob [103]. Among the topics relating to Chapters 3 and 4, but (regrettably) not developed there, one should mention processes living on domains with a non-empty boundary and related boundary problems for PDEs and ΨDEs. The literature on the boundary-value problem for parabolic equations is of course enormous. For much less studied generators having both a diffusive and a jump part, one could consult e.g. Taira [233] or Kolokoltsov [130] for recent results and a short review.

Chapter 5 presented a systematic development of the Lyapunov or barrier function method for processes with Lévy–Khintchine-type pseudo-differential generators having unbounded coefficients (previously used mainly for diffusions and jump-type processes; see e.g. the monographs of Freidlin [78] or Ethier and Kurtz [74])). Special attention was paid to the description of appropriate function spaces where the corresponding semigroup is strongly continuous (yielding a useful extension of the notion of Feller semigroups) and to the invariant domains of the generator. These aspects are seemingly new, even in the well-developed theory of diffusions with unbounded coefficients.

Chapters 6 and 7 were devoted to the rigorous mathematical construction of solutions to kinetic equations. The literature on this subject is extensive; the review given below is not meant to be exhaustive.

The results in Section 6.2 are mostly well known. However, we put together various approaches on the level of generality that unify a number of particular results on bounded generators that serve as starting points for the analysis of unbounded-coefficient extensions of particular models. Sections 6.3–6.6

followed the author's paper [132], presenting the unification and extension of a variety of particular situations analyzed previously by many authors (see the review below). The existence result of Theorem 6.7 can be essentially improved. Namely, the assumption of the existence of a finite moment $(1 + E^\beta, \mu)$ with $\beta > 1$ is not needed. To work without this assumption, two ideas are used. First, if $(1+E, \mu) < \infty$, there exists an increasing smooth function G on $\mathbf{R}_+$ such that $(G(1+E), \mu) < \infty$ (this is a well-known, and not very difficult, general measure-theoretic result; see e.g. [188] for a proof); second, one has to work with the barrier function $G(1 + E)$ instead of the moment L^β. Mishler and Wennberg [188] give a corresponding treatment of the Boltzmann equation and Laurencot and Wrzosek [157] consider a coagulation model. The results in Sections 6.7 and 6.8 are new.

The results in Sections 7.1 and 7.2 are also new; they complement the constructions of nonlinear stable-like processes obtained in Kolokoltsov [134] and extend, to rather general Lévy–Khintchine-type generators, results previously available for nonlinear diffusions. In Section 7.3 we simply formulated the results obtained in [137], yielding a universal approach to the probabilistic interpretation of nonlinear evolutions by means of distribution-dependent SDEs driven by nonlinear Lévy noise. In Section 7.4 we indicated how some classes of sufficiently regular evolutions with unbounded coefficients can be treated using the direct nonlinear duality approach applied earlier in the bounded-coefficient case.

The main streams of developments in nonlinear kinetic equations in the literature are those devoted to nonlinear diffusions, the Boltzmann equation and coagulation–fragmentation processes, some basically identical techniques having been developed independently for these models. The interplay between these directions of research is developing quickly, especially in connection with spatially nontrivial models of interaction. The theory of nonlinear diffusions, pioneered by McKean [181], is well established; see Gärtner [82] for a review of early developments. New advances are mainly in the study of the Landau–Fokker–Planck equation, (1.82), which presents a certain diffusive limit for the Boltzmann equation. This nonlinear diffusion equation has a degeneracy of a kind that is intractable by a direct application of the McKean approach. The probabilistic analysis of this equation is based on an ingenious solution in terms of SDEs driven by space–time white noise due to Guérin [89], [90]; see also [91] for further regularity analysis based on the Malliavin calculus. Namely, the starting point of this analysis is a nice observation that if $A = \sigma\sigma^T$ in (1.82), σ being a $d \times n$ matrix, the Landau equation is solved by the distributions of the solutions to the stochastic differential equation

$$X_t = X_0 + \int_0^t \int_0^1 \sigma(X_s - Y_s(\alpha)) W^n(d\alpha\, ds) + \int_0^t \int_0^1 b(X_s - Y_s(\alpha))\, d\alpha\, ds,$$

where W^n is the n-dimensional space–time white noise (i.e. a zero-mean Gaussian $\mathbf{R}^n$-valued random measure on $\mathbf{R}_+ \times [0, 1]$ with $\mathbf{E} W_i^n W_j^n(d\alpha\, ds) = \delta_i^j\, d\alpha\, ds$), the pair of processes (X, Y) is defined on the product probability space $(\Omega, \mathcal{F}, \mathcal{F}_t, P) \times ([0, 1], \mathcal{B}([0, 1]), d\alpha)$ and the distributions of X and Y are taken as coinciding. Analytically, a rigorous treatment of the Landau–Fokker–Planck equation (1.82) was initiated by Arseniev and Buryak [13]; see Desvillettes and Villani [65] and Goudon [85], and references therein, for a wealth of further development.

There is of course plenty of related activity on various nonlinear parabolic equations, reaction–diffusion equations, etc. arising in various areas in the natural sciences (nonlinear wave propagation, super-processes, gravity etc.), which we have not aimed to review; see e.g. Smoller [224], Maslov and Omel'yanov [176], Dynkin [69] and Biler and Brandolese [37]. Nonlinear SPDEs driven by space–time white noise are also being actively studied (see e.g. Crisan and Xiong [55], Kurtz and Xiong [152], [153]), often using approximations of the Ito type. In general, the branching-particle mechanism for Monte-Carlo-type approximations to the solutions of stochastic equations have become very popular in applications; see e.g. the monograph Del Moral [64] and references therein.

The literature on the Boltzmann equation is immense, even if one concentrates on the mathematical analysis of the spatially homogeneous model (and the related mollified equation), which are particular cases of the general model analyzed in Chapter 6. Let us review only the development of the key well-posedness facts. The first was obtained by Carleman [45], for the class of rapidly decreasing continuous functions, by means of quite specific representations that are available for the Boltzmann equation. The L_1-theory was developed by Arkeryd [10] using an approach similar to that used later for coagulation models; it is incorporated in the general results of Sections 6.2 and 6.4. The L^∞-theory of solutions was developed by Arkeryd [11] and the L^p-theory by Gustafsson [92], [93]. The above-mentioned works are mostly concerned with the so-called hard potential interactions with a cutoff, where, roughly speaking, the collision kernel $B(|v|, \theta)$ is bounded by $|v|^\beta$, $\beta \in (0, 1]$. Further extensions under various growth and singularity assumptions have been developed by many authors; see e.g. Mishler and Wennberg [188], Lu and Wennberg [164] and references therein. The propagation of smoothness in integral norms, i.e. well-posedness theory in Sobolev spaces, was developed by Mouhot and Villani [191], and the propagation of smoothness in uniform norms, i.e. well-posedness in the spaces of continuously differentiable functions, was developed in Kolokoltsov [133] under the rather general assumption of polynomially growing collision kernels. Namely, suppose that

$$B(|v|, \theta) = |v|^{\beta} \cos^{d-2} \theta h(\theta)$$

in (1.52) for a bounded function h. Then, for any β, d, the energy- and mass-preserving solution f_t to the Boltzmann equation is well defined for any initial non-negative function f (the density of the initial measure μ) having finite energy and entropy. It was shown in [133] that there exist bounds, which are global in time, for the uniform moments $\sup_x [f_t(x)(1 + |x|)^r]$ of the solution f_t (and also for its spatial derivatives) in terms of the initial integral moments $\int [f(x)(1 + |x|)^s] \, dx$. These bounds depend in a nontrivial way on the relations between the parameters β, d, s, r, but can be obtained in a more or less explicit form.

The mathematical analysis of Smoluchovski's coagulation–fragmentation equation was initiated by Ball and Carr in [17], where an equation with a discrete mass distribution was analyzed subject to additive bounds for the rates. Various discrete models of interactions were unified in Kolokoltsov [131]: discrete versions of the results of Section 6.4 were given. Related results were obtained by Lachowicz [154]. Well-posedness for weak measure-valued kinetic equations for coagulation, in continuous time and in a general measurable space, was obtained by Norris [194]. When coagulation alone is taken into consideration, Theorems 6.10 and 6.12 yield the Norris well-posedness result. The method used by Norris was a little different, as it relied heavily on the monotonicity built into the coagulation model. Using the methods of complex analysis, Dubovskii and Stewart [68] obtained well-posedness for coagulation–fragmentation processes in the class of measures with exponentially decreasing continuous densities. As coagulation evolution does not preserve the number of particles, the underlying nonlinear process, unlike in the Boltzmann case, is sub-Markov. However, since the mass is preserved one can "change the variable" by considering the process to be one of mass evolution, which then becomes Markovian. This idea was exploited by Deaconu, Fournier and Tanré [62] to give an alternative probabilistic interpretation of Smoluchovski evolution yielding naturally a well-posedness result for infinite initial measure (the scaled number of particles) but finite total mass. Much modern literature is devoted to the qualitative behavior of coagulation–fragmentation models, i.e. their gelation (involving the non-conservativity of mass), self-similarity and long-time behavior, which are not studied in this monograph; see e.g. Ernst and Protsinis [72], Fournier and Laurencot [77], Lushnikov and Kulmala [166], da Costa, Roessel and Wattis [52] and references therein.

Another stream of activity is connected with spatially nontrivial models, mainly with coagulating particles that move in space according to a Brownian motion with a parameter depending on the mass; see e.g. Collet and Poupaud [51] and Wrzosek [249] for discrete mass distributions and [5], [156], [195] for continuous masses. Finally, we refer to Eibeck and Wagner [70], Kolodko, Sabelfeld and Wagner [122] and references therein for the extensive work on numerical solutions to the Smoluchovski equation. Various special classes of coagulation and fragmentation processes are dealt with in the monograph Bertoin [34].

The systematic development of the theory of smoothness of solutions to kinetic equations with respect to the initial data given in Chapter 8 is a novelty of the book. The exposition in this chapter in some places extends and in some places complements results from the author's papers [133], [134], [136]. In Bailleul [14] the smoothness results of Section 8.3 were extended to treat smoothness with respect to a parameter in the coagulation kernel. This sensitivity of the Smoluchovski equation is important in numerical analysis; see Kraft and Vikhansky [149] and Bailleul [14].

Among the topics related to Part II, but not touched upon, we mention the development of nonlinear evolutions on lattices (by Zegarlinski in [254] and by Olkiewicz, Xu and Zegarlinski in [197]), the evolution of nonlinear averages, in particular with applications to financial market analysis, by Maslov (see [174], [175]) and the theory of nonlinear Dirichlet forms (see Sipriani and Grillo [222] and Jost [112]).

Chapter 9 dealt with the dynamic LLN. Section 9.3 contained a unified exposition of the traditional approach to proving the LLN (one proves the tightness of particle approximation systems and picks up a converging subsequence) for rather general jump-type interactions. It included the basic LLNs for binary coagulations (see Norris [194]) and for Boltzmann collisions (see e.g. Sznitman [231] and [232]). Other sections in Chapter 9 were devoted to estimating the rates of convergence in the LLN for interactions with unbounded rates, using the smoothness of the kinetic equations with respect to the initial data. They were based mostly on [134], [136]. The results in Section 9.5 are new. Convergence estimates in the LLN for Boltzmann-type collisions with bounded rates supplemented by a spatial movement (the mollified Boltzmann equation) were obtained by Graham and Méléard [87], [88] using the method of Boltzmann trees.

Turning to Chapter 10, let us mention first of all that the CLT for interacting diffusions has been thoroughly developed; see Dawson [58] and Giné and Wellner [83] and references therein. The CLT for coagulation processes with finite rates and discrete mass was established by Deaconu, Fournier and

Tanré [63]. For the Boltzmann equation with bounded rates it was obtained by Méléard [185] in the more general setting of collisions supplemented by a spatial movement (the mollified Boltzmann equation). The method in these papers is different from ours; instead of working out analytical estimates of the smoothness of kinetic equations with respect to the initial data, it is based on the direct analysis of infinite-dimensional SDEs underlying the limiting Ornstein–Uhlenbeck (OU) process.

The results on the CLT for stable-like processes from Section 10.2 complement and improve the results of the author's paper [134], where non-degenerate stable-like processes (described by Theorem 4.25 and hence having additional regularity properties) were analyzed. The results on the Smoluchovski equation with unbounded rates were based on the author's paper [136]; the estimates of convergence in the CLT for the Boltzmann equation with unbounded rates seem to be new.

Interesting applications of the circle of ideas around the CLT limit for evolutionary games can be found in Mobilia, Georgiev and Tauber [189] and Reichenbach, Mobilia and Frey [208].

The analysis of infinite-dimensional OU processes just touched upon in Section 10.4 represents a vast area of research, for which one can consult the above-mentioned papers of Méléard as well as for example Lescot and Roeckner [161], van Neerven [244], Dawson *et al.* [60] and references therein.

In Chapter 11 we have considered the various developments, indicating possible directions and perspectives of further research. Its results are due mostly to the author.

The controlled nonlinear Markov processes touched upon in a general way in Section 11.2 are well under investigation in the setting of McKean nonlinear diffusions; see [100] and references therein.

In Section 11.4 we applied the methods of Chapter 3 to the construction of Markov processes on manifolds and also extended to Lévy-type noise the analytical construction of stochastic geodesic flows with Brownian noise suggested in [124]. Geometrical constructions of Ornstein–Uhlenbeck processes on manifolds driven by Browninan noise were given in [67], [117].

Section 11.5 dealt with a nonlinear counterpart of the notion of conditional positivity developed in Kolokoltsov [134]. A characterization of "tangent vectors" in the space of probability laws at the Dirac point measures, see Proposition 11.19, which follows from the Courrège representation of linear conditionally positive functionals, was given by Stroock [227]. In this book Stroock posed the question of characterizing "tangent vectors" to arbitrary measures. Theorem 11.20, taken from Kolokoltsov [134], solves this problem for polynomial vector fields, which can be extended to analytic vector fields. The discussion in Section 11.5 established formal links between general positivity-preserving evolutions and mean field limits for interacting particles.

The appendices contain some technical material used in the main body of the text. The results are well known apart from, possibly, some in Appendices E, G and H.

Appendices

Appendix A Distances on measures

The properties of separability, metrizability, compactness and completeness for a topological space S are crucial for the analysis of S-valued random processes. Here we shall recall the basis relevant notions for the space of Borel measures, highlighting the main ideas and examples and omitting lengthy proofs.

Recall that a topological (e.g. metric) space is called *separable* if it contains a countable dense subset. It is useful to have in mind that separability is a topological property, unlike, say, completeness, which depends on the choice of distance. (For example, an open interval and the line $\mathbf{R}$ are homeomorphic, but the usual distance is complete for the line and not complete for the interval). The following standard examples show that separability cannot necessarily be assumed.

Example A.1 The Banach space l_∞ of bounded sequences of real (or complex) numbers $a = (a_1, a_2, \ldots)$ equipped with the sup norm $\|a\| = \sup_i |a_i|$ is not separable, because its subset of sequences with values in $\{0, 1\}$ is not countable but the distance between any two such (not coinciding) sequences is 1.

Example A.2 The Banach spaces $C(\mathbf{R}^d)$, $L_\infty(\mathbf{R}^d)$, $\mathcal{M}^{\text{signed}}(\mathbf{R}^d)$ are not separable because they contain a subspace isomorphic to l_∞.

Example A.3 The Banach spaces $C_\infty(\mathbf{R}^d)$, $L_p(\mathbf{R}^d)$, $p \in [1, \infty)$, are separable; this follows from the Stone–Weierstrass theorem.

Recall that a sequence of finite Borel measures μ_n is said to converge *weakly* (resp. *vaguely*) to a measure μ as $n \to \infty$ if (f, μ_n) converges to (f, μ) for any $f \in C(S)$ (resp. for any $f \in C_c(S)$). If S is locally compact, the Riesz–Markov theorem states that the space of finite signed Borel measures is the Banach dual to the Banach space $C_\infty(S)$. This duality specifies the *$\star$-weak topology* on measures, where the convergence μ_n to μ as $n \to \infty$ means that (f, μ_n) converges to (f, μ) for any $f \in C_\infty(S)$.

Example A.4 Let $S = \mathbf{R}$. The sequence $\mu_n = n\delta_n$ in $\mathcal{M}(\mathbf{R})$ converges vaguely but not $\star$-weakly. The sequence $\mu_n = \delta_n$ converges $\star$-weakly but not weakly.

Proposition A.5 *Suppose that p_n, $n \in \mathbf{N}$, and p are finite Borel measures in $\mathbf{R}^d$.*

(i) If the p_n converge vaguely to p, and the $p_n(\mathbf{R}^d)$ converge to $p(\mathbf{R}^d)$, as $n \to \infty$ then the p_n converge to p weakly (in particular, if p_n and p are probability laws, the vague and the weak convergence coincide).

(ii) $p_n \to p$ $\star$-weakly if and only if $p_n \to p$ vaguely and the sequence p_n is bounded.

Proof (i) Assuming that p_n is not tight (see the definition before Theorem A.12) leads to a contradiction, since then $\exists\, \epsilon$: $\forall$ compact set K $\exists n$: $\mu_n(\mathbf{R}^d \setminus K) > \epsilon$, implying that

$$\liminf_{n\to\infty} p_n(\mathbf{R}^d) \geq p(\mathbf{R}^d) + \epsilon.$$

(ii) This is straightforward. □

Proposition A.6 *If S is a separable metric space then the space $\mathcal{M}(S)$ of finite Borel measures is separable in the weak (and hence also in the vague) topology.*

Proof A dense countable set is given by linear combinations, with rational coefficients, of the Dirac masses δ_{x_i} where $\{x_i\}$ is a dense subset of S. □

The following general fact from functional analysis is important for the analysis of measures.

Proposition A.7 *Let B be a separable Banach space. Then the unit ball $B_1^\star$ in its dual Banach space $B^\star$ is $\star$-weakly compact and there exists a complete metric in $B_1^\star$ compatible with this topology.*

Proof Let $\{x_1, x_2, \ldots\}$ be a dense subset in the unit ball of B. The formula

$$\rho(\mu, \eta) = \sum_{k=1}^{\infty} 2^{-k} |(\mu - \eta, x_k)|$$

specifies a complete metric in $B_1^\star$ that is compatible with the $\star$-weak convergence, i.e. $\rho(\mu_n, \mu) \to 0$ as $n \to 0$ if and only if $(\mu_n, x) \to (\mu, x)$ for any $x \in B$. Completeness and compactness follow easily. To show compactness, for example, we have to show that any sequence μ_n has a converging subsequence. To this end, we first choose a subsequence μ_n^1 such that (μ_n^1, x_1) converges, then a further subsequence μ_n^2 such that (μ_n^2, x_2) converges, etc. Finally, the diagonal subsequence μ_n^n converges on any of the x_i and is therefore converging. □

Remark A.8 The unit ball can be seen to be compact without assuming the $B_1^\star$ separability of B (the so-called Banach–Alaoglu theorem).

Proposition A.9 *If S is a separable locally compact metric space then the set $\mathcal{M}_M(S)$ of Borel measures with norm bounded by M is a complete separable metric compact set in the vague topology.*

Proof This follows from Propositions A.6 and A.7. □

To metricize the weak topology on measures one needs other approaches. Let S be a metric space with distance d. For $P, Q \in \mathcal{P}(S)$ define the *Prohorov distance*

$$\rho_{\text{Proh}}(P, Q) = \inf\{\epsilon > 0 : P(F) \leq Q(F^\epsilon) + \epsilon \,\forall \text{ closed } F\},$$

where $F^\epsilon = \{x \in S : \inf_{y \in F} d(x, y) < \epsilon\}$.

It is not difficult to show that

$$P(F) \leq Q(F^\epsilon) + \beta \quad \Longleftrightarrow \quad Q(F) \leq P(F^\epsilon) + \beta,$$

leading to the conclusion that ρ_{Proh} is actually a metric.

Theorem A.10 *(i) If S is separable then $\rho(\mu_n, \mu) \to 0$ as $n \to \infty$ for $\mu, \mu_1, \mu_2, \ldots \in \mathcal{P}(S)$ if and only if $\mu_n \to \mu$ weakly.*

(ii) If S is separable (resp. complete and separable) then $(\mathcal{P}(S), \rho_{\text{Proh}})$ is separable (resp. complete and separable).

Proof See e.g. Ethier and Kurtz [74]. □

It is instructive to have a probabilistic interpretation of this distance.

One says that a measure $\nu \in \mathcal{P}(S \times S)$ is a *coupling* of the measures $\mu, \eta \in \mathcal{P}(S)$ if the margins of ν are μ and η, i.e. if $\nu(A \times S) = \mu(A)$ and $\nu(S \times A) = \eta(A)$ for any measurable A or, equivalently, if

$$\int_{S \times S} [\phi(x) + \psi(y)]\, \nu(dxdy) = (\phi, \mu) + (\psi, \eta) \tag{A.1}$$

for any $\phi, \psi \in C(S)$.

Theorem A.11 *Let S be a separable metric space and $P, Q \in \mathcal{P}(S)$. Then*

$$\rho_{\text{Proh}}(P, Q) = \inf_\nu \inf\{\epsilon > 0 : \nu(x, y : d(x, y) \geq \epsilon) \leq \epsilon\},$$

where $\inf_\nu$ *is taken over all couplings of* P, Q.

Proof See Ethier and Kurtz [74]. □

As in the usual analysis, for the study of the convergence of probability laws the crucial role belongs to the notion of compactness. Recall that a subset of a metric space is called *relatively compact* if its closure is compact.

A family Π of measures on a complete separable metric space S is called *tight* if for any ϵ there exists a compact set $K \subset S$ such that $P(S \setminus K) < \epsilon$ for all measures $P \in \Pi$. The following fact is fundamental (a proof can be found e.g. in [74], [220] or [114]).

Theorem A.12 (Prohorov's compactness criterion) *A family* Π *of measures on a complete separable metric space* S *is relatively compact in the weak topology if and only if it is tight.*

Another convenient way to metricize the weak topology of measures is using Wasserstein–Kantorovich distances. Namely, let $\mathcal{P}^p(S)$ be the set of probability measures μ on S with finite pth moment, $p > 0$, i.e. such that

$$\int d^p(x_0, x)\mu(dx) < \infty$$

for some (and hence clearly for all) x_0.

The *Wasserstein–Kantorovich metrics* W_p, $p \geq 1$, on the set of probability measures $\mathcal{P}^p(S)$ are defined as

$$W_p(\mu_1, \mu_2) = \left(\inf_\nu \int d^p(y_1, y_2)\nu(dy_1 dy_2) \right)^{1/p}, \tag{A.2}$$

where inf is taken over the class of probability measures ν on $S \times S$ that couple μ_1 and μ_2. Of course, W_p depends on the metric d. It follows directly from the definition that

$$W_p^p(\mu_1, \mu_2) = \inf \mathbf{E} d^p(X_1, X_2), \tag{A.3}$$

where inf is taken over all random vectors (X_1, X_2) such that X_i has the law μ_i, $i = 1, 2$. One can show (see e.g. [246]) that the W_p are actually metrics on $\mathcal{P}^p(S)$ (the only point that is not obvious being that the triangle inequality holds) and that they are complete.

Proposition A.13 *If* S *is complete and separable, the infimum in* (A.2) *is attained.*

Proof In view of Theorem A.12, in order to be able to pick out a converging subsequence from a minimizing sequence of couplings, one needs to know that the set of couplings is tight. But this is straightforward: if K is a compact set in S such that $\mu_1(S \setminus K) \le \delta$ and $\mu_2(S \setminus K) \le \delta$ then

$$\nu(S \times S \setminus (K \times K)) \le \nu(S \times (S \setminus K)) + \nu((S \setminus K) \times S) \le 2\delta,$$

for any coupling ν. □

The main result connecting weak convergence with the Wasserstein metrics is as follows.

Theorem A.14 *If S is complete and separable, $p \ge 1$, and $\mu, \mu_1, \mu_2, \ldots$ are elements of $\mathcal{P}^p(S)$ then the following statements are equivalent:*

(i) $W_p(\mu_n, \mu) \to 0$ as $n \to \infty$;
(ii) $\mu_n \to \mu$ weakly as $n \to \infty$ and for some (and hence any) x_0

$$\int d^p(x, x_0)\mu_n(dx) \to \int d^p(x, x_0)\mu(dx).$$

Proof See e.g. Villani [246]. □

Remark A.15 If d is bounded then, of course, $\mathcal{P}^p(S)=\mathcal{P}(S)$ for all p. Hence, changing the distance d to the equivalent $\tilde{d} = \min(d, 1)$ allows us to use Wasserstein metrics as an alternative way to metricize the weak topology of probability measures.

In the case $p = 1$ the celebrated *Monge–Kantorovich theorem* states that

$$W_1(\mu_1, \mu_2) = \sup_{f \in Lip} |(f, \mu_1) - (f, \mu_2)|,$$

where Lip is the set of continuous functions f such that $|f(x) - f(y)| \le \|x - y\|$ for all x, y; see [246] or [202].

We shall need also the wasserstein distance between the distributions in the spaces of the paths (curves) $X : [0, T] \mapsto S$. Its definition depends on the way in which the distance between paths is measured. The most natural choices are the uniform and integral measures leading to the distances

$$\begin{aligned} W_{p,T,\mathrm{un}}(X^1, X^2) &= \inf\left(\mathbf{E} \sup_{t \le T} d^p(X_t^1, X_t^2)\right)^{1/p}, \\ W_{p,T,\mathrm{int}}(X_1, X_2) &= \inf\left(\mathbf{E} \int_0^T d^p(X_t^1, X_t^2)\, dt\right)^{1/p}, \end{aligned} \tag{A.4}$$

where inf is taken over all couplings of the distributions of the random paths X_1, X_2. The estimates in $W_{p,T,\mathrm{int}}$ are usually easier to obtain, but those in $W_{p,T,\mathrm{un}}$ are stronger. In particular, uniform convergence is stronger than

Skorohod convergence, implying that the limits in $W_{p,T,\mathrm{un}}$ preserve the Skorohod space of càdlàg paths while the limits in $W_{p,T,\mathrm{int}}$ need not necessarily do so.

Appendix B Topology on càdlàg paths

The main classes of stochastic processes, i.e. martingales and sufficiently regular Markov process, have modifications with *càdlàg* paths, meaning that they are right continuous and have left limits (the word càdlàg is a French acronym). This is a quite remarkable fact, in view of the general Kolmogorov result on the existence of processes on the space of all (even non-measurable) paths. Suppose that (S, ρ) is a complete separable metric space. The set of S-valued càdlàg functions on a finite interval $[0, T]$, $T \in \mathbf{R}_+$ or on the half-line $\mathbf{R}_+$ is usually denoted by $D = D([0, T], S)$ or $D = D(\mathbf{R}_+, S)$ and is called the *Skorohod path space*. We shall often write $D([0, T], S)$ for both these cases, meaning that T can be finite or infinite.

Proposition B.1 *If $x \in D([0, T], S)$ then for any $\delta > 0$ there can exist only finitely many jumps of x on $[0, T]$ of a size exceeding δ.*

Proof If this were not so then jumps exceeding δ would have an accumulation point on $[0, T]$. □

In the analysis of continuous functions a useful characteristic is the *modulus of continuity*

$$w(x, t, h) = \sup\{\rho(x(s), x(r)) : r - h \le s \le r \le t\}, \qquad h > 0.$$

As one can easily see, a function $x \in D([0, t], S)$ is continuous on $[0, t]$ if and only if $\lim_{h\to 0} w(x, t, h) = 0$. In the analysis of càdlàg functions a similar role belongs to the *modified modulus of continuity*, defined as

$$\tilde{w}(x, t, h) = \inf_{\Delta} \max_{k} \sup_{t_k \le r, s < t_{k+1}} \rho(x(r), x(s)),$$

where the infimum extends over all partitions $\Delta = (0 = t_0 < t_1 < \cdots < t_l < t)$ such that $t_{k+1} - t_k \ge h$ for $k = 1, \ldots, l - 1$.

Proposition B.2 *(i) The definition of $\tilde{w}(x, t, h)$ is not be changed if the infimum is extended only over partitions with $h \le t_{k+1} - t_k < 2h$. In particular $\tilde{w}(x, t, h) \le \tilde{w}(x, t, 2h)$ for all x. (ii) If $x \in D([0, t], S)$ then* $\lim_{h\to 0} \tilde{w}(x, t, h) = 0$.

Proof By Proposition B.1, for an arbitrary δ there exists a partition $0 = t_0^0 < t_1^0 < \cdots < t_k^0 = t$ of $[0, t]$ such that inside the intervals $I_l = [t_l^0, t_{l+1}^0)$ of the partition there are no jumps with sizes exceeding δ. Let us make a further partition of each I_l, defining recursively

$$t_l^{j+1} = \min(t_{l+1}^0, \inf\{s > t_l^j : |x(s) - x(t_l^{j-1})| > 2\delta\}),$$

with as many j as one needs to reach t_{l+1}^0. Clearly the new partition thus obtained is finite and all the differences $t_l^j - t_l^{j+1}$ are strictly positive, so that on the one hand

$$h = \min_{l,j}(t_l^j - t_l^{j+1}) > 0.$$

On the other hand, $\tilde{w}(x, t, h) \leq 4\delta$. □

The appropriate topology on D is not obvious. Our intuition arising from the study of random processes suggests that, in a reasonable topology, the convergence of the sizes and times of jumps should imply the convergence of paths. For example, the sequence of step functions $\mathbf{1}_{[1+1/n,\infty)}$ should converge to $\mathbf{1}_{[1,\infty)}$ as $n \to \infty$ in $D([0, T], \mathbf{R}_+)$ for $T > 1$. However, the usual uniform distance

$$\|\mathbf{1}_{[1+1/n,\infty)} - \mathbf{1}_{[1,\infty)}\| = \sup_y |\mathbf{1}_{[1+1/n,\infty)}(y) - \mathbf{1}_{[1,\infty)}(y)|$$

equals 1 for all n not allowing such a convergence in the uniform topology. The main idea is to make $\mathbf{1}_{[1+1/n,\infty)}$ and $\mathbf{1}_{[1,\infty)}$ close by introducing a time change that connects them. Namely, a *time change* on $[0, T]$ or $\mathbf{R}_+$ is defined as a monotone continuous bijection of $[0, T]$ or $\mathbf{R}_+$ onto itself. One says that a sequence $x_n \in D([0, T], S)$ converges to $x \in D([0, T], S)$ in the *Skorohod topology* J_1 if there exists a sequence of time changes λ_n of $[0, T]$ such that

$$\sup_s |\lambda_n(s) - s| + \sup_{s \leq t} \rho(x_n(\lambda_n(s)), x(s)) \to 0, \qquad n \to \infty,$$

for $t = T$ in the case where T is finite or for all $t > 0$ in the case where $T = \infty$.

For example, for

$$\lambda_n = \begin{cases} (1 + 1/n)t, & t \leq 1, \\ (1 - 1/n)(t - 1) + (1 + 1/n), & 1 \leq t \leq 2, \\ t, & t \geq 2, \end{cases} \tag{B.1}$$

one has $\mathbf{1}_{[1+1/n,\infty)}(\lambda_n(t)) = \mathbf{1}_{[1,\infty)}(t)$ for all t, so that

$$\sup_{s \leq t} \left(|\lambda_n(s) - s| + |\mathbf{1}_{[1+1/n,\infty)}(\lambda_n(s)) - \mathbf{1}_{[1,\infty)}(s))|\right) = 1/n \to 0$$

as $n \to \infty$ for all $t \geq 2$. Thus the step functions $\mathbf{1}_{[1+1/n,\infty)}$ converge to $\mathbf{1}_{[1,\infty)}$ in the Skorohod topology of $D(\mathbf{R}_+, \mathbf{R}_+)$ as $n \to \infty$.

Proposition B.3 *(i) Let*

$$J(x, T) = \sup_{t \leq T} d(x(t), x(t-))$$

for $T < \infty$ *and*

$$J(x, \infty) = \int_0^\infty e^{-u} \min(1, J(x, u))\, du.$$

Clearly $J(x, T) = 0$ *if and only if* x *is continuous. We claim that the function* $J(x, T)$ *is continuous on* $D([0, T], S)$.

(ii) If $x_n \to x$ *in* $D([0, T], S)$ *and the limiting curve* x *is continuous then* $x_n \to x$ *point-wise.*

Proof This is left as an exercise. □

It is more or less obvious that the functions

$$d_S(x, y) = \inf_\lambda \sup_{s \leq T} \Big(|\lambda(s) - s| + \rho(x(\lambda(s)), y(s)) \Big)$$

for finite T and

$$d_S(x, y) = \inf_\lambda \left[\sup_{s \geq 0} |\lambda(s) - s| + \sum_{n=1}^{\infty} 2^{-n} \min\left(1, \sup_{s \leq n} \rho\Big(x(\lambda(s)), y(s)\Big)\right) \right]$$

for infinite T (where inf extends over the set of all time changes) specify a metric on the sets $D([0, T], S)$, called *Skorohod's metric*, that is compatible with J_1-topology in the sense that x_n converges to x in this topology if and only if $d_S(x_n, x) \to 0$ as $n \to \infty$. This metric is not very convenient for the analysis, since the space $D([0, T], S)$ is not complete in this metric, as the following shows:

Example B.4 Consider a sequence of indicators $\mathbf{1}_{[1-1/n,1)}$. This sequence is fundamental (or Cauchy) with respect to the metric d_S on $D([0, T], \mathbf{R})$ for any $T \geq 1$ (because the time changes

$$\lambda_{m,n} = \begin{cases} \frac{t(1-1/n)}{(1-1/m)}, & t \leq 1 - 1/m, \\ 1 - \frac{1}{n} + \frac{m}{n}[t - (1 - 1/m)], & 1 - 1/m \leq t \leq 1 \\ t, & t \geq 1 \end{cases} \tag{B.2}$$

transform $\mathbf{1}_{[1-1/m,1)}$ to $\mathbf{1}_{[1-1/n,1)}$) but is not converging. Convince yourself that this sequence is not fundamental in the metric d_P introduced below.

However, one can improve this situation by measuring the distance between any time change and the identity map not by a uniform norm as in d_S but rather by the distance of the corresponding infinitesimal increments. More precisely, one measures the distance between a time change and the identity map by the quantity

$$\gamma(\lambda) = \sup_{0\leq s<t} \left| \log \frac{\lambda(t)-\lambda(s)}{t-s} \right|,$$

The corresponding distance,

$$d_P(x, y) = \inf_{\lambda} \left(\gamma(\lambda) + \sup_{t\leq T} \rho\Big(x(\lambda(s)), y(s)\Big) \right)$$

for finite T and

$$d_P(x, y) = \inf_{\lambda} \left[\gamma(\lambda) + \sum_{n=1}^{\infty} 2^{-n} \min\left(1, \sup_{s\leq n} \rho\left(x(\lambda(s)), y(s)\right)\right)\right]$$

for $T = \infty$ (where inf is over all time changes with a finite $\gamma(\lambda)$) is called *Prohorov's metric* on $D([0, T], S)$. Clearly this is again a metric on $D([0, T], S)$.

Proposition B.5 *The Prohorov metric d_P is compatible with the J_1-topology on $D([0, T], S)$, so that $x_n \to x$ if and only if $d_P(x_n, x) \to 0$ as $n \to 0$.*

Proof In one direction this is clear, since $\gamma(\lambda_n) \to 0$ implies $\sup_{0\leq s\leq t} |\lambda_n(s)-s| \to 0$ for all t (where one takes into account that $\lambda(0) = 0$). Suppose now that $d_S(x, x_n) \to 0$ for $x_n, x \in D([0, T], S)$. Assume that T is finite (the modification for $T = \infty$ being more or less straightforward). Then for any $\delta \in (0, 1/4)$ one can choose a partition $0 = t_0 < t_1 < \cdots$ of $[0, T]$, with, for $t_{k+1} - t_k \geq \delta$,

$$\sup_{t_k\leq r,s<t_{k+1}} \rho(x(r), x(s)) < \tilde{w}(x, T, \delta) + \delta$$

and time change λ with

$$\sup_{0\leq t\leq T} |\lambda(t) - t| < \delta^2, \qquad \sup_{0\leq t\leq T} \rho\left(x(\lambda^{-1}(t), x_n(t))\right) < \delta.$$

Let $\tilde{\lambda}$ be the time change obtained from λ by linear interpolation between its values at the points of the partition t_k, $k = 0, 1, 2, \ldots$ Then

$$\gamma(\tilde{\lambda}) = \max_k \left| \log \frac{\lambda(t_{k+1}) - \lambda(t_k)}{t_{k+1} - t_k} \right| \leq \max(\log(1 + 2\delta), -\log(1 - 2\delta)) \leq 4\delta.$$

Moreover, as the composition $\lambda^{-1} \circ \tilde{\lambda}$ maps each interval $[t_k, t_{k+1})$ to itself, one has

$$\rho\Big(x(t), x_n(\lambda(t))\Big) \le \rho\left(x(t), x(\lambda^{-1} \circ \tilde{\lambda}(t))\right) + \rho\left(x(\lambda^{-1} \circ \tilde{\lambda}(t)), x_n(\lambda(t))\right)$$
$$\le \tilde{w}(x, T, \delta) + 2\delta.$$

This implies that $d_P(x, x_n) \to 0$, as δ can be chosen arbitrarily small (and one takes into account Proposition B.2). □

The following fact is fundamental.

Theorem B.6 (Skorohod–Kolmogorov–Prohorov) *(i) The metric space $(D([0, T], S), d_P)$ is complete and separable.*

(ii) The Borel σ-algebra of $D([0, T], S)$ is generated by the evaluation maps $\pi_t : x \mapsto x(t)$ for all $t \le T$ (or $t < \infty$ in the case $T = \infty$).

(iii) A set $A \subset D([0, T], S)$ is relatively compact in the J_1-topology if and only if $\pi_t(A)$ is relatively compact in S for each t and

$$\lim_{h \to 0} \sup_{x \in A} \tilde{w}(x, t, h) = 0, \qquad t > 0. \tag{B.3}$$

A (by now standard) proof can be found e.g. in [107], [38] or [74].

Proposition B.7 *If S is locally compact, then condition (iii) of the above theorem implies that there exists a compact set Γ_T such that $\pi_t(A) \subset \Gamma_T$ for all $t \in [0, T]$.*

Proof If $\tilde{w}(x, T, h) < \epsilon$ then let Γ be the union of a finite number of compact closures of $\pi_{hk/2}(A)$, $hk/2 \le T$, $k \in \mathbf{N}$. Then all intervals of any partition with $[t_k, t_{k+1}] \ge h$ contain a point $hk/2$, so that the whole trajectory belongs to the compact set $\cup \Gamma^{\epsilon}_{kh/2}$. □

We conclude with further remarks on the space D.

An interesting feature of J_1-topology is the fact that addition is not continuous in this topology, i.e. D is not a linear topological space. In fact, let $x = -y = \mathbf{1}_{[1,\infty)}$ be a step function in $D(\mathbf{R}_+, \mathbf{R})$. Consider the approximating step functions from the left and from the right, namely $x_n = \mathbf{1}_{[1-1/n,\infty)}$, $y_n = -\mathbf{1}_{[1+1/n,\infty)}$. Then $x + y = 0$ and $x_n \to x$, $y_n \to y$ as $n \to \infty$ in the J_1-topology. However, $x_n + y_n = \mathbf{1}_{[1-1/n,1+1/n)}$ does not converge to zero.

Another important feature of J_1-topology is in the fact that one cannot approximate discontinuous functions by continuous functions. For instance, if one tries to approximate the step function $x = \mathbf{1}_{[1,\infty)}$ by the broken-line continuous paths

$$x_n(t) = n(t - 1 + 1/n)\mathbf{1}_{[1-1/n,1)} + \mathbf{1}_{[1,\infty)} \tag{B.4}$$

then x_n converges to x pointwise and monotonically but not in J_1. In fact, one can easily see that $d_S(x_n, x) = 1$ for all n.

Let us mention that J_1 is not the only reasonable topology on D. In fact, in his seminal paper in 1956, Skorohod introduced four different topologies on D: the so-called J_1, J_2, M_1 and M_2 topologies. The topology M_2, for example, arises from a comparison of the epigraphs $\{(t, y) : y \geq x(t)\}$ of paths by means of the Hausdorff distance. In this topology, the sequence (B.3) converges to the step function $x = \mathbf{1}_{[1,\infty)}$. Though these topologies are of interest sometimes, J_1 is by far the most important. Only J_1-topology is used in this book.

Appendix C Convergence of processes in Skorohod spaces

Everywhere in this appendix, S denotes a complete separable metric space with distance d.

The results of this section are quite standard, though a systematic exposition of the main tools needed for the study of convergence in Skorohod spaces is not easy to find in a single textbook. Such a systematic exposition would take us far from the content of this book, so we have just collected everything that we need, giving precise references to where the proofs can be found. The basic references are the books Jacod and Shiryaev [108], Kallenberg [115], Ethier and Kurtz [74]; see also Dawson [59] and Talay, and Tubaro [239].

Let X^α be a family of S-valued random processes, each defined on its own probability space with a fixed filtration $\mathcal{F}_t^\alpha$ with respect to which it is adapted. One says that the family X^α possesses *the compact containment condition* if for any $\eta, T > 0$ there exists a compact set $\Gamma_{\eta,T} \subset S$ such that

$$\inf_\alpha P\{X_\alpha(t) \in \Gamma_{\eta,T} \,\forall t \in [0, T]\} \geq 1 - \eta. \tag{C.1}$$

The following is the basic criterion of compactness for distributions on Skorohod spaces (see Appendix B for the notation $\tilde{w}$).

Theorem C.1 *Let X^α be a family of random processes with sample paths in $D(\mathbf{R}_+, S)$. Then $\{X^\alpha\}$ is relatively compact if and only if*

(i) for every $\eta > 0$ and a rational $t \geq 0$ there exists a compact set $\Gamma_{\eta,t} \subset S$ such that

$$\inf_\alpha P\{X_\alpha(t) \in \Gamma_{\eta,t}^\eta\} \geq 1 - \eta \tag{C.2}$$

(where $\Gamma_{\eta,t}^\eta$ is the η-neighborhood of $\Gamma_{\eta,t}$),

(ii) for every $\eta > 0$ and $T \geq 0$ there exists a $\delta > 0$ such that

$$\sup_\alpha P\{\tilde{w}(X_\alpha, \delta, T) \geq \eta\} \leq \eta.$$

Moreover, if (i) and (ii) hold then the compact containment condition holds also.

Proof See Ethier and Kurtz [74]. □

The following result (often referred to as the *Jakubovski criterion of tightness*) reduces the problem of compactness to real-valued processes.

Theorem C.2 *Let X^α be a family of random processes with sample paths in $D(\mathbf{R}_+, S)$ possessing the compact containment condition. Let H be a dense subspace of $C(S)$ in the topology of uniform convergence on compact sets. Then $\{X_\alpha\}$ is relatively compact if and only if the family $\{f \circ X_\alpha\}$ is relatively compact for any $f \in H$.*

Proof See Ethier and Kurtz [74], Theorem 3.9.1, or Jakubovski [108]. □

As the conditions of Theorem C.1 are not easy to check, more concrete criteria have been developed.

A sequence X^n of S-valued random processes (each defined on its own probability space with a fixed filtration $\mathcal{F}_n$ with respect to which it is adapted) is said to enjoy the one of *Aldous conditions*, (A), (A′) or (A″), if

(A) For each $N, \epsilon, \eta > 0$ there exist a $\delta > 0$ and n_0 such that, for any sequence of $\mathcal{F}_n$-stopping times $\{\tau_n\}$ with $\tau_n \le N$,

$$\sup_{n\ge n_0}\sup_{\theta\le\delta} \mathbf{P}^n\{d(X^n_{\tau_n}, X^n_{\tau_n+\theta}) \ge \eta\} \le \epsilon.$$

(A′) For each $N, \epsilon, \eta > 0$ there exist a $\delta > 0$ and an n_0 such that, for any sequence of pairs of $\mathcal{F}_n$-stopping times $\{\sigma_n, \tau_n\}$ with $\sigma_n \le \tau_n \le N$,

$$\sup_{n\ge n_0} \mathbf{P}^n\{d(X^n_{\sigma_n}, X^n_{\tau_n}) \ge \eta, \tau_n < \sigma_n + \delta\} \le \epsilon.$$

(A″) One has that $d(X^n_{\tau_n}, X^n_{\tau_n+h_n}) \to 0$ in probability as $n \to \infty$ for any sequence of bounded $\mathcal{F}_n$-stopping times $\{\tau_n\}$ and any sequence of positive numbers $h_n \to 0$.

Theorem C.3 (Aldous criterion for tightness) *Conditions (A), (A′), (A″) are equivalent and imply the basic tightness condition (ii) of Theorem C.1.*

Proof See Joffe and Métivier [111], Kallenberg [114], Jacod and Shiryaev [107] or Ethier and Kurtz [74]. □

As an easy consequence of this criterion we obtain the following crucial link between semigroup convergence and Skorohod convergence for Feller processes.

Theorem C.4 *Let S be locally compact and X, X^1, X^2, ... be S-valued Feller processes with corresponding Feller semigroups T_t, T_t^1, T_t^2, ... If the compact containment condition holds and the semigroups T^n converge to T_t strongly and uniformly for bounded times then the sequence $\{X^n\}$ is tight and the distributions of X^n converge to the distribution of X.*

Proof Using Theorems C.1 and C.3 and the strong Markov property of Feller processes, one only needs to show that $d(X_0^n, X_{h_n}^n) \to 0$ in probability as $n \to \infty$ for any initial distributions μ_n that may arise from the optional stopping of X^n and for any positive constants $h_n \to 0$. By the compact containment condition (and Prohorov's criterion for tightness) we may assume that μ_n converges weakly to a certain μ, the law of X_0. By the assumed uniformity in time of semigroup convergence, $T_{h_n}^n g \to g$ for any $g \in C_\infty(S)$, implying that

$$\mathbf{E}[f(X_0^n)g(X_{h_n}^n)] = \mathbf{E}(f T_{h_n}^n g)(X_0^n) \to \mathbf{E}(fg)(X_0)$$

for $f, g \in C_\infty(S)$. Consequently, $(X_0^n, X_{h_n}^n) \to (X_0, X_0)$ in distribution. Then $d(X_0^n, X_{h_n}^n) \to d(X_0, X_0) = 0$ in distribution and hence also in probability.

For applications to the dynamic law of large numbers for interacting particles the following slight modification of this result is often useful. □

Theorem C.5 *Let S be locally compact, S_n be a family of closed subsets of S and X, X^1, X^2, $\cdots$ be S, S_1, S_2, ... -valued Feller processes with corresponding Feller semigroups T_t, T_t^1, T_t^2, ... Suppose that the compact containment condition holds for the family X^n and that the semigroups T^n converge to T_t in the sense that*

$$\sup_{s \le t} \sup_{x_n \in S_n} |T_s^n f(x_n) - T_s f(x_n)| \to 0, \qquad n \to \infty$$

for any $t > 0$ and $f \in C_\infty(S)$. Finally, assume that the sequence $x_n \in S_n$ converges to an $x \in S$. Then the sequence of processes $\{X^n(x_n)\}$ (with initial conditions x_n) is tight and the distributions of $X^n(x_n)$ converge to the distribution of $X(x)$.

Proof It is the same as for the previous theorem. Notice that after choosing a subsequence of laws μ_n in S_n converging to a law μ in S one has

$$(T_t^n f, \mu_n) - (T_t f, \mu) = ((T_t^n - T_t) f, \mu_n) + (T_t f, \mu_n - \mu),$$

implying that $(T_t^n f, \mu_n) \to (T_t f, \mu)$ because the first term in the above expression tends to zero by our assumption on the convergence of T_n and the second by the weak convergence of μ_n. □

The following result allows us to obtain convergence for the solutions of the martingale problem without assuming the convergence of semigroups and the local compactness of S.

Theorem C.6 *Let X_α be a family of random processes with sample paths in $D(\mathbf{R}_+, S)$ and let C_a be a subalgebra in $C(S)$. Assume that for any f from a dense subset of C_a there exist càdlàg adapted processes Z^α such that*

$$f(X_t^\alpha) - \int_0^t Z_f^\alpha(s)\,ds$$

is an $\mathcal{F}_t^\alpha$-martingale and, for any $t > 0$,

$$\sup_\alpha \mathbf{E}\left(\int_0^t |Z_f^\alpha(s)|^p\,ds\right)^{1/p} < \infty$$

for a $p > 1$. Then the family $f \circ X^\alpha$ is tight in $D(\mathbf{R}_+, \mathbf{R})$ for any $f \in C_a$.

Proof See Theorem 3.9.4 in Ethier and Kurtz [74]. □

To formulate other criteria we need the notion of quadratic variation. For two real-valued processes X and Y the *mutual variation* or *covariation* is defined as the limit in probability

$$[X, Y]_t$$

$$= \lim_{\max_i(s_{i+1}-s_i)\to 0} \sum_{i=1}^n (X_{\min(s_{i+1},t)} - X_{\min(s_i,t)})(Y_{\min(s_{i+1},t)} - Y_{\min(s_i,t)}) \tag{C.3}$$

(the limit is over finite partitions $0 = s_0 < s_1 < \cdots < s_n = t$ of the interval $[0, t]$). In particular, $[X, X]_t$ is often denoted for brevity by $[X]_t$ and is called the *quadratic variation* of X.

Proposition C.7 *If $[X, Y]$ is well defined for two martingales X_t, Y_t then $(XY)_t - [X, Y]_t$ is a martingale.*

Proof For a partition $\Delta = (s = s_0 < s_1 < \cdots < s_n = t)$ of the interval $[s, t]$ write

$$[X, Y]^\Delta = \sum_{j=1}^n (X_{s_j} - X_{s_{j-1}})(Y_{s_j} - Y_{s_{j-1}}).$$

Then

$$(XY)_t - (XY)_s = \sum_{j=1}^n (X_{s_j} - X_{s_{j-1}})Y_{s_j} + \sum_{j=1}^n (Y_{s_j} - Y_{s_{j-1}})X_{s_{j-1}}$$

$$= [X, Y]^{\Delta} + \sum_{j=1}^{n} (Y_{s_j} - Y_{s_{j-1}}) X_{s_{j-1}} + \sum_{j=1}^{n} (X_{s_j} - X_{s_{j-1}}) Y_{s_{j-1}}, \qquad (C.4)$$

and the expectation of the last two terms vanishes. □

Remark C.8 Formula (C.4) suggests a stochastic integral representation for XY as

$$(XY)_t - (XY)_0 = [X, Y]_t + \int_0^t X_{s-} \, dY_s + \int_0^t Y_{s-} \, dX_s,$$

since the integrals in the above relation are clearly the only reasonable notation for the limit of the last two terms in (C.4). In stochastic analysis such integrals are studied systematically.

By a straightforward generalization of Proposition C.7, one obtains that if

$$X_t^1 = M_t^1 + \int_0^t b_s^1 \, ds, \qquad X_t^2 = M_t^2 + \int_0^t b_s^2 \, ds,$$

where M_t^1, M_t^2 are martingales and b_s is a bounded measurable process, then

$$(X^1 X^2)_t - [X^1, X^2]_t - \int_0^t (X_s^1 b_s^2 + X_s^2 d_s^1) \, ds \qquad (C.5)$$

is a martingale.

One can deduce from (C.3) that if X and Y have locally finite variations ΔX_s and ΔY_s then a.s.

$$[X, Y]_t = \sum_{s \le t} \Delta X_s \Delta Y_s \qquad (C.6)$$

(see e.g. Kallenberg [114] or Talay and Tubaro [237] for a proof).

It is also known that if $[X, Y]_t$ is locally integrable then there exists a unique predictable process with paths of finite variation, denoted by $\langle X, Y \rangle$ and called the *predictable covariation*, such that $[X, Y]_t - \langle X, Y \rangle_t$ is a local martingale; $\langle X, X \rangle_t$ is often denoted for brevity as $\langle X \rangle_t$.

The following two results are usually referred to as the Rebolledo criteria for tightness (see e.g. [204] or [73] for the first and [203] or [111] for the second).

Theorem C.9 *Let X_t^n be a family of square integrable processes such that*

$$X_t^n = M_t^n + V_t^n,$$

where the V_t^n are predictable finite-variation processes and the M_t^n are martingales. Then the family X_t^n satisfies the Aldous condition (A) whenever V_t^n and the quadratic variation $[M_t^n]$ satisfy this condition.

Theorem C.10 *In the above theorem, the quadratic variation $[M_t^n]$ and the martingales M_t^n themselves satisfy the Aldous condition (A) whenever the predictable quadratic variation $\langle M_t^n \rangle_t$ satisfies this condition.*

To apply Rebolledo's criteria, one has to be able to calculate either the covariation or the predictable covariation. In the case of a jump process, one can often use (C.6) effectively. For a Markov process it is often easy to find the predictable covariation, as we now explain. Let X_t be a Markov process in $\mathbf{R}^d$ solving the martingale problem for an operator L. Let $\phi_i(x)$ denote the ith coordinate of $x \in \mathbf{R}^d$, and suppose that ϕ^i and $\phi^i \phi^j$ belong to the domain of L, so that

$$M^i = X_t^i - X_0^i - \int_0^t L\phi^i(X_s)\, ds$$

and

$$M^{ij} = X_t^i X_t^j - X_0^i X_0^j - \int_0^t L(\phi^i \phi^j)(X_s)\, ds$$

are well-defined martingales. Notice first that

$$[X^i, X^j]_t = [M^i, M^j]_t$$

by (C.6). In the next two lines M_t denotes an arbitrary martingale. From equation (C.5) one has

$$\begin{aligned} &X_t^i X_t^j - X_0^i X_0^j \\ &\quad = [X^i, X^j]_t + \int_0^t \left[X_s^i L(\phi^j)(X_s) + X_s^j L(\phi^i)(X_s) \right] ds + M_t, \end{aligned}$$

implying that

$$\begin{aligned} &[X^i, X^j]_t \\ &\quad = \int_0^t L(\phi^i \phi^j)(X_s)\, ds - \int_0^t \left[X_s^i L(\phi^j)(X_s) + X_s^j L(\phi^i)(X_s) \right] ds + M_t. \end{aligned}$$

Consequently, by the definition of the predictable covariation,

$$\langle X^i, X^j \rangle_t = \int_0^t [L(\phi^i \phi^j)(X_s) - X_s^i L(\phi^j)(X_s) - X_s^j L(\phi^i)(X_s)]\, ds, \quad \text{(C.7)}$$

which is the required formula.

Appendix D Vector-valued ODEs

For the sake of completeness we present here in a concise form the basic result on the smoothness of solutions to Banach-space-valued ODEs (more precisely,

ODEs on functions with values in a Banach space) with respect to a parameter. We recall in passing the notion of Gateaux differentiation (see e.g. Martin [170] for a detailed exposition).

Let B_1 and B_2 denote Banach spaces with norms $\|.\|_1$, $\|.\|_2$. We assume that M is a closed convex subset of B_1. In the examples we have in mind, M stands for the set of positive elements in B_1 (say, $M = \mathcal{M}(X)$ in the Banach space B_1 of signed measures on X).

One says that a mapping $F : M \mapsto B_2$ is *Gateaux differentiable* if, for any $Y \in M$ and $\xi \in B_1$ such that there exists an $h > 0$ with $Y + h\xi \in M$, the limit

$$D_\xi F(Y) = \lim_{h \to 0_+} \frac{1}{h} [F(Y + h\xi) - F(Y)]$$

exists (in the norm topology of B_2); $D_\xi F(Y)$ is called the *Gateaux derivative* of $F(Y)$ in the direction ξ. From the definition it follows that

$$D_{a\xi} F(Y) = a D_\xi F(Y)$$

whenever $a > 0$ and that

$$F(Y + \xi) = F(Y) + \int_0^1 D_\xi F(Y + s\xi)\, ds \qquad \text{(D.1)}$$

whenever $Y + \xi \in M$.

Exercise D.1 Show that if F is Gateaux differentiable and $D_\xi F(Y)$ depends continuously on Y for any ξ then the mapping $D_\xi F(Y)$ depends linearly on ξ. Hint: use (D.1) to show additivity and also to show homogeneity with respect to negative multipliers.

Exercise D.2 Deduce from (D.1) that if $D_\xi F(Y)$ is Lipschitz continuous, in the sense that

$$\|D_\xi \Omega(Y_1) - D_\xi \Omega(Y_2)\|_2 = O(1)\|\xi\|_1 \|Y_1 - Y_2\|_1$$

uniformly on Y_1, Y_2 from any bounded set, then

$$\|F(Y + \xi) - F(Y) - D_\xi F(Y)\|_2 = O(\|\xi\|_1^2) \qquad \text{(D.2)}$$

uniformly for Y, ξ from any bounded set.

From now on let $B_1 = B_2 = B$ with norm denoted by $\|.\|$.

Theorem D.1 (Differentiation with respect to the initial data) *Let Ω_t be a family of continuous Gateaux differentiable mappings $M \mapsto B$ depending continuously on t, such that, uniformly for finite t,*

$$\|D_\xi \Omega_t(Y)\| \le c(\|Y\|)\|\xi\|,$$
$$\|D_\xi \Omega_t(Y_1) - D_\xi \Omega_t(Y_2)\| \le c(\|Y_1\| + \|Y_2\|)\|\xi\|\|Y_1 - Y_2\| \tag{D.3}$$

for a continuous function c on $\mathbf{R}_+$. *Let the Cauchy problem*

$$\dot{\mu}_t = \Omega_t(\mu_t), \qquad \mu_0 = \mu, \tag{D.4}$$

be well posed in M in the sense that for any $\mu \in M$ *there exists a unique continuous curve* $\mu_t = \mu_t(\mu) \in B$, $t \ge 0$, *such that*

$$\mu_t = \mu_0 + \int_0^t \Omega_s(\mu_s)\, ds \tag{D.5}$$

and, uniformly for finite t and bounded μ_0^1, μ_0^2,

$$\|\mu_t(\mu_0^1) - \mu_t(\mu_0^2)\| \le c\|\mu_0^1 - \mu_0^2\| \tag{D.6}$$

for all pairs of initial data μ_0^1, μ_0^2. *Finally, let* $\xi_t = \xi_t(\xi, \mu)$ *be a continuous curve in B satisfying the equation*

$$\xi_t = \xi + \int_0^t D_{\xi_s} \Omega_s(\mu_s)\, ds \tag{D.7}$$

for a given solution $\mu_t = \mu_t(\mu)$ *of* (D.5). *Then*

$$\sup_{s \le t} \frac{1}{h} \|\mu_s(\mu + h\xi) - \mu_s(\mu) - h\xi_s\| \le \kappa(t) h \|\xi\|^2 \tag{D.8}$$

uniformly for bounded ξ *and* μ, *implying in particular that* $\xi_t = D_\xi \mu_t(\mu)$.

Proof Subtracting from the integral equations for $\mu_t(\mu + h\xi)$ the integral equations for $\mu_t(\mu)$ and $h\xi_t$ yields

$$\mu_t(\mu + h\xi) - \mu_t(\mu) - h\xi_t$$
$$= \int_0^t [\Omega_s(\mu_s(\mu + h\xi)) - \Omega_s(\mu_s(\mu)) - h D_{\xi_s} \Omega_s(\mu_s(\mu))]\, ds.$$

However, it follows from (D.2) and (D.6) that

$$\|\Omega_s(\mu_s(\mu+h\xi)) - \Omega_s(\mu_s(\mu)) - D_{\mu_s(\mu+h\xi)-\mu_s(\mu)} \Omega_s(\mu_s(\mu))\| = O(h^2)\|\xi\|^2.$$

Hence, writing $\phi_t = [\mu_t(\mu + h\xi) - \mu_t(\mu) - h\xi_t]/h$ and using Exercise D.1 implies that

$$\phi_t = \int_0^t D_{h\phi_s} \Omega_s(\mu_s(\mu))\, ds + O(h)\|\xi\|^2,$$

which in turn implies (D.8), by Gronwall's lemma, and the first estimate in (D.3). □

Theorem D.2 (Differentiation with respect to a parameter) *Let Ω_t be a family of Gateaux differentiable mappings from M to the Banach space $\mathcal{L}(B)$ of bounded linear operators in B satisfying* (D.3)*, where the norm on the l.h.s. of the inequalities is now understood as the operator norm. Then the solution to the linear Cauchy problem*

$$\dot{Z}_t = \Omega_t(Y)Z_t, \qquad Z_0 = Z \in B,$$

is Gateaux differentiable with respect to Y, and the derivative $D_\xi Z_t(Y)$ is the unique solution of the problem

$$\dot{D}_\xi Z_t(Y) = \Omega_t(Y)D_\xi Z_t(Y) + D_\xi \Omega_t(Y)Z_t, \qquad D_\xi Z_0(Y) = 0.$$

Proof This is the same as for the previous theorem and is left as an exercise. □

Appendix E Pseudo-differential operator notation

The *Fourier transform*

$$Ff(p) = \frac{1}{(2\pi)^{d/2}} \int e^{-ipx} f(x)\,dx = \frac{1}{(2\pi)^{d/2}}(e^{-ip\cdot}, f)$$

is known to be a bijection on the Schwarz space $S(\mathbf{R}^d)$, the inverse operator being the inverse Fourier transform

$$F^{-1}g(x) = \frac{1}{(2\pi)^{d/2}} \int e^{ipx} g(p)\,dp.$$

As one easily sees, the Fourier transform takes the differentiation operator to the multiplication operator, i.e. $F(f')(p) = (ip)F(f)$. This property suggests a natural definition for fractional derivatives. Namely, one defines a *symmetric fractional operator* in $\mathbf{R}^d$ of the form

$$\int_{S^{d-1}} |(\nabla, s)|^\beta \mu(ds),$$

where $\mu(ds)$ is an arbitrary centrally symmetric finite Borel measure on the sphere S^{d-1}, as the operator that multiplies the Fourier transform of a function by

$$\int_{S^{d-1}} |(p, s)|^\beta \mu(ds),$$

i.e. via the equation

$$F\left(\int_{S^{d-1}} |(\nabla, s)|^\beta \mu(ds) f\right)(p) = \int_{S^{d-1}} |(p, s)|^\beta \mu(ds) Ff(p).$$

Well-known explicit calculations (see e.g. [124] or [217]) show that

$$\int_{S^{d-1}} |(\nabla, s)|^\beta \mu(ds)$$
$$= \begin{cases} c_\beta \int_0^\infty \int_{S^{d-1}} \left[f(x+y) - f(x)\right] \dfrac{d|y|}{|y|^{1+\beta}} \mu(ds), \\ \qquad\qquad\qquad\qquad\qquad\qquad\qquad\qquad \beta \in (0,1), \\ c_\beta \int_0^\infty \int_{S^{d-1}} \left[f(x+y) - f(x) - (y, \nabla f(x))\right] \dfrac{d|y|}{|y|^{1+\beta}} \mu(ds), \\ \qquad\qquad\qquad\qquad\qquad\qquad\qquad\qquad \beta \in (1,2), \end{cases} \tag{E.1}$$

with certain explicit constants c_β and with

$$\int_{S^{d-1}} |(\nabla, s)| \mu(ds) = -\frac{2}{\pi} \lim_{\epsilon \to 0} \int_\epsilon^\infty \int_{S^{d-1}} \left[f(x+y) - f(x)\right] \frac{d|y|}{|y|^2} \mu(ds). \tag{E.2}$$

Fractional derivatives constitute a particular case of the so-called *pseudo-differential operators* (ΨDOs). For a function $\psi(p)$ in $\mathbf{R}^d$, called in this context a *symbol*, one defines the ΨDO $\psi(-i\nabla)$ as the operator taken by Fourier transform to multiplication by ψ; this is expressed in the equation

$$F(\psi(-i\nabla)f)(p) = \psi(p)(Ff)(p). \tag{E.3}$$

Comparing this definition with the above fractional derivatives, one sees that $\int_{S^{d-1}} |(\nabla, s)|^\alpha \mu(ds)$ is the ΨDO with symbol $\int_{S^{d-1}} |(p, s)|^\alpha \mu(ds)$.

The explicit formula for F^{-1} yields an explicit integral representation for the ΨDO with symbol ψ:

$$\psi(-i\nabla)f(x) = \frac{1}{(2\pi)^{d/2}} \int e^{ipx} \psi(p)(Ff)(p)\, dp.$$

This expression suggests the following further extension. For a function $\psi(x, p)$ on $\mathbf{R}^d$ one defines the ΨDO $\psi(x, -i\nabla)$ *with symbol* ψ via the formula

$$\psi(x, -i\nabla)f(x) = \frac{1}{(2\pi)^{d/2}} \int e^{ipx} \psi(x, p)(Ff)(p)\, dp. \tag{E.4}$$

Appendix F Variational derivatives

We recall here basic definitions for the variational derivatives of functionals on measures, deduce some of their elementary properties and finally specify the natural class of functional spaces, on which the derivatives on the dual space can be defined analogously. Suppose that X is a metric space. For a function F on $\mathcal{M}(X)$, the *variational derivative* $\delta F(Y)/\delta Y(x)$ is defined as the Gateaux derivative of $F(Y)$ in the direction δ_x:

$$\frac{\delta F(Y)}{\delta Y(x)} = D_{\delta_x} F(Y) = \lim_{s \to 0_+} \frac{1}{s} [F(Y + s\delta_x) - F(Y)],$$

where $\lim_{s \to 0_+}$ means the limit over positive s and D denotes the Gateaux derivative. The higher derivatives $\delta^l F(Y)/\delta Y(x_1) \cdots \delta Y(x_l)$ are defined inductively.

As follows from the definition, if $\delta F(Y)/\delta Y(.)$ exists for an $x \in X$ and depends continuously on Y in the weak topology of $\mathcal{M}(X)$ then the function $F(Y + s\delta_x)$ of $s \in \mathbf{R}_+$ has a continuous right derivative everywhere and hence is continuously differentiable, implying that

$$F(Y + \delta_x) - F(Y) = \int_0^1 \frac{\delta F(Y + s\delta_x)}{\delta Y(x)} ds. \tag{F.1}$$

From the definition of the variational derivative it follows that

$$\lim_{s \to 0_+} \frac{1}{s} [F(Y + sa\delta_x) - F(Y)] = a \frac{\delta F}{\delta Y(x)}$$

for a positive a, allowing us to extend equation (F.1) to

$$F(Y + a\delta_x) - F(Y) = a \int_0^1 \frac{\delta F}{\delta Y(x)} (Y + sa\delta_x) \, ds. \tag{F.2}$$

It is easy to see that this still holds for negative a provided that $Y + a\delta_x \in \mathcal{M}(X)$.

We shall need an extension of this identity for measures that are more general than the Dirac measure δ_x. Let us introduce some useful notation. We shall say that F belongs to $C^k_{\text{weak}}(\mathcal{M}(X)) = C^k(\mathcal{M}(X))$, $k = 1, 2, \ldots$, if, for all $l = 1, \ldots, k$, $\delta^l F(Y)/\delta Y(x_1) \cdots \delta Y(x_l)$ exists for all $x_1, \ldots, x_k \in X^k$, $Y \in \mathcal{M}(X)$, and represents a continuous mapping of $k + 1$ variables (when measures are equipped with the weak topology) that are uniformly bounded on the sets of bounded Y. If X is locally compact, one similarly defines the spaces $C^k_{\text{vague}}(\mathcal{M}(X))$, which differ from the spaces $C^k_{\text{weak}}(\mathcal{M}(X))$, by assuming that there is continuity in the vague topology and that the derivatives $\delta^l F(Y)/\delta Y(x_1) \cdots \delta Y(x_l)$ belong to $C_\infty(X^l)$ uniformly for bounded sets of Y. If $X = \mathbf{R}^d$, we shall say that F belongs to $C^{1,l}_{\text{weak}}(\mathcal{M}(X))$, $l = 1, 2, \ldots$, if, for any $m = 0, 1, \ldots, l$, the derivatives

$$\frac{\partial^m}{\partial x^m} \frac{\delta F(Y)}{\delta Y(x)}$$

exist for all $x \in X$, $Y \in \mathcal{M}(X)$, are continuous functions of their variables (when measures are equipped with the weak topology) and are uniformly bounded on the sets of bounded Y.

Lemma F.1 *If $F \in C^1(\mathcal{M}(X))$ then F is Gateaux differentiable on $\mathcal{M}(X)$ and*

$$D_\xi F(Y) = \int \frac{\delta F(Y)}{\delta Y(x)} \xi(dx). \tag{F.3}$$

In particular, if $Y, Y + \xi \in \mathcal{M}(X)$ then

$$F(Y+\xi) - F(Y) = \int_0^1 \left(\frac{\delta F(Y+s\xi)}{\delta Y(.)}, \xi \right) ds \tag{F.4}$$

(note that ξ can be a signed measure).

Proof Using the representation

$$F(Y + s(a\delta_x + b\delta_y)) - F(Y)$$
$$= F(Y + sa\delta_x) - F(Y) + b \int_0^s \frac{\delta F}{\delta Y(y)}(Y + sa\delta_x + hb\delta_y)\, dh$$

for arbitrary points x, y and numbers $a, b \in \mathbf{R}$ such that

$$Y + a\delta_x + b\delta_y \in \mathcal{M}(X)$$

and also the uniform continuity of $\delta F(Y + s\delta_x + h\delta_y; y)$ in s, h allows us to deduce from (F.1) the existence of the limit

$$\lim_{s \to 0_+} \frac{1}{s} \left[F(Y + s(a\delta_x + b\delta_y)) - F(Y) \right] = a \frac{\delta F}{\delta Y(x)} + b \frac{\delta F}{\delta Y(y)}.$$

Extending similarly to an arbitrary number of points, one obtains (F.4) for ξ a finite linear combination of the Dirac measures.

Assume now that $\xi \in \mathcal{M}(X)$ and $\xi_k \to \xi$ as $k \to \infty$ weakly in $\mathcal{M}(X)$, where the ξ_k are finite linear combinations of the Dirac measures with positive coefficients. We now pass to the limit $k \to \infty$ in equation (F.4) written for $\xi = \xi_k$. As $F \in C(\mathcal{M}(X))$, one has

$$F(Y + \xi_k) - F(Y) \to F(Y + \xi) - F(Y), \qquad k \to \infty.$$

Next, the difference

$$\int_0^1 \left(\frac{\delta F}{\delta Y(.)}(Y + s\xi_k), \xi_k \right) ds - \int_0^1 \left(\frac{\delta F}{\delta Y(.)}(Y + s\xi), \xi \right) ds$$

can be written as

$$\int_0^1 \left(\frac{\delta F}{\delta Y(.)}(Y + s\xi_k), \xi_k - \xi \right) ds$$
$$+ \int_0^1 \left[\left(\frac{\delta F}{\delta Y(.)}(Y + s\xi_k), \xi \right) - \left(\frac{\delta F}{\delta Y(.)}(Y + s\xi), \xi \right) \right] ds.$$

By our assumptions,

$$\frac{\delta F}{\delta Y(.)}(Y + s\xi_k) - \frac{\delta F}{\delta Y(.)}(Y + s\xi)$$

converges to zero as $k \to \infty$ uniformly for x from any compact set. However, as $\xi_k \to \xi$ this family is tight, so that the integrals in the above formula can be made arbitrarily small outside any compact domain. Hence both integrals converge to zero.

To complete the proof of (F.4) let us note that if $\xi \notin \mathcal{M}(X)$ (i.e. ξ is a signed measure) then the same procedure is applied, but one has to approximate both Y and ξ. Clearly (F.3) follows from (F.4) and the definition of a Gateaux derivative. □

Corollary F.2 *If* $Y, Y+\xi \in \mathcal{M}(X)$ *and* $F \in C^2(\mathcal{M}(X))$ *or* $F \in C^3(\mathcal{M}(X))$, *the following Taylor expansions hold respectively:*

$$F(Y+\xi) - F(Y) = \left(\frac{\delta F(Y)}{\delta Y(.)}, \xi\right) + \int_0^1 (1-s)\left(\frac{\delta^2 F(Y+s\xi)}{\delta Y(.)\delta Y(.)}, \xi \otimes \xi\right) ds, \tag{F.5a}$$

$$F(Y+\xi) - F(Y) = \left(\frac{\delta F}{\delta Y(.)}, \xi\right) + \frac{1}{2}\left(\frac{\delta^2 F(Y)}{\delta Y(.)\delta Y(.)}, \xi \otimes \xi\right) + \frac{1}{2}\int_0^1 (1-s)^2 \left(\frac{\delta^3 F(Y+s\xi)}{\delta Y^3(.,.,.)}, \xi^{\otimes 3}\right) ds. \tag{F.5b}$$

Proof This follows straightforwardly from the usual Taylor expansion. □

Lemma F.3 *If* $t \mapsto \mu_t \in \mathcal{M}(X)$ *is continuously differentiable in the weak topology then, for any* $F \in C^1(\mathcal{M}(X))$,

$$\frac{d}{dt}F(\mu_t) = (\delta F(\mu_t; \cdot), \dot{\mu}_t). \tag{F.6}$$

Proof This requires the chain rule of calculus in an infinite-dimensional setting. The details are left as an exercise. □

Another application of the chain rule that is of importance is the following.

Lemma F.4 *If* $F \in C^{1,l}_{\text{weak}}(\mathcal{M}(\mathbf{R}^d))$ *then*

$$\frac{\partial}{\partial x_i}F(h\delta_{\mathbf{x}}) = h\frac{\partial}{\partial x_i}\frac{\delta F(Y)}{\delta Y(x_i)}, \qquad Y = h\delta_{\mathbf{x}} = h\delta_{x_1} + \cdots + h\delta_{x_n}.$$

Exercise F.1 Prove Lemma F.4. Hint: formally, this is a chain rule, where one uses the fact that δ'_x is the generalized function (distribution) acting as $(f, \delta'_x) = f'(x)$. For a more rigorous footing one represents the increments of F by (F.4).

One often needs also the derivatives of Banach-space-valued functions on $\mathcal{M}(X)$. Let us say that a mapping $\Phi : \mathcal{M}(X) \mapsto \mathcal{M}(X)$ has a *strong variational derivative* $\delta\Phi(\mu, x)$ if, for any $\mu \in \mathcal{M}(X)$, $x \in X$, the limit

$$\frac{\delta\Phi}{\delta Y(x)} = \lim_{s\to 0_+} \frac{1}{s}\,[\Phi(Y + s\delta_x) - \Phi(Y)]$$

exists in the norm topology of $\mathcal{M}(X)$ and is a finite signed measure on X. Higher derivatives are defined inductively. We shall say that Φ belongs to $C^l(\mathcal{M}(X); \mathcal{M}(X))$, $l = 0, 2, \ldots$, if for all $k = 1, \ldots, l$ the strong variational derivative $\delta^k\Phi(Y; x_1, \ldots, x_k)$ exists for all $x_1, \ldots, x_k \in X^k$, $Y \in \mathcal{M}(X)$, and represents a mapping $\mathcal{M}(X) \times X^k \mapsto \mathcal{M}^{\mathrm{sign}}(X)$ that is continuous in the sense of the weak topology and is bounded on the bounded subsets of Y.

Lemma F.5 *Let $\Phi \in C^1(\mathcal{M}(X); \mathcal{M}(X))$ and $F \in C^1(\mathcal{M}(X))$; then the composition $F \circ \Phi(Y) = F(\Phi(Y))$ belongs to $C^1(\mathcal{M}(X))$, and*

$$\frac{\delta F}{\delta Y(x)}(\Phi(Y)) = \int \left.\frac{\delta F(Z)}{\delta Z(y)}\right|_{Z=\Phi(Y)} \frac{\delta\Phi}{\delta Y(x)}(Y, dy). \tag{F.7}$$

Proof By (F.4),

$$\frac{\delta F \circ \Phi}{\delta Y(x)}(Y) = \lim_{h\to 0_+} \frac{1}{h}\,[F(\Phi(Y + h\delta_x)) - F(\Phi(Y))]$$
$$= \lim_{h\to 0_+} \frac{1}{h} \int_0^1 ds \left[\frac{\delta F}{\delta Z(.)}(\Phi(Y) + s[\Phi(Y + h\delta_x) - \Phi(Y)], .)\,\Phi(Y + h\delta_x) \right.$$
$$\left. - \Phi(Y)\right]$$
$$= \lim_{h\to 0_+} \frac{1}{h} \int_0^1 ds \int_X \left[\frac{\delta F}{\delta Z(y)}\Big(\Phi(Y) + sh\,[\delta\Phi(Y, x) + o(1)]\Big)\frac{\delta\Phi}{\delta Y(x)}(Y, dy)\right],$$

yielding (F.7). □

The above lemmas are derived under the strong assumption of boundedness. In practice one often uses their various extensions, in which functions and their variational derivatives are unbounded or defined on a subset of $\mathcal{M}(X)$. The matter is often complicated by the necessity to work in different weak or strong topologies. In all these situations the validity of the calculations should be justified, of course.

Moreover, variational derivatives can be defined also for other functional spaces. The next statement identifies a natural class.

Proposition F.6 *Let a Banach space B with norm $\|.\|_B$ be a subspace of the space of continuous functions on a complete locally compact metric space*

X (we shall apply it in the case where X is a subset of $\mathbf{R}^d$) such that, for any compact set $K \subset X$, the mapping $f \mapsto f\mathbf{1}_K$ is a contraction and a continuous embedding $B \to C(K)$ with a closed image and, for any $f \in B$, one has $f = \lim(f\mathbf{1}_{B_R})$ in B as $R \to \infty$, where B_R is a ball with any fixed center. Then the Dirac measures δ_x belong to $B^\star$ and their linear combinations are $\star$-weakly dense there. Consequently, the variational derivatives $\delta F/\delta Y$ can be defined in the usual way and possess all the above properties.

Proof By the Hahn–Banach theorem, any element $\phi \in B^\star$ can be lifted to an element of $(C(K))^\star = \mathcal{M}(K)$ and hence approximated by linear combinations of Dirac measures. Finally, by the assumptions made in the proposition, any (ϕ, f) is approximated by $(\phi_R, f) = (\phi, f\mathbf{1}_{B_R})$. □

Spaces satisfying the conditions of Proposition F.6 are numerous. As examples, one can take the spaces $C_\infty(X)$, $C_f(X)$, $C_{f,\infty}(X)$, their subspaces of differentiable functions with uniform or integral norms and various weighted Sobolev spaces.

Appendix G Geometry of collisions

We recall here two points on Boltzmann's collision geometry. Namely, we shall deduce the collision inequalities and the Carleman representation for the collision kernel, in both cases for an arbitrary number of dimensions.

Recall that the Boltzmann equation in weak form, (1.50), can be written as

$$\frac{d}{dt}(g, \mu_t) = \frac{1}{2} \int_{n \in S^{d-1}:(n, w-v) \geq 0} \int_{\mathbf{R}^{2d}} \mu_t(dv)\mu_t(dw) \times [g(v') + g(w') - g(v) - g(w)]B(|v_2 - v_1|, \theta), \qquad \text{(G.1)}$$

where $\theta \in [0, \pi/2]$ is the angle between $n = (v' - v)/|v' - v|$ and $w - v$, and

$$v' = v - n(v - w, n), \qquad w' = w + n(v - w, n), \qquad n \in S^{d-1}, \qquad (n, v - w) \geq 0; \qquad \text{(G.2)}$$

this is a natural parametrization of collisions under the conservation laws (1.48)

$$v + w = v' + w', \qquad v^2 + w^2 = (v')^2 + (w')^2.$$

Unlike in (1.52), which is an adaptation of the general jump-type kinetic equation, we denote here the input and output pairs of velocities by (v, w) and (v', w'), which is more usual in the literature on the Boltzmann equation. Clearly, equation (G.1) can be rewritten as

$$\frac{d}{dt}(g, \mu_t)$$
$$= \frac{1}{4} \int_{S^{d-1}} \int_{R^{2d}} [g(v') + g(w') - g(v) - g(w)] B(|v - w|, \theta) dn \mu_t(dv) \mu_t(dw), \tag{G.3}$$

if B is extended to the angles $\theta \in [\pi/2, \pi]$ by using $B(|v|, \theta) = B(|v|, \pi - \theta)$. Also, the symmetry allows us to rewrite (G.1) as

$$\frac{d}{dt}(g, \mu_t)$$
$$= \int_{n \in S^{d-1}:(n, w-v) \geq 0} dn \int_{\mathbf{R}^{2d}} \mu_t(dv) \mu_t(dw) [g(v') - g(v)] B(|v_2 - v_1|, \theta). \tag{G.4}$$

For measures with density $\mu(dx) = f(x)dx$, this can be given in the following strong form:

$$\frac{d}{dt} f_t(v)$$
$$= \int_{n \in S^{d-1}:(n, w-v) \geq 0} \int_{\mathbf{R}^d} [f_t(w') f_t(v') - f_t(w) f_t(v)] B(|w - v|, \theta) \, dn \, dw. \tag{G.5}$$

A useful tool for the analysis of collisions is supplied by the following elementary fact.

Proposition G.1 *If v', w' are the velocities attained after a collision of particles with velocities $v, w \in \mathbf{R}^d$, so that*

$$v' = v - n(v - w, n), \qquad w' = w + n(v - w, n),$$
$$n \in S^{d-1}, \qquad (n, v - w) \geq 0,$$

then

$$\begin{aligned} |v'|^2 &= |v|^2 \sin^2\theta + |w|^2 \cos^2\theta - \sigma |v| \, |w| \sin\theta \cos\theta \cos\phi \sin\alpha, \\ |w'|^2 &= |v|^2 \cos^2\theta + |w|^2 \sin^2\theta + \sigma |v| \, |w| \sin\theta \cos\theta \cos\phi \sin\alpha, \end{aligned} \tag{G.6}$$

where $\alpha \in [0, \pi]$ is the angle between the vectors v and w, $\theta \in [0, \pi/2]$ is the angle between $v' - v$ and $w - v$ and $\phi \in [0, \pi]$ is the angle between the planes generated by v, w and $v' - v, w - v$.

Exercise G.1 Prove this statement. Hints: clearly the second equation in (G.6) is obtained from the first by symmetry, i.e. by changing θ to $\pi/2 - \theta$. Furthermore, though (G.6) is formulated in $\mathbf{R}^d$, it is effectively a three-dimensional statement involving three vectors, v, w, v' (or the four points

$0, v, w, v'$ in the corresponding affine space). Finally, it can be shown by lengthy but straightforward calculations that the main point is that the vectors $v' - v$ and $v' - w$ are perpendicular (because, as follows from (G.2), the end points of all possible vectors v' lie on the sphere whose poles are the end points of v, w).

The form of the *collision inequality* given below (and its proof) is taken from Lu and Wenberg [164]. This form is the result of much development and improvement, the main contributors being, seemingly, Povsner and Elmroth (see [71]).

Proposition G.2 *With the notation of the previous proposition, the following collision inequality holds for* $\theta \in (0, \pi/2)$ *and any* $c > 0$*:*

$$\begin{aligned}(c+|v'|^2)^{s/2} &+ (c+|w'|^2)^{s/2} - (c+|v|^2)^{s/2} - (c+|w|^2)^{s/2} \\ &\le 2^{s+1}[(c+|v|^2)^{(s-1)/2}(c+|w|^2)^{1/2} \\ &\qquad + (c+|w|^2)^{(s-1)/2}(c+|v|^2)^{1/2}]\cos\theta\sin\theta \\ &- \min\left(\tfrac{1}{4}s(s-2), 2\right)\cos^2\theta\sin^2\theta[(c+|v|^2)^{s/2} + (c+|w|^2)^{s/2}]. \quad \text{(G.7)}\end{aligned}$$

Proof By (G.6)

$$\begin{aligned}|v'|^2 &\le (|v|\sin\theta + |w|\cos\theta)^2, \\ |w'|^2 &\le (|v|\cos\theta + |w|\sin\theta)^2.\end{aligned}$$

Hence

$$\begin{aligned}c+|v'|^2 &\le c(\sin^2\theta + \cos^2\theta) + (|v|\sin\theta + |w|\cos\theta)^2 \\ &\le \left(\sqrt{c+|v|^2}\sin\theta + \sqrt{c+|w|^2}\cos\theta\right)^2\end{aligned}$$

and

$$c+|w'|^2 \le \left(\sqrt{c+|v|^2}\cos\theta + \sqrt{c+|w|^2}\sin\theta\right)^2.$$

Using inequality (6.15) one can write

$$\begin{aligned}(c+|v'|^2)^{s/2} &+ (c+|w'|^2)^{s/2} \\ &\le [(c+|v|^2)^{s/2} + (c+|w|^2)^{s/2}](\sin^s\theta + \cos^s\theta) \\ &\quad + 2^s\sin\theta\cos\theta(\sin^{s-2}\theta + \cos^{s-2}\theta) \\ &\qquad \times [(c+|v|^2)^{(s-1)/2}(c+|w|^2)^{1/2} + (c+|w|^2)^{(s-1)/2}(c+|v|^2)^{1/2}],\end{aligned}$$

the last term being bounded by

$$2^{s+1}\sin\theta\cos\theta\Big[(c+|v|^2)^{(s-1)/2}(c+|w|^2)^{1/2}+(c+|w|^2)^{(s-1)/2}(c+|v|^2)^{1/2}\Big]$$

owing to the assumption $s \geq 2$. Hence, it remains to show the elementary inequality

$$\sin^s\theta+\cos^s\theta \leq 1-\min\left[\tfrac{1}{4}s(s-2),2\right]\cos^2\theta\sin^2\theta, \qquad s>2.$$

Exercise G.2 Prove the above inequality, thus completing the proof of the proposition. Hint: for $s \geq 4$ it reduces to the obvious inequality

$$\sin^s\theta+\cos^s+2\cos^2\theta\sin^2\theta \leq 1;$$

for $s \in [2,4]$ one simply needs to prove that

$$a^\beta+(1-a)^\beta+\beta(\beta-1)a(1-a) \leq 1, \qquad a\in[0,1/2],\ \beta\in[1,2].$$

□

As a second topic in this appendix, we shall discuss a couple of other representations for the Boltzmann equation that are crucial for its qualitative analysis.

Namely, writing

$$n=\frac{w-v}{|w-v|}\cos\theta+m\sin\theta, \qquad dn=\sin^{d-2}\theta\,d\theta\,dm,$$

with $m \in S^{d-2}$ and dm the Lebesgue measure on S^{d-1}, allows us to rewrite (G.5) as

$$\frac{d}{dt}f_t(v_1)=\int_0^{\pi/2}d\theta\int_{S^{d-2}}dm\int_{\mathbf{R}^d}dw[f_t(w')f_t(v')-f_t(w)f_t(v)]$$
$$\times\sin^{d-2}\theta B(|w-v|,\theta). \qquad \text{(G.8)}$$

Under the symmetry condition

$$\sin^{d-2}\theta B(|z|,\theta)=\sin^{d-2}(\pi/2-\theta)B(|z|,\pi/2-\theta)$$

the integral in (G.8) is invariant under the transformation $\theta \mapsto \pi/2-\theta, m \mapsto -m$ (or equivalently, $v' \mapsto w', w' \mapsto v'$). Hence, decomposing the domain of θ into two parts, $[0,\pi/4]$ and $[\pi/4,\pi/2]$, and making the above transformation in the second integral allows us to represent the Boltzmann equation in the following reduced form:

$$\frac{d}{dt}f_t(v)=2\int_{n\in S^{d-1}:\theta\in[0,\pi/4]}\int_{\mathbf{R}^d}[f_t(w')f_t(v')-f_t(w)f_t(v)]$$
$$\times B(|w-v|,\theta)\,dn\,dw. \qquad \text{(G.9)}$$

Next, let us denote by $E_{v,z}$ the $(d-1)$-dimensional plane in $\mathbf{R}^d$ that passes through v and is perpendicular to $z-v$ and by $d_{E_{v,v'}}$ the Lebesgue measure on it. Changing the variables w, n in (G.5) to $v' \in \mathbf{R}^d$, $w' \in E_{v,v'}$, so that

$$dn\, dw = dn\, d_{E_{v,v'}}w'\, d|v'-v| = \frac{1}{|v'-v|^{d-1}} dv'\, d_{E_{v,v'}}w',$$

leads to the following *Carleman representation* of the Boltzmann equation:

$$\frac{d}{dt} f_t(v) = \int_{\mathbf{R}^d} dv' \int_{E_{v,v'}} d_{E_{v,v'}}w'[f_t(w')f_t(v') - f_t(w)f_t(v)]\frac{B(|w-v|,\theta)}{|v-v'|^{d-1}}. \tag{G.10}$$

Finally, using the same transformation for (G.9) leads to the following reduced form of the Carleman representation, the Carleman–Gustafsson representation, which was proposed in [92]:

$$\frac{d}{dt} f_t(v) = 2\int_{\mathbf{R}^d} dv' \int_{w' \in E_{v,v'}:|w'| \le |v-v'|} d_{E_{v,v'}}w'$$
$$\times [f_t(w')f_t(v') - f_t(w)f_t(v)]\frac{B(|w-v|,\theta)}{|v-v'|^{d-1}}. \tag{G.11}$$

Carleman's representation and its modification are useful for obtaining pointwise estimates for solutions to the Boltzmann equation; see e.g. the end of Section 6.2.

Appendix H A combinatorial lemma

Clearly, for any $f \in C_{\mathrm{sym}}(X^2)$ and $\mathbf{x} = (x_1, \ldots, x_n) \in X^n$,

$$\sum_{I \subset \{1,\ldots,n\}, |I|=2} f(\mathbf{x}_I)$$
$$= \tfrac{1}{2} \iint f(z_1, z_2)\delta_{\mathbf{x}}(dz_1)\delta_{\mathbf{x}}(dz_2) - \tfrac{1}{2}\int f(z,z)\delta_{\mathbf{x}}(dz). \tag{H.1}$$

The following is a generalization for functions of k variables.

Proposition H.1 *For any natural k, $f \in C_{\mathrm{sym}}(X^k)$ and $\mathbf{x} = (x_1, \ldots, x_n) \in X^n$,*

$$\sum_{I \subset \{1,\ldots,n\}, |I|=k} f(\mathbf{x}_I) = \frac{1}{k!}[(f, \delta_{\mathbf{x}}^{\oplus k}) - \sigma(f)] \tag{H.2}$$

where $\sigma(f)$ is a positive linear mapping on $C_{\mathrm{sym}}(X^k)$ given by

$$\sigma(f) = \sum_{m=1}^{k-1} m! \sum_{I\subset\{1,\dots,n\},|I|=m} \sum_{l\in I} \int_{X^{k-m-1}} f(x_l, x_l, \mathbf{x}_{I\setminus l}, \mathbf{z}) \times \delta_{\mathbf{x}}^{\oplus(k-m-1)}(d\mathbf{z}). \quad \text{(H.3)}$$

Proof Clearly

$$(f, \delta_{\mathbf{x}}^{\oplus k}) = \sum_{i_1,\dots,i_k=1}^{n} f(x_{i_1},\dots,x_{i_k}) = k! \sum_{I\subset\{1,\dots,n\},|I|=k} f(\mathbf{x}_I) + \sigma(f),$$

where $\sigma(f)$ is the sum of the terms $f(x_{i_1},\dots,x_{i_k})$ in which at least two indices coincide. Let $i, j, i < j$, denote the numbers of the first repeated indices there. Then, by the symmetry of f,

$$\sigma(f) = \sum_{j=2}^{k}\sum_{i=1}^{j-1}\sum_{l=1}^{n} \sum_{J\subset\{1,\dots,n\}\setminus l,|J|=j-2} (j-2)!$$
$$\times \sum_{i_1,\dots,i_{k-j}=1}^{n} f(x_l, x_l, \mathbf{x}_J, x_{i_1},\dots,x_{i_{k-j}})$$
$$= \sum_{j=2}^{k}(j-1)! \sum_{l=1}^{n} \sum_{J\subset\{1,\dots,n\}\setminus l,|J|=j-2} \int_{X^{k-j}} f(x_l, x_l, \mathbf{x}_J, \mathbf{z})\delta_{\mathbf{x}}^{\oplus(k-j)}(d\mathbf{z}).$$

Setting $m = j-1$ and $I = J \cup l$ yields (H.3). □

Corollary H.2 *For $h > 0$, $f \in C_{\text{sym}}(X^k)$ and $\mathbf{x} = (x_1,\dots,x_n) \in X^n$,*

$$h^k \sum_{I\subset\{1,\dots,n\},|I|=k} f(\mathbf{x}_I) = \frac{1}{k!}[(f, (h\delta_{\mathbf{x}})^{\oplus k}) - \sigma_h(f)], \quad \text{(H.4)}$$

where $\sigma_h(f)$ is a positive linear mapping on $C_{\text{sym}}(X^k)$ possessing the estimates

$$\left|\frac{\sigma_h(f)}{k!}\right| \le \frac{h}{2(k-2)!}\int_{X^{k-1}} Pf(\mathbf{y})(h\delta_{\mathbf{x}})^{\otimes(k-1)}(d\mathbf{y}), \quad \text{(H.5)}$$

where $Pf(y_1,\dots,y_{k-1}) = f(y_1, y_1, y_2, y_3,\dots,y_{k-1})$, and

$$\left|\frac{\sigma_h(f)}{k!}\right| \le \frac{h}{2(k-2)!}\|f\|\|h\delta_{\mathbf{x}}\|^{k-1}. \quad \text{(H.6)}$$

Proof Multiplying (H.2) by h^k yields (H.4) with

$$\sigma_h(f) = \sum_{m=1}^{k-1} h^{m+1} m! \sum_{I\subset\{1,\ldots,n\},|I|=m} \sum_{l\in I} \int_{X^{k-m-1}} f(x_l, x_l, \mathbf{x}_{I\setminus l}, \mathbf{z}) \times (h\delta_{\mathbf{x}})^{\oplus(k-m-1)}(d\mathbf{z}). \tag{H.7}$$

Hence, estimating the sum over the subsets I by the sum over all combinations yields

$$|\sigma_h(f)| \le h \sum_{m=1}^{k-1} h^{m+1} m \sum_{x_1,\ldots,x_m=1}^{n} \int f(x_1, x_1, x_2, \ldots, x_m, \mathbf{z}) \times (h\delta_{\mathbf{x}})^{\oplus(k-m-1)}(d\mathbf{z})$$

$$= \sum_{m=1}^{k-1} hm \int Pf(\mathbf{y})(h\delta_{\mathbf{x}})^{\oplus(k-1)}(d\mathbf{y}),$$

implying (H.5) and hence (H.5). □

Exercise H.1 Show that

$$h^k \sum_{I\subset\{1,\ldots,n\},|I|=k} f(\mathbf{x}_I) = \frac{1}{k!}(f, (h\delta_x)^{\otimes k}) + \sum_{l=1}^{k-1} (-h)^l (\Phi_l^k[f], (h\delta_{\mathbf{x}})^{\otimes(k-l)}), \tag{H.8}$$

where the $\Phi_l^k[f]$ are positive bounded operators taking $C_{\text{sym}}(X^k)$ to $C_{\text{sym}}(X^{k-l})$. Write down an explicit formula for $\Phi_l^k[f]$.

Appendix I Approximation of infinite-dimensional functions

Let B and B^* be a real separable Banach space and its dual, with duality denoted by $(\cdot,\cdot)$ and the unit balls by B_1 and B_1^*. It follows from the Stone–Weierstrass theorem that finite-dimensional (or cylindrical) functions of the form $F_f(v) = f((g_1, v), \ldots, (g_m, v))$ with $g_1, \ldots, g_m \in B$ and $f \in C(\mathbf{R}^m)$ are dense in the space $C(B_1^*)$ of $*$-weakly continuous bounded functions on the unit ball in B^*. We need a more precise statement (which to date the author has not found in the literature) that these approximations can be chosen in such a way that they respect differentiation. It is sufficient for us to discuss the case $B = C_\infty(X)$, where X is $\mathbf{R}^n$ or a submanifold of $\mathbf{R}^n$, and we will focus our attention on this case.

A family $P_1, P_2, \ldots$ of linear contractions in B of the form given by

$$P_j v = \sum_{l=1}^{L_j} (w_j^l, v)\phi_j^l, \tag{I.1}$$

where ϕ_j^l and w_j^l are finite linearly independent sets from the unit balls B_1^* and B_1 respectively, is said to form an *approximative identity* if the sequence P_j converges strongly to the identity operator as $j \to \infty$.

To see the existence of such a family, let us choose a finite $1/j$ net $x_1, x_2, \ldots, x_{L_j}$ in the ball $\{\|x\| \leq j\}$. Let ϕ_j^l be a collection of continuous non-negative functions such that $\phi_j(x) = \sum_l \phi_j^l(x)$ belongs to $[0, 1]$ everywhere, equals 1 for $\|x\| \leq j$ and vanishes for $\|x\| \geq j+1$ and such that each ϕ_j^l equals 1 in a neighborhood of x_j^l and vanishes for $\|x - x_j^l\| \geq 2/j$. Then the operators given by

$$P_j f(x) = \sum_{l=1}^{L_j} f(x_j^l)\phi_j^l(x) = \sum_{l=1}^{L_j} (f, \delta_{x_j^l})\phi_j^l(x)$$

form an approximative identity in $B = C_\infty(X)$, which one may check first for $f \in C^1(\mathbf{R}^d)$ and then, by approximating for all $f \in C_\infty(\mathbf{R}^d)$.

Proposition I.1 *Suppose that the family $P_1, P_2, \ldots$ of finite-dimensional linear contractions in B given by* (I.1) *form an approximative identity in B. Then:*

(i) for any $F \in C(B_1^*)$ the family of finite-dimensional (or cylinder) functionals $F_j = F(P_j^*)$ converges to F uniformly (i.e. in the norm topology of $C(B_1^*)$);

(ii) if F is k-times continuously differentiable, i.e. $\delta^k F(\mu)(v_1, \ldots, v_k)$ exists and is a $*$-weakly continuous function of $k+1$ variables, then the derivatives of F_j of order k converge to the corresponding derivatives of F uniformly on B_1^*.

Proof (i) Notice that

$$P_j^*(\mu) = \sum_{l=1}^{L_j} (\phi_j^l, \mu) w_j^l.$$

The required convergence for functions of the form $F_g(\mu) = \exp[(g, \mu)]$, $g \in C_\infty(X)$, follows from the definition of the approximative identity

$$F_g(P_j^*(\mu)) = \exp\left[(P_j g, \mu)\right].$$

For arbitrary $F \in C(B_1^*)$ the statement is obtained through its approximation by linear combinations of exponential functions F_g (which is possible owing to the Stone–Weierstrass theorem).

(ii) If F is k-times continuously differentiable then

$$\frac{\delta F_j(\mu)}{\delta\mu(x)} = \sum_{l=1}^{L_j} \phi_j^l(x) \frac{\delta F}{\delta\mu(x_j^l)}(P_j^\star \mu)$$
$$= \left(\frac{\delta F}{\delta\mu(x_j^l)}(P_j^\star \mu),\ P_j^\star \delta_x \right) = P_j \left(\frac{\delta F(P_j^\star \mu)}{\delta\mu(.)} \right)(x),$$

and similarly

$$\frac{\delta^k F_j(\mu)}{\delta\mu(v_1)\cdots\delta\mu(v_k)} = \left(\frac{\delta^k F}{\delta\mu(v_1)\cdots\delta\mu(v_k)}(P_j^\star\mu),\ P_j^\star\delta_{v_1} \otimes \cdots \otimes P_j^\star\delta_{v_k} \right);$$

the result then follows from (i). □

From Proposition I.1 one easily deduces the possibility of approximating $F \in C(B_1^*)$ by polynomials or exponential functionals together with their derivatives.

Corollary I.2 *For any $F \in C(B_1^*)$ there exist two families $F_n \in C(B_1^*)$, of finite linear combinations of the analytic functions of $\mu \in B^*$, the first of which is of the form*

$$\exp\left\{ -\epsilon_j \sum_{l=1}^{L_j} \left[(\phi_j^l, \mu) - \xi_l \right]^2 \right\}, \qquad \epsilon_j > 0, \quad \xi_l \in \mathbf{R}^d,$$

and the second of which is of the form

$$\int g(x_1, \ldots, x_k)\mu(dx_1)\cdots\mu(dx_n), \qquad g \in S(\mathbf{R}^d),$$

such that $\Pi_j(F)$ converges to F uniformly on B_1^*; if F is differentiable then the corresponding derivatives also converge.

Proof For any j, clearly the functional $F_j(\mu) = F(P_j^*(\mu))$ can be written in the form $F_j(\mu) = f_j(y(\mu))$ where $y(\mu) = \{(\phi_j^1, \mu), \ldots, (\phi_j^{L_j}, \mu)\}$ and the f_j are bounded continuous functions of L_j variables. Approximating f_j first by $e^{\epsilon\Delta} f_j$ and the latter function by a linear combination h_j of Gaussian functions of the type $\exp[-\epsilon_j \sum_{l=1}^{L_j} (y_l - \xi_l)^2]$, we then define $\Pi_j(F(\mu)) = h_j(y(\mu))$, which has the required property. One obtains the polynomial approximation in a similar way, invoking the finite-dimensional Weierstrass theorem. □

Appendix J Bogolyubov chains, generating functionals and Fock-space calculus

This appendix is an addendum to Section 1.9 indicating an alternative method of deducing the basic kinetic equations (historically one of the first methods; see the comments in Section 11.6) that uses so-called Bogolyubov chains, called also the BBGKY hierarchy.

The discussion in Section 1.9 suggests that it could be useful to look at the evolution (1.68) in terms of correlation functions. This is actually the method normally used in statistical physics. The corresponding equations are called *Bogolyubov chains*. A neat and systematic way to obtain these equations is via Fock-space calculus, using creation and annihilation operators; this also yields a fruitful connection with similar problems in quantum mechanics.

For an arbitrary $Y \in \mathcal{M}(X)$ the *annihilation operator* $a_-(Y)$ is defined by

$$(a_-(Y)f)(x_1, \ldots, x_n) = \int_{x_{n+1}} f(x_1, \ldots, x_n, x_{n+1})Y(dx_{n+1})$$

on $C(S\mathcal{X})$ and, for an arbitrary $h \in C(X)$, the *creation operator* $a_+(h)$ acts on $C(S\mathcal{X})$ as follows:

$$(a_+(h)f)(x_1, \ldots, x_n) = \sum_{i=1}^{n} f(x_1, \ldots, \check{x}_i, \ldots, x_n)h(x_i), \qquad n \neq 0,$$
$$(a_+(h)f)^0 = 0$$

(the inverted caret on x_i indicates that it is removed from the list of arguments of f). Their duals are the creation and annihilation operators on $\mathcal{M}_{\text{sym}}(\mathcal{X})$, defined as

$$(a_-^\star(Y)\rho)(dx_1 \cdots dx_n) = \frac{1}{n}\sum_{i=1}^{n} \rho(dx_1 \cdots \check{dx}_i \cdots dx_n)Y(dx_i),$$
$$(a_+^\star(h)\rho)(dx_1 \cdots dx_n) = (n+1)\int_{x_{n+1}} \rho_{n+1}(dx_1 \cdots dx_{n+1})h(x_{n+1}).$$

These operators satisfy the *canonical commutation relations*

$$[a_-(Y), a_+(h)] = \int h(x)Y(dx)\mathbf{1}, \qquad [a_+^\star(h), a_-^\star(Y)] = \int h(x)Y(dx)\mathbf{1}.$$

For brevity, we shall write a_+ and $a_+^\star$ for the operators $a_+(\mathbf{1})$ and $a_+^\star(\mathbf{1})$ respectively. The powers of $a_+(h)$ and its dual are clearly given by the formulas

$$\left[(a_+(h))^m f\right]^n (x_1, \ldots, x_n) = m! \sum_{I \subset \{1,\ldots,n\}, |I|=m} f^{n-m}(x_{\bar{I}}) \prod_{i \in I} h(x_i),$$

$$\left[(a_+^\star(h))^m \rho\right]_n (dx_1 \cdots dx_n) = \frac{(n+m)!}{n!} \int_{x_{n+1},\ldots,x_{n+m}} \rho_{n+m}(dx_1 \cdots dx_{n+m}) \prod_{j=1}^m h(x_{n+j})$$

(the first formula holds only for $m \leq n$ and, for $m > n$, the power $\left[(a_+(h))^m f\right]^n$ vanishes).

In terms of canonical creation and annihilation operators, the transformations $\rho \mapsto \mu^h$ and $g \to S^h g = f$ from Section 1.9 are described by the formulas

$$\rho \mapsto \nu = N! \, h^N e^{a_+^\star} \rho \tag{J.1}$$

and

$$f = \{f^j\} \mapsto g = \{g_j\} = (N!)^{-1} h^{-N} e^{-a_+} f. \tag{J.2}$$

Next, for an operator D on $C(X)$, the density number operator $n(D)$ (or *gauge operator* or *second-quantization* operator for D) in $C(S\mathcal{X})$ is defined by the formula

$$n(D) f(x_1, \ldots, x_n) = \sum_{i=1}^n D_i f(x_1, \ldots, x_n), \tag{J.3}$$

where D_i acts on $f(x_1, \ldots, x_n)$ regarded as a function of x_i. One easily checks that

$$[n(A), n(D)] = n([A, D]),$$
$$[n(D), a_+(h)] = a_+(Dh), \qquad [a_-(Y), n(D)] = a_-(D^\star Y).$$

In particular, if D is a finite-dimensional operator of the form

$$D = \sum_{j=1}^l h_j \otimes Y_j, \qquad h_j \in C(X), \qquad Y \in \mathcal{M}(X), \tag{J.4}$$

which acts as $Df = \sum_{j=1}^l (f, Y_j) h_j$, then $n(D) = \sum_{j=1}^l a_+(h_j) a_-(Y_j)$. In general, however, one cannot express $n(D)$ in terms of a_+ and a_-.

The tensor power of the operator n is defined for a possibly unbounded operator $L^k : C_{\rm sym}(\mathcal{X}) \to C_{\rm sym}(X^k)$ as the operator $n^{\otimes k}(L^k)$ in $C_{\rm sym}(\mathcal{X})$ given by the formula

$$(n^{\otimes k}(L^k)f)(x_1, \ldots, x_n) = \sum_{I \subset \{1,\ldots,n\}, |I|=k} (L^k f_{x_{\bar{I}}})(x_I), \tag{J.5}$$

where we write

$$f_{x_1,\ldots,x_n}(y_1, \ldots, y_m) = f(x_1, \ldots, x_n, y_1, \ldots, y_m),$$

for an arbitrary $f \in C_{\rm sym}(\mathcal{X})$. This is precisely the transformation from a B_k of the form (1.65) to the generator of k-ary interaction I_k given by (1.66), so that

$$I_k[P^k, A^k] = n^{\otimes k}(B^k). \tag{J.6}$$

Using the formalism of the creation, annihilation and gauge operators, one may deduce the Bogolyubov-chain equations, i.e. the equations for the correlation functions ν^h that correspond to the evolution of ρ^h dual to equation (1.68), i.e.

$$\dot{\rho}^h(t) = (I^h)^\star[P, A]\rho^h(t), \qquad I^h[P, A] = \frac{1}{h}\sum_{l=1}^{k} h^l I_l[P^l, A^l]. \tag{J.7}$$

Although we will not carry out this calculation (see the comments in Section 11.6), let us note that it turns out that the limiting equations, as $h \to 0$, to the equations for ν and g expressed by (J.1) and (J.2) are given by the chain of equations (1.91) and (1.92), yielding the method of deducing kinetic equations from Bogolyubov chains for correlation functions.

Finally, let us mention a method of storing information on many-particle evolutions using generating functionals. The *generating functionals* for an observable $f = \{f^n\} \in C(\mathcal{X})$ and for a state $\rho = \{\rho_n\} \in \mathcal{M}(\mathcal{X})$ are defined respectively as functionals Φ_f of $Y \in \mathcal{M}(X)$ and $\tilde{\Phi}_\rho$ of $Q \in C(X)$, where

$$\Phi_f(Y) = (f, Y^{\tilde{\otimes}}), \qquad \tilde{\Phi}_\rho(Q) = (Q^{\otimes}, \rho). \tag{J.8}$$

Clearly generating functionals represent an infinite-dimensional analogue of the familiar generating functions for discrete probability laws.

As follows from the above discussion, the kinetic equation (1.71) yields the characteristic equation for the evolution of the generating functional Φ_{g_t} of the observable g_t satisfiying (1.92) (the dual limiting equation for correlation functions). Namely, if Y_t solves (1.71) with initial condition Y_0 and g_t solves (1.92) with initial condition g_0 then

$$\Phi_{g_t}(Y_0) = \Phi_{g_0}(Y_t).$$

This equation represents a natural nonlinear extension of the usual duality between measures and functions.

Appendix K Infinite-dimensional Riccati equations

The construction of the Ornstein–Uhlenbeck (OU) semigroups from Section 10.4 is very straightforward. However, the corresponding process is Gaussian; hence it is also quite natural and insightful to construct infinite-dimensional OU semigroups and/or propagators alternatively, via the completion from its action on Gaussian test functions. In analyzing the latter, the Riccati equation appears. We shall sketch here this approach to the analysis of infinite-dimensional OU semigroups, starting with the theory of differential Riccati equations on symmetric operators in Banach spaces.

Let B and $B^\star$ be a real Banach space and its dual, duality being denoted as usual by $(.,.)$. Let us say that a densely defined operator C from B to $B^\star$ (that is possibly unbounded) is *symmetric* (resp. *positive*) if $(Cv, w) = (Cw, v)$ (resp. if $(Cv, v) \geq 0$) for all v, w from the domain of C. By $SL^+(B, B^\star)$ let us denote the space of bounded positive operators taking B to $B^\star$. Analogous definitions are applied to the operators taking $B^\star$ to B. The notion of positivity induces a (partial) order relation on the space of symmetric operators.

The (time-nonhomogeneous differential) *Riccati equation* in B is defined as the equation

$$\dot{R}_t = A(t)R_t + [A(t)R_t]^\star - R_t C(t) R_t, \tag{K.1}$$

where $A(t)$ and $C(t)$ are given families of possibly unbounded operators from B to B and from B to $B^\star$ respectively, and where the solutions R_t are sought in the class $SL^+(B^\star, B)$. The literature on the infinite-dimensional Riccati equation is extensive (see e.g. Curtain [53] and McEneaney [181] and references therein), but it is usually connected with optimal control problems and has a the Hilbert space setting. We shall give here a short proof of well-posedness for an unbounded family $C(t)$ in a Banach space using an explicit formula arising from the "interaction representation" and bypassing any optimization interpretations and tools, and we then discuss the link with infinite-dimensional Ornstein–Uhlenbeck processes describing fluctuation limits.

Before approaching (K.1) we will summarize the main properties of a simpler reduced equation with vanishing $A(t)$, namely the equation

$$\dot{\pi}(t) = -\pi(t)C(t,s)\pi(t), \qquad t \geq s. \tag{K.2}$$

Proposition K.1 *Suppose that $C(t,s)$, $t \geq s$, is a family of densely defined positive operators taking B to $B^\star$ that are bounded and strongly continuous in t for $t > s$, their norms being an integrable function, i.e.*

$$\int_s^t \|C(\tau,s)\| d\tau \leq \kappa(t-s) \tag{K.3}$$

for a continuous $\kappa : \mathbf{R}_+ \mapsto \mathbf{R}_+$ vanishing at the origin. Then, for any $\pi_s \in SL^+(B^\star, B)$, there exists a unique global strongly continuous family of operators $\pi(t,s) \in SL^+(B^\star, B)$, $t \geq s$, such that

$$\pi(t,s) = \pi_s - \int_s^t \pi(\tau,s)C(\tau,s)\pi(\tau,s)\, d\tau \tag{K.4}$$

(the integral is defined in the norm topology). Moreover:

(i) $\pi(t,s) \leq \pi_s$ and the image of $\pi(t,s)$ coincides with that of π_s for all $t \geq s$;

(ii) the family $\pi(t,s)$ depends continuously on t,s and Lipschitz continuously on the initial data in the uniform operator topology (defined by the operator norm);

(iii) equation (K.2) *holds in the strong operator topology for $t > s$;*

(iv) if π_s has a bounded inverse π_s^{-1} then all $\pi(t,s)$ are invertible and

$$\|\pi^{-1}(t,s)\| \leq \|\pi_s^{-1}\| + \kappa(t-s).$$

Proof The existence of a positive solution for times $t-s$ such that $\kappa(t-s)\|\pi_s\| < 1$ follows from the explicit formula

$$\pi(t,s) = \pi_s \left(\mathbf{1} + \int_s^t C(\tau,s)\, d\tau\, \pi_s\right)^{-1} = \left(\mathbf{1} + \pi_s \int_s^t C(\tau,s)\, d\tau\right)^{-1} \pi_s$$

(notice that the series representations for both these expressions coincide), which is obtained from the observation that, in terms of the inverse operator, equation (K.2) takes the simpler form $(d/dt)\pi^{-1}(t,s) = C(t,s)$. This implies also the required bound for the inverse operator. From (K.4) and the positivity of $\pi(t,s)$ it follows that $\pi(t,s) \leq \pi_s$. Consequently, for large $t-s$ one can construct solutions by iterating the above formula. Lipschitz continuity and uniqueness follow from Gronwall's lemma, since (K.4) implies that

$$\|\pi^1(t,s)-\pi^2(t,s)\| \le \|\pi_s^1-\pi_s^2\| + \int_s^t \left[\|\pi^1(\tau,s)\|+\|\pi^2(\tau,s)\|\right] \times \|C(\tau,s)\|\|\pi^1(\tau,s)-\pi^2(\tau,s)\|\,d\tau. \tag{K.5}$$

□

Now we return to equation (K.1).

Proposition K.2 *(i) Let the domains of all $A(t)$ and $C(t)$ contain a common dense subspace D in B and depend strongly continuously on t as bounded operators taking D to B and D to $B^\star$ respectively (we assume that D is itself a Banach space with a certain norm);*

(ii) let $A(t)$ generate a bounded propagator $U^{t,s}$ with common invariant domain D so that, for any $\phi \in D$, the family $U^{t,s}\phi$ is the unique solution in D of the Cauchy problem

$$\frac{d}{dt}U^{t,s}\phi = A(t)U^{t,s}\phi, \qquad U^{s,s}\phi = \phi$$

and $U^{t,s}$ is strongly continuous both in B and D;

(iii) let $C(t)$ be positive and let the norm of $C(t,s) = (U^{t,s})^\star C(t)U^{t,s} \in SL^+(B, B^\star)$ for $t > s$ satisfy (K.3).

Then, for any $R \in SL^+(B^\star, B)$ with image belonging to D, the family

$$R_t = U^{t,s}\pi(t,s)(U^{t,s})^\star, \qquad t \ge s, \tag{K.6}$$

where $\pi(t,s)$ is the solution to (K.2) *given by Proposition K.1 with $\pi_s = R$ and $C(t,s) = (U^{t,s})^\star C(t)U^{t,s}$, is a continuous function $t \mapsto SL^+(B^\star, B)$, $t \ge s$, in the strong operator topology; the images of the R_t belong to D and R_t depends Lipschitz continuously on R and satisfies the Riccati equation weakly, i.e.*

$$\frac{d}{dt}(R_t v, w) = (A(t)R_t v, w) + (v, A(t)R_t w) - (R_t C(t) R_t v, w) \tag{K.7}$$

for all $v, w \in B^\star$. Finally, if R is compact or has a finite-dimensional range then the same holds for all R_t.

Proof Everything follows by inspection from the explicit formula given (it is straightforward to check that if π satisfies the required reduced Riccati equation then the "dressed" operator (K.6) satisfies (K.7)) and Proposition K.1.

□

Remark K.3 The obstruction to the strong form of the Riccati equation lies only in the second term of (K.7) and can be easily removed by an appropriate assumption.

Remark K.4 The above results were formulated for the usual forward Cauchy problem. However, in stochastic processes one often has to deal with the inverse Cauchy problem. Of course, everything remains the same for the inverse-time Riccati equation

$$\dot{R}_s = -A(s)R_s - [A(s)R_s]^\star + R_s C(s) R_s, \qquad s \le t, \tag{K.8}$$

if $A(t)$ is assumed to generate the backward propagator $U^{s,t}$, $s \le t$, in B. Then the family $R_s = U^{s,t}\pi(s,t)(U^{s,t})^\star$ solves (K.8) with initial condition $R_t = R$, where π solves the reduced Riccati equation in terms of the inverse time $\dot{\pi}_s = \pi_s C(s,t)\pi_s$ with the same initial condition $\pi_t = R$ and $C(s,t) = (U^{s,t})^\star C(s) U^{s,t}$.

Now we point out how the Riccati equation appears from an analysis of the backward propagators in $C(B^\star)$, which are specified by formal generators of the form

$$O_t F(Y) = \left(A(t)\frac{\delta F}{\delta Y(.)}, Y \right) + \tfrac{1}{2} L_{C(t)} \frac{\delta^2 F}{\delta Y^2(.,.)},$$

where $A(t)$ and $C(t)$ are as above and $L_{C(t)}$ is the bilinear form corresponding to $C(t)$, i.e.

$$L_{C(t)}(f \otimes g) = (C(t)f, g),$$

so that

$$\|C(t)\|_{B \mapsto B^\star} = \sup_{\|f\|,\|g\| \le 1} (C(t)f, g).$$

In order to give meaning to the above operator O_t it is convenient to reduce the analysis to the functional Banach spaces B introduced in Proposition F.6. If $R \in SL^+(B^\star, B)$ for such a space B then R can be specified by its integral kernel (which, with some abuse of notation, we shall denote by the same letter R). Thus

$$R(x, y) = (R\delta_x, \delta_y) = (R\delta_x)(y),$$

so that

$$(Rf)(x) = \int (R\delta_x)(y) f(y)\, dy = \int R(x, y) f(y)\, dy.$$

Now let

$$F_t(Y) = F_{R_t,\phi_t,\gamma_t}(Y) = \exp\left[-\tfrac{1}{2}(R_t Y, Y) + (\phi_t, Y) + \gamma_t \right], \qquad Y \in B^\star, \tag{K.9}$$

with $R_t \in SL^+(B^\star, B)$, $\phi_t \in B$. Then

$$O_t F_t(Y) = \Big(A(t)(-R_t Y + \phi_t), Y\Big) + \tfrac{1}{2}\Big(C(t)(R_t Y - \phi_t), R_t Y - \phi_t\Big) - \tfrac{1}{2} L_{C(t)} R_t,$$

implying the following.

Proposition K.5 *If a Banach space B is of the kind introduced in Proposition F.6, if R_s, $s \le t$, satisfies the weak inverse-time Riccati equation*

$$\frac{d}{ds}(R_s v, v) = -2(A(s)R_s v, v) + (R_s C(s) R_s v, v), \qquad s \le t, \qquad \text{(K.10)}$$

and if ϕ_s, γ_s satisfy the system

$$\begin{aligned} \dot{\phi}_s &= -A(s)\phi_s + R_s C(s)\phi_s, \\ \dot{\gamma}_s &= \tfrac{1}{2} L_{C(s)}(R_s - \phi_s \otimes \phi_s) \end{aligned} \qquad \text{(K.11)}$$

then the Gaussian functional F_s solves the equation $\dot{F}_s = -O_s F_s$. If the assumptions of Proposition K.2 hold for the inverse equation (K.8) *(see Remark K.4) and the first equation in* (K.11) *is well posed in B, the resolving operators of the equation $\dot{F}_s = -O_s F_s$ form a backward propagator of positivity-preserving contraction in the closure (in the uniform topology) of the linear combinations of the Gaussian functionals* (K.9) *it R_t has images from D.*

Proof This is straightforward from Proposition K.2. The positivity of the propagators obtained follows from the conditional positivity of O_t or alternatively can be deduced from the finite-dimensional approximations. □

Remark K.6 The propagator constructed in Proposition K.5 does not have to be strongly continuous. Continuity in time holds only in the sense of convergence on bounded sets.

For example, in the case of the operator $O_t F$ from (10.16) under the assumptions of Theorem 8.14 (the pure coagulation model) the operator $A(t)$ is given by

$$A(t)f(x) = \int \big[f(x+z) - f(x) - f(z)\big] K(x,z)\mu_t(dz),$$

and generates the propagator $U^{t,r}$ constructed in Chapter 7. Hence one can apply Proposition K.5 for an alternative construction of the Ornstein–Uhlenbeck semigroup from Section 10.4.

References

[1] L. Accardi, F. Fagnola (eds.). Quantum interacting particle systems. In: *Proc. Volterra–CIRM Int. School, Trento, 2000, QP–PQ: Quantum Probability and White Noise Analysis*, vol. 14, World Scientific, 2002.

[2] S. Albeverio, A. Hilbert and V. Kolokoltsov. Sur le comportement asymptotique du noyau associé á une diffusion dégénéré. *C.R. Math. Rep. Acad. Sci. Canada* **22:4** (2000), 151–159.

[3] S. Albeverio, B. Rüdiger. Stochastic integrals and the Lévy–Ito decomposition theorem on separable Banach spaces. *Stoch. Anal. Appl.* **23:2** (2005), 217–253.

[4] D. J. Aldous. Deterministic and stochastic models for coalescence (aggregation and coagulation): a review of the mean-field theory for probabilists. *Bernoulli* **5:1** (1999), 3–48.

[5] H. Amann. Coagulation–fragmentation processes. *Arch. Ration. Mech. Anal.* **151** (2000), 339–366.

[6] W. J. Anderson. *Continuous-Time Markov Chains. Probability and its Applications.* Springer Series in Statistics. Springer, 1991.

[7] D. Applebaum. *Probability and Information.* Cambridge University Press, 1996.

[8] D. Applebaum. *Lévy Processes and Stochastic Calculus.* Cambridge Studies in Advanced Mathematics, vol. 93. Cambridge University Press, 2004.

[9] O. Arino, R. Rudnicki. Phytoplankton dynamics. *Comptes Rendus Biol.* **327** (2004), 961–969.

[10] L. Arkeryd. On the Boltzmann equation. Parts I and II. *Arch. Ration. Mech. Anal.* **45** (1972), 1–35.

[11] L. Arkeryd. L^∞ Estimates for the spatially-homogeneous Boltzmann equation. *J. Stat. Phys.* **31:2** (1983), 347–361.

[12] A. A. Arseniev. Lektsii o kineticheskikh uravneniyakh (in Russian) (Lectures on kinetic equations). *Nauka*, Moscow, 1992.

[13] A. A. Arseniev, O. E. Buryak. On a connection between the solution of the Boltzmann equation and the solution of the Landau–Fokker–Planck equation (in Russian). *Mat. Sb.* **181:4** (1990), 435–446; English translation in *Math. USSR Sb.* **69:2** (1991), 465–478.

[14] I. Bailleul. Sensitivity for Smoluchovski equation. Preprint 2009. http://www.statslab.cam.ac.uk/ismael/files/Sensitivity.pdf.

[15] A. Bain, D. Crisan. *Fundamentals of Stochastic Filtering*. Stochastic Modelling and Applied Probability, vol. 60. Springer, 2009.

[16] R. Balescu. *Statistical Dynamics. Matter out of Equilibrium*. Imperial College Press, 1997.

[17] J. M. Ball, J. Carr. The discrete coagulation–fragmentation equations: existence, uniqueness and density conservation. *J. Stat. Phys.* **61** (1990), 203–234.

[18] R. F. Bass. Uniqueness in law for pure jump type Markov processes. *Prob. Theory Relat. Fields* **79** (1988), 271–287.

[19] R. F. Bass, Z.-Q. Chen. Systems of equations driven by stable processes. *Prob. Theory Relat. Fields* **134** (2006), 175–214.

[20] P. Becker-Kern, M. M. Meerschaert, H.-P. Scheffler. Limit theorems for coupled continuous time random walks. *Ann. Prob.* **32:1B** (2004), 730–756.

[21] V. P. Belavkin. Quantum branching processes and nonlinear dynamics of multi-quantum systems (in Russian). *Dokl. Acad. Nauk SSSR* **301:6** (1988), 1348–1352.

[22] V. P. Belavkin. Multiquantum systems and point processes I. *Rep. Math. Phys.* **28** (1989), 57–90.

[23] V. P. Belavkin, V. N. Kolokoltsov. Stochastic evolutions as boundary value problems. In: *Infinite Dimensional Analysis and Quantum Probability, RIMS Kokyuroku* **1227** (2001), 83–95.

[24] V. P. Belavkin, V. N. Kolokoltsov. Stochastic evolution as interaction representation of a boundary value problem for Dirac type equation. *Inf. Dim. Anal., Quantum Prob. Relat. Fields* **5:1** (2002), 61–92.

[25] V. P. Belavkin, V. N. Kolokoltsov. On general kinetic equation for many particle systems with interaction, fragmentation and coagulation. *Proc. Roy. Soc. London A* **459** (2003), 727–748.

[26] V. P. Belavkin, V. P. Maslov. Uniformization method in the theory of nonlinear hamiltonian systems of Vlasov and Hartree type (in Russian). *Teoret. i Matem. Fizika* **33:1** (1977), 17–31. English translation in *Theor. Math. Phys.* **43:3** (1977), 852–862.

[27] R. E. Bellman. *Dynamic Programming*. Princeton University Press and Oxford University Press, 1957.

[28] G. Ben Arous. Developpement asymptotique du noyau de la chaleur sur la diagonale. *Ann. Inst. Fourier* **39:1** (1989), 73–99.

[29] A. Bendikov. Asymptotic formulas for symmetric stable semigroups. *Exp. Math.* **12** (1994), 381–384.

[30] V. Bening, V. Korolev, T. Suchorukova, G. Gusarov, V. Saenko, V. Kolokoltsov. Fractionally stable distributions. In: V. Korolev, N. Skvortsova (eds.), *Stochastic Models of Plasma Turbulence* (in Russian), Moscow State University, 2003, pp. 291–360. English translation in V. Korolev, N. Skvortsova (eds.), *Stochastic Models of Structural Plasma Turbulence*, VSP, 2006, pp. 175–244.

[31] V. Bening, V. Korolev, V. Kolokoltsov. Limit theorems for continuous-time random walks in the double array limit scheme. *J. Math. Sci.* (NY) **138:1** (2006), 5348–5365.

[32] J. Bennett, J.-L. Wu. Stochastic differential equations with polar-decomposed Lévy measures and applications to stochastic optimization. *Fron. Math. China* **2:4** (2007), 539–558.

[33] J. Bertoin. *Lévy Processes*. Cambridge Tracts in Mathematics, vol. 121, Cambridge University Press, 1996.

[34] J. Bertoin. *Random Fragmentation and Coagulation Processes*. Cambridge Studies in Advanced Mathematics, vol. 102, Cambridge University Press, 2006.

[35] K. Bichteler. *Stochastic Integration with Jumps*. Encyclopedia of Mathematics and Applications, Cambridge University Press, 2002.

[36] K. Bichteler, J.-B. Gravereaux, J. Jacod. *Malliavin Calculus for Processes with Jumps*. Stochastic Monographs, vol. 2, Gordon and Breach, 1987.

[37] P. Biler, L. Brandolese. Global existence versus blow up for some models of interacting particles. *Colloq. Math.* **106:2** (2006), 293–303.

[38] P. Billingsley. *Convergence of Probability Measures*. Wiley, 1968.

[39] H. Bliedtner, W. Hansen. *Potential Theory – An Analytic Approach to Balayage*. Universitext, Springer, 1986.

[40] R. M. Blumenthal, R. K. Getoor. Some theorems on stable processes. *Trans. Amer. Math. Soc.* **95** (1960), 263–273.

[41] A. V. Bobylev. The theory of the nonlinear spatially uniform Boltzmann equation for Maxwell molecules. *Sov. Sci. Rev. C, Math. Phys. Rev.* **7** (1988), 111–233.

[42] N. N. Bogolyubov. *Problems of the Dynamic Theory in Statistical Physics*. Moscow, 1946 (in Russian).

[43] J.-M. Bony, Ph. Courrège, P. Priouret. Semi-groupes de Feller sur une variété a bord compacte et problèmes aux limites intégro-différentiels du second ordre donnant lieu au principe du maximum. *Ann. Inst. Fourier, Grenoble* **18:2** (1968), 369–521.

[44] Yu. D. Burago, V. A. Zalgaller. *Geometric Inequalities*. Springer, 1988.

[45] T. Carleman. *Problèmes mathématique dans la théorie cinétique des gaz*. Almquist and Wiksells, 1957.

[46] R. A. Carmona, D. Nualart. *Nonlinear Stochastic Integrators, Equations and Flows*. Stochatic Monographs, vol. 6, Gordon and Breach, 1990.

[47] C. Cercognani, R. Illner, M. Pulvirenti. *The Mathematical Theory of Dilute Gases*. Springer, 1994.

[48] A. M. Chebotarev. A priori estimates for quantum dynamic semigroups (in Russian). *Teoret. Mat. Fiz.* **134:2** (2003), 185–190; English translation in *Theor. Math. Phys.* **134:2** (2003), 160–165.

[49] A. M. Chebotarev, F. Fagnola. Sufficient conditions for conservativity of minimal quantum dynamic semigroups. *J. Funct. Anal.* **118** (1993), 131–153.

[50] A. M. Chebotarev, F. Fagnola. Sufficient conditions for conservativity of minimal quantum dynamic semigroups. *J. Funct. Anal.* **153** (1998), 382–404.

[51] J. F. Collet, F. Poupaud. Existence of solutions to coagulation-fragmentation systems with diffusion. *Transport Theory Statist. Phys.* **25** (1996), 503–513.

[52] Ph. Courrège. Sur la forme integro-différentiélle du générateur infinitésimal d'un semi-groupe de Feller sur une variété. In: *Sém. Théorie du Potentiel*, 1965–66. Exposé 3.

[53] F. P. da Costa, H. J. Roessel, J. A. D. Wattis. Long-time behaviour and self-similarity in a coagulation equation with input of monomers. *Markov Proc. Relat. Fields* **12** (2006), 367–398.

[54] D. Crisan, J. Xiong. Approximate McKean–Vlasov representations for a class of SPDEs. To appear in *Stochastics*.

[55] R. F. Curtain. Riccati equations for stable well-posed linear systems: the generic case. *SIAM J. Control Optim.* **42:5** (2003), 1681–1702 (electronic).

[56] E. B. Davies. *Quantum Theory of Open Systems*. Academic Press, 1976.

[57] E. B. Davies. *Heat Kernels and Spectral Theory*. Cambridge University Press, 1992.

[58] D. Dawson. Critical dynamics and fluctuations for a mean-field model of cooperative behavior. *J. Stat. Phys.* **31:1** (1983), 29–85.

[59] D. Dawson. Measure-valued Markov processes. In: P. L. Hennequin (ed.), *Proc. Ecole d'Eté de probabilités de Saint-Flour XXI, 1991*. Springer Lecture Notes in Mathematics, vol. 1541, 1993, pp. 1–260.

[60] D. Dawson *et al.* Generalized Mehler semigroups and catalytic branching processes with immigration. *Potential Anal.* **21:1** (2004), 75–97.

[61] A. de Masi, E. Presutti. *Mathematical Methods for Hydrodynamic Limits*. Springer, 1991.

[62] M. Deaconu, N. Fournier, E. Tanré. A pure jump Markov process associated with Smoluchovski's coagulation equation. *Ann. Prob.* **30:4** (2002), 1763–1796.

[63] M. Deaconu, N. Fournier, E. Tanré. Rate of convergence of a stochastic particle system for the Smoluchovski coagulation equation. *Methodol. Comput. Appl. Prob.* **5:2** (2003), 131–158.

[64] P. Del Moral. *Feynman–Kac Formulae. Genealogical and Interacting particle Systems with Applications*. Probability and its Application. Springer, 2004.

[65] L. Desvillettes, C. Villani. On the spatially homogeneous Landau equation for hard potentials. Part I. *Comm. Partial Diff. Eq.* **25** (2000), 179–259.

[66] S. Dharmadhikari, K. Joag-Dev. *Unimodality, Convexity, and Applications*. Academic Press, 1988.

[67] B. Driver, M. Röckner. Constructions of diffusions on path spaces and loop spaces of compact riemannian manifolds. *C.R. Acad. Sci. Paris, Ser. I* **320** (1995), 1249–1254.

[68] P. B. Dubovskii, I. W. Stewart. Existence, uniqueness and mass conservation for the coagulation–fragmentation equation. *Math. Meth. Appl. Sci.* **19** (1996), 571–591.

[69] E. B. Dynkin. *Superdiffusions and Positive Solutions of Nonlinear Partial Differential Equations*. University Lecture Series, vol. 34, American Mathematical Society, 2004.

[70] A. Eibeck, W. Wagner. Stochastic particle approximation to Smoluchovski's coagulation equation. *Ann. Appl. Prob.* **11:4** (2001), 1137–1165.

[71] T. Elmroth. Global boundedness of moments of solutions of the Boltzmann equation for forces of inifinite range. *Arch. Ration. Mech. Anal.* **82** (1983), 1–12.

[72] F. O. Ernst, S.E. Protsinis. Self-preservation and gelation during turbulance induced coagulation. *J. Aerosol Sci.* **37:2** (2006), 123–142.

[73] A. M. Etheridge. *An Introduction to Superprocesses*. University Lecture Series, vol. 20, American Mathematical Society, 2000.

[74] S. N. Ethier, Th. G. Kurtz. *Markov Processes – Characterization and Convergence*. Wiley Series in Probability and Mathematical Statistics, Wiley, 1986.

[75] K. Evans, N. Jacob. Feller semigroups obtained by variable order subordination. *Rev. Mat. Comput.* **20:2** (2007), 293–307.

[76] W. Feller. *An Introduction to Probability. Theory and Applications*, second edition, vol. 2. John Wiley and Sons, 1971.

[77] N. Fournier, Ph. Laurencot. Local properties of self-similar solutions to Smoluchowski's coagulation equation with sum kernels. *Proc. Roy. Soc. Edinburgh. A* **136:3** (2006), 485–508.

[78] M. Freidlin. *Functional Integration and Partial Differential Equations*. Princeton University Press, 1985.

[79] T. D. Frank. Nonlinear Markov processes. *Phys. Lett. A* **372:25** (2008), 4553–4555.

[80] B. Franke. The scaling limit behavior of periodic stable-like processes. *Bernoulli* **21:3** (2006), 551–570.

[81] M. Fukushima, Y. Oshima, M. Takeda. *Dirichlet Forms and Symmetric Markov Processes*. de Gruyter, 1994.

[82] J. Gärtner. On the McKean–Vlasov limit for interacting diffusions. *Math. Nachri.* **137** (1988), 197–248.

[83] E. Giné, J. A. Wellner. Uniform convergence in some limit theorem for multiple particle systems. *Stochastic Proc. Appl.* **72** (1997), 47–72.

[84] H. Gintis. *Game Theory Evolving*. Princeton University Press, 2000.

[85] T. Goudon. Sur l'equation de Boltzmann homogène et sa relation avec l'equation de Landau–Fokker–Planck. *C.R. Acad. Sci. Paris* **324**, 265–270.

[86] S. Graf, R. D. Mauldin. A classification of disintegrations of measures. In: *Measures and Measurable Dynamics*. Contemporary Mathematics, vol. 94, American Mathematical Society, 1989, 147–158.

[87] G. Graham, S. Méléard. Chaos hypothesis for a system interacting through shared resources. *Prob. Theory Relat. Fields* **100** (1994), 157–173.

[88] G. Graham, S. Méléard. Stochastic particle approximations for generalized Boltzmann models and convergence estimates. *Ann. Prob.* **25:1** (1997), 115–132.

[89] H. Guérin. Existence and regularity of a weak function-solution for some Landau equations with a stochastic approach. *Stoch. Proc. Appl.* **101** (2002), 303–325.

[90] H. Guérin. Landau equation for some soft potentials through a probabilistic approach. *Ann. Appl. Prob.* **13:2** (2003), 515–539.

[91] H. Guérin, S. Méléard, E. Nualart. Estimates for the density of a nonlinear Landau process. *J. Funct. Anal.* **238** (2006), 649–677.

[92] T. Gustafsson. L^p-properties for the nonlinear spatially homogeneous Boltzmann equation. *Arch. Ration. Mech. Anal.* **92** (1986), 23–57.

[93] T. Gustafsson. Global L^p-properties for the spatially homogeneous Boltzmann equation. *Arch. Ration. Mech. Anal.* **103** (1988), 1–38.

[94] O. Hernandez-Lerma. *Lectures on Continuous-Time Markov Control Processes*. Aportaciones Matematicas, vol. 3, Sociedad Matematica Mexicana, Mexico, 1994.

[95] O. Hernandez-Lerma, J. B. Lasserre, J. Bernard. *Discrete-Time Markov Control Processes. Basic Optimality Criteria*. Applications of Mathematics, vol. 30. Springer, 1996.

[96] J. Hofbauer, K. Sigmund. *Evolutionary Games and Population Dynamics*. Cambridge University Press, 1998.

[97] W. Hoh. The martingale problem for a class of pseudo differential operators. Math. Ann. **300** (1994), 121–147.

[98] W. Hoh, N. Jacob. On the Dirichlet problem for pseudodifferential operators generating Feller semigroups. *J. Funct. Anal.* **137:1** (1996), 19–48.

[99] A. S. Holevo. Conditionally positive definite functions in quantum probability (in Russian). In: *Itogi Nauki i Tekniki*. Modern Problems of Mathematics, vol. 36, 1990, pp. 103–148.

[100] M. Huang, R. P. Malhame, P. E. Caines. Large population stochastic dynamic games: closed-loop McKean–Vlasov systems and the Nash certainty equivalence principle. *Commun. Inf. Syst.* **6:3** (2006), 221–251.

[101] T. J. R. Hughes, T. Kato, J.E. Marsden. Well-posed quasi-linear second-order hyperbolic systems with applications to nonlinear elastodynamics and general relativity. *Arch. Ration. Mech. Anal.* **63:3** (1976), 273–294.

[102] S. Ito, Diffusion equations. Translations of Mathematical Monographs, vol. 114. American Mathematical Society, 1992.

[103] N. Jacob. *Pseudo-Differential Operators and Markov Processes*, vols. I, II, III. Imperial College London Press, 2001, 2002, 2005.

[104] N. Jacob, R. L. Schilling. Lévy-type processes and pseudodifferential operators. In: O. E. Barndorff-Nielsen *et al.* (eds), *Lévy Processes, Theory and Applications*, Birkhäuser, 2001, pp. 139–168.

[105] N. Jacob *et al.* Non-local (semi-)Dirichlet forms generated by pseudo differential operators. In: Z. M. Ma *et al.* (eds.), *Dirichlet Forms and Stochastic Processes, Proc. Int. Conf. Beijing 1993*, de Gruyter, 1995, pp. 223–233.

[106] J. Jacod, Ph. Protter. *Probability Essentials*. Springer, 2004.

[107] J. Jacod, A. N. Shiryaev. *Limit Theorems for Stochastic Processes*. Springer, 1987. Second edition, 2003.

[108] A. Jakubowski. On the Skorohod topology. *Ann. Inst. H. Poincaré* **B22** (1986), 263–285.

[109] I. Jeon. Existence of gelling solutions for coagulation–fragmentation equations. *Commun. Math. Phys.* **194** (1998), 541–567.

[110] E. Joergensen. Construction of the Brownian motion and the Orstein–Uhlenbeck Process in a Riemannian manifold. *Z. Wahrsch. verw. Gebiete* **44** (1978), 71–87.

[111] A. Joffe, M. Métivier. Weak convergence of sequence of semimartingales with applications to multitype branching processes. *Adv. Appl. Prob.* **18** (1986), 20–65.

[112] J. Jost. Nonlinear Dirichlet forms. In: *New Directions in Dirichlet Forms*, American Mathematical Society/IP Studies in Advanced Mathematics, vol. 8, American Mathematical Society, 1998, pp. 1–47.

[113] M. Kac. *Probability and Related Topics in Physical Science*. Interscience, 1959.

[114] O. Kallenberg. *Foundations of Modern Probability*, second edition. Springer, 2002.

[115] I. Karatzas, S. Shreve. *Brownian Motion and Stochastic Calculus*. Springer, 1998.

[116] T. Kato. Quasi-linear equations of evolution, with applications to partial differential equations. In: *Spectral Theory and Differential Equations, Proc. Symp. Dundee, 1974*, Lecture Notes in Mathematics, vol. 448, Springer, 1975, pp. 25–70.

[117] T. Kazumi. Le processes d'Ornstein–Uhlenbeck sur l'espace des chemins et le probleme des martingales. *J. Funct. Anal.* **144** (1997), 20–45.

[118] A. Khinchine. Sur la crosissance locale des prosessus stochastiques homogènes à acroissements indépendants. *Isvestia Akad. Nauk SSSR, Ser. Math.* (1939), 487–508.

[119] K. Kikuchi, A. Negoro. On Markov processes generated by pseudodifferential operator of variable order. *Osaka J. Math.* **34** (1997), 319–335.

[120] C. Kipnis, C. Landim. *Scaling Limits of Interacting Particle Systems*. Grundlehren der Mathematischen Wissenschaften, vol. 320, Springer, 1999.

[121] A. N. Kochubei. Parabolic pseudo-differentiable equations, supersingular integrals and Markov processes (in Russian). *Izvestia Akad. Nauk, Ser. Matem.* **52:5** (1988), 909–934. English translation in *Math. USSR Izv.* **33:2** (1989), 233–259.

[122] A. Kolodko, K. Sabelfeld, W. Wagner. A stochastic method for solving Smoluchowski's coagulation equation. *Math. Comput. Simulation* **49** (1999), 57–79.

[123] V. N. Kolokoltsov. On linear, additive, and homogeneous operators in idempotent analysis. In: V. P. Maslov and S. N. Samborski: (eds.), *Idempotent Analysis*, Advances in Soviet Mathematics, vol. 13, 1992, pp. 87–101.

[124] V. N. Kolokoltsov. *Semiclassical Analysis for Diffusions and Stochastic Processes*. Springer Lecture Notes in Mathematics, vol. 1724, Springer, 2000.

[125] V. N. Kolokoltsov. Symmetric stable laws and stable-like jump-diffusions. *Proc. London Math. Soc.* **3:80** (2000), 725–768.

[126] V. N. Kolokoltsov. Small diffusion and fast dying out asymptotics for superprocesses as non-Hamiltonian quasi-classics for evolution equations. *Electronic J. Prob.*, http://www.math.washington.edu/ ejpecp/ **6** (2001), paper 21.

[127] V. N. Kolokoltsov. Measure-valued limits of interacting particle systems with k-nary interactions I. *Prob. Theory Relat. Fields* **126** (2003), 364–394.

[128] V. N. Kolokoltsov. On extension of mollified Boltzmann and Smoluchovski equations to particle systems with a k-nary interaction. *Russian J. Math. Phys.* **10:3** (2003), 268–295.

[129] V. N. Kolokoltsov. Measure-valued limits of interacting particle systems with k-nary interactions II. *Stoch. Stoch. Rep.* **76:1** (2004), 45–58.

[130] V. N. Kolokoltsov. On Markov processes with decomposable pseudo-differential generators. *Stoch. Stoch. Rep.* **76:1** (2004), 1–44.

[131] V. N. Kolokoltsov. Hydrodynamic limit of coagulation–fragmentation type models of k-nary interacting particles. *J. Stati. Phys.* **115: 5/6** (2004), 1621–1653.

[132] V. N. Kolokoltsov. Kinetic equations for the pure jump models of k-nary interacting particle systems. *Markov Proc. Relat. Fields* **12** (2006), 95–138.

[133] V. N. Kolokoltsov. On the regularity of solutions to the spatially homogeneous Boltzmann equation with polynomially growing collision kernel. *Advanced Stud. Contemp. Math.* **12** (2006), 9–38.

[134] V. N. Kolokoltsov. Nonlinear Markov semigroups and interacting Lévy type processes. *J. Stat. Phys.* **126:3** (2007), 585–642.

[135] V. N. Kolokoltsov. Generalized continuous-time random walks (CTRW), subordination by hitting times and fractional dynamics. arXiv:0706.1928v1[math.PR] 2007. *Probab. Theory Appl.* **53:4** (2009), 594–609.

[136] V. N. Kolokoltsov. The central limit theorem for the Smoluchovski coagulation model. arXiv:0708.0329v1[math.PR] 2007. *Prob. Theory Relat. Fields* **146:1** (2010), 87. Published online, http://dx.doi.org/10.1007/s00440-008-0186-2.

[137] V. N. Kolokoltsov. The Lévy–Khintchine type operators with variable Lipschitz continuous coefficients generate linear or nonlinear Markov processes and semigroupos. To appear in *Prob. Theory. Relat. Fields.*

[138] V. N. Kolokoltsov, V. Korolev, V. Uchaikin. Fractional stable distributions. *J. Math. Sci. (N.Y.)* **105:6** (2001), 2570–2577.

[139] V. N. Kolokoltsov, O. A. Malafeyev. *Introduction to the Analysis of Many Agent Systems of Competition and Cooperation (Game Theory for All).* St Petersburg University Press, 2008 (in Russian).

[140] V. N. Kolokoltsov, O. A. Malafeyev. *Understanding Game Theory*. World Scientific, 2010.

[141] V. N. Kolokoltsov, V. P. Maslov. *Idempotent Analysis and its Application to Optimal Control.* Moscow, Nauka, 1994 (in Russian).

[142] V. N. Kolokoltsov, V. P. Maslov. *Idempotent Analysis and its Applications.* Kluwer, 1997.

[143] V. N. Kolokoltsov, R. L. Schilling, A. E. Tyukov. Transience and non-explosion of certain stochastic newtonian systems. *Electronic J. Prob.* **7** (2002), paper no. 19.

[144] T. Komatsu. On the martingale problem for generators of stable processes with perturbations. *Osaka J. Math.* **21** (1984), 113–132.

[145] V. Yu. Korolev, V. E. Bening, S. Ya. Shorgin. *Mathematical Foundation of Risk Theory.* Moscow, Fismatlit, 2007 (in Russian).

[146] V. Korolev *et al.* Some methods of the analysis of time characteristics of catastrophes in nonhomogeneous flows of extremal events. In: I. A. Sokolov (ed.), *Sistemi i Sredstva Informatiki. Matematicheskie Modeli v Informacionnich Technologiach*, Moscow, RAN, 2006, pp. 5–23 (in Russian).

[147] M. Kostoglou, A. J. Karabelas. A study of the nonlinear breakage equations: analytical and asymptotic solutions. *J. Phys. A* **33** (2000), 1221–1232.

[148] M. Kotulski. Asymptotic distribution of continuous-time random walks: a probabilistic approach. *J. Stat. Phys.* **81:3/4** (1995), 777–792.

[149] M. Kraft, A. Vikhansky. A Monte Carlo method for identification and sensitivity analysis of coagulation processes. *J. Comput. Phys.* **200** (2004), 50–59.

[150] H. Kunita. *Stochastic Flows and Stochastic Differential Equations.* Cambridge Studies in Advanced Mathematics, vol. 24, Cambridge University Press, 1990.

[151] T. G. Kurtz, J. Xiong. Particle representations for a class of nonlinear SPDEs. *Stochastic Proc. Appl.* **83:1** (1999), 103–126.

[152] T. G. Kurtz, J. Xiong. Numerical solutions for a class of SPDEs with application to filtering. In: *Stochastics in Finite and Infinite Dimensions*, Trends in Mathematics, Birkhäuser, 2001, pp. 233–258.

[153] A. E. Kyprianou. *Introductory Lectures on Fluctuations of Lévy Processes with Applications.* Universitext, Springer, 2006.

[154] M. Lachowicz. Stochastic semigroups and coagulation equations. *Ukrainian Math. J.* **57:6** (2005), 913–922.

[155] M. Lachowicz, Ph. Laurencot, D. Wrzosek. On the Oort–Hulst–Savronov coagulation equation and its relation to the Smoluchowski equation. *SIAM J. Math. Anal.* **34** (2003), 1399–1421.

[156] P. Laurencot, S. Mischler. The continuous coagulation–fragmentation equations with diffusion. *Arch. Ration. Mech. Anal.* **162** (2002), 45–99.

[157] P. Laurencot, D. Wrzosek. The discrete coagulation equations with collisional breakage. *J. Stat. Phys.* **104: 1/2** (2001), 193–220.

[158] R. Leandre. Uniform upper bounds for hypoelliptic kernels with drift. *J. Math. Kyoto University* **34:2** (1994), 263–271.

[159] J. L. Lebowitz, E. W. Montroll (eds.). *Non-Equilibrium Phenomena I: The Boltzmann Equation.* Studies in Statistical Mechanics, vol. X, North-Holland, 1983.

[160] M. A. Leontovich. Main equations of the kinetic theory from the point of view of random processes (in Russian). *J. Exp. Theoret. Phys.* **5** (1935), 211–231.

[161] P. Lescot, M. Roeckner. Perturbations of generalized Mehler semigroups and applications to stochastic heat equation with Lévy noise and singular drift. *Potential Anal.* **20:4** (2004), 317–344.

[162] T. Liggett. *Interacting Particle Systems.* Reprint of the 1985 original. Classics in Mathematics, Springer, 2005.

[163] G. Lindblad. On the Generators of quantum dynamic semigroups. *Commun. Math. Phys.* **48** (1976), 119–130.

[164] X. Lu, B. Wennberg. Solutions with increasing energy for the spatially homogeneous Boltzmann equation. *Nonlinear Anal. Real World Appl.* **3** (2002), 243–258.

[165] A. A. Lushnikov. Some new aspects of coagulation theory. *Izv. Akad. Nauk SSSR, Ser. Fiz. Atmosfer. i Okeana* **14:10** (1978), 738–743.

[166] A. A. Lushnikov, M. Kulmala. Singular self-preserving regimes of coagulation processes. *Phys. Rev. E* **65** (2002).

[167] Z.-M. Ma, M. Röckner. *Introduction to the Theory of Non-Symmetric Dirichlet Forms.* Springer, 1992.

[168] P. Mandl. *Analytic Treatment of One-Dimensional Markov Processes.* Springer, 1968.

[169] A. H. Marcus. Stochastic coalescence. *Technometrics* **10** (1968), 133–143.

[170] R. H. Martin. *Nonlinear Operators and Differential Equations in Banach Spaces.* Wiley, 1976.

[171] N. Martin, J. England. *Mathematical Theory of Entropy.* Addison-Wesley, 1981.

[172] V. P. Maslov. *Perturbation Theory and Asymptotical Methods.* Moscow State University Press, 1965 (in Russian). French Translation, Dunod, Paris, 1972.

[173] V. P. Maslov. *Complex Markov Chains and Functional Feynman Integrals.* Moscow, Nauka, 1976 (in Russian).

[174] V. P. Maslov. Nonlinear averaging axioms in financial mathematics and stock price dynamics. *Theory Prob. Appl.* **48:04** (2004), 723-733.

[175] V. P. Maslov. *Quantum Economics.* Moscow, Nauka, 2006 (in Russian).

[176] V. P. Maslov, G. A. Omel'yanov. *Geometric Asymptotics for Nonlinear PDE. I.* Translations of Mathematical Monographs, vol. 202, American Mathematical Society, 2001.

[177] V. P. Maslov, C. E. Tariverdiev. Asymptotics of the Kolmogorov–Feller equation for systems with a large number of particles. Itogi Nauki i Techniki. Teoriya veroyatnosti, vol. 19, VINITI, Moscow, 1982, pp. 85–125 (in Russian).

[178] N. B. Maslova. Existence and uniqueness theorems for the Boltzmann equation. In: Ya. Sinai (ed.), *Encyclopaedia of Mathematical Sciences*, vol. 2, Springer, 1989, pp. 254–278.

[179] N. B. Maslova. *Nonlinear Evolution Equations: Kinetic Approach*. World Scientific, 1993.

[180] W. M. McEneaney. A new fundamental solution for differential Riccati equations arising in control. *Automatica (J. IFAC)* **44:4** (2008), 920–936.

[181] H. P. McKean. A class of Markov processes associated with nonlinear parabolic equations. *Proc. Nat. Acad. Sci.* **56** (1966), 1907–1911.

[182] H. P. McKean. An exponential formula for solving Boltzmann's equation for a Maxwellian gas. *J. Combin. Theory* **2:3** (1967), 358–382.

[183] M. M. Meerschaert, H.-P. Scheffler. *Limit Distributions for Sums of Independent Random Vectors*. Wiley Series in Probability and Statistics, John Wiley and Son, 2001.

[184] M. M. Meerschaert, H.-P. Scheffler. Limit theorems for continuous-time random walks with infinite mean waiting times. *J. Appl. Prob.* **41** (2004), 623–638.

[185] S. Méléard. Convergence of the fluctuations for interacting diffusions with jumps associated with Boltzmann equations. *Stocha. Stoch. Rep.* **63: 3–4** (1998), 195–225.

[186] R. Metzler, J. Klafter. The random walk's guide to anomalous diffusion: a fractional dynamic approach. *Phys. Rep.* **339** (2000), 1–77.

[187] P.-A. Meyer. *Quantum Probability for Probabilists*. Springer Lecture Notes in Mathematics, vol. 1538, Springer, 1993.

[188] S. Mishler, B. Wennberg. On the spatially homogeneous Boltzmann equation. *Ann. Inst. H. Poincaré Anal. Non Linéaire* **16:4** (1999), 467–501.

[189] M. Mobilia, I. T. Georgiev, U. C. Tauber. Phase transitions and spatio-temporal fluctuations in stochastic lattice Lotka–Volterra models. *J. Stat. Phys.* **128: 1–2** (2007), 447–483.

[190] E. W. Montroll, G. H. Weiss. Random walks on lattices, II. *J. Math. Phys.* **6** (1965), 167–181.

[191] C. Mouhot, C. Villani. Regularity theory for the spatially homogeneous Boltzmann equation with cut-off. *Arch. Ration. Mech. Anal.* **173:2** (2004), 169–212.

[192] A. Negoro. Stable-like processes: construction of the transition density and the behavior of sample paths near $t = 0$. *Osaka J. Math.* **31** (1994), 189–214.

[193] J. Norris. *Markov Chains*. Cambridge University Press, 1998.

[194] J. Norris. Cluster coagulation. *Commun. Math. Phys.* **209** (2000), 407–435.

[195] J. Norris. Notes on Brownian coagulation. *Markov Proc. Relat. Fields* **12:2** (2006), 407–412.

[196] D. Nualart. *The Malliavin Calculus and Related Topics. Probability and its Applications*, second edition. Springer, 2006.

[197] R. Olkiewicz, L. Xu, B. Zegarlin'ski. Nonlinear problems in infinite interacting particle systems. *Inf. Dim. Anal. Quantum Prob. Relat. Topics* **11:2** (2008), 179–211.

[198] S. Peszat, J. Zabczyk. *Stochastic Partial Differential Equations with Lévy Noise*. Encyclopedia of Mathematics, Cambridge University Press, 2007.

[199] D. Ya. Petrina, A. K. Vidibida. Cauchy problem for Bogolyubov's kinetic equations. *Trudi Mat. Inst. USSR Acad. Sci.* **136** (1975), 370–378.

[200] N. I. Portenko, S. I. Podolynny. On multidimensional stable processes with locally unbounded drift. *Random Oper. Stoch. Eq.* **3:2** (1995), 113–124.

[201] L. Rass, J. Radcliffe. *Spatial Deterministic Epidemics*. Mathematical Surveys and Monographs, vol. 102, American Mathematical Society, 2003.

[202] S. Rachev, L. Rüschendorf. *Mass Transportation Problems*, vols. I, II. Springer, 1998.

[203] R. Rebolledo. La methode des martingales appliquée l'etude de la convergence en loi de processus (in French). *Bull. Soc. Math. France Mem.* **62**, 1979.

[204] R. Rebolledo. Sur l'existence de solutions certains problemes de semimartingales (in French). *C. R. Acad. Sci. Paris A–B* **290:18** (1980), A843–A846.

[205] M. Reed, B. Simon. *Methods of Modern Mathematical Physics*, vol. 1, *Functional Analysis*. Academic Press, 1972.

[206] M. Reed, B. Simon. *Methods of Modern Mathematical Physics*, vol. 2, *Harmonic Analysis*. Academic Press, 1975.

[207] M. Reed, B. Simon. *Methods of Modern Mathematical Physics*, vol. 4, *Analysis of Operators*. Academic Press, 1978.

[208] T. Reichenbach, M. Mobilia, E. Frey. Coexistence versus extinction in the stochastic cyclic Lotka–Volterra model. *Phys. Rev. E (3)* **74:5** (2006).

[209] D. Revuz, M. Yor. *Continuous Martingales and Brownian Motion*. Springer, 1999.

[210] Yu. A. Rozanov. *Probability Theory, Stochastic Processes and Mathematical Statistics* (in Russian). Moscow, Nauka, 1985. English translation: *Mathematics and its Applications*, vol. 344, Kluwer, 1995.

[211] R. Rudnicki, R. Wieczorek. Fragmentation–coagulation models of phytoplankton. *Bull. Polish Acad. Sci. Math.* **54:2** (2006), 175–191.

[212] V. S. Safronov. *Evolution of the Pre-Planetary Cloud and the Formation of the Earth and Planets*. Moscow, Nauka, 1969 (in Russian). English translation: Israel Program for Scientific Translations, Jerusalem, 1972.

[213] A. I. Saichev, W. A. Woyczynski. *Distributions in the Physical and Engineering Sciences* vol. 1, Birkhäuser, Boston, 1997.

[214] A. I. Saichev, G. M. Zaslavsky. Fractional kinetic equations: solutions and applications. *Chaos* **7:4** (1997), 753–764.

[215] S. G. Samko. *Hypersingular Integrals and Applications*. Rostov-na-Donu University Press, 1984 (in Russian).

[216] S. G. Samko, A. A. Kilbas, O. A. Marichev. *Fractional Integrals and Derivatives and Their Applications*. Naukla i Teknika, Minsk, 1987 (in Russian). English translation Harwood Academic.

[217] G. Samorodnitski, M. S. Taqqu. *Stable Non-Gaussian Random Processes, Stochastic Models with Infinite Variance*. Chapman and Hall, 1994.

[218] R. L. Schilling. On Feller processes with sample paths in Besov spaces. *Math. Ann.* **309** (1997), 663–675.

[219] R. Schneider. *Convex Bodies: The Brunn–Minkowski Theory*. Cambridge University Press, 1993.

[220] A. N. Shiryayev. *Probability*. Springer, 1984.

[221] Ja. G. Sinai, Ju. M. Suchov. On an existence theorem for the solutions of Bogoljubov's chain of equations (in Russian). *Teoret. Mat. Fiz.* **19** (1974), 344–363.

[222] F. Sipriani, G. Grillo. Nonlinear Markov semigroups, nonlinear Dirichlet forms and applications to minimal surfaces. *J. Reine Angew. Math.* **562** (2003), 201–235.

[223] A. V. Skorohod. *Stochastic Equations for Complex Systems*. Translated from the Russian. Mathematics and its Applications (Soviet Series), vol. 13, Reidel, 1988.

[224] J. Smoller. *Shock Waves and Reaction–Diffusion Equations*. Springer, 1983.

[225] H. Spohn. *Large Scaling Dynamics of Interacting Particles*. Springer, 1991.

[226] D. W. Stroock. Diffusion processes associated with Lévy generators. *Z. Wahrsch. verw. Gebiete* **32** (1975), 209–244.

[227] D. W. Stroock. *Markov Processes from K. Ito's Perspective*. Annals of Mathematics Studies. Princeton University Press, 2003.

[228] D. Stroock, S. R. S. Varadhan. On degenerate elliptic–parabolic operators of second order and their associated diffusions. *Commun. Pure Appl. Math.* **XXV** (1972), 651–713.

[229] D. W. Stroock. S. R. S. Varadhan. *Multidimensional Diffusion Processes*. Springer, 1979.

[230] A.-S. Sznitman. Nonlinear reflecting diffusion process and the propagation of chaos and fluctuation associated. *J. Funct. Anal.* **56** (1984), 311–336.

[231] A.-S. Sznitman. Equations de type de Boltzmann, spatialement homogènes. *Z. Wahrsch. verw. Gebeite* **66** (1984), 559–592.

[232] A.-S. Sznitman. Topics in propagation of chaos. In: *Proc. Ecole d'Eté de probabilités de Saint-Flour XIX-1989*. Springer Lecture Notes in Mathematics, vol. 1464, Springer, 1991, pp. 167–255.

[233] K. Taira. On the existence of Feller semigroups with boundary conditions. *Mem. Ameri. Math. Soc.* 99 (1992), 1–65.

[234] K. Taira. On the existence of Feller semigroups with Dirichlet conditions. *Tsukuba J. Math.* **17** (1993), 377–427.

[235] K. Taira. Boundary value problems for elliptic pseudo-differential operators II. *Proc. Roy. Soc. Edinburgh* **127A** (1997), 395–405.

[236] K. Taira, A. Favini and S. Romanelli. Feller semigroups and degenerate elliptic operators with Wentzell boundary conditions. *Stud. Math.* **145: 1** (2001), 17–53.

[237] D. Talay, L. Tubaro (eds.). Probabilistic Models for Nonlinear Partial Differential Equations. In: *Proc. Conf. at Montecatini Terme, 1995*, Springer Lecture Notes in Mathematics, vol. 1627, Springer, 1996.

[238] H. Tanaka. Purely discontinuous Markov processes with nonlinear generators and their propagation of chaos (in Russian). *Teor. Verojatnost. i Primenen* **15** (1970), 599–621.

[239] H. Tanaka. On Markov process corresponding to Boltzmann's equation of Maxwellian gas. In: *Proc. Second Japan–USSR Symp on Probability Theory, Kyoto, 1972*, Springer Lecture Notes in Mathematics, vol. 330, Springer, 1973, pp. 478–489.

[240] H. Tanaka, M. Hitsuda. Central limit theorems for a simple diffusion model of interacting particles. *Hiroshima Math. J.* **11** (1981), 415–423.

[241] V. V. Uchaikin, V.M. Zolotarev. *Chance and Stability: Stable Distributions and their Applications*. VSP, 1999.

[242] V. V. Uchaikin. Montroll–Weisse problem, fractional equations and stable distributions. *Int. J. Theor. Phys.* **39:8** (2000), 2087–2105.

[243] K. Uchiyama. Scaling limit of interacting diffusions with arbitrary initial distributions. *Prob. Theory Relat. Fields* **99** (1994), 97–110.

[244] J. M. van Neerven. Continuity and representation of Gaussian Mehler semigroups. *Potential Anal.* **13:3** (2000), 199–211.

[245] C. Villani. On a new class of weak solutions to the spatially homogeneous Boltzmann and Landau equations. *Arch. Ration. Mech. Anal.* **143** (1998), 273–307.

[246] C. Villani. *Topics in Optimal Transportation*. Graduate Studies in Mathematics vol. 58, American Mathematical Society, 2003.

[247] W. Whitt. *Stochastic-Process Limits*. Springer, 2002.

[248] E.T. Whittaker, G.N. Watson. *Modern Analysis*, third edition. Cambridge University Press, 1920.

[249] D. Wrzosek. Mass-conservation solutions to the discrete coagulation–fragmentation model with diffusion. *Nonlinear Anal.* **49** (2002), 297–314.

[250] K. Yosida. *Functional Analysis*. Springer, 1980.

[251] M. Zak. Dynamics of intelligent systems. *Int. J. Theor. Phys.* **39:8** (2000), 2107–2140.

[252] M. Zak. Quantum evolution as a nonlinear Markov process. *Found. Phys. Lett.* **15:3** (2002), 229–243.

[253] G.M. Zaslavsky. Fractional kinetic equation for Hamiltonian chaos. *Physica D* **76** (1994), 110–122.

[254] B. Zegarlinski. Linear and nonlinear phenomena in large interacting systems. *Rep. Math. Phys.* **59:3** (2007), 409–419.

[255] V.M. Zolotarev. *One-Dimensional Stable Distributions*. Moscow, Nauka, 1983 (in Russian). English translation: Translations of Mathematical Monographs, vol. 65, American Mathematical Society, 1986.

Index

Aldous condition, 330
Aldous criterion, 330
annihilation operator, 352

backward propagator, xiv
Bogolyubov chain, 352
Boltzmann collisions, 21
Boltzmann equation, 33
  mollified, 33
  spatially trivial, 22

càdlàg paths, 324
canonical commutation relations, 352
Carleman representation, 347
chain rule, xiv
Chapman–Kolmogorov equation, 54
  nonlinear, 4
coagulation kernel, 21, 22
collision breakage, 22
collision inequality, 345
collision kernel, 22, 23
compact containment condition, 136, 329
complete positivity, 286
conditional positivity, 5, 60, 173, 300, 302
  local, 60
conservativity, 5
contraction, 44
  conservative, 54
convergence of measures
  vague, 319
  weak, 319
  $\star$-weak, 319
correlation functions, 35
coupling, 321
  of Lévy processes, 96
Courrège theorem, 61
covariation, 332
  predictable, 333
creation operator, 352
curvilinear Ornstein–Uhlenbeck process, 297

decomposable generator, 125
decomposable measures, 15
Dirichlet form, 129
du Hamel principle, 70
duality, 49
  nonlinear, 355
  quantum, 285
duality formula, 277
dynamic law of large numbers, 11
Dynkin's formula, 62

epidemic, 7
evolutionary games, 24
  weak CLT, 267

Feller process, 55
Feller semigroup, 55
  conservative, 56
  minimal extension, 55
C-Feller semigroup, 56
fractional derivative, 337
fragmentation, 22
fragmentation kernel, 22

Gateaux derivative, 335
gauge operator, 353
generating functional, 354
generator, conservative, 61
generator of k-ary interaction, 28
generators of order at most one, 102, 114, 232
  LLN, 233
  nonlinear, 180, 193
  nonlinear, Feller property, 198
  weak CLT, 267
geodesic flow, 294
Green function, 68

Hamiltonian, 294
heat kernel, 68
  for stable-like processes, 126
Hille–Yosida theorem, 129

integral generators, 104
intensity of interaction, 16
interacting stable-like processes
  LLN, 236
  weak CLT, 267
interaction of kth order, 16

Jakubovski criterion, 330

kinetic equation, 39
  k-ary interaction, 20
  binary interaction, 19
  discrete, 11

Landau–Fokker–Planck equation, 34
Lévy kernels, xiv
Lévy–Khintchine generator, 12
Lévy process, 12
  nonlinear, 13, 178
  time-nonhomogeneous, 175
Lindblad theorem, 287
linear operator, 43
  bounded, 43
  closable, 44
  closed, 44
  closure of, 44
  core, 44
  densely defined, 43
  dissipative, 60
  domain, 43
  norm of, 43
  positive, 54
Lotka–Volterra equations, 6

Malliavin calculus, 130
Markov semigroup of deterministic
    measure-valued process, 197
Markov transition probability family, 54
martingale problem, 63
  well-posed, 63
mass-exchange process, 8
  of order k or k-ary, 9
  profile, 9
mean field interaction, 30
minimal propagator, 108
mixed states, 14
mixed strategy, 24
modulus of continuity, 324
  modified, 324
mollifier, 33
moment measures, 35
  scaled, 35
Monge–Kantorovich theorem, 323
multiple coagulation, 22
mutual variation, 332

Nash equilibrium, 25
nonlinear Lévy semigroup, 13, 177
nonlinear Markov chain, 2
nonlinear Markov process, 40
nonlinear Markov semigroup, 1, 4, 39
  generator, 4
  stochastic representation, 4, 5, 148
  transition probabilities, 4
nonlinear martingale problem, 40
  well-posed, 40
nonlinear quantum dynamic semigroup, 292
nonlinear random integral, 91
nonlinear stable-like processes, 184
  Feller property of, 200
  localized, 187
  smoothness with respect to initial data, 200
  unbounded coefficients, 187
nonlinear transition probabilities, 2

observables, 14
  decomposable, 15
Ornstein–Uhlenbeck (OLL)
    infinite-dimensional processes, 268

k-person game, 24
  profile of, 24
  symmetric, 24
perturbation theory, 46
positive maximum principle (PMP), 60
principle of uniform boundedness, 48
probability kernel, xiv
T-product, 103, 151, 170
Prohorov's criterion, 322
propagation of chaos, 37, 244
propagator, xiv
  generator, 48
  Markov, 39, 54
  sub-Markov, 39, 54
propagator equation, xiv
pseudo-differential operators (ΨDOs), 338
pure coagulation, 9
pure fragmentation, 9

quadratic variation, 91, 332
quantum dynamic semigroup, 287

random measure, xiv
randomization lemma, 74
Rebolledo criterion, 333
relative entropy, 28
replicator dynamics, 6, 25
resolvent, 45
Riccati equation, 355

second quantization, 29, 353
semiclassical asymptotics, 129
semigroup, xiv
  Markov, 54
  sub-Markov, 54
Skorohod space, 324
Skorohod topology, 325
Smoluchovski's equation, 8, 21
Sobolev spaces, 120, 124
  weighted, 271
stable density, 68
stable-like processes, 82
  heat kernel, 126
  regularity, 95
  truncated, 142
  unbounded coefficients, 137, 142
stochastic differential equation (SDE)
  driven by nonlinear Lévy noise, 79
  with nonlinear noise, 79, 84
stochastic geodesic flow, 297
  induced by embedding, 298
stochastic integral with nonlinear noise, 78
stochastic matrix, 2
  infinitesimally stochastic matrix, 5
strong convergence, 44
strongly continuous propagator, 48
strongly continuous semigroup, 44
  generator, 45
subordination, 129
symbol of ΨDO, 338
symmetric function, 14

tight family of measures, 322
transition kernel, xiv, 113
  additively bounded, 155
  critical, 155
  dual, 113
  multiplicatively bounded, 155
  E-preserving, 155
  strongly multiplicatively bounded, 240
  subcritical, 155

variational derivative, 338
Vlasov's equation, 33

Wasserstein–Kantorovich distance, 322
weak CLT
  for Boltzmann collisions, 266
  for Smoluchovski coagulation, 265

Printed in the United States
by Baker & Taylor Publisher Services